Fundamentals of
Energy Production

Alternate Energy: ————————————

A WILEY SERIES

Series Editors:

MICHAEL E. McCORMICK
Department of Naval Systems Engineering
U.S. Naval Academy
Annapolis, Maryland

DAVID L. BOWLER
Department of Engineering
Swarthmore College
Swarthmore, Pennsylvania

Solar Selective Surfaces
O. P. Agnihotri and B. K. Gupta

Energy, The Biomass Options
Henry Bungay

Ocean Wave Energy Conversion
Michael E. McCormick

Fundamentals of Energy Production
Edwin L. Harder

Fundamentals of Energy Production

EDWIN L. HARDER

A WILEY-INTERSCIENCE PUBLICATION

JOHN WILEY & SONS

New York   Chichester   Brisbane   Toronto   Singapore

Library of Congress Cataloging in Publication Data:

Harder, Edwin L. (Edwin Leland), 1905–
  Fundamentals of energy production.

  (Alternate energy)
  "A Wiley-Interscience publication."
  Includes index.
  1. Power (Mechanics) 2. Power resources. I.
Title. II. Series.

TJ163.9.H37   333.79   81-16257
ISBN 0-471-08356-9   AACR2

Printed in the United States of America

10 9 8 7 6 5 4 3 2 1

Series Preface

During the 1970s it became clear that the world's known nonrenewable energy resources are decreasing rapidly and may be exhausted within the foreseeable future. In response to this disturbing prospect, the technologically advanced countries began to focus attention on renewable resources. As a result, there have been significant advances in such areas as solar heating and cooling, photovoltaics, wind power, bioenergy, and ocean wave energy.

The purpose of Alternate Energy: A Wiley Series is to discuss solutions to the technological and economic problems associated with the widespread use of renewable energy resources. The series is intended to introduce readers to the range and potential of these resources, to describe currently available and anticipated methods for conversion and delivery, and to consider economic aspects. The authors published in this series are well known in their fields and have made significant contributions. We have planned the series to meet the needs of all those interested in alternate energy, both practitioners and students.

MICHAEL E. MCCORMICK
DAVID L. BOWLER

Annapolis, Maryland
Swarthmore, Pennsylvania

Preface

The factors affecting energy, and man's use of it, are sharply divided into two categories. First are the physical facts regarding energy, the laws of nature, unchanging and timeless, the historical facts of what has happened up to the present, the factual information about energy reserves, and the technical information concerning the conversion of energy to useful forms and its transportation and storage.

Second are the economic and political factors and the attitudes of people concerning pollution, the environment, lifestyle, and social and national goals that are continually changing. This book deals with the first category. While many of the problems arise from the second category, their solutions must inevitably be based on the first.

The book is an encapsulation of information about the various fields of energy. Anyone working in the energy field will of course know more about his special areas than is contained in the corresponding chapters of this book. However, he will find a useful encapsulation of the essential information about the other areas with which he is less familiar. Prior to the several years of research involved in writing the book, the author, after 45 years in the electric power field, was relatively unfamiliar with many of the other fields.

Most of the book can be read and understood by any intelligent layman. Parts, particularly the three more theoretical chapters on chemistry, physics (energy conversion), and nuclear physics of energy, would require a college education. Parts of these chapters would require an engineering or science education. However, most of the concepts are explained simply. The treatment is in no way comparable to the usual theoretical treatment in texts on these subjects.

Nature has bounteously provided the earth with enough energy for everyone for as long as man can live on the earth. Fission energy is adequate for many centuries. The energy available from nuclear fusion has been likened to a Pacific Ocean full of oil that could be refilled 500 times, enough to last 64 billion years at the current rate of use. Solar energy will last as long as man can live on the earth, or vice versa.

Nature has also equipped man with ample creative ability to learn to use this energy safely, cleanly, and economically. There is enough energy to supply every living human being far more than is now used in the most advanced countries of the world. There is energy for all.

Many of the problems standing in the way of universal use of energy lie in political and social areas. However, the goal of this book is to explain the fundamental laws that nature has attached to this energy, the basic facts, historical or physical, about it, and the progress that man has made in processes and cycles for adapting it to his use. This should help to provide a clearer understanding of our opportunities and guidance as we endeavor to solve the much more difficult human problems* that now stand in the way of universal use of one of nature's richest blessings—energy.

The historical information about energy costs and cost comparisons, while factual, is of little value unless it can be updated to the period of interest. Had this book been published in 1958, there would have been 10 years of constant energy prices. Costs and cost comparisons would have remained essentially valid for 10 years. With the current (1981) high inflation and even higher energy cost escalation, this presents a serious problem. The solution adopted has been to date all costs and cost comparisons. Cost–time indices are given in Chapter 1. Also the costs vs year are given for major energy components such as power plants of various types and fuels of all kinds at various levels. Would that we could have another 10 years of negligible inflation and constant energy prices; but that being unlikely, this solution has been adopted.

*CONAES Report, *Energy in Transition—1985–2010,* Washington, D.C., NRC, 1979.

The heart of the book is included in nine chapters on energy sources: five conventional sources—oil, coal, gas, hydro, and nuclear—and four "alternate" sources—solar, wind, geothermal, and current growth or biomass. The chapters on oil, coal, and gas also include shale oil, tar sands and coal conversion technology (synfuels). An initial chapter presents the resources and rates of use of energy and the general cost structure, including the indices mentioned earlier. Except for resources, this chapter applies mainly to the United States.

The remaining five chapters appeared necessary for a variety of reasons. The chemistry of energy, mainly organic chemistry of hydrocarbons, is included since this science is not usually a part of electrical or mechanical engineering curricula, but is essential in understanding petroleum, gas, and coal conversion. This chapter also covers combustion and heating values. The chapter on physics of energy covers energy conversion cycles and processes and associated thermodynamic concepts, from steam turbines to refrigerators or solar cells.

The chapter on nuclear physics of energy contains essentially the briefing received by the author when placed in a position requiring knowledge of the nuclear developments in two great companies, Westinghouse and Siemens. It is the briefing that anyone should have to understand the nuclear fission development. The chapter on transportation and storage covers the five principal modes of energy transport over 50 miles—ship, barge, rail, pipeline, and transmission line. It covers several storage concepts not covered elsewhere. Pumped storage is included with hydro, batteries with energy conversion, and LNG with gas. Finally, the chapter on environmental coordination gives a very brief encapsulation of four areas of environment—air, water, "the good earth," and radiation.

For the preparation of this book, I am deeply indebted to scores of my friends and associates, but particularly to Gene Whitney for review and suggestions regarding hydro (I know of no better authority); to Prof. Fletcher Osterle, whose timely suggestions were invaluable in areas of thermodynamics; to C. van Mook of the Dravo Corp. for authentic information about towboats, barges, and waterways; to Bob Creagan for my briefing on nuclear energy; and to my wife, Esther, whose constant support and cooperation has made this book possible and whose love and affection has made it all worthwhile.

EDWIN L. HARDER

Pittsburgh, Pennsylvania
February 1982

Contents

Reference Information

1

Energy Reserves, Current Use, and General Cost Structure

INTRODUCTION

It is the purpose of this chapter to answer three fundamental questions:

1. What energy reserves are available in the United States and in the world?
2. What is the current rate of production and use of energy in the United States and in the world?
3. What is the general cost structure of energy in the United States?

Energy Sources

The complete list of energy sources that will be considered in this book is as follows:

Percent of 1978 U.S. use

Current Energy Sources		Alternate Energy Sources
Petroleum	48.5	Geothermal
Natural gas	25.6	Breeder reactor fuel
Coal and Lignite	17.9	Shale oil
Water power	4.1	Peat
Fission uranium	3.8	Tar sands
Other	0.1	Nuclear fusion fuel
		Solar energy
Total	100.0	Wind power
		Tidal and wave power
		Current growth, biomass
		Food

Of these, the first five provided practically all the energy used commercially in the United States in 1978. They are referred to as the "conventional energy sources," in distinction to the other "alter-

nate" or "unconventional" sources. Geothermal, next in order, supplied about 0.1% of U.S. electric energy in 1978. Breeder reactors and shale oil may soon enter the picture. Lignite and brown coal are more important in other countries of the world. Tar sands, negligible in the United States with present technology, are emerging as important elsewhere.

All of these energy sources are covered in this book. For each source, an encapsulation is given covering the principal facts about it and the cycles or processes involved in using it. However, for the purposes of this chapter, the first eight, excluding "other", are principally considered.

The plan of the chapter is to give first the reserves available of each of these principal resources. The U.S. reserves are then to be compared with the U.S. use of energy, current and projected. This will bring out the major energy problems facing the country in 1981 and the major developments that must be completed in order to solve them. The energy cost structure is then given.

Reserves. It would be logical to define what is meant by reserves first, before discussing them. But this would entail a detailed discussion of each source and how it is treated. Instead, the reader is asked to accept at this point that the reserves as given represent reasonably well what can be economically recovered from the earth in the coming years. Successful development of breeder reactors, coal conversion, and oil from shale will be required to utilize these reserves fully.

Costs. The whole use of energy, now and in the future, depends on economics. This affects both the amount used and the choice of sources and methods. Hence, it was felt that this chapter must answer the third question, "What is the relative cost?" Here again, the reader is asked to accept

that the cost of energy in the United States in the future will be a small percentage of the personal income, as it is now (under 5% in 1974, under 10% in 1979). Hence, *relative* cost will affect the choice of energy alternatives, just as it does now, but *need* will primarily determine the amount used. Future requirements are postulated on this basis.

This premise is borne out by the fact that a quadrupling of energy prices, a doubling in constant dollars, during the 1970s, did not reduce the amount of energy used in the United States. It merely reduced the rate of increase (see Price Indices and Table 1.2).

U.S. AND WORLD RESERVES

Information of sufficient accuracy for understanding the energy situation in the United States, in broad outline, is given in the tables and figures of this chapter. Table 1.1 gives the total reserves reasonably recoverable and is used to determine the adequacy of reserves for expected use of different energy sources in the future. The primary information in metric units is from the *United Nations Statistical Yearbook*,[1] with supplementary information from many sources. These data have been converted to the commonly used large energy units—quads (10^{15} Btu) and billions of barrels of oil equivalent—by factors shown in the footnotes.

The lignite and brown coal energy reserves are based on 1750 kcal/kg (3140 Btu/lb), the lower limit considered as reserves. Heating values vary up to two or three times this much, and the values shown are therefore lower limits.

The conversion of fission uranium from millions of metric tons to quads is based on plutonium recycle, and the amount would be much less if the plutonium in spent fuel is not used. The uranium production shown is for all purposes, military and civilian.

The relatively huge reserves of breeder reactor fuel are due to the usability of all the uranium instead of tenths of 1%, as well as to the very much larger amount recoverable at the price that could then be afforded.

For the unconventional sources, reserves have not been defined generally. There are about 2100 billion barrels of oil in the Green River shale deposits of Colorado, Utah, and Wyoming, one of the largest known deposits in the world.[2] The

Athabasca tar sands of Alberta, Canada, contain 626 billion barrels of crude bitumen, of which 38 billion barrels can be reached by open cut mining, the only currently developed process. This would yield about 26 billion barrels of a syncrude, comparable to petroleum, which is shown in the table (Table 1.1).

Estimates of geothermal reserves vary from the limitless energy available 10 miles down in the earth, currently uneconomic to tap, to little more than the currently operating fields. The estimate shown,[3] comparable to coal reserves, involves unproved economics and undeveloped technology.

Reserves of unconventional exhaustible fuels are explained more fully in the chapters devoted to those energy sources.

Other Resources

Resources are defined as all material available to man in the future, known and unknown. For the conventional sources of energy, reserves are defined as those resources that are known and recoverable with existing technology and under existing economic conditions.

For the coal series of fuels, this is defined as including only veins at least 12 in. (30 cm) thick, and not over 4000 ft (1200 m) below the surface, or 1640 ft (500 m) for lignite and brown coal.

For oil and gas, only discovered deposits are included as reserves. Estimates are made of future discoveries, but again these assume existing technology and economics. "Potential" reserves are the sum of discovered and undiscovered reserves.

Known and estimated resources greatly exceed reserves as defined above. As fossil fuel reserves run out and fuel prices rise relative to other costs, new technology will no doubt be developed to use these additional resources.

Thus in addition to the 29,000 quads of coal reserves in the United States there are an additional 41,000 quads at less than 3000 ft that cannot be recovered economically at present.

In addition to the 1000 quads of potential oil reserves, there are 5900 quads locked up in shale oil, 87 quads in tar sands, and 174 quads in heavy oil.[16]

In addition to the 1200 odd quads of potential natural gas reserves, there are 600 quads in Devonian shale, 500 quads in coal seams, 3000 quads in geopressurized zones, and 800 quads in "tight sands."[16]

Table 1.1. World and U.S. Reserves, Annual Production, and Use of Principal Exhaustible Energy Resources—1972[a]

Resource	Reserves				Production[b] (quads)
	Millions of Metric Tons	Billions of Cubic Meters	Quads	Billions of Barrels of Oil Equivalent	
Crude Petroleum					
World	76,800		3,310	568	106
U.S.	4,899		211	36	19.5
U.S. Potential	23,900		1,030	178	
Natural Gas					
World		54,100	1,975	340	44
U.S.		7,535	275	47.4	23.2
U.S. Potential		32,500	1,186	205	
Coal					
World	6,641,000		175,000	30,200	56.6
U.S.	1,100,000		29,000	5,010	14.1
Lignite, Brown Coal					
World	2,041,000		14,100	2,430	5.5
U.S.	406,000		2,800	484	0.04
Fission Uranium					
World	0.869[d]		70,000[c]	12,300	(19,185 Metric tons[b])
U.S.	0.259[d]		16,000[c]	2,830	(10,514 Metric tons[b])
Breeder Uranium					
World			420,000,000	74,000,000	
U.S.			96,000,000	17,000,000	
Shale Oil[2]					
World	3,885,000		166,750	28,750	
U.S.	311,000		13,340	2,300	
Fusion Fuel	- - - - Virtually unlimited - - - -				
Tar Sands					
Canada, Athabasca	3,500		151	26	0.10
U.S.	Negligible				
Geothermal					
World					0.045
U.S.			35,800	6,350	0.009

[a]Conversion Data: 1 barrel = 42 U.S. gallons; 1 quad = 10^{15} Btu.; Oil (sp. gr. 0.85), Higher Heating Value = 19,590 Btu/lb, 7.4 barrels/metric ton, 43.1 million Btu/metric ton, 5.8 million Btu/barrel, 5.8 quads per billion barrels. Natural gas, 1035 Btu/Standard Cubic Foot, 36,500 Btu/m³; coal, 12,000 Btu/lb, 26.4 million Btu/metric ton; lignite and brown coal (minimum value used; see text), 1750 kcal/kg, 3140 Btu/lb, 6.91 million Btu/metric ton.

[b]World use is approximately equal to production, except for nuclear fuels.

[c]Based on plutonium recycle.

[d]1972 Data shown are uranium contents of ore reasonably assured at a market price of $10.00/lb U_3O_8. In 1978, the corresponding figures were 1.650 and 0.523 metric megatons for world and United States, respectively, reasonably assured at a market price of less than $30.00/lb U_3O_8. (Sources: Nuclear Energy Agency of OECD and International Atomic Energy Agency.[1])

ENERGY USE IN THE UNITED STATES

The energy balance of the United States is given in Table 1.2, showing the use of the five principal energy sources by the five recognized energy markets. The end use of energy is also shown. This "structure" conforms to the record keeping by industry and government and hence is adhered to in this book. A further breakdown by type of load is given later (Fig. 4.9). Note that the electric utility, while a market for the prime energy sources, is itself a secondary energy source, distributing energy in turn to the ultimate-use sectors.

As noted in Table 1.2A, 49% of the oil was imported in 1978. The relative amount of oil imported has increased from about 25% in 1960 to about 50% in 1977–1980, as shown in Fig. 1.1. Since 1976 the total oil used in the United States has remained a little over 6 billion barrels per year. Fig. 1.1 shows the amount of oil imported and produced and the amount of imported oil that is in the form of products.

More than half of the oil is used for transportation. Oil supplies over half of all residential heating. A large part of the oil used in industry is for feedstocks. Oil is essential for combustion turbines for electric utility peak loads. The production of liquid fuels is of highest priority in the synfuels program.

Table 1.2B shows the sources of U.S. energy consumed over the last 25 years. The total has almost doubled. Natural gas peaked about 1973. Nuclear energy, negligible in 1965, reached 3.8% in 1978.

Both the distribution of electric energy and the total energy use by the three end-use sectors will be

Table 1.2. U.S. Sources of Energy Consumed, Measured in Quads (10^{15} Btu)[a]

A. *1978 by Source, Market, and End-Use Sector (Population 218,000,000)*

Prime Energy Source	Transportation	Industrial[c]	Commercial	Residential	Electric Utility	Total Quads[4]	%
Coal		3.5	0.5	0.2	9.8	13.97	17.9
Natural gas		8.2	2.5	4.6	4.7	20.00	25.6
Oil	20.5	6.9	1.5	5.8	3.0	37.96[d]	48.5[d]
Water power					3.2	3.17	4.1
Nuclear					3.0	2.98	3.8
All other					0.1	.07	0.1
Total[5]	20.5	18.6	15.1		23.8	78.15	100.0
End use[4]	20.62	28.72	28.81		—	78.15	

B. *Total Consumption by Source of Energy and Year*

Source	1955	1960	1965	1970	1973	1974	1975	1976	1977	1978	1979	1980
Coal	11.703	10.285	12.358	12.698	13.292	12.935	12.837	13.732	13.980	13.977	15.106	15.550
Natural gas	9.232	12.736	16.079	22.029	22.512	21.732	19.948	20.345	19.931	20.000	20.546	20.394
Oil	17.524	20.035	23.419	29.537	34.840	33.455	32.731	35.175	37.176	37.965	37.135	34.248
Water power	1.497	1.626	2.057	2.650	3.010	3.309	3.219	3.066	2.519	3.168	3.163	3.105
Nuclear			0.038	0.229	0.910	1.272	1.900	2.111	2.702	2.977	2.748	2.727
All other					0.046	0.056	0.072	0.081	0.082	0.068	0.089	0.116
Total	39.956	44.816	53.969	67.143	74.609	72.759	70.707	74.509	76.390	78.154	78.787	76.140

[a]Multiply by 0.472 for millions of barrels of oil per day equivalent.

[b]Distribution of fuels among markets is approximate.

[c]Part used for feedstocks (approx.): coal, 2.2 quads; natural gas, 1.0 quad; oil, 4.2 quads; total, 7.4 quads.

[d]49% imported.

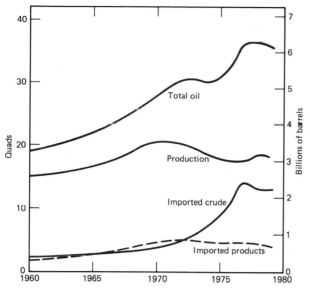

Figure 1.1. U.S. oil supply.

needed in later chapters. However, the use of the prime energy sources, as given in Table 1.2A, is sufficient for a consideration of the corresponding reserves. Note the total use—78 quads, or 13.4 billion barrels of oil equivalent in 1978; this is 37 million barrels of oil per day equivalent.

Future Use of Energy in the United States

It is not the function of this book to predict the future, but rather to provide data and methods for determining the consequences of any future course of action. The historical rate of increase in the use of energy, 4%/yr, is used in the next section to get a very preliminary feel for the significance of the energy reserve situation. While the rate has been less in recent years (1979), there have been deviations before for other reasons. Decades are required to determine average rates. In view of the uncertainties in the reserves themselves, the use of the historical rate appears justified in this case.

The U.S. Department of Commerce has projected a 1.9%/yr energy growth rate from 1977 to 2000.[6] A growth rate of 2%/yr would halve the cumulative requirements in six decades, compared with the 4%/yr rate. The effect of this assumption is also shown in Fig. 1.2. The numbers in parentheses are for 2% increase per year as stated.

Additional summary data on energy use and costs in the United States are given in Ref. 7, and energy use, in Ref. 8.

U.S. Use of Reserves

The U.S. reserves and rates of use have now been given. What does this show? The energy reserve situation is depicted graphically in Fig. 1.2. It is certainly not to scale; this would be impossible and is unnecessary since the figures are all shown. But it does show the principal energy "facts of life" in the United States in 1981.

Each energy source is shown as a reservoir, with its size marked in quads, and a pipe leading to the market, with valves open or closed. The U.S. energy needs are indicated by the small table, with both annual and cumulative requirements in quads. Thus, the needs can easily be compared with the supplies.

We have huge breeder fuel reserves, enough for many thousands of years. At a 4% growth rate, these reserves are 1000 times our cumulative needs for the next 100 years. But the valve is closed until the breeder reactor development is complete—hopefully in the 1980s.

We have very large reserves of coal and fission uranium fuel, enough for seven and five decades, respectively, at 4%/yr, together enough for over eight decades, of all our needs. But the coal pipeline has suffered severe restrictions; in fact, it has been shut off to many former uses, as is well known. Arbitrarily low, government-fixed gas prices at the wells for many years have made a serious dent in the coal pipe. Low imported-oil prices until 1973 effectively shut off coal from U.S. railroads. "Smoke control" eliminated coal as a domestic fuel. And we even increased the use of oil in power plants. Environmental restrictions have further impeded the flow of this plentiful fuel—coal. This pipe is so badly damaged that not until a new pipe, with new valves, namely, coal conversion, clean fuels from coal, and stack gas cleaning or fluidized-bed combustion, are fully developed will energy from this plentiful source really flow freely again.

We have another large reservoir—shale oil— which was little talked of in the past. But it is larger than the whole world reserve of petroleum. The United States, not the OPEC countries, is the oil-rich land! Here again, the valve is closed until a huge development—oil from shale—is completed, hopefully within a decade or two.

In the meantime we have large pipes, with wide-open valves, draining our most precious and scarce reserves, natural gas and petroleum, as though we were trying to get rid of them. Worse than that, we

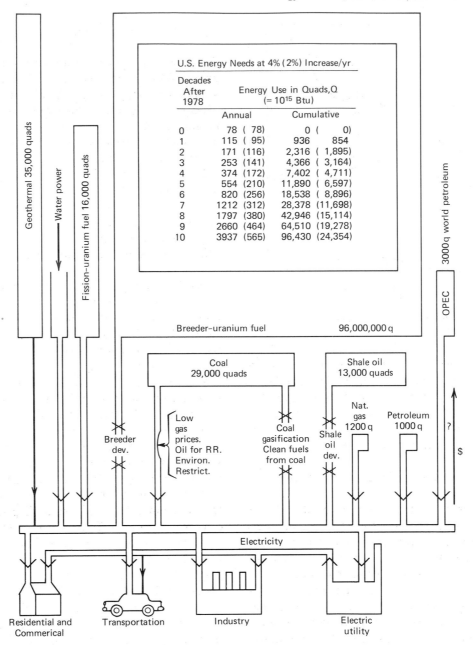

U.S. Energy Needs at 4% (2%) Increase/yr			
Decades After 1978	Energy Use in Quads,Q (= 10^15 Btu)		
	Annual	Cumulative	
0	78 (78)	0 (0)	
1	115 (95)	936	854
2	171 (116)	2,316 (1,895)	
3	253 (141)	4,366 (3,164)	
4	374 (172)	7,402 (4,711)	
5	554 (210)	11,890 (6,597)	
6	820 (256)	18,538 (8,896)	
7	1212 (312)	28,378 (11,698)	
8	1797 (380)	42,946 (15,114)	
9	2660 (464)	64,510 (19,278)	
10	3937 (565)	96,430 (24,354)	

Figure 1.2. The U.S. energy picture, 1978.

are bending every effort (1981) to enlarge these pipes and drain our precious reserves sooner. Together, in 1978, domestic gas and oil supplied about 48% of our energy.

Imported Petroleum. Figure 1.2 also shows a pipeline from world petroleum that is now (1981) a very controversial pipeline. In 1978, it carried nearly 50% of our oil requirements, 24% of our total energy. But there is a dollar pipeline out of the country to compensate. This we could tolerate until 1973. But then OPEC raised the price by 4:1, four times as many dollars out per barrel as oil in. By 1980 world oil had increased from $2.50 per barrel to $32, 13:1. This had a disastrous effect on U.S. balance of payments abroad, weakened the dollar, and added to serious domestic inflation and business recession at the same time. It also enriched the OPEC nations and placed in their hands a powerful diplomatic and economic weapon.

This then is our reason for depleting our own scarce oil and gas reserves even more rapidly than we have been. Meanwhile, we hasten to make temporary repairs to the old coal pipe, increase nuclear power with reservations, and strive to reduce our energy consumption, increase the use of alternative energy sources, and get the new channels opened into our vast stores of coal.

Initial enthusiasm (1974) for shale oil and geothermal has considerably dampened (1980) and while extensive preparations are being made, large-scale production must await economic conditions and assurances that favor their commercial development. Both the use of fission uranium and development of breeder nuclear sources are currently (1980) set back by adverse public opinion in the United States, although these sources are obviously badly needed. Nuclear use and development is proceeding or being accelerated elsewhere in the world.

Electricity. As shown in Fig. 1.2, electricity forms a secondary pipeline into the user markets. It accounts for 28% of the nation's energy in 1978. But by the year 2005, it could become over 65% and be a major factor, in fact, an essential factor, in getting energy from our largest reservoirs—coal, breeder fuel, and geothermal (if and when)—to the users in a highly usable form.

Why Not Hydro or Geothermal? Figure 1.2 also shows an open-ended water power reservoir, which is unfortunately being refilled rather slowly, and which we can tap for only 4.1% of our energy in 1978. In spite of some increases in hydro, this is a steadily lowering percentage as our needs rapidly mount.

It also shows a tiny pipe into a very large geothermal reservoir, about the size of our coal reserves by some estimates. This tiny pipe supplies $\frac{1}{10}$ of 1% of the nation's electric energy in 1978; and while it is being expanded, the process is long and slow. If the supply gets to be 1% by 1995, it will be very surprising.

Why Not Solar Energy or Winds and Tides? Certainly, solar energy powers satellite communications and heats buildings and water in favorable locations. But the tremendous visions of using it commercially on a large scale for generating electricity involve huge developments with tremendous unknowns. This must be compared with the three unopened valves of Fig. 1.2. These are straightforward avenues of development into very large energy reserves. Economic solar generation of electricity on a large scale is yet to be proven,[17] in spite of remarkable progress in photovoltaics. Substantially the same is true of the winds and tides—possible but uneconomic—on a large scale. Hence, these sources have been omitted from Table 1.1 and Fig. 1.2. They are however covered in detail later in this book, as is energy from biomass. Their contribution, though small, is extremely important nationally in reducing dependence on imported energy.

Energy Production

The world production and consumption for each fossil fuel are practically the same, since variations in storage from year to year are generally negligible. Both the world production and approximate use of fossil fuels are given in the last column of Table 1.1.

For uranium, only a tiny part of that produced is used for commercial energy, both because of military uses and storage and because a reactor burns only a very small part of its uranium fuel. The production listed is therefore in metric tons for all purposes.

For the United States, the differing figures for energy production in Table 1.1, and annual use in Table 1.2, reflect the fact that about 50% of the oil is imported (1978), as is 5% of the gas.[9] About 10% of the coal produced is exported. Also, the data in the two tables are for different years.

GENERAL COST STRUCTURE

The energy cost structure from producer to user in 1969 or 1970 and in 1979 is shown in Table 1.3. In 1969, 78% of the cost of gasoline at the pump was in refining, distribution, and taxes. Only 22% of the cost was for crude oil. By 1979, the composite (domestic and import) cost of crude had risen 500%, and crude comprised 44% of the cost of gasoline at the pump.

Natural gas at the well was 16% of the cost of residential gas in 1970, 84% of the cost being in the distribution system. In 1979, 41% of the cost was wellhead cost.

For electric energy, fuel was only 17% of the residential cost in 1970 but 33% in 1979. Two thirds of the cost was still in the fixed system and services.

Table 1.3.   Energy Cost Structure in the United States

	Oil (Cents/Gallon)		Natural Gas ($/MCF or Approx. $/10^6 Btu)			Electricity (¢/kWh)		
	1969	1979		1970	1979		1970	1979
Gasoline with State and Fed. tax	34.8	85	Residential	1.09	2.78	Residential	2.3	5.5
Gasoline at service sta. (without tax)	23.9	73				Industrial	1.0	3.0 est.
Refinery output (gasoline or distillate)	11.9	50 (approx.)	Elec. util. fuel	0.292	1.63	Busbar	0.84	2.5 est.
Composite crude	7.5	37	Well	0.171	1.14	Fuel	0.4	1.8

These relationships represent national averages and vary considerably from area to area of the country. They are of particular importance in considering alternate energies.

For example, ethanol for gasohol would involve the same distribution costs as gasoline. It must therefore compete with the refinery output cost of gasoline on an energy basis. Having 65% of the energy per gallon (Table 10.13), its cost must be 65% of that of gasoline to be equivalent. Necessary subsidies must take this into account if alcohol from current growth is to contribute substantially to lowered energy imports.

Similarly the substitution of wind power or solar energy for residential or farm electricity may displace only the fuel cost of the utility if *all* of the same equipment and services are required for backup. The lowest cost to the economy, and lowest cost to the residential user are quite different. However, if wind or solar energy displace oil fuel, delivered by truck, the alternate energy competes with the *delivered price* in both cases.

Change in Conventional/Alternate Cost Ratio. As shown in Fig. 1.3, the inflation in fuels and energy has been over twice the inflation in all commodities. While general costs have doubled, conventional energy-related costs have quadrupled. This has two very important consequences:

1. Energy conservation is much more valuable. Larger investment in insulation and in energy-efficient devices is economic. It is also strongly in the national interest for other reasons.

2. The costs of alternate energies not based on fossil fuels, and not having large environ-

mental and safety costs have been reduced by about 2:1 relative to conventional energies. The areas in which alternate energies are economical have been enormously expanded.

Updating Energy-Related Costs

Throughout this book, many costs and cost comparisons are given in current dollars. Some date back to 1936 (Hoover Dam plant cost, Table 5.1); some are more recent (electric generating options, year end 1978, Table 3.6). The high general and energy inflation rates (Fig. 1.3) and the enormous increase in the cost of world oil, from $2.50 to $32 per barrel in six years, renders all but the most recent costs obsolete, unless updated in some manner. Cost *comparisons* are valid under certain conditions to be discussed later.

Price Indices

In reporting historical energy prices to the U.S. Congress, the Energy Information Administration

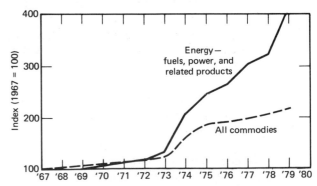

Figure 1.3.   Producer price indices. *Source.*   Reference 5.

Table 1.4. GNP Price Deflators

Year	Index	Year	Index	Year	Index
1950	1.864	1971	1.042	1976	0.747
1955	1.640	1972	1.000	1977	0.706
1960	1.456	1973	0.945	1978	0.657
1965	1.346	1974	0.862	1979	0.604
1970	1.094	1975	0.786		

Consumers Price Index^a Ratio, CPI_{1972}/CPI

Year	Index	Year	Index	Year	Index
1950		1971	1.03	1976	0.73
1960	1.412	1973	0.94	1978	0.64
1970	1.08	1975	0.78		

Source: Ref. 5.

^{a}U.S. Bureau of Labor Statistics. CPI = Consumer Price Index.

(EIA) has given additionally the prices in constant dollars,[9] using the Gross National Product (GNP) price deflators[5] based on 1972 = 1.0, Table 1.4.

While this places past and present real costs in better perspective, it would not be a proper basis for updating the costs of energy or energy facilities, since these have inflated at about double the rate of all commodities, or of the GNP. The Consumer Price Index Ratios are not greatly different from the GNP deflators, as shown by the supplement to Table 1.4.

The purchasing power of the dollar for fuels, power, and related products is best given by the energy curve[5] of Fig. 1.3, which shows that current prices for the same energy commodity have advanced as indicated in Table 1.5, in the United States (1967 = 100). Prior to 1970, energy inflation was *less* than general inflation.

Both consumer and producer price indices[5] are broken down into subclasses such as "nonresidential structures," "durable equipment," and "imports," which provide updating ratios for specific costs. All costs are dated, and many are broken

Table 1.5. Producer Energy Price Index^{ab}

Year	Index	Year	Index	Year	Index
1950	87.1	1971	115.2	1976	265.6
1955	91.2	1972	118.6	1977	302.2
1960	96.1	1973	134.3	1978	322.5
1965	95.5	1974	208.3	1979	408.1
1970	106.2	1975	245.1	May 1980	571.9

^{a}Table may be updated from the latest issue of Ref. 5.
^{b}1967 = 100

down into feedstocks, capital cost, and operating cost to facilitate revision.

From 1972 to 1980, the world oil price increased 12.8 times; the U.S. residential electric rates increased 2.3 times. Thus, the use of *any* single index for all fuels and energy facilities would be very misleading.

Also, a power plant built in 1960 for $150/kWe would cost in 1979, according to the producer energy price index given above, approximately 150 (408.1/96.1) = $637/kWe, or 4.25 times as much, *for the same plant*. However, the same plant could no longer be built. Extensive environmental controls would be required for coal-fired plants, such as stack-gas scrubbers, cooling towers, and many other features not formerly required. Refinery costs have more than doubled in constant dollars, largely because of environmental control requirements.

For all of these reasons, cost inflation data are given for various fuels and for typical power plants in the following sections. These will serve as a guide in updating corresponding costs in the later chapters.

Cost of Fuels

Tables 1.6 and 1.7 show the increase in the cost of the major fuels from 1950 to 1981, in current dollars.[9] Table 1.6 shows the market prices in customary units, tons of coal, barrels or gallons of oil, and thousands of standard cubic feet (MCF) of natural gas. These can be converted to constant dollars, if needed, by multiplying by the GNP price deflators.

For utilities, there are some further costs beyond the point of FOB. Also the heating value per ton, barrel, or MCF varies with many factors from the typical values used earlier in this chapter, that is, coal 24 million Btu/ton, oil 5.8 million Btu/barrel, and natural gas 1035 Btu/MCF. The actual cost in cents per million Btu, as burned, is shown in Table 1.7,[10] through 1977, and as delivered, 1977–1981.

The costs of electricity[4,11] are shown in Table 1.8.

Coal-Fired Steam Electric Plant Costs

Many coal-fired plants over 500 MWe were built during the 1960s for $100 to $150/kWe. For example, the Keystone station in western Pennsylvania, 1872 MWe, was built in 1967 for $101/kWe. The Homer City, Pennsylvania, plant, 1319 MWe, was built in 1969 for $144/kWe. The costs of a number of plants chosen at random are shown in Fig. 1.4.

Table 1.6. Cost of Fuels in Current Dollars[a]

Year	Bituminous Coal ($/ton) FOB Mines	FOB Utility	Oil $/barrel Wellhead	Import	Refinery Acquisition Average	Oil ¢/gallon Heating Oil[b]	Gasolene, retail Regular Unleaded	Average All Grades	Natural Gas (¢/MCF) Wellhead	To Electric Utility	Residential
1950	4.84		2.51						6.5		
1955	4.50	6.07	2.77						10.4		
1960	4.69	6.26	2.88						14.0		
1965	4.44	5.71	2.86						15.6		
1967	4.62	5.85	2.92						16.0	28.0	104.4
1970	6.26	7.13	3.18						17.1	29.2	109.0
1971	7.07	8.00	3.39						18.2	32.3	114.9
1972	7.66	8.84	3.39						18.6	34.0	121.4
1973	8.53	9.01	3.89						21.6	37.8	128.7
1974	15.75	15.46	6.87		9.07				30.4	51.0	143.1
1975	19.23	17.63	7.67	12.45	10.38	37.7	56.2		44.5	77.3	170.8
1976	19.43	18.38	8.19	13.34	10.89	40.6	58.7		58.0	98.0	197.8
1977	19.82	20.37	8.57	14.31	11.96	46.0	62.6		79.0	132.4	234.9
1978	21.78	23.75	9.00	14.38	12.46	49.4	63.9	65.2	90.5	147.8	256.3
1979	23.50	27.30	12.64	18.62	17.72	65.6	85.3	88.2	117.8	180.3	323.1
1980			21.19	37	28.07	97.8		122.1			
May 1981			32.71		37.86	122.7		137.0			

[a]Table can be updated from latest issue of Ref. 4.
[b]Average selling price to residential customers.

Units may be added to plants over many years. The cumulative cost[10] per kWe and the in-service date and size of each unit[12] are given in DOE/EIA reports. Generally, most of the station capacity is included in one or two most recent units. An equivalent date is used for Fig. 1.4, usually not more than one year before the last large unit.

Costs are plotted against in-service date. Electric Power Research Institute (EPRI)-1978[13] range of expected costs is plotted at 1980. This is arbitrary,

as these costs are based on a plant going into service in 1978 with 1978 year-end costs, but conforming to the 1979 NSPS, hence with costs that would be incurred a few years after 1978.

Ebasco, about 1979, projected[15] a cost of $1266/ kWe for a three-unit coal-fired plant with service dates of 1988, 1989, and 1990.

It may be noted that the actual cost increase of these plants in current dollars is approximately that indicated by the producer energy price index,

Table 1.7. Cost of Fuels as Burned by Electric Utilities in the United States

	1967	1968	1969	1970	1971	1972	1973	1974	1975	1976	1977	1977	1978	1979	1980	1981
	Cents per Million Btu (As Burned)											(as delivered to plant)				(May)
Coal	25.2	25.5	26.6	31.1	36.0	38.1	40.5	71.0	81.5	84.8	94.7	94.7	111.6	122.4	135.2	146.7
Gas	24.7	25.1	25.4	27.0	28.0	30.3	33.8	47.9	74.2	103.6	130.2	130.0	143.8	175.4	212.9	282.7
Oil	32.2	32.8	31.9	36.6	51.5	58.8	80.3	192.2	202.4	198.9	222.4	220.4	212.3	299.7	427.9	552.8
Weighted average	25.7	26.1	26.9	30.7	36.4	39.9	47.5	90.8	104.8	112.4	129.7	127.7	139.3	162.1	189.3	250.8

Source: 1967–1977, Ref. 10; 1977–1980, Ref. 4.

Table 1.8. Average Electricity Prices in Cents/kWh[a]

Year	Residential		Commercial[b]		Industrial[c]		Total	
	I	II	I	II	I	II	I	II
1950	2.88	2.88						
1955	2.65	2.65						
1960	2.47	2.47						1.69
1965	2.25	2.25						1.59
1967	2.17	2.17						1.56
1970	2.10	2.10						1.59
1971	2.19	2.19						1.69
1972	2.29	2.29						1.77
1973	2.54	2.38	2.41	2.30	1.25	1.17	1.96	1.86
1974	3.10	2.83	3.04	2.85	1.69	1.55	2.49	2.30
1975	3.51	3.21	3.45	3.23	2.07	1.92	2.92	2.70
1976	3.73	3.45	3.69	3.46	2.21	2.07	3.09	2.89
1977	4.05	3.78	4.09	3.84	2.50	2.33	3.42	3.20
1978	4.31	4.03	4.36	4.10	2.79	2.59	3.69	3.46
1979	4.63		4.67		3.03		3.97	
1980	5.36		5.48		3.69		4.73	
June 1981	6.48		6.48		4.36		5.59	

Source. References 4 and 11.
[a]I = Class A and B privately owned utilities[4]; II = total electricity industry.[11]
[b]Edison Electric Institute designates this "small light and power."
[c]Edison Electric Institute designates this "large light and power."

namely, 3.9 times from 1960 to 1979. However, there is a wide spread in costs due to various factors.

Oil- or Gas-Fired Plants

The costs of a number of oil- and gas-fired plants, selected at random, are shown in Fig. 1.5, with data

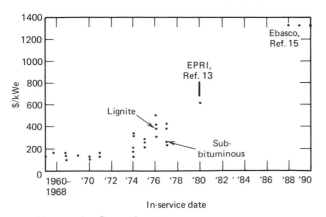

Figure 1.4. Costs of coal-fired steam-electric plants.

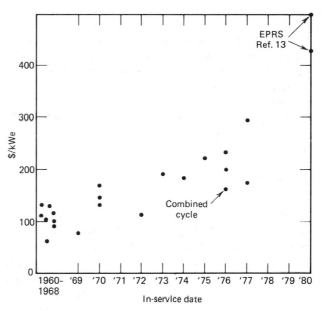

Figure 1.5. Costs of oil- and gas-fired steam-electric plants.

from the same sources.[10,12] As before, the EPRI 1978 "overnight construction" costs have been arbitrarily plotted at 1980—$425/kWe with distillate, $500/kWe with residual fuel. Here again, the producer energy price index is indeed representative of the trend.

Nuclear Plant Costs in the United States

Randomly selected nuclear plant costs from the same data sources[10,12] are plotted in Fig. 1.6. While no new nuclear plants are being priced in 1980, informed estimates are about $1500/kWe, for delivery several years hence. Since the Connecticut Yankee, 600 MWe, was $156/kWe in 1967, later increased to

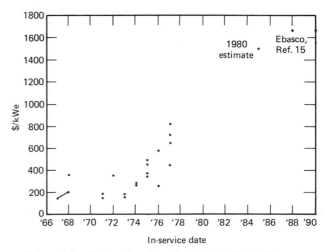

Figure 1.6. Costs of nuclear plants in the United States.

$195/kWe by modifications, it is clear that the cost of nuclear plants has far outstripped the producer energy price index. Increased design complexity, added environmental and safety requirements, and licensing and construction delays are among the reasons cited.[10]

Ebasco, about 1979, projected[15] a cost of $1648/kWe for a two-unit nuclear plant with service dates of 1988 and 1990.

Hydro Plant Costs[14]

Hydro plant costs have increased at least in proportion to the producer energy price index. However, the conditions are too variable to plot costs, particularly for multipurpose dams, where costs are simply "allocated." For three plants first reported in the 1978 supplement to Ref. 14, the costs were $388 to $503/kWe, but some costs have been considerably higher.

Cost Comparisons

In general, cost comparisons can be updated by considering separately the changes in feedstocks, capital costs, and operating costs. Recent comparisons, such as the EPRI study in Chapter 3, (Table 3.6 and 3.7 and Fig. 3.2), should require a minimum of modification for some years. The assumptions are clearly stated. Ten or 15 years is a short time to establish the relative economics of large power innovations.

REFERENCES

1. *United Nations Statistical Yearbook,* New York: Publishing Services, United Nations, 1973, 1979.

2. J. S. Hutchins, "Shale Oil in the 1980's," *Proc. 6th Annu.*

3. P. Kruger and C. Otte, Eds., *Geothermal Energy,* Stanford, CA: Stanford University Press, 1973.

4. *Monthly Energy Review,* DOE/EIA, 0035/08/80.

5. *U.S. Bureau of the Census, Statistical Abstract of the United States,* 101st ed., Washington, D.C.: Supt. of Documents, 1980, p. 476.

6. *Projected Annual Resource Requirements—Energy Forecast 1985 and 2000,* U.S. Department of Commerce, C1.2:R31, Washington, D.C.: Supt. of Doc., 1978. See also *Energy Projections to the Year 2000,* DOE/PE-0029, Washington, D.C.: Supt. of Documents, 1981.

7. U.S. Department of Energy, *Secretary's Annual Report to Congress,* Washington, D.C.: Supt. of Documents, 1980.

8. *Coal Data Book* (President's Commission on Coal), Pr. 39.8: C63/C63/2/980, Washington, D.C.: Supt. of Documents 1980.

9. U.S. Department of Energy, EIA, *Report to Congress, Vol. 2, Data,* Washington, D.C.: Supt. of Documents 1979.

10. *Steam-Electric Plant Construction Costs and Annual Production Expenses, 1977,* DOE/EIA, 0033/3, 1977, Washington, D.C.: Supt. of Documents, 1978.

11. *Edison Electric Institute, Statistical Yearbook of the Electric Utility Industry,* Washington, D.C.: EEI, 1978. And *Historical Statistics of the Electric Utility Industry* EEI. Pub. 62–69, Washington, D.C.: EEI, 1978.

12. *Inventory of Power Plants in the United States,* DOE/EIA, 0095(79), April 1979, Dec. 1979, Washington, D.C.: Supt. of Documents, 1979, 1980.

13. *Technical Assessment Guide,* Palo Alto: EPRI, July 1979.

14. *Hydro-Electric Plant Construction Cost and Annual Production Expenses,* DOE/EIA, 0171-78, Washington, D.C.: Supt. of Documents, 1978.

15. *Navy Energy Fact Book,* 0584-LP-200-1420, Philadelphia: Navy Publication Forms Center, 1979, p. 232.

16. "Energy," *National Geographic,* February 1981, special issue.

17. S. Baron, "Solar Energy: Economics vs. Energetics," *Mechanical Engineering,* May 1981, p. 35.

Int. Conf. on Coal Gasification, University of Pittsburgh, 1979.

NOTE: Most of these references are published annually or monthly. The latest issues can be used to update the tables in this chapter.

2

Petroleum, Shale Oil, and Tar Sands

PETROLEUM

While the term petroleum can be applied to all hydrocarbons, both liquid and gaseous, that occur in the earth, it is more generally applied to the liquid that is recovered from wells. It is also called crude oil. The gas component is called natural gas. Together, they constitute the principal prime energy sources of earth in current use. In 1978 oil, an inclusive name for petroleum and its products, provided 48.5% of all prime energy in the United States, and natural gas provided 25.6%, together 74.1% of the prime energy (Table 1.2).

Because of its great importance in the affairs of men, detailed information is available in handbooks and references on every phase of petroleum, from its origin nearly half a billion years ago to the thousands of present-day products derived from it for use as energy or materials.

What are the questions that should be answered by this chapter? In keeping with the premise of this book, they must be restricted to the physical facts that are timeless and unchanging and to historical facts, in distinction to forecasts or predictions, or to the attitudes of people concerning pollution, the environment, and social and national goals, which are continually changing. The problems to be solved in the future are of course intimately related to these latter questions, but sound solutions must start with the inexorable laws of nature and the facts of history, the subject matter of this book.

The broad scope of the subject and the ready availability of detailed information in the references dictate that only the most basic and summary information should be addressed here, that information which a person seeking a fundamental grounding in energy would want first. These questions are the following:

1. What is petroleum?
2. What are the ultimate products to be derived from petroleum, especially energy products?
3. What are the principal processes in refining the raw petroleum into its market products?

If these three questions can be answered so that the reader has a clear picture of the possibilities and limitations of one of our greatest resources, the chapter will certainly have served its purpose.

The energy resources, rates of use, and general cost structure have already been given in Chapter 1. The chemical composition of the hydrocarbons in petroleum and, of some importance, their names are given in Chapter 11. Even an encyclopaedic reference uses the chemical terms, alkane or paraffin, isomer, isooctane, aromatics, saturated, and so on. If this nomenclature is unfamiliar, a perusal of Chapter 11 will clarify it. The transportation and storage of energy are covered in Chapter 14.

What Is Petroleum?

Petroleum and the associated natural gas are mainly a mixture of hydrocarbons of from 1 to 30 carbon atoms per molecule. As the molecules become larger, there is a regular increase in boiling point with increased number of carbon atoms in the molecules, 20–30°C (36–54°F) per carbon atom, except for the first few, which have larger differences.

Hydrocarbons having one to four carbon atoms per molecule[1] boil below 0°C (32°F) at 1 atm and are thus gases at normal temperature and pressure. Hydrocarbons with 5 to 17 carbon atoms boil above 36°C (97°F) and are liquids under normal atmospheric conditions. Note that isomers of hydrocarbons with five carbon atoms may boil at as low as 12°C (54°F). Hydrocarbons with 18 or more carbon atoms are solids under normal conditions.

Of the five families of hydrocarbons described in Chapter 11, the three most stable families are the principal constituents of crude oil: (1) alkanes or paraffins, both normal and isoalkanes; (2) cycloalkanes, also called naphthenes; and (3) the aromatics. Some crude oils, such as Pennsylvania, consist mainly of alkanes (paraffins). Others, such as the heavy Mexican and Venezuelan crudes, are predominantly naphthenic and are rich in asphalt, a high-boiling semisolid material. Wax is usually but not always associated with paraffin-base crudes.

In addition to the hydrocarbons, petroleum contains compounds of sulfur, nitrogen, and oxygen in small amounts. These impurities are harmful unless removed. There are also usually traces of vanadium, nickel, chlorine, and arsenic.

The U.S. Bureau of Mines classifies crude oil as paraffin base, naphthene base, or mixed base, according to the properties of key fractions distilled from the oil. Naphthene base oils are also known as "asphalt base." Every intermediate gradation occurs in various oil fields throughout the world.

In addition to differences in the predominant families of hydrocarbons present, there is a wide spread in the percentages of different boiling point hydrocarbons present. For example, if samples from various wells throughout the world are separated by simple distillation into three fractions— those having boiling points below 330°F, called gasolines and naphthas, those between 330 and 650°F, called middle distillates (kerosene, gas oil, and diesel fuel), and those above 650°F, called fuel oils—variations are found even exceeding those shown in Table 2.1.

Thus, petroleum includes a very wide diversity of hydrocarbon mixtures, depending on the particular oil field, depth of burial, and other factors.

The products needed are equally varied. In Europe, a large percentage of petroleum is used for fuel oil. In the United States, with plentiful natural gas, the largest use of petroleum is for gasoline. But the proportion needed for gasoline or for fuel oil, in either case, of course varies from winter to summer.

The sophisticated methods now available for converting or "refining" from the highly varied petroleum input to the multitude of needed products are described later.

Specific Gravity

Reference has been made to "light" and "heavy" crudes. The heavier molecules do not occupy a correspondingly larger space in a liquid or solid; consequently the heavy crudes, with more larger molecules, have a higher specific gravity. The lighter crudes, with their greater proportion of gasoline or light fuel oil, are somewhat more valuable.

Specific gravity is generally measured by a hydrometer calibrated in an easily read scale, such as 0 to 100. The American Petroleum Institute (API) scale is commonly used. Its relation to the usual units (water = 1.0000) is shown in Table 2.2.

A gravity of 30° API is a rather common basis of reference. Most of the world production of petroleum lies between 25 and 35° API. A crude with a higher API gravity is a lighter crude with more of the valuable lighter products such as gasoline, or one that requires less refining to obtain the same yield of gasoline.

Heating Values of Petroleum and Products

Since petroleum is a highly variable mixture of hydrocarbons, it cannot be assigned a specific heating

Table 2.1.   Variation in Composition of Crudes

Fraction and Boiling Range	Heavy Crude (%)		Light Crude (%)
Gasolines and naphthas below 330°F (166°C)	6	to	33
Middle distillates (kerosene, gas oil, etc.) 330–650°F (166–343°C)	21	to	37
Fuel oils, above 650°F (343°C)	63	to	30
	100		100

Table 2.2.   Specific Gravity and Higher Heating Value (HHV) of Petroleum and Products

Specific Gravity[a] at 60/60°F	API[a] (deg)	Approx. HHV (Btu/lb) from Eq. 2.1
1.0000 (Times water)	10	18,540
0.9340	20	19,020
0.8762	30 (Most common)	19,420
0.8251	40	19,750
0.7796	50	20,020
0.7389	60	20,260
0.7022	70	20,460
0.6690	80	20,630
0.6388	90	20,777
0.6112	100	20,908

[a]Degrees API $= \dfrac{141.5}{\text{sp. gr. } 60/60°F} - 131.5$.

value like the specific compounds in Table 11.3. However, its heating value varies with its specific gravity and is very closely given (within 1 or 2%) by the expression

$$H = 22{,}320 - 3780d^2 \quad \text{(Btu/lb)} \quad (2.1)$$

where H is the higher heating value (See Chap. 11) and d is the specific gravity. These values are shown in Table 2.2 and apply as well to fractions of petroleum such as gasoline, kerosene, or fuel oil as to the complete mixture.

The specific gravity of petroleum is generally between 0.80 and 0.97, and the API gravity is between 45 and 15°. Its HHV therefore lies between 19,900 and 18,763. In this book, unless otherwise stated, oil is considered to have the U.S. average specific gravity of 0.85 (API 35°). This corresponds to 19,589 Btu/lb, or to 5.836 million Btu per 42-gal barrel. For general conversions from Btu to barrels of oil equivalent, this figure is rounded to 5.8 million Btu per barrel, or to 5.8 quads per billion barrels oil.

Just why should the heating values of the heavier hydrocarbons be less *per pound* than the lighter? Evidently, more energy is expended in breaking the larger molecules apart before their atoms can combine with oxygen to form CO_2 and H_2O.

Typical specific gravities and heating values of petroleum products are given in Table 2.3. The No.

1 fuel oil is for vaporizing-type burners. The No. 2 distillate is for general purpose (such as residential) heating in atomizing-type burners that do not require the No. 1 oil. The No. 4 oil is a light residual or heavy distillate for burners that can atomize heavier oils than domestic burners, but requires no preheating except in extremely cold weather. No. 5 may require preheating. No. 6, also called "Bunker C," generally requires heat in the storage tank to permit pumping and preheating at the burner to permit atomizing.

Carbon/Hydrogen Ratio of Petroleum

For the cycloalkanes (naphthenes), of formula C_nH_{2n}, the carbon/hydrogen ratio is 6:1 by weight. For the alkanes (paraffins), there are two more hydrogen atoms, C_nH_{2n+2}. But this increase is slight for the liquid and solid compounds.

For example, the carbon/hydrogen ratio of C_8H_{18} is 96:18 = 5.33; that of $C_{30}H_{62}$ is 360:62 = 5.81. Only for the aromatics is the ratio much different: 72:6 = 12 for C_6H_6 (benzene), and 96:10 = 9.6 for C_8H_{10} (xylene).

The aromatics are generally present in small amounts, so that the carbon/hydrogen ratio is never far from 6:1. In fact, almost all crude oils range between 82 and 87% carbon and between 13 and 18% hydrogen by weight.

Table 2.3. Typical Heating Values of Petroleum Products

Product	API Gravity (deg)	Sp. Gr.	Density (lb/gal)	HHV (Btu/lb)[a]
Gasoline	65.3	0.719	6.00	20,336
Kerosene	41.3	0.819	6.84	19,785
Military turbine fuel				
JP-4				18,725
JP-5				18,515
Commercial jet fuel				
A				18,589
B				18,744
Fuel oil Grade No.				
1	38–45	0.83–0.80	6.95–6.68	19,710–19,910
2	30–38	0.87–0.83	7.30–6.96	19,440–19,710
4	20–28	0.93–0.89	7.78–7.40	19,020–19,350
5L	17–22	0.95–0.92	7.94–7.69	18,810–19,100
5H	14–18	0.97–0.95	8.08–7.89	18,810–18,940
6	8–15	1.01–0.96	8.45–8.05	18,450–18,790
Crude oil, U.S. average	35	0.85	7.10	19,589

[a]Btu/gal = Btu/lb × lb/gal. Heating Oil, U.S. (Aug. 1980): 140,000 Btu/gal.

Petroleum Products

What are the products to be manufactured from petroleum? In a typical year, 1969, there were 3,371,751 thousand barrels of crude oil produced in the United States, and about 516,000 barrels imported (about 13%). The input and output of refineries in the United States[2] in the same year are given in Table 2.4. In subsequent years, both the total amount and the imports have increased greatly, but the percentage distribution of products has changed very little.

From Table 2.4, the following observations may be made:

1. Practically all of the crude oil produced or imported goes to refineries.

2. The principal product, gasoline, accounted for 49.5% of the product in 1969. It was 48% in 1959, 49.2% in 1949, 48.5% in 1939, and 42% (about equal to total fuel oil) in 1929. For many years, it has constituted about half the total product.

3. There is an "overage." In spite of obvious losses, the total of the products exceeds the input by 122 million barrels, or about 3%. As the heavy hydrocarbons are cracked in some of the refinery processes, the volume expands. Note that "still gas," used in the refinery, is accounted for as one of the products.

4. The kerosene includes kerosene-type jet fuels as used in commercial aircraft. Jet fuel includes only naphtha-type jet fuel, primarily for military aircraft.

5. This is a gross subdivision of the products as used in the petroleum industry. Each of these in turn is subdivided into more specialized products. For example, gasolines of different octane ratings must be manufactured, leaded and unleaded. Fuels for aircraft must have low freezing temperatures. There are three grades of distillate fuel oil and two grades of residual fuel oil, of varying viscosity (pour point) and impurities specifications. Similar subdivision by quality or characteristics occur in every category.

6. As mentioned earlier, over half the oil used in industrial plants goes into feedstocks, some 11.4% of the national use of oil in 1975. To reconcile this with the 3.5% listed for pet-

rochemical feedstocks in Table 2.4, we must include every nonfuel use. The total "feedstocks" are then in the same range, namely, 11 to 12% of total oil.

7. Naphtha does not appear in the list of major products, although it is an important fraction of the petroleum, as will appear later. In the United States, it is mostly used for naphtha-type jet fuels, as mentioned above, reformed into high-octane gasoline or hydrocracked into LPG gases, as discussed later. In many countries without natural gas, hydrogen, methanol, ammonia, and oxochemicals are derived primarily from naphtha

Table 2.4. Refinery Input and Output in the United States in 1969

Product	Thousands of Barrels	%
Input		
Crude petroleum		
Domestic	3,363,602	
Foreign	516,003	
Natural gas liquids	268,801	
Total	4,148,406	
(Earlier totals)		
1959	3,070,984	
1949	2,029,678	
1939	1,277,446	
1929	1,034,165	
1919	364,477	
Output		
Gasoline	2,050,804	49.5
Kerosene	318,690	7.7
Distillate fuel oil	846,863	20.4
Jet fuel	104,748	2.5
Residual fuel oil	265,906	6.4
Lubricating oil	65,080	1.6
Wax	6,049	0.1
Coke	102,868	2.5
Asphalt	135,691	3.3
Still gas	160,363	3.9
Road oil	9,086	0.2
Petrochemical feedstocks	146,268	3.5
Other finished products	92,798	2.2
Unfinished oil	− 34,346	− 0.8
Overage	− 122,412	− 3.0
Total	4,148,406	100.0

Source. Petroleum Facts and Figures, 1971.[2]

(Chapter 4, section on hydrogen manufacture), and LPG gases are derived largely from refineries (Chapter 4, section on LPG). Note: Syngas from coal is also widely used for these products (Chapter 4, section on methanol).

Refining Methods

We have now answered two of the three questions: What is petroleum? What products are manufactured from it? We now come to the third question: What are these refining methods?

As we have seen in Table 2.1, a heavy crude may contain only 6% of gasolines and naphthas combined, whereas in the United States 50% gasoline is needed. Also, if we simply distill the products from most crudes the gasoline obtained has an octane rating of only about 40. It is entirely inadequate for modern cars requiring 90 to 100 octane (see Chapter 11 for definition of octane). Obviously, something has to be done. The hydrocarbon molecules must actually be broken up or combined in some way to obtain more of the lighter products such as gasoline and changed to types of molecules that have higher octane ratings or other desirable characteristics.

Thus, refining involves three general steps:

1. Chemically altering the molecules.
2. Physically separating the products.
3. Purifying—may be chemical or physical.

The processes are not necessarily carried out in this order, but a refinery will involve all three steps, in various combinations, depending on the crude oil being used and the particular products being manufactured. Actually there is a fourth process—blending components into the desired useful products.

The general principles of the processes used in the first three steps will be briefly explained, the fourth step being fairly obvious.

Physical Techniques of Petroleum Refining

Fractional Distillation. While the compounds of different boiling points could of course be separated by simple distillation, the process is carried out much more efficiently in a continuous process using a fractional distillation column. The general principle is illustrated in Fig. 2.1, although there are many variations. An actual tower may be 150 ft high, process 100,000 or 200,000 barrels crude oil per day, and operate continuously for years.

The crude oil heated to 600 to 700°F, depending on the crude oil used and the products desired, is fed in part way up the column of trays. There may be 30 or 40 trays spaced at regular intervals, each having a large number of small holes, either "bubble caps" or simple perforations $\frac{3}{16}$ to $\frac{1}{4}$ in. diameter.

The products that are gaseous at the input temperature and pressure (approximately atmospheric) rise through the column and are condensed in a water-cooled condenser at the top. Except for a small amount of gas that does not condense and water which is separated out, the balance of the condensate—a mixture of gasolines, naphthas, kerosenes, and gas oils—is pumped back into the column at the top. It is "refluxed." It settles down through the column. Its lighter products reevaporate, and from the uprising stream it condenses out heavier products.

The uncondensed gas at the top is piped off to be used for heating purposes (still gas) or for further processing. A valve in this line controls the pressure in the column, usually to about 1 atm. When a steady state is reached, there is a gradual gradation of temperature and liquid products, from the point of entry to the top. Product sidestreams, or "fractions," are drawn off as shown.

Flexibility and control are provided by varying the location and magnitude of the various sidestreams and the ratio of the reflux to the product drawn off. Usually, the reflux is from one to three times the amount of "overhead" product.

The heavier hydrocarbons, fuel oils, and residuals that are liquid at the entrance temperature settle down through the lower set of trays (stripping trays) and are drawn off at the bottom. Steam is admitted near the bottom and assists in the separation of some further light products as it rises through the column, thereby increasing the yield of the more valuable lighter products.

Typical boiling point ranges for the products are:

Product	Temperature	
	°F	°C
Light gasoline (overhead)	75–200	25– 95
Naphtha (#1 sidestream)	200–300	95–150
Kerosene (#2 sidestream)	300–450	150–230
Gas oil (#3 sidestream)	450–650	230–340

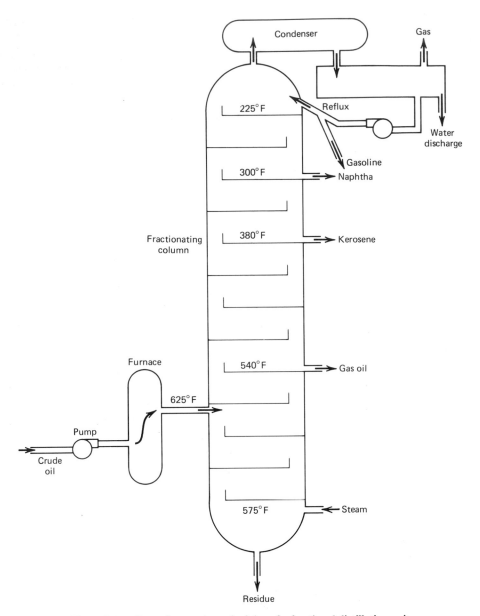

Figure 2.1. General operating principles of a fractional distillation unit.

But these ranges can be varied within wide limits by controls, as mentioned above.

The residue may be blended directly into fuel oil or subjected to vacuum distillation, as described next.

Vacuum Distillation. In order to separate the heavy hydrocarbons such as the residue from the fractional distillation above without at the same time cracking them, vacuum distillation is used, with the temperature limited to about 750°F (400°C). The vacuum is produced by steam ejectors. The

column is similar in most respects to the fractionating column described above, but of larger diameter in order to limit the vapor velocities at the reduced pressure.

The sidestreams in this case provide stock for manufacturing lubricating oil and feedstocks for catalytic cracking (below), whereas the residue is asphalt or bitumen.

Superfractionation. By increasing the number of trays in the column, for example, to 100, and increasing the reflux ratio to about 5, the fractions can be limited to very few or even to one single hydrocarbon. Ninety percent pure isopentane and also isohexane and isoheptane concentrates, all of very high octane number, can be produced for aviation gasoline.

Other Physical Processes. These include:

1. Absorption and stripping—used for example to recover the LPG gases from the gas at the top of a fractionating column.
2. Solvent extraction to remove particular undesired components—as aromatics from kerosene.
3. Adsorption (in zeolites) to remove particular molecular shapes—as normal paraffins from gasolene.
4. Crystallization—used, for example, to remove wax from lubricating oil.

Changing Molecular Structure

By the methods described above, petroleum can be separated into groups of compounds of different boiling point ranges. Also, families or homologous groups can be separated. However, this provides neither the proper proportion of products nor the required quality of products to meet the needs. The molecular structure must actually be altered to mold the crude petroleum into the desired mix of products. The processes used for changing molecular structure are the following:

Cracking (by heating) converts large molecules into smaller ones suitable for gasolines or fuel oils. Thermal cracking is used for cracking kerosene or light oil into gasoline of about 70 octane. Catalytic cracking performs a similar function at higher efficiency. Typical catalysts are clay or molecular sieves.

In hydrocracking with catalysts, excess hydrogen is used in cracking heavy oils into middle distillates and gasoline, residuals into lighter oils, and straight-run naphthas into liquefied petroleum gases—all without the simultaneous formation of coke and large quantities of gas. Thus, the catalyst does not need to be continually regenerated to burn off the carbon.

Vis breaking is the cracking of heavy crude to fuel oil. Steam cracking is used, for example, to obtain ethylene from naphtha.

Polymerization is used to combine two small molecules into a larger one in the gasoline range, such as propylene to gasoline.

Alkylation is used to combine two small molecules into one of high octane.

Reforming is the process of changing the structure of particular molecules. Naphtha of less than 40 octane may be reformed to a gasoline of 70 to 80 octane.

Petroleum Purification

Objectionable impurities in petroleum include sulfur, oxygenated compounds such as alkyl phenols and naphthenic acids, nitrogen bases, gums and gum-forming constituents, and unstable compounds that can lead to color formation in the product. Of these, sulfur is the most objectionable.

The principal purification processes used are sulfuric-acid treatment, which removes most impurities into an acid tar; sweetening (oxidizing the mercaptans); mercaptan extraction (with the aid of solutizers); activated clay treatment (absorbs impurities); hydrogen treatment (to remove sulfur, nitrogen, and oxygen); and molecular sieves, which have a strong affinity for polar compounds such as water, CO_2, H_2S, and mercaptans.

SHALE OIL

What is shale oil? What are the processes in getting from the shale in the ground to a "syncrude" refinery input equivalent to petroleum? From there on the processing closely parallels that of petroleum and will not be repeated.

What Is Shale Oil?

Shales are thinly layered sedimentary rocks widely dispersed throughout the surface of the earth. In certain of these shale deposits, organic material has been laid down, intermixed with the inorganic material. It has partially decomposed and metamorphosed over the ages until it exists now as a mixture of organic substances locked within the shale. It cannot move about as does petroleum and natural gas. This organic material contained in shale is called *kerogen*.

The shale contains no oil as such. However, if an "oil-bearing" shale is broken up and retorted (pyrolyzed) at 900°F (482°C) or higher, the kerogen decomposes and a large part of it is given off as an oily vapor which condenses to a viscous liquid shale oil. Also some gas is given off, about 6% of the energy, which does not condense, and a carbonaceous residue remains on the rock, about 14% of the energy.

Unlike petroleum, this liquid shale oil contains large amounts of compounds of nitrogen and large amounts of sulfur. These must be removed by a process such as hydrogenation (high-temperature hydrogen treatment) before a syncrude refinery stock similar to crude oil results.

The energy content of oil shale is equivalent to 0 to 100 gallons oil (petroleum) per ton of shale, 25 gallons per ton being a representative figure for the richer beds in the United States (parts of Colorado, Wyoming, and Utah) that will be used for some decades at least.

The Recovery Problem

Now we can begin to visualize the mining problem involved in recovering oil from shale. A ton of coal, at 12,000 Btu/lb, provides 24 million Btu. A gallon of 0.92 sp. gr. shale oil, 18,460 Btu/lb, weighs 7.68 lb and contains 142,000 Btu. Thus, 25 gallons from a ton of shale provides 3.55 million Btu. Some 6.8 tons of good oil shale must be mined to get as much energy as from 1 ton of coal.

In 1978, oil supplied 2.72 times as much energy as coal in the United States (Table 1.2). Had this *all* come from U.S. shale oil, the amount of shale to be mined and processed would have been over 18 times, by weight, the amount of coal actually mined (about half that by volume). Furthermore, the main deposits are concentrated in about $\frac{1}{30}$ of the area of Colorado. Thus, while the U.S. shale oil reserves far exceed the whole world petroleum reserves (Fig. 1.1), there are formidable problems in its use. There has been no large-scale development or use through 1980.

The Project Independence Task Force on Oil Shale[3] (1974) estimated that shale oil production would be economic with world oil at $7.00 per barrel or higher and placed possible U.S. production at 450,000 to 1,600,000 barrels a day by 1990 (15 years), depending on the degree of stimulation. While either of these figures represents huge mining operations, involving tonnage of 47 to 167% of 1978 U.S. coal tonnage, they still represent but 2.4 to 8.5% of probable U.S. oil requirements in 1990 and would require invested capital of $2.5 to $8.2 billions.

Even at these production levels, air pollution would be at or beyond acceptable limits, depending on the particular state or federal limits considered. Water requirements would require additional dams at the higher level.

In spite of high world oil prices, $32 per barrel in 1980, the high general inflation, uncertain energy policy, environmental problems, and conditions unfavorable to risk capital have provided negative "stimulation"[4] and have effectively stifled this development to date (1980). Many limited production projects are, however, in progress.

The Green River Formation of Oil Shale

The largest known oil shale deposit in the United States, the Green River Formation, covers a 16,500 square mile area in contiguous parts of Colorado, Utah, and Wyoming and contains 2100 billion barrels of oil. The Colorado reserves are contained within 3000 square miles (3% of the state area) and are the richest, thickest, and best defined. They contain 84% of the oil, with 10% in Utah and 6% in Wyoming.

One of the richest and most accessible portions is in western Colorado near the town of Rifle. In this area, the oil shale formation lies in three zones, one above the other:

Main Zone	480 to 630 ft thick
Middle Zone	230 to 270 ft thick
Lower Zone	205 to 220 ft thick

These are separated generally by 50 to 150 ft of transitional beds with little or no oil. However, in some areas the beds are contiguous, forming an oil-shale "measure" of about 1300 ft in thickness.

The Mahogany Ledge, the richest portion of the entire Green River formation, constitutes the bottom 73 ft, average, of the top or main zone and has an average yield of 30 gal/ton. It extends for about 1000 square miles and contains about 100 billion barrels of oil, some three times the nation's proved petroleum reserves.

Assaying Oil Shale[5]

The oil content of oil shale is usually determined by the modified Fischer retort method. A 100-g sample of crushed shale is heated in 49 minutes to 900°F and held there for 20 minutes, by which time essentially all of the oil obtainable under these conditions

Table 2.5. Average Composition of Kerogen in Colorado Green River Shale[a]

Element	Percent
Carbon	76.1
Hydrogen	10.5
Nitrogen	2.6
Sulfur	1.3
Oxygen	9.5
Total	100.0

[a]C/H ratio 7.2, avg. sp. gr. 1.05.
Source. Reference 8.

has exited, and is condensed in a vial. The gases escape. The water is centrifuged out and the oil is weighed.

Composition of Oil Shale and Products

Throughout a given formation, such as the Green River oil shale formation of the United States, the mineral and kerogen compositions are nearly constant. For the Green River formation, the mineral content has specific gravity of about 2.7, and the kerogen, of about 1.05. The kerogen composition is shown in Table 2.5. The oil produced per ton is directly proportional to the amount of kerogen per ton.

When the kerogen is decomposed in a commercial or assay retort, the oil, gas, and residue account for the fractions of weight and heating value shown in Table 2.6. The weight of oil is 66.3%, or two thirds of the weight of kerogen, for the Green River formation. It varies from 10 to 80% for various formations throughout the world.

The shale oil has the properties shown in Table 2.7, HHV 18,460 Btu/lb and sp. gr. 0.916 (API

Table 2.6. Products of Decomposition of Kerogen by Retorting

Product	Percent by Weight	Percent by Heating Value
Oil	66.3	79.4
Gas	9.1	6.4
Organic residue	24.6	14.2
Total	100.0	100.0

Source. Reference 8.

Table 2.7. Average Composition of Shale Oil, Mahogany Ledge

Element Percent by Weight		Other Properties
Carbon	84.59	C/H Ratio = 7.3
Hydrogen	11.53	HHV = 18,460 Btu/lb
Nitrogen	1.96	Sp. gr. 60/60°F − 0.916
Sulfur	0.61	Kinematic viscosity
		= 20 centistokes

Source. Ref. 8.

23). It must be hydrogenated to form a high-quality syncrude refinery input.

Approximate Assay[6]

Since the composition and specific gravity of kerogen and minerals are nearly constant over the entire Green River formation, approximate assays can be made by simply measuring the specific gravity, bulk density, or HHV of the shale.

For example, 25 gal/ton corresponds to $25 \times 8.35 \times 0.916 = 191$ lb oil/ton, or 191/0.663 = 288 lb kerogen/ton, and 1712 lb minerals/ton. Hence, the specific gravity of the shale is $(288/2000)1.05 + (1712/2000)2.70 = 2.46$. The bulk density is $62.4 \times 2.46 = 153.5$ lb/ft^3, and the HHV (since 79.4% of the energy is in the oil) is $(191/2000)(18460/0.794) = 2220$ Btu/lb shale. Similarly, each oil content corresponds to a particular value of specific gravity, bulk density, and HHV.

At 100 gal/ton (the upper limit) there would be 1152 lb kerogen per ton, or 57.6% by weight (78% by volume).

Nitrogen and Sulfur

The nitrogen in Colorado shale varies directly with the kerogen content and appears to be associated entirely with the organic material. The sulfur is distributed in the organic and inorganic material.

Specific Heat

The specific heat of the shale varies with the kerogen content and the temperature range but is of the order of 0.33 Btu/lb °F. The following comparison is therefore of interest: At 25 gal/ton shale, the gas contains about 285,000 Btu, the organic residue contains about 631,000 Btu.

To heat one ton of shale from 60 to 900°F, neglecting losses, requires $2000(900 - 60)0.33 = 555,000$ Btu. Thus, for the rich shale, yielding 25 gal/ton, the gas and organic residue contain enough energy to retort the shale. This is used in some of the processes to be described.

Oil Shale Production

As mentioned, the richest and most accessible oil shale deposits are in western Colorado. The mining and processing of these is of greatest interest. As of 1974, the following processes of mining and retorting were being considered and were the basis of economic estimates and growth estimates given earlier.

Mining

The majority of oil shale mining was expected to be underground room and pillar mining, confined initially to the 30- to 90-ft-thick beds of the Mahogany Ledge in Colorado. Some 60 to 75% of the mined strata can be recovered using 60-ft square rooms, with pillars 60 to 75 feet high. Shale would be brought to the surface for retorting.

While surface mining can be used, it is applicable to only 15 to 20% of the minable shale in Colorado. It would be limited to relatively small overburden, since the underground deposits are very thick and economically mined.

Retorting

The three commercially acceptable above-ground retorting methods as of 1974 all involve crushing the shale and heating it to 900°F (482°C), or higher, to decompose the organic material. These methods yield shale oil and gas. They differ in the manner of heating. Some combination of these retort methods will probably be used.

Tosco II Process. Externally heated balls are fed through a rotary kiln containing preheated shale crushed to $-\frac{1}{2}$ in. After transferring part of their heat, the balls are separated out and reheated. The oily vapors are collected and condensed externally.

Union Oil Co. Process. Air is fed down through a column of shale that is being forced upward by a "rock pump." The downdraft of air burns off the carbonaceous residue in a combustion zone just below the surface, providing the necessary heat. The high temperature combustion products move on down the column, pyrolyzing the shale. The shale in this case is crushed to about 3 in. in size, with "fines" removed. By using oxygen instead of air, gas of high Btu content can be produced. Part of the vapors condense on the shale and exit as liquids. The remainder exit with the gas and are condensed externally.

Gas Combustion Process. Recycled gases flow up through a gravity downfeed of shale. They are heated by the hot spent shale. About a third of the way up, air and more recycle gas are fed in through a distributor. There is a combustion zone just above this, creating a retorting temperature near the top. Carbonaceous material from the shale is burned off as well as the recycle gas. There are several variations of this general principle.

Paraho Process.[10]   An example is the Paraho technology, which has been developed over the last 10 years (1981) in a research facility at Anvil Points, Rifle, Colorado, the earlier BuMines facility. The Paraho retort is a vertical steel vessel with a firebrick lining. Shale crushed to $\frac{1}{4}$ inch to 3 inches moves downward by gravity through four zones as it descends in the retort: mist formation, retorting, combustion, and cooling.

Recycle gas rising through the lower part of the bed cools the spent shale so that no water is required for cooling or transport to the disposal site. The combustion zone is near the center of the bed, air and additional recycle gas admitted through distributors at two levels. Residual carbon on the shale and recycle gas provide all the necessary heat. Above the combustion zone retorting takes place, converting the kerogen into shale oil and gas, which become a mist as they are cooled by the incoming shale in the upper zone. These products are drawn off near the top of the retort, separated and stored.

Over 4.6 million gallons of Paraho shale oil have been produced, more than by any other process (May 1981). It has been refined into all types of civilian and military oil products which have met practically all specifications. Shales from Israel and Morocco have been processed and preliminary tests made on shales from 12 other countries. While 4.6 million gallons is a large sample, 110,000 bbls, it should be noted that an oil refinery processes 100,000 to 200,000 barrels per day. Thus the Anvil Points plant is still a pilot/research facility. Paraho plans a commercial facility at a site in Utah.

In Situ Production

In situ, or in place, processing of shale oil was in an experimental stage in 1976. Research interest stems from four potential advantages it offers:

1. Two thirds fewer people to operate the process.
2. One half to two thirds the amount of water needed by some above-ground retorts.
3. One third to no disposal of spent shale.
4. Increased recovery, since lower-grade shales can be processed.

However, there are many difficult problems to be solved before these benefits can be realized. Research is being carried out on two concepts, one fully in situ, the other partially, as follows:

Borehole, or Horizontal Sweep, Processing— Fully In Situ. A pattern of boreholes is drilled into the shale to be processed. Communication between the holes is established hydraulically or with chemical explosives. Heat is then applied either by establishing a fire in the shale or by injecting hot gases or liquids. The resulting retorted shale oil is driven to "producer" boreholes for removal.

Modified In Situ Processing. A vertical underground retort is prepared by mining a cavity below the shale to be retorted. The shale is then fragmented by drilling and blasting, filling the cavity, which is then fired to retort the shale. The oil is pumped from a previously prepared sump below the retort cavity.

The mined shale, about 25% of the total, would be retorted above ground by methods described previously.

Occidental-Tenneco.[11] An example of this process is the Occidental-Tenneco in situ retort being prepared on the c-b (Colorado-B) lease in the Piceance Creek Basin near Rio Blanco, Colorado, since approval by the Secretary of the Interior on June 30, 1977. Occidental holds a joint lease with Ashland Oil. Three shafts were being sunk from the surface, main, service, and ventilation, and were at about 1700 ft depth in May 1981.

Occidental has been producing small quantities of high quality shale oil on Mt. Logan, near DeBeque,

Colorado, for several years. The company "intends to move forward immediately toward full-scale production of 50,000 barrels a day" at the c-b tract location. That level should be reached in 1983.

Problems inherent in the in situ technique include (1) obtaining uniform fracturing, (2) obtaining uniform heating as the flame zone moves through the fracture zone, and (3) self healing of the fractures. Only practical experience can answer these questions and provide the basis for improvements in techniques.

Nuclear In Situ Processing. The nuclear method envisioned is similar to the modified in situ method, except that the initial cavity is produced by a nuclear explosion below the shale bed. A "chimney" of fractured shale above collapses into this void and is retorted. The nuclear charge is placed in a large borehole, 1 to 2 ft in diameter, and no mining as such is required. This concept has been discussed since 1958. However, there have been no nuclear experiments in actual oil shale beds up to the present (1980), and none is being actively considered.

As mentioned earlier, most production through 1990 is planned using underground mining and above-ground retorts. Practically all major U. S. oil companies hold leases in the Green River Formation[12] and are either involved in development or have plans for future production.

TAR SANDS (OR BITUMINOUS SANDS OR OIL SANDS)

In certain sand beds, the interstices between the grains of sand are largely filled with a thick tarry hydrocarbon mixture that can be extracted and used like petroleum. This bitumen can be processed to yield 65 to 75% synthetic crude oil. By present methods, the sand must be mined and transported to a separation plant. Only open-pit mining is practical, and since most of the deposits are buried too deeply for this, there are relatively few places where commercial, large-scale operations are possible at present.

Athabasca Tar Sands

One of these deposits is the Athabasca Tar Sands in northern Alberta, Canada, along both sides of the Athabasca River. Here, midway between Hudson

Bay and the Pacific Coast, in the foothills of the Rocky Mountains, is an area of about 780 square miles, out of the 9000 square miles extent of the sand beds, where the overburden has been eroded to a workable depth (0 to 150 ft). The area in which surface mining is economic contains proved reserves of 38 billion barrels bitumen in sands yielding 5% bitumen by weight or more, and with overburden under 150 ft and an overburden/pay-sand ratio of 1.0 or less. At an estimated conversion ratio of 0.7, some 26.5 billion barrels synthetic crude oil (syncrude) can be produced.

There are about 626 billion barrels crude bitumen in sands yielding 2% by weight or more. Of this, 74 billion barrels have overburden under 150 ft, and 552 billion barrels have overburden of 150 to 2000 ft. In situ methods for recovery of bitumen from sands yielding less than 5%, or having more than 150 ft overburden, are being studied.

In 1964, the Great Canadian Oil Sands Ltd. started construction in the Mildred-Ruth area on a large-scale operation to produce 50,000 barrels per day (BPD) of syncrude. This installation is now complete, and for several years it has produced 45,000 to 50,000 BPD, with plans to increase to 58,000 BPD by 1981. By 1978, a consortium, Syncrude Ltd. was bringing into operation a 130,000 BPD facility. In mid-1980, the combined output of these two plants was 179,000 BPD, up from 94,000 BPD average in 1979. A third plant by Shell Canada and others was planned by 1985, 125,000 BPD; and with future plans, a combined output of 800,000 BPD is forecast by 1995. In 30 years, 800,000 BPD would use up 8.8 billion barrels of the 26.5 billion-barrel prime resource.

By-products of the original 50,000 BPD plant were 2900 tons/day coke and about 300 tons/day sulfur.

Most of the Athabasca tar sands, as well as those elsewhere, including the Utah tar sands, must await in situ methods of oil recovery. The energy density, even for medium-grade oil sands, is greater than in most oil fields, so that in situ appears to be a promising development for the future.

Since the Athabasca development is the most advanced, the characteristics of these oil sands and the associated recovery operation will be described.

Formation of Tar Sands

Geologists are not agreed on how the oil got into the sand. However, this much seems clear. These sands were at one time along the shore of a sea, blown by winds, washed by waters—some salt seas, some fresh waters. Dunes of all shapes and descriptions were formed. Some were clean sands of quartz granulations, with fine material, clay and shale, largely washed out. These have been most fully impregnated with oil. Others contain more fine material. These have been less fully impregnated with oil. Finally, there is sand quite full of fine material, and much shale, which has been only slightly impregnated with oil. The oil impregnation does not extend over the full extent of the sand beds but stops short of it along one side.

All of these dunes and beds are intermixed haphazardly, so that an extensive deposit of good-grade oil sand with small overburden is rare. The Mildred-Ruth workings is one of the few.

Classification of Oil Sands

The Athabasca sands have been explored quite extensively by boreholes and the results classified as follows:

Good-Grade Oil Sand. The -200-mesh material is less than 20% by weight, usually under 10%. Oil is over 10% by weight, average 13.5 to 15%. Clay is about 1%, water, 2 to 5%. In these sands, the natural sorting action during bar formation removed most of the fines.

Interbedded Oil Sand and Shale. The -200-mesh material is of medium percentage. Oil is 4 to 10% (by definition), average 6.5%. Water averages 9%.

Shale. There is a high percentage of -200-mesh material. Oil is less than 4% (by definition), usually 1 to 2%. Water is 10 to 15%.

Out of 90 holes bored on the west side of the Athabasca River, totaling 14,000 ft of section, the proportions of different grades of oil sand were found to be as follows:

	Avg. Percent Oil by Weight	Percent of Section
Good-grade oil sands in beds at least 5 ft thick	13.5	40
Interbedded oil sand and shale	7	35
Shale	2	25

Table 2.8. Characteristics of Good-Grade Oil Sand—Athabasca

	Mean	Range
Specific gravity (water = 1.0)	1.9	1.75–2.09
Porosity (%)	40.5	34–46
Oil saturation (%)	65	45–90
Water saturation (%)	20	1–40
Total saturation (%)	85	65–100
Oil yield (gal/yd³)	53	31–84

Source. Reference 7.

Characteristics of Oil Sand

The approximate characteristics of good-grade oil sand are shown in Table 2.8. The permeability of the mineral content is high—20 darcys or more. With 30% silt content, it drops to about 250 millidarcys.

The mean yield of good grade oil sand, 53 gal/yd³, at a mean sp. gr. of 1.9, is 33 gal/ton, the same order as good grade oil shale.

Processes of Extraction and Treatment

The oil obtained before treatment is naphthenic and heavy. Its specific gravity at 25°C (77°F) is 1.002 to 1.027. The lighter oil will flow at ordinary temperatures; the heavier oil can hardly be said to flow. It contains about 5% sulfur.

The process that has evolved as most satisfactory and economic for the open-cut Athabasca (McMurray Formation) oil sands involves the following steps. The cost estimates, 1957 basis, for a production of 20,000 BPD are shown below at the right to give an idea of the relative costs of the different operations.[7] The operation of the Great Canadian Oil Sands Ltd. was based on 50,000 BPD.

	Cost ($/Barrel)
Large rotary cutters, loading into belts / Conveyor belts to the separation plant	0.55
Hot water separation to separate the oil from most of the sand.	0.36
Fluidized-bed "coker" retort to complete the separation from sand and water and to coke the residue.	0.36
Desulfurization by mild hydrogenation	0.81
Pipeline to Great Lakes and associated storage	1.02
Total	3.10

At these costs, syncrude was competitive with petroleum in the Great Lakes marketing area as a refinery input. It consisted of about 12% gasoline and 88% gas oil quite similar to the No. 2 fuel oil. Its sulfur content was about 0.5%. It was a high-grade refinery input.

U.S. Tar Sands

In the state of Utah, there are 24 tar sand deposits that have been explored enough to determine that they contain about 28 billion barrels of oil (U.S. consumption in 1978 was 6.4 billion barrels). Seventeen of these deposits are in the Uinta Basin in N.E. Utah and contain some 10 to 11.3 billion bar-

Table 2.9. Properties of P.R. Springs Deposits of Tar Sands in N.E. Utah

	Asphalt Wash	Three Mile Canyon	South Seep Ridge Area
Number of zones (one above the other)	2	1	4
Total net thickness (ft)	23–29	23.8	71–112
Occurrences, depth (ft)	56–263	189 avg	27–254
Sand avg. porosity (% of bulk volume)	24.7	29.2	25.6
Avg oil saturation (% of pore volume)	58.1	67.4	36.5
Oil gravity (API)	10.9	10.5	8.2–16.5
Oil sp. gr.	0.9937	0.9966	
Zone thickness (ft)			13–24
Avg. oil yield per yd³ of sand (gal)	29	40	19

Source. Ref. 9.

rels of viscid (sticky), "sweet" oil with less than 0.5% sulfur.

The P.R. Spring deposits in this basin, some 350 square miles, contain about 4.5 billion barrels and have been quite thoroughly explored by BuMines.[9] While there are outcroppings of low overburden, in general the overburden is far too great for open-pit mining. Recovery awaits the development of in situ methods. Some characteristics of these sands are given in Table 2.9. Note that the oil yield may be low due to low porosity, low saturation, or both. The Three Mile Canyon yield approaches 1 barrel per yd^3.

REFERENCES

1. R. T. Morrison and R. N. Boyd, *Organic Chemistry*, 3rd ed., Boston: Allyn and Bacon, 1973.

2. American Petroleum Institute, *Petroleum Facts and Figures*, Washington, D.C.: API, 1971.

3. Project Independence Task Force, *Potential Future Role of Oil Shale: Prospects and Constraints,* Washington, D.C.: Federal Energy Administration, 1974.

4. F. A. L. Holloway, "Tar Sands and Oil Shale," *The Bridge,* Washington, D.C.: *Nat. Acad. Eng.,* Summer 1977, p. 7.

5. K. E. Stanfield and I. C. Frost, *Method of Assaying Oil Shale by a Modified Fischer Retort*, Rept. of Invest. 4477, Pittsburgh: BuMines, 1949.

6. J. W. Smith, *Theoretical Relationship Between Density and Oil Yield for Oil Shales*, Rept. of Invest. 7248, Pittsburgh: BuMines, 1969.

7. K. A. Clark, *Athabasca Oil Sands. Historical Review and Summary of Technical Data*, Contrib. No. 69, Edmonton, Alberta, Canada: Research Council of Alberta, 1957.

8. K. E. Stanfield, I. C. Frost, W. S. McAuley, and H. N. Smith, *Properties of Colorado Oil Shale*, R.I. 4825, Pittsburgh: BuMines, 1951.

9. L. A. Johnson, L. C. Marchant, C. Q. Cupps, *Properties of Utah Tar Sands, Asphalt Wash Area, P.R. Spring Deposit*, R.I. 8030, Pittsburgh: BuMines, 1975.

10. Oil Shale Brochure, Grand Junction, CO.: Paraho Development Corp., May 1981.

11. P. J. Ognibene, "Oil From Shale—Is It Worth the Price?", *Parade*, Nov. 13, 1977, p. 28.

12. *Energy Fact Book*, Arlington, Va.: Tetratech, 1975, p. G-5.

3

Coal, Lignite, and Peat

We are now near the end of the age of oil and gas and near the beginning of the age of coal and nuclear energy. While large-scale use of coal preceded that of gas and oil, and coal was in fact the fuel of the industrial revolution, the immense reserves of coal have scarcely been tapped. The more convenient gas and oil, once discovered, have been used at such prodigious rates that they can last but a few more decades.

However, the days of direct burning of high-sulfur coal are also numbered, and ways must be found to use this abundant fuel without polluting the atmosphere. These include gasification, liquefaction, and various ways of cleaning either the coal or the stack gases.

The information presented here provides an understanding of this changing role of coal. Since the literature on coal is vast and readily accessible, our task is to encapsulate the most important facts, characteristics, and processes upon which this transformation will be based.

The facts about reserves, rates of use, and general cost structure of fuels are given in Chapter 1. The gasification of coal is covered in Chapter 4, the transportation and storage of fuels in Chapter 14, and the chemistry of the hydrocarbons involved and their combustion in Chapter 11.

Coal is a highly variable substance. In addition to the variations from lignite to bituminous and anthracite, there are vast differences in its heating value, amount of volatiles, caking properties, sulfur, moisture, and so on. In the past selected coals have been used for selected applications. However, as the use of coal expands to substitute for diminishing oil and gas supplies, with conversion to liquid and gaseous fuels increasing by orders of magnitude, more universal burning and conversion methods are being developed which can use a wide range of coals.

THE COMPOSITION AND PROPERTIES OF COAL

Coal is thought of as a "pure coal substance" of organic origin which has been intimately mixed with various impurities and with moisture. The six items included in an ultimate analysis of coal are: carbon, hydrogen, oxygen, nitrogen, sulfur, and ash. Carbon, hydrogen, and oxygen are combined in the original glucose molecules, $C_6H_{12}O_6$, of plant life, which form the carbohydrates, cellulose, and so on. Nitrogen is taken up by living plants to form proteins, essential parts of both plant and animal life. The sulfur and ash are mainly mineral, although part of the sulfur is organic. Proteins sometimes contain sulfur and phosphorous. Values of these six items are shown in Table 3.1 for five random coal samples ranging from lignite to anthracite.

The properties of coal of particular importance are its *heating value,* the amounts of *fixed carbon* and *volatile material,* and the amounts of *moisture, sulfur,* and *ash.* The coal is of course solid and contains no volatile material. However, when it is heated above 400°C (752°F), some of the coal material decomposes and is given off as gas. This is referred to as the volatile material in the coal.

The heating value is determined in a bomb calorimeter (Chapter 11), as with other solid or liquid fuels. The sulfur is determined by special tests (later) or by ultimate analysis. The other four items are determined by a special test known as the "proximate analysis," D 3172-73 (D items refer to the applicable ASTM standards).

Proximate Analysis

In this analysis, the pulverized coal is first heated to just above boiling, 104 to 110°C, to drive off the moisture M, which is weighed by difference,

Table 3.1. Sources and Analyses of Various Ranks of Coal as Received

Classification by Rank	State	Bed	Proximate (%)				Ultimate (%)					Calorific Value (Btu/bb)
			Moisture	Volatile Matter	Fixed Carbon	Ash	Sulfur	Hydrogen	Carbon	Nitrogen	Oxygen	
Anthracite	Pennsylvania	Clark	4.3	5.1	81.0	9.6	0.8	2.9	79.7	0.9	6.1	12,880
Medium-volatile bituminous coal	Pennsylvania	Upper Kittanning	2.1	24.4	67.4	6.1	1.0	5.0	81.6	1.4	4.9	14,310
High-volatile-C	Illinois	No.5	1.4	34.5	40.6	9.6	3.8	5.8	59.7	1.0	20.1	10,810
Sub-bituminous-B	Wyoming	Monarch	22.2	33.2	40.3	4.3	0.5	6.9	53.9	1.0	33.4	9,610
Lignite	North Dakota	Unnamed	36.8	27.8	29.5	5.9	0.9	6.9	40.6	0.6	45.1	7,000

Source. Ref. 17.

D 3173-73. The volatile material *VM* is then driven off by heating to 950°C for 7 minutes in a closed crucible and weighing by difference, D 3175-73. The remainder is then burned in air at 700 to 750°C, and the remaining ash *A* is weighed, D 3174-73. The balance burned off is the fixed carbon, $FC = 100 - (M + VM + A)$, all quantities in percentages, D 3172-73.

Parr Formulas

The classification of coal is based on the pure coal substance, without the moisture and minerals, which are chance inclusions. The proximate analysis is based on the coal as received. After much research, the Parr formulas, D 388-77, have been standardized for correcting the results of the proximate analysis to the properties of the pure coal substance. They are

$$\text{dry, Mm-free } FC = \frac{(FC - 0.15\,S)\,100}{100 - (1.08\,A + 0.55\,S)} \tag{3.1}$$

$$\text{dry, Mm-free } VM = 100 - \text{dry, Mm-free } FC \tag{3.2}$$

$$\text{moist, Mm-free Btu} = \frac{(\text{Btu} - 50\,S)}{100 - (1.08\,A + 0.055\,S)} \tag{3.3}$$

with all materials in % and Btu/lb. S is the percent sulfur and Mm is "mineral matter." Simpler "approximate" formulas have been adopted for all but exact legal definition, D 300-77.

Classification of Coals by Rank

Coal is classified into ranks as shown in Table 3.2. The ranks correspond to the various stages of coalification but are precisely defined in terms of the properties of the coal. Higher rank coals are identified by either fixed carbon or volatile material, and the lower ranks are identified by calorific (heating) value. The agglomerating character is also given when it is a factor in identification. Coals are considered agglomerating if, in the test to determine the volatile matter, they produce either a coherent button that will support a 500 g weight without pulverizing or a button that shows swelling or cell structure.

The variation of fixed carbon, volatile material, and moisture with rank is shown in Fig. 3.1, the volatile becoming very small in anthracite and about equal to the fixed carbon in lignite.

Plastic Properties of Coal

When coal is heated through the range of 400 to 500°C (752 to 932°F), it passes through a plastic stage. It partially melts or fuses and loses its original coal structure. As its temperature is raised, it reaches its state of greatest fluidity. Both chemical and physical changes take place in the coal material. Then, at a higher temperature, still under 500°C, it solidifies again after losing most of its volatile material. The initial stage of plasticity may or may not be accompanied by swelling, and the latter stage involves contraction as part of the solid material is volatilized and given off. The remaining solid material may stick tightly together, "glued" as it were by

Table 3.2. Classification of Coals by Rank

Class	Group	Fixed-Carbon Limits (%) (Dry, Mineral-Matter-Free Basis)		Volatile-matter limits (%) (Dry, Mineral-Matter-Free Basis)		Calorific value limits (Btu/lb) (Moist, Mineral-Matter-Free Basis)		Agglomerating Character
		Equal or Greater Than	Less Than	Greater Than	Equal or Less Than	Equal or Greater Than	Less Than	
I. Anthracite	1. Meta-anthracite	98			2			
	2. Anthracite	92	98	2	8			
	3. Semianthracite	86	92	8	14			Nonagglomerating
II. Bituminous	1. Low-volatile bituminous coal	78	86	14	22			
	2. Medium volatile bituminous coal	69	78	22	31			Commonly Agglomerating
	3. High-volatile A bituminous coal		69	31		14,000		
	4. High-volatile B bituminous coal					13,000	14,000	
	5. High-volatile C bituminous coal					11,500	13,000	
						10,500	11,500	Agglomerating
III. Subbituminous	1. Subbituminous A coal					10,500	11,500	Nonagglomerating
	2. Subbituminous B coal					9,500	10,500	
	3. Subbituminous C coal					8,300	9,500	
IV. Lignite	1. Lignite A					6,300	8,300	
	2. Lignite B						6,300	

Source. ASTM Standard D 388, which see for footnotes.

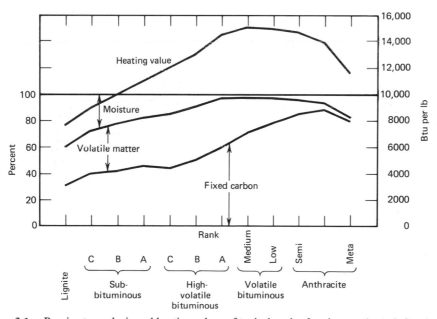

Figure 3.1. Proximate analysis and heating values of typical coals of various ranks (ash-free basis).

the tar and pitch before the volatiles were lost, or it may be noncaking.

These great variations in plastic properties have a profound influence on the various processes using coal: direct combustion, coke making, liquefaction, and gasification. Very limited swelling can be tolerated without wrecking a coke oven. However, unless coals agglomerate or stick together, they will not produce a metallurgical coke. If they do stick or agglutinate appreciably, they are unsuitable for any of the fixed-bed processes of combustion or conversion. They form an impervious mass in the firebed that blocks the flow of air.

Gieseler Plastometer

Several indices of agglomerating tendency are used, such as the coke-button test described earlier (Table 3.2) or the "free swelling index," D 720-77. However, the Gieseler plastometer, D 2639-74, actually measures the shear properties of the plastic coal. Movements of the torque-loaded rabble arms in small crucible of ground coal as the temperature is raised register the initial softening temperature, the maximum fluidity and temperature, and the final solidification temperature.

The several ways of dealing with the agglomerating characteristic, such as pretreatment or fluidized or entrained beds, are treated in later sections.

Characteristics of Coal and Its Constituents

Strength of Coal and Coke. Larger lump sizes are less susceptible to moisture absorption and spontaneous combustion. The breaking up of coal into smaller pieces, or even "fines," in handling or weathering is most undesirable. The strength of coal is measured by a Drop Shatter Test, D 440, or a Tumbler Test, D 441.

Grindability of Coal. The power required for auxiliaries may be 5% in a bituminous coal power plant and 10% in an anthracite coal plant (before scrubbers), the difference being largely the increased grinding energy. Grinding energy is determined by the Hardgrove Machine, a miniature ball mill in which the coal is compared with standard samples provided by the ASTM. D409-71.

Moisture in Coal. Starting with the inherent "bed moisture," the coal may either absorb moisture or dry out in handling and storage, particularly with small or slack sizes. Large lumps vary little from the bed moisture. See Table 1, Fig. 1, and Proximate Analysis.

Ash in Coal. The softening temperature of ash varies from 1830 to 3010°F (1000 to 1654°C). Slag tapping requires a temperature slightly above the softening temperature for adequate fluidity.

The standard test, D 1857-68, for fusibility of coal and coke ash involves slowly heating a $\frac{3}{4}$ in. high pyramid of ground ash (with binder) and noting the temperatures of (1) initial softening, (2) softening temperature (spherical lump), (3) hemispherical temperature (height = $\frac{1}{2}$ of base), and (4) fluid temperature (depth $\frac{1}{16}$ in.) Softening temperature is frequently used to determine the suitability for slag tapping.

Heating Values—DuLong's Formula. Heating values are determined by the bomb calorimeter (See Chapter 11). The HHV of coal may also be approximated by DuLong's formula:

$$\text{Btu/lb} = 14,544\, C + 62,028\left(H - \frac{O}{8}\right) + 4050\, S$$

$$(3.9)$$

where C, H, O, and S are weight fractions from the ultimate analysis. For coals from anthracite through bituminous ranks, the values usually agree with the bomb calorimeter within 1.5%. Sub-bituminous coals and lignite may deviate 4 or 5%.

Sulfur in coal may occur as pyritic (FeS_2), organic, or sulfate. Organic sulfur may constitute 20 to 85% of the total. In low-sulfur coal, the sulfur is mainly organic. In high-sulfur coal, the added sulfur is mainly mineral (pyritic). Sulfates of iron or calcium are found in weathered coal, there being relatively little in freshly mined coal.

The sulfur can be determined without a full ultimate analysis. For example, a weighed sample of the coal can be mixed with Eschka mixture and burned. The sulfur is then precipitated from the resulting solution as barium sulfate, $BaSO_4$, which is then filtered, ashed, and weighed.

The three forms of sulfur can then be separated by successive extractions with hydrochloric acid (for sulfate) and nitric acid (for pyritic). The remainder is organic.

The percentage of sulfur in coal ranges from a trace to 12%. Most commercial coals have between

0.3 and 5.5%. In 1975, electric utility coal averaged 2.9% sulfur.

Nitrogen. Runs from a trace to 3% in coals of various ranks, entirely in the organic material.

Weight and Bulk of Coal and Coke. Coal is a rock, since minerals are defined as inorganic with fixed chemical composition. It is about half as heavy as most rocks or minerals, bituminous coal having a specific gravity of about 1.4. Most of the minerals in its ash have a specific gravity of about 2.7, except pyrite, which is about 5. Due to the increased degree of compaction, there is a regular increase in specific gravity with rank, as shown in Table 3.3.

The bulk density of crushed coal depends on the degree of crushing, whether the fines are removed, the moisture content, and the degree of compaction. For coal as received, the bulk density is determined by shoveling loosely into an 8 ft³ box and weighing. Coal in piled storage may be compressed as much as 25% and may absorb up to 25% moisture. Consequently, bulk densities can only be given as rather broad ranges.

Other Properties of Coal. The structure or lithological classification of coal is treated in detail by Rose.[1] Coals have been classified according to use, considering some 33 selection factors.[2] Hardness, strength, and grindability are treated thoroughly by Yancey and Geer, including a scratch test for hardness.[3]

Summary of Composition and Properties of Coal

The composition, classification, and principal characteristics of coal have now been presented as follows:

1. Its rank and the associated fixed carbon, volatile matter, heating value, and agglomerating characteristics.
2. It plastic properties when heated above 400°C (752°F).
3. Its physical characteristics, hardness, strength, grindability, weight, and bulk.
4. The characteristics of its principal constituents, fixed carbon, volatile material, moisture, and ash or minerals.

Table 3.3. Weight and Bulk of Coal and Coke

| | Specific Gravity | | |
Substance	Solid	Piled	Bulk Density, Piled[a] (lb/ft³)
Graphite	1.64–2.7		
Anthracite	1.4–1.8	0.75–0.93	47–58
Bituminous	1.2–1.5	0.64–0.87	40–54
Lignite	1.1–1.4		
Peat, dry	0.65–0.85	0.32–0.42	20–26
Coke	1.0–1.4	0.37–0.51	23–32

Source. Courtesy BuMines.
[a]Before compaction.

5. Its heating values.
6. It sulfur and nitrogen.

The uses of coal are profoundly influenced by the very wide range of characteristics that are found in all of these categories.

THE USE OF COAL

In the past, most coal has been burned directly, although over the past century and a half an increasing amount has been converted to gaseous and liquid fuels and to petrochemicals. About 15% has been made into metallurgical coke. The current conversion technology is the basis of many very large industries supplying manufactured gas, ammonia, refinery products, and petrochemicals. Over 100 coal conversion methods have been developed[4] or are in process of development at the present time (1980). With a few exceptions, the developed methods have been tailored or limited to coals of particular characteristics.

However, none of the fully developed methods are suitable for the coal conversion requirements of the next quarter century. Processes for large-scale production of high-Btu pipeline-quality gas, of low- and medium-Btu gases for electric utilities and industry, of liquid and solid clean fuels from coal, of clean combustion techniques, and of coal or stack-gas cleaning are all under intensive development. Few have developed beyond the pilot plant stage (1980).

The reason for this is twofold: (1) Events of the last 10 years have only recently brought a full awareness of the impending exhaustion of oil and gas reserves and the consequences. (2) During the same period, environmental concerns have resulted in the institution of rigid control of sulfur and nitrogen emissions in all new facilities. While both factors have greatly stimulated the developments needed for clean use, and also greatly expanded use of our large coal reserves, the time has been too short to bring any of them to a successful conclusion.

The use of coal in the United States in 1975 and that expected in the year 2000 is given in Table 3.4. Compared with this projected fourfold increase in coal production in the next quarter century, the increase in the last quarter century was less than 15% overall.

The coal used in the United States in 1975 was overwhelmingly of the high-sulfur variety. It ran about 12,000 Btu/lb, and that used by electric utilities had an average sulfur content of 2.4 lb/10^6 Btu. This represents a potential SO_2 emission of 4.8 lb/10^6 Btu, since the weight of SO_2 is twice the weight of atomic sulfur.

Low-sulfur coal is considered to have less than 0.5% sulfur by weight; and since it runs about 8000 Btu/lb, its sulfur content is about 0.6 lb/10^6 Btu. Its potential SO_2 emission is 1.2 lb/10^6 Btu.

New Source Performance Standards (NSPS)

The New Source Performance Standard in effect in 1977 limited the SO_2 emission of new plants to 1.2 lb/10^6 Btu. Thus, high-sulfur coal required about 75% sulfur removal, on the average, and low-sulfur coal required little or no sulfur removal.

New Source Performance Standards—June 11, 1979[5]

The federal NSPS in effect since June 1979 requires that SO_2 emissions from new coal-fired plants be limited to 1.2 lb/10^6 Btu (520 ng/J). Also, a 90% reduction from the potential SO_2 emissions is required except when emissions are less than 0.6 lb/10^6 Btu (260 ng/J). In this case, a minimum reduction of 70% from the potential emissions is required.

The NO_x (nitrogen oxides) emissions must be limited to 0.6 lb/10^6 Btu (260 ng/J), except 0.5 lb/10^6 Btu (210 ng/J) for subbituminous coal. Somewhat more NO_x is permitted, 0.8 lb/10^6 Btu (340 ng/J), if over 25% by weight of the fuel is lignite from North Dakota, South Dakota, or Montana.

No_x emissions are dependent primarily on combustion conditions, particularly temperature, rather than on the coal composition.

Table 3.4. Production and Use of Coal in the United States

	Millions of Tons	
	1975	2000 (Estimated)
Electric utilities	400	1500
Other	200	850
Total	600	2350

Source. EPRI Journal, Nov. 1976, p. 8.

NSPS Examples

Application of the 1979 NSPS to new plants may be illustrated by two examples: (1) the use of high-sulfur coal having potentially 4.8 lb/10^6 Btu of SO_2 emissions, and (2) the use of a low-sulfur coal having potentially 1.2 lb/10^6 Btu of SO_2 emissions.

For the high-sulfur coal, a 90% reduction would leave 0.48 lb/10^6 Btu of SO_2 emission. Since this is less than 0.6 lb/10^6 Btu, the emission need be limited only to 0.6 lb/10^6 Btu. Note that this results in an 88% reduction and therefore meets the 70% requirement also.

For the low-sulfur coal, a 70% reduction from the potential SO_2 would leave 0.36 lb/10^6 Btu. Since this is well below 0.6 lb/10^6 Btu, the low-sulfur coal is governed entirely by the 70% reduction requirement.

A 9% sulfur coal of 12,000 Btu/lb would have potentially 15 lb/10^6 Btu of SO_2 emissions. A 90% reduction would leave 1.5 lb/10^6 Btu. It would have to be further reduced to 1.2 lb/10^6 Btu, the maximum allowed in any case. This is a 92% reduction.

Future Emission Limitations

While the requirements of the year 2000 are not yet fixed, a consideration of the possible regulatory actions leads the electric utility industry to expect that SO_2 emission from high-sulfur coal may be controlled to 0.4 to 0.8 lb/10^6 Btu by the year 2000. Thus, high-sulfur coal, with a potential of 4.8 lb/10^6 Btu, would require 83 to 92% sulfur removal to meet these limits.

By the year 2000, it is expected that 350 to 650 million tons of low-sulfur coal will be available to the electric utilities, requiring at least 70% sulfur removal.

Thus, it can be seen that the coal conversion, or alternate sulfur-removing techniques, envisioned for the next quarter century bear little relation to those of the past, being of the order of 50 to 100 times greater in magnitude and applying to over 90% of all coal used in the United States.

COAL CONVERSION TECHNOLOGY

Coal conversion methods may be roughly divided into three groups: (1) those used by electric utilities and large industries for sulfur control; (2) the production of SNG, synthetic or substitute natural gas; and (3) syngas, largely CO and H_2, for the manufacture of a wide variety of fuels, chemicals, and refinery products. In countries where oil and natural gas are not produced, syngas from coal has been widely used as a feedstock. It has largely replaced coal tars because of its simpler composition, the greater quantities available, and its greater flexibility for synthesis. As oil and natural gas become exhausted, syngas will no doubt be used more in the United States also. Over 10% of the natural gas and over 50% of the oil used by industrials in the United States is used for feedstocks (see Table 1.2).

The plan of treatment in this book is as follows. All methods of using coal in electric utilities, solid, liquid, and gas are first covered broadly in this chapter, in order to give a rough economic comparison. Then the solid and liquid phases will be treated, as well as some methods that are general and produce a variety of fuels. All gasification methods are treated in Chapter 4. However, since these processes are all under intensive development at the present time (1980) and hence subject to many changes, only the basic concepts of the more important techniques are outlined.

The processes treated later in this chapter include three direct combustion and seven conversion processes, as follows:

1. Atmospheric fluidized-bed combustion process.
2. Wet alkali scrubbers.
3. Regenerative scrubbers.
4. Solvent-refined coal (SRC), Gulf and others, liquid or solid (prills).
5. Exxon Donor Solvent (EDS), a petroleum-type fuel.

6. H-Coal, Hydrocarbon Research, a petroleum-type fuel, ebullated catalyst.
7. Synthoil, BuMines/ERDA, a petroleum-type fuel, packed bed system.
8. Fischer–Tropsch synthesis, the conversion of syngas into variety of hydrocarbon products.
9. Char Oil Energy Development (COED), F.M.C. Corp., the production of a variety of fuels by multistage pyrolysis.
10. SRC-II, producing a range of liquid and gaseous products.

Sulfur Control Measures in Electric Utilities

Of the many sulfur control measures under development in 1980, those receiving principal consideration by the electric utility industry are indicated in Table 3.5. All of these have sulfur-removal capabilities adequate for the 1979 standards, NSPS, and most of them can meet more stringent standards in the future, at increased cost. These include fluidized-bed combustion, incorporating sulfur removal, the direct firing of coal with stack-gas scrubbers, and the use of clean solids and liquids derived from coal. They also include gasification with sulfur removal. MHD and fuel cells in the future will also use clean fuels derived from coal.

In addition to the methods listed in Table 3.5, a number of others—coal cleaning, dry alkali bag filters, and several regenerative scrubbing processes—have been omitted because of their early stage of development. Also, only those processes

---

Table 3.5. Nine Clean Ways to Burn Coal

1. Direct firing of pulverized coal with flue gas desulfurization (FGD) (wet alkali scrubbing)
2. Direct firing of pulverized coal with regenerative FGD.
3. Atmospheric fluidized-bed combustion (AFBC). (Dry limestone is added to the bed)
4. Pressurized fluidized-bed combustion (PFBC).
5. Direct firing of pulverized solvent-refined coal (SRC).
6. Liquid firing of a petroleum-type fuel derived from coal.
7. Gasification-combined-cycle plant (GCC).
8. Magnetohydrodynamic plant (MHD). (Sulfur combines with carbonate seed and is removed from stack gases.)
9. Fuel cell using liquid fuel derived from coal.

have been included which have adequate sulfur removal potentiality to meet the likely standards of the year 2000.

Future Generating Plant Options

For the nine clean coal-burning systems of Table 3.5 and several other options—oil, gas, nuclear, batteries, and underground pumped storage—the power plant costs and performance data are shown in Table 3.6.[6] These are based on end-of-year 1978 dollars for a plant going into service at that time.

Actually, some of the options shown require extensive development before first commercial service. These developments are currently (1981) underway, and the dates of possible first commercial service, pending success of ongoing developments, are shown.

The coal burning plant of the past (before FGD) would be most comparable to the solid SRC-steam plant shown in the table, $675/kW capital cost (1978 basis), 9680 Btu/kWh heat rate. Oil- and gas-fired plants of 1980 would correspond roughly to the oil-steam plant, $440 to $515/kW (1978 basis), 9680 Btu/kWh heat rate.

Table 3.6. Power Plant Cost and Performance Data (All Costs in End-of-Year 1978 Dollars)

Plant	Total Capital ($kW)	First Commercial Service	O&M Costs Fixed ($/kW-yr)	O&M Costs Variable (mils/kWh)	Mature Plant Forced Outage (1)	Average Heat Rate (Btu/kWh)
Nuclear (1000 MW)	820	Current	3.1	1.5	18.2	10,700
Coal-limestone FGD						
500 MW	800	Current	12.8	3.6	17.6	10,150
1000 MW	730	Current	12.9	3.5	19.5	9,735
Coal-regenerative FGD						
(500 MW)	845	Current	14.2	2.9	17.4	10,600
Gasification-combined cycle (1000 MW)						
Texaco—2000°F GT	815	1990	14.4	1.5	5.5	9,250
BGC Slagger—2600°F GT	655	2000	19.9	1.8	7.1	8,155
Atmospheric fluid bed						
(500 MW)	700	1992	10.8	5.2	12.8	10,250
Pressurized fluid bed						
(600 MW)	680	1994	13.3	3.6	12.1	8,720
MHD (1000 MW)	1010	2001	29.0	2.8	28.6	7,640
Solid SRC-steam (500 MW)	675	1989	2.8	2.7	16.3	9,680
Oil-steam (500 MW)						
Distillate	440	Current	1.4	1.5	16.3	9,680
Residual	515	Current	1.4	1.5	16.3	9,680
Combustion turbine (75 MW)	190	Current	0.25	2.7	8.0	13,800
Combined cycle (250 MW)						
distillate	340	Current	4.4	1.1	6.9	8,600
Advanced combined cycle (250 MW)						
Distillate	365	1989	4.4	1.1	5.0	7,520
Residual	425	1989	5.5	1.35	6.9	7,620
Fuel cell (10 MW)						
1st Gen. (distillate)	445	1986	3.3	3.2	5.6	9,000
Advanced (distillate)	475	1992	3.3	3.2	5.6	7,300
Batteries—5 hour						
advanced	630	1992	0.3	2.0	3.5	72% eff.
Underground pumped						
Hydro—10 hour	575	1991	1.5	—	1.8	72% eff.

Source. Ref. 6, copyright 1980, published by Government Institutes, Inc., Rockville, Md., U.S.A.

Table 3.7. Fuel Price Scenarios—East Central Region ($/10⁶ BTU, Constant End-of-Year 1978 Dollars)

	1978	1990	2000	2020
Uranium (with reprocessing)	0.45	0.51	0.56	0.68
Bituminous coal (4% S)	1.43	1.60	1.70	2.05
Petroleum				
Distillate	4.50	5.35	6.85	9.00
Residual (0.4% S)	3.73	4.45	5.90	7.90
Coal-derived fuels				
Upgraded liquids	5.10	5.35	5.60	6.05
Full-range liquids	4.40	4.60	4.80	5.20
Solid SRC	3.55	3.75	3.90	4.20

Source. Ref. 6, copyright 1980, published by Government Institutes, Inc., Rockville, Md., U.S.A.

Note the low capital cost of the combustion turbine, $190/kW, that justifies its use for peaking service, even with the 13,800 Btu/kWh heat rate and expensive fuel.

The expected fuel costs for 1978 to 2020 are shown in Table 3.7, coal-derived liquids becoming cheaper than petroleum fuels after 1990.[6] Refined fuel for the gas turbine (upgraded liquid) is over three times the cost of bituminous coal.

Most Likely Choice

Based on these data, computer studies have been run[6] to determine which type of unit would be most economic as a function of date of installation, region of the country, and number of hours of operation per year. Figure 3.2 shows the results of such a study for the East Central Region (Wisconsin, Illinois, Indiana, Ohio, and Kentucky). Studies were made both with and without nuclear as an option. If nuclear is an option, it is uniformly most economic for all base load in the future (3000 hr/yr or more). However, even if current (1981) uncertainties surrounding the nuclear option are resolved, it may not be selected exclusively. The chart shows in dashed lines which type of plant would be *next* most economic.

Thus, in the East Central Region, coal with scrubbers or oil would be the economic selections for intermediate or peaking duty, respectively, until 1986. If nuclear is ruled out, coal with scrubbers will also be the economic selection for base load. About 1986, a plant based on coal liquids becomes more economic over its lifetime. Coal liquids are used both for peaking (combustion turbine) and intermediate (steam), until underground pumped storage and batteries become commercial. After that, coal liquids continue to have a narrow field of application.

For base load (above 3000 hr/yr), coal gasification–combined cycles are the economic choice after 1990, when this system becomes commercial (Table 3.6), if nuclear is not an option. Pressurized fluidized-bed combustion would be most economic for installation in the period 1993 to

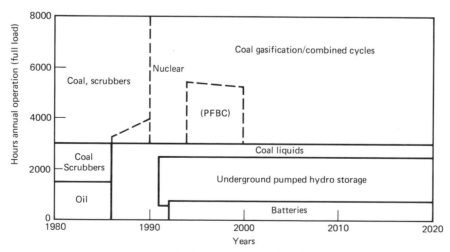

Figure 3.2. Preferred technologies for East Central Region, life-cycle cost. *Source.* R. A. Loth, O. D. Gildersleeve, Jr., S. A. Vejtasa, and R. W. Zeren, Reference 6, Copyright 1980, published by Government Institutes Inc., Rockville, Md., USA.

2000 for plants running 3000 to 5500 hr/yr if nuclear is not an option.

Based on the findings of these studies and similar studies in other regions, it appears that scrubbers and coal liquids are clearly economic options for the future. Coal gasification–combined cycles, when commercial, will be the economic choice for base load if the nuclear option is not available. Both atmospheric and pressurized fluidized-bed combustion have substantial fields of application in various regions, barring nuclear. These are therefore the primary coal-related developments being supported by the electric power industry and being given principal focus in this book.

The general principles of the gasification schemes for electric generation are covered in Chapter 4, along with gasification for other purposes. The various coal-burning and liquid-fuel schemes are outlined in the following paragraphs.

Direct Burning of Coal with Sulfur Removal

The amount of sulfur dioxide emitted to the atmosphere from combustion of fuels can be reduced (a) by removing the sulfur before combustion, (b) by fixing the sulfur by direct reaction in the fuel bed, or (c) by scrubbing the exhaust gases after combustion.

Coal cleaning or washing can at most remove only those sulfur compounds in segregated form, mainly pyrites. Since these constitute about 41 to 79% of the sulfur in high-sulfur coal[7] and since at least 90% removal will generally be needed, this measure is inadequate by itself. It may however be used as a preliminary step to reduce the overall cost.

Fluidized-Bed Combustion[8]

Direct reaction in the fuel bed is best accomplished by fluidized-bed combustion (FBC). If compressed gas is passed upward through a bed of inert particles, at a sufficient velocity to overcome gravity, each particle will just float on the gas stream in a boiling turbulent mass. The surface appears as a boiling fluid. This is a fluidized bed. It has been used since the 1920s for certain chemical reactors, the reactants being added to the inert particles of the bed. Since 1944, it has been used for combustion to produce steam and to burn low-quality fuels, as in incinerator boilers.

In the case of fluidized-bed combustion, the fuel particles are added to the inert mass. They may constitute only 1% of the mass and thus cannot adhere to each other or agglomerate. The inert particles may be sand or ash. The entire bed is at a red-hot combustion temperature and is continually mixed. Almost any solid or liquid fuel can be fed in at bottom, middle, or top and *usually* results in a residence time adequate for complete combustion. A gaseous fuel can be fed at the bottom. The fluidizing gas is the combustion air in slightly over-stoichiometric proportion plus fuel gas if used.

Limestone or dolomite can be added in proper proportion to combine with the sulfur in the fuel and produce solid sulfate particles, which can be removed with the ash.

The boiler tubes are immersed in the bed, where the excellent heat transfer from the fluidized solids permits a lower combustion temperature, less scaling, and a smaller and hopefully lower-cost boiler. The production of oxides of nitrogen is also greatly reduced at the lower combustion temperature.

At this combustion temperature, the ash does not melt. Ash is tapped off as it accumulates to keep the bed level constant.

The limestone or dolomite is proportioned from 2.5 to 4.0 times the theoretical combining weight for the sulfur present and results in 90 to 95% sulfur removal. However, there is a substantial increase in the amount of waste for disposal, much more than is required for wet-alkali scrubbing of stack gas. A reduction to 1.5 times the theoretical minimum would be most desirable both because of cost and waste disposal. However, disposal in solid form is less of a problem than sludge from a scrubber. The calcium and magnesium sulfates may be useful for land conditioning as crops remove sulfur.

Thus, FBC has several advantages: (a) direct removal of sulfur during combustion, (b) low nitrogen emission, (c) ability to burn a variety of fuels, including low-grade fuels, (d) lower combustion temperature and less slagging of boiler tubes, and (e) smaller size.

Atmospheric Fluidized-Bed Combustion (AFBC).[8] If the gas is not pressurized beyond the small pressure required for fluidizing, the system is termed "atmospheric." The clean, hot exhaust gas passes to the stack in the normal manner. This process simply replaces the conventional boiler of the steam cycle.

However, a dual-cycle arrangement is also under investigation at the Oak Ridge National Laboratory.

The compressed air between the recuperator and the gas turbine is heated by boiler tubes in the fluidized bed instead of by fuel in the gas turbine combustor. Waste heat from the gas turbine is then used to raise steam for the steam turbine (see Fig. 13.9—recuperated gas turbine).

Pressurized Fluidized-Bed Combustion (PFBC)[8]. There has always been an interest in a pressurized coal-fired boiler, so that the exhaust gases could drive a gas turbine before giving up their heat to a steam cycle. However, the flyash is highly abrasive and quickly erodes the turbine blades. Cyclones can remove 95% of the flyash, electric precipitators 99%. But this is not adequate with the hard abrasive flyash produced at normal combustion temperatures.

With fluidized-bed combustion, because of the excellent heat transfer, the combustion temperature can be much lower. The flyash is much softer and less abrasive. Again, the dual cycle is being tried, with good chance of success. The fluidized bed is pressurized, typically to 16 atm for a 16:1 compression ratio gas turbine. Exhaust gas supplies heat to the steam cycle.

State of the Art—1980. While fluidized-bed technology dates from the 1920s, and fluidized-bed combustion from 1944, there was little interest in new coal-burning technology until 1974. At first, this interest centered in coal gasification and liquefaction, but soon turned to direct burning as well, as the cost and difficulties of the other alternatives became apparent. Thus, while there has been remarkable progress in the last few years, it will no doubt be 1984 before any full-scale atmospheric or pressurized FBC power plants are in operation (DOE goal). Some of the most significant developments are the following.

Fluidized-bed technology is well established for small boilers, 10,000 to 100,000 lb/hr, equivalent to 1.2 to 12 MWe, and for incinerator boilers, heat treatment, ore roasters, and a variety of other uses. Hundreds are in operation.

In 1977, seven large-scale industrial prototypes were under construction in the United States and Britain. Atmospheric FBC boilers up to 600,000 lb/hr (75 MWe equiv.) were being offered with warranty by several firms.

It is generally believed that atmospheric FBC will be most suitable for industrial applications but that dual-cycle pressurized FBC will be needed to pro-

vide any real advantage for the larger electric utility units. While more complex, they offer higher thermal efficiency and other advantages. Feasibility and design studies are underway for American Electric Power Co. by B&W Ltd. and the British affiliate of Stal-Laval, which could lead to commercial orders for a demonstration plant of 170 MWe about 1980.[8] Advantages cited are (a) direct sulfur removal, (b) higher efficiency, (c) smaller size and lower cost of boiler, and (d) the spent sorbent does not significantly increase the waste disposal problem.

DOE is sponsoring a Curtiss–Wright PFBC dual-cycle pilot plant of 13 MWe (7 MWe gas turbine), using crushed dolomite and a 1650°F bed temperature. One third of the compressed air, used for combustion, will be cleaned. Two thirds will pass through boiler tubes. In addition, the blades will bleed air to cool and to deflect any residual flyash.

Several conceptual design studies have been commissioned for both AFBC and PFBC concepts of 500 to 600 MWe full-scale plants. DOE has contracted with Curtiss–Wright and others for the design of a 200 MWe demonstration PFBC plant.

DOE is proceeding with extensive Component Test and Integration Units (CTIUs) at the Morgantown, West Virginia, Energy Research Center and at Argonne National Laboratories.

The largest operating FBC units are:

1. The B&W boiler at Renfrew, Scotland, converted in 1975, 40,000 lb/hr, equivalent to about 5 MWe.
2. The OCR (now DOE) financed 30 MWe boiler on the Monongahela Power Co. at Rivesville, West Virginia, operating since 1977. Already made obsolete by later developments, it has been discontinued (1981).
3. The municipal heating boiler at Enkoeping, Sweden, 25 MWt, equivalent to about 8 MWe.

All are atmospheric. There are many smaller test units, both AFBC and PFBC, in the United States and elsewhere.

Flue Gas Desulfurization (Wet Alkali Scrubbing)[7,9]

Since coal washing is inadequate and fluidized-bed combustion has not yet reached the demonstration plant stage, most utilities are turning to wet alkali scrubbers for sulfur removal in new plants (1976–1980). In 1976, the 10 companies finalizing selections for new plants practically all decided on direct lime–limestone scrubbing in spite of several nasty unresolved problems. Only a few companies favored regenerative or other processes that are less developed. The possible variations in scrubber systems are limitless, and only a few principal developments can be mentioned here.

Basic Principle of Wet Alkali Scrubbing. If sulfur dioxide gas, SO_2, is brought into contact with a lime and water mixture (calcium hydroxide), the sulfur dioxide reacts with the solution to form a calcium salt, a sulfate, $CaSO_4$, or a sulfite, $CaSO_3$. These precipitate out as solids, forming a sludge.

The necessary intimate contact can be achieved by running the alkaline solution over closely spaced plates through which the flue gas flows.[7] However, in recent large installations[21,22] the alkaline solution is sprayed directly into the flue gas stream, as well as wetting the surfaces. Particulate removal may be combined with the sulfur removal in a two-stage scrubber,[21] or an initial precipitator may be used to remove most of the particulates followed by a scrubber to remove the sulfur and remaining particulates.[22]

Bruce Mansfield Plant. At the 2.4 GW Bruce Mansfield plant at Shippingport, Pennsylvania both of these methods are in use. The Chemico[21] two-stage scrubber is used on unit 1, in service 1976, and on unit 2, in service 1977. A precipitator and Kellogg-Weir scrubber[22] is used for unit 3, in service 1980. Both systems are designed to remove 92.1% of the sulfur dioxide and practically all of the particulates.

In the Chemico system the flue gases are admitted at the top of the scrubber and raised to high velocity in a variable venturi, where they are thoroughly saturated with the lime-water solution. The heavy droplets containing most of the pollutants fall to the bottom. The rest of the gas turns 180° upward and passes through a secondary spray of the lime-water "liquor." Altogether some 70% of the SO_2 and nearly all of the particulates have been removed at this point. The flue gas then passes upward through a mist eliminator and out of stage 1. It then passes through an induced draft fan and enters the top of the second stage, referred to as the absorber. The absorber is similar to the scrubber, except that its venturi is fixed. The absorber removes any remaining particulates and brings the cleaning up to 92.2% SO_2 removal. The flue gas is then heated by oil-fired heaters from 120°F to 160°F to give it adequate buoyancy, and passes up the stack.

Six parallel cleaning trains, converging to two parallel heaters, are used for each unit. Units 1 and 2 use a common 950 ft stack.

Liquor is recirculated within the scrubbers and maintained at a pH of 7 by added lime. It is bled to maintain 8–10% solids, the outflow passing to a large thickener tank. Here the overflow is recirculated back to the scrubbers. The underflow, maintained at about 30% solids, is treated with a hardening agent, calcilox, a Dravo product, and pumped 6 miles to a 1400 acre disposal site. The disposal site is a complete valley, with a hydraulic dam 400 ft high and 2200 ft long, the largest rock-filled dam in Eastern United States. An estimated 200 million tons of sludge from the three units will fill this impoundment in 20–25 years. The calcilox hardener causes the sludge to solidify in the reservoir to the consistency of normal earth, able to support 4½ tons/ft².

The air quality control system, AQCS, for unit 3 differs primarily in using a precipitator to remove about 95% of the particulates, and a Weir horizontal cross-flow scrubber, using Kellogg modified chemistry. Soluble magnesium sulfate is added to the scrubber liquid, resulting in rapid absorption of the SO_2. This liquor, containing SO_2 and flyash, passes into a reactor below the scrubber, where it is reacted with lime-water to precipitate the sulfate/sulfites and regenerate the scrubber liquor. The scrubber has six absorption stages each with 50 spray nozzles arranged along the top, and draining separately into the reactor. A chevron type mist eliminator is used.

The SO_2 laden slurry is bled from the reactor into a thickener tank from which a 30% solids underflow is treated with hardener and goes to the waste disposal system common with units 1 and 2. Ambient air heated by oil-fired heaters to 1700°F is blown into the flue gas emanating from the scrubber to raise its temperature about 35°F before passing up the stack.

The AQCS uses over 5% of the energy generated, nearly doubling the station service power. Including the cooling towers, environmental controls added over 50% to the cost of the plant without controls. More than $1 out of $3 of both capital and operating costs are spent for environmental controls. Altogether the 2340 MW plant (3–780 MW units) cost

$1.3 billion or $555/kW, which could be plotted on Fig. 1.4 at a mean date of 1979.

Indirect Lime–Limestone Scrubbing. Because of the plugging and scaling problems in direct lime–limestone scrubbing, an indirect process is also under investigation. The SO_2 is scrubbed by a clear solution of water or alkali, and the effluent solution is reacted with lime or limestone outside of the scrubber to precipitate calcium sulfate and sulfite. One such "double alkali" system was selected for a major installation in 1976. Recent progress in solving the plugging and scaling problems has lessened the need for this alternative.

Regenerative Scrubbing. If ammonia or sodium scrubbing are used, the sulfates and sulfur resulting are salable by-products, and the sludge pond is eliminated. This is particularly desirable where space is limited. The use of ammonium salts or ammonia followed by oxidation results in ammonium sulfate and sulfur in about a 90:10 ratio. The ammonium sulfate is a fertilizer, and the sulfur is a marketable product.

In the regenerative sodium sulfate process, the sulfur is recovered, after several intermediate steps, as sulfur dioxide gas, which can be converted to sulfur in a standard Claus unit.

State of the Art—1980. While scrubbing processes involve major unsolved problems—plugging and scaling as mentioned, mist elimination, flue gas reheat, and sludge disposal—direct lime–limestone scrubbing is farther advanced than any alternative. Indirect and regenerative processes have distinct advantages. Their development is continuing, and they may well play a larger role in the future.

The following are a few of the key current developments[9]:

1. Southern Services test facility at Gulf Power's Sholz station in Florida. Two 20 MWe units are equipped with indirect lime-limestone scrubbers. The Chiyoda unit uses weak sulfuric acid with lime–limestone precipitant, and the ADL-CEA scrubber uses sodium salt scrubbing with lime regenerant.

2. Work is being carried out on six other variations of the sodium–lime regenerative process at different locations, and on one process using an aluminum sulfate solution.

3. EPA is sponsoring five demonstration regenerative units of 100 MWe or over.

4. In addition several other processes are being investigated by BuMines, TVA, and others.

As noted in the EPRI study (Fig. 3.2), alkali scrubbing, the most nearly developed of the sulfur-removing processes, also results in a busbar cost both for base-load and intermediate-load plants as low or lower than any other coal option. It is likely to be the selected coal option for some time.

Fluidized-bed combustion, including PFBC, must await experience on the first full-scale plants in the late 1980s before firm comparisons with alkali scrubbing can be made. However, as shown in Fig. 3.2, it is expected to be a viable coal option after 1993 if ongoing developments are successful.

Within the last few years (1981) there has been substantial development and application of dry flue gas desulphurization utilizing an alkali spray and baghouses[18], both for industrial and utility boilers.

Coal Refining and Liquefaction

The conversion of coal to a liquid or gas requires the addition of hydrogen. The carbon/hydrogen ratio of a typical bituminous coal is given in Table 3.1. It is $81.6/5.0 = 16.3$ and corresponds to the chemical formula $CH_{0.74}$. Lignite ($40.6/6.9 = 5.9$) corresponds to $CH_{1.4}$. Saturated hydrocarbons C_nH_{2n+2}, in the range of petroleum-type fuels with many carbon atoms per molecule, corresponds to CH_{2+}, whereas methane, the principal constituent of natural gas, is CH_4. Thus, conversion is largely a matter of hydrogenation.

The hydrogen added is least in solvent-refined coal, where it serves primarily to dissolve and purify. The product is a solid at normal temperature.

Liquefaction to petroleum-type fuels requires more hydrogen, with larger amounts needed for the lighter synthetic crudes and less for the heavier fuel oils.

Conversion to a gas requires the most added hydrogen. Part of the coal can be converted to liquid or gas, using the hydrogen already present in the coal. However, full conversion requires large amounts of hydrogen.

Brief descriptions are given of four coal liquefaction systems in this chapter, all under development in 1977, namely, Exxon Donor Solvent (EDS),

H-Coal, Synthoil, and SCR-II. Solvent-refined coal, which can also be used as a liquid at elevated temperatures, is also included. It is normally used in solid form as prills. The earlier Bergius processes of coal liquefaction of I.G. Farben, used in Germany for about 15 years, used relatively high pressure, about 400 bars. Pressure vessels for this process are expensive and limited in diameter. Also, the cost of hydrogen compression is high. Processes now under development are directed to more economic and higher capacity units.

However, at lower pressures such as 100 bars, a catalyst is generally required, either within the reactor (H-Coal, Synthoil) or indirectly operating through a recycle solvent (EDS).

The EDS research, to be described, has demonstrated that liquids boiling up to 540°C (100°F) can be consistently separated from heavy bottoms (higher boiling liquids, unconverted coal, and ash) by vacuum separation.

Temperatures used for liquefaction at the lower pressures (100 to 150 bars) range from 400 to 470°C (750 to 880°F).

The degree of hydrogenation required in a donor solvent for a high degree of conversion of coal to liquid is greatly reduced if molecular hydrogen is used in addition. Thus, in the EDS process, the relative amounts of hydrogen used in hydrogenation of the solvent or supplied as molecular hydrogen are optimized.

Solvent Refined Coal[10] *(SRC-I)*[11]

Solvent-refined coal (SRC) is being developed by several companies (1977). The process being developed by Gulf Oil Co. (Pittsburgh and Midway Coal Mining Co.) is described below.

The finely pulverized coal is dissolved in a coal-based liquid at high temperature and pressure (800 to 900°F, 1000 to 2000 psi) with the addition of hydrogen. The purpose of the hydrogen is to make the coal dissolve more readily (depolymerize) and, as the coal molecules break up, to prevent their reforming. Some of the hydrogen also combines with the sulfur in the coal to form hydrogen sulfide gas. Almost all of the pyritic and up to 75% of the organic sulfur is removed from the coal product and recovered as a solid.

After the pressure is reduced, the coal solution is passed through a rotary filter which screens out undesired coal solids. The solvent is removed by flash evaporation and reused. The product is still a liquid which can be "rained" down through a cooling tower into "prills" that solidify as they fall. It can be maintained and used as a liquid at 300 to 400°F or shipped and used as prills.

In the resulting product, the ash content is less than 0.1%, sulfur is less than 1%, moisture is 0, oxygen is about 4%, and calorific value is at least 16,000 Btu/lb (about one third higher than bituminous coal).

The use of SRC in a "coal"-fired plant without scrubbers is fully compliant. Three thousand tons were used in an 18-day test at Plant Mitchell, Georgia. Both SO_2 and NO_x emissions were well under EPA limits.[11] Compared with the cost when high-sulfur coal and flue gas scrubbers are used, the plant capital cost is much lower (about 77%) and fuel cost is much higher. The advantage of SRC therefore increases as the load factor drops. In the EPRI study (Fig. 3.2), SCR is not a selected option. However, other studies[11] have shown the costs to be within 10% over a wide range, with SRC more economic for an intermediate-load plant.

SRC also has potential as a building block[11] or intermediate in the production of premium liquid and solid products. For example, it can be hydrotreated catalytically to produce low-sulfur and low-nitrogen liquid fuels and aromatic chemicals, using technology already proved on petroleum residuals. It can also be coked and calcined to produce anode-quality coke for the anodes of aluminum smelters.

A 50 tons/day SRC pilot plant has been in operation near Tacoma, Wash., since 1975. See also SRC-II later.

Exxon Donor Solvent[4,12]

As indicated in the name, the unique feature of the Exxon Donor Solvent (EDS) method is the use of a hydrogenated solvent which acts as a hydrogen donor, improving the liquefaction, in addition to its role as a slurry solvent to carry the crushed coal and promote dissolution. This solvent is a midrange fraction of the product oil.

A typical donor molecule is tetralin, which changes to naphthalene as it "donates" hydrogen in the process. This is illustrated in Fig. 3.3, in which the tetralin molecule donates four hydrogens to free radicals formed by the thermal cracking of the coal "molecules" during liquefaction. A direct hydrogen feed is also used, as shown in Fig. 3.4.

Also unique to the EDS process is the use of

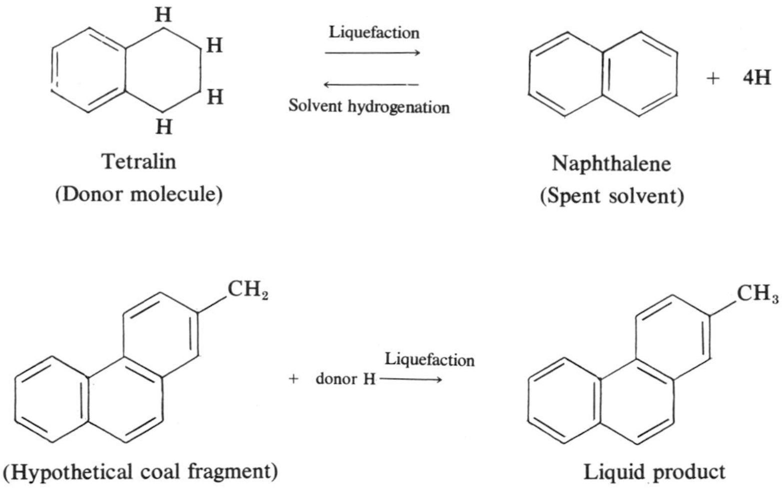

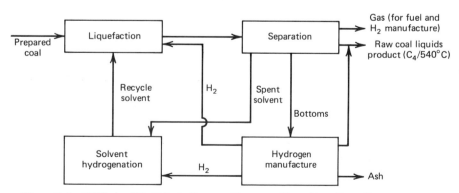

Figure 3.3. Exxon Donor Solvent coal liquefaction process. Action of donor molecule. Courtesy Chem. Eng. Prog.[12] and authors.

component processes requiring minimal modification from counterparts already proved in refinery or chemical service and the absence of any solids separation as such.

The four major elements of the EDS process are shown in Fig. 3.4. In the liquefaction element, a catalytic reactor operating at 800°F, 2000 psi, coal, recycle solvent, and hydrogen are reacted to produce gas, raw coal liquids, and heavy bottoms containing unreacted coal and mineral matter. These are separated by distillation. The spent solvent goes to "solvent hydrogenation" and is then recycled. The heavy bottoms, including solids, are used to produce hydrogen and further raw coal liquids.

In addition to these four major steps, four additional processes are required in a pilot or full-scale plant. These are:

1. The coal feed and slurry preparation system. Coal is ground and screened to −30 mesh for a one ton/day pilot plant, or −8 mesh typically for a full-scale plant.

2. The solvent fractionation system, which removes lower and higher boiling-range components from the solvent.

3. The recycle hydrogen acid–gas removal and compression system.

4. The low-pressure gas scrubbing system to remove heavier hydrocarbons such as C_5 through C_8.

From a typical Illinois No. 6 coal of 10,700 Btu/lb (24.9 MJ/kg) as received, or 12,814 Btu/lb (29.8 MJ/kg) dry, raw liquid in the heavy naphtha boiling

Figure 3.4. EDS simplified block diagram. Courtesy *Chem. Eng. Prog.*[12] and authors.

range (70 to 200°C) is obtained having a HHV of 18,327 Btu/lb (42.6 MJ/kg) and raw liquid in the fuel oil range (200 to 540°C) of 17,122 Btu/lb (39.8 MJ/kg). These heating values can be increased 5 to 6% by hydrotreatment before distillation.

In producing 18,000 Btu/lb oil from 12,814 Btu/lb coal (dry), the weight of oil produced can obviously not exceed 12,814/18,000 = 71% of the weight of coal. This represents 100% thermal efficiency and is reduced in a practical process by the energy required for hydrogen production and for process heat and power. Thus as shown in Table 3.8, only 65 to 75% of the original coal energy is actually available to be included in the product oil. Taking 71% of this results in 46 to 54% liquid yield by weight. This amounts to 2.7 to 3.1 barrels liquid per ton coal.

The EDS coal liquefaction process is designed to produce only naphtha and fuel oil products and to be a balanced, self-sufficient, process. It is capable of closely approaching the liquid yields and process efficiencies indicated in Table 3.8.

The course of a development process is well illustrated by the EDS process. The research began in 1966, and the development was at the 1 ton/day pilot plant stage in 1975, earlier work having been conducted in units ranging from 100 cc batch units to a $\frac{1}{2}$ ton/day integrated pilot plant. The next step, a 250 tons/day pilot plant, with joint industry and government financing, was under construction in 1977 in Baytown, Tex. At 3 barrels/ton, this amounts to 750 barrels/day, or about 1½% of a full-scale 50,000 barrels/day facility. This intermediate step, 250 tons/day, is necessary to establish equipment reliability, to obtain engineering scale-up data, to confirm process performance, and to establish

Table 3.8. Maximum Practical Yields in EDS Self-Sufficient Liquefaction Plant

	Percent of Dry Coal Feed
Feed coal energy	100
Hydrogen production	10–15
Process heat and power	15–20
Feed energy available for liquids production	65–75
Overall process efficiency (%)	65–75
Maximum practical liquid yield (wt %)	46–54
Barrels/ton	2.7–3.1

Source. Courtesy *Chem. Eng. Prog.*[12] and authors.

operability limits. It also provides large product samples for testing in anticipated applications.

To summarize, the donor solvent process of coal liquefaction was selected by Exxon for development after years of research for three main reasons:

1. It can produce high yields of low-sulfur liquid products.

2. Its chance of commercial success is enhanced since the processing elements require minimal modification from similar units in petroleum processing and chemical industries. Solids separation is not required. Products are separated by flashes and vacuum distillation.

3. The cost of products should be at least competitive with other coal liquefaction systems under development.

H-Coal (Hydrocarbon Research, Inc.)[4]

In the H-coal process, the coal is dried, pulverized, and formed into a slurry with a coal-derived oil. This slurry is mixed with compressed hydrogen and heated, and the two are fed upward through an ebullated-bed catalytic reactor, operating at 850°F, 2250 to 2700 psi, shown in Fig. 3.5. Most of the coal is catalytically hydrogenated into the desired hydrocarbon oils and some gases and is carried off. The catalyst, which is bubbled up, or ebullated, rises only to the indicated level because of its size and is thus separated from the through product. By adjustments in the operating conditions, such as space velocity, the product can be varied from a syncrude refinery input, using about 18.6 MSCF hydrogen per ton coal, to a low-sulfur fuel oil requiring about 12.2 MSCF hydrogen per ton coal. About 94% of the coal is reacted yielding a product 74 to 78% by weight of the dry, ash-free coal feed.

Synthoil Process[13]

The Synthoil process is being developed by DOE, formerly BuMines, ERDA (1977). The process is shown conceptually in Fig. 3.6 (not all loops shown).

Coal is pulverized and slurried in about 35% coal mixture with some of the product oil. It is preheated to about 850°F and mixed with recycle and fresh hydrogen; and after a residence time of about 20 minutes, it is introduced into the bottom of a reactor

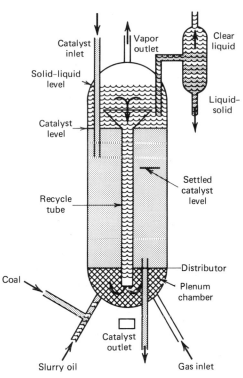

Figure 3.5. Reactor used in H-coal process. Courtesy C. C. Kang and P. H. Kydd.[16]

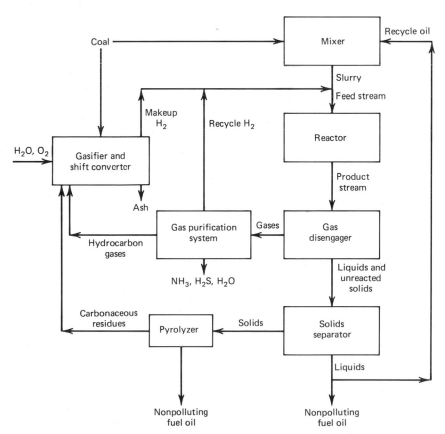

Figure 3.6. Synthoil process. *Source.* BuMines/DOE.

packed with cobalt–molybdenum catalyst pellets, accomplishing over 94% conversion to oil in about 2 minutes of reaction time. The reactor temperature is controlled by a quenching stream of recycle product oil, not shown.

The hydrogen flow rate is high enough to create a highly turbulent flow through the packed bed, which prevents the ash from settling out, promotes mass and heat transfer, and maintains the catalyst surface in a clean condition.

After passing out of the reactor, the gases, liquids, and solids are separated from the product stream, the latter by centrifuge. The gases are purified to remove the NH_3, H_2S, H_2O, the C_1–C_5 hydrocarbons, and other gases from the recycle hydrogen stream. These are mostly useful by-products.

The liquid is divided into recycle, quench, and product oil.

The solids are pyrolyzed, yielding additional low-sulfur fuel oil and residue. The residue can be used, along with some fresh coal, in a steam–oxygen gasifier to produce makeup hydrogen for the process.

Typically, a Kentucky high-volatile coal, having 5.5% sulfur and 16% ash, is converted to an oil that flows at room temperature (sp. gr. 1.02 to 1.08) and has about 0.2% sulfur and 0.2% ash. It has about 17,400 Btu/lb.

Process conditions are 4000 psi (or 2000 psi with somewhat less yield and more sulfur) and 850°F. Oil yields are over 3 barrels oil per ton coal (Kentucky coal with 11,200 Btu/lb).

As of 1977, work had been completed on a $\frac{5}{16}$ in. ID pilot unit (48 lb coal/day), and a larger 1.1 in. ID reactor (400 lb coal/day) was in operation. Based on the very satisfactory results with this unit, a 10 ton/day development unit was under construction at the DOE Energy Research Center at Bruceton, Pennsylvania.

The thermal efficiency of the process is about 75%.

General Conversion Methods

Fischer–Tropsch Synthesis[4]

Fischer and Tropsch made extensive studies of the catalytic reduction of carbon monoxide to various hydrocarbon liquids in 1923 to 1933. Their work was followed by pilot and full-scale plants in Germany and France for the production of gasoline, diesel fuel, paraffins, and other products from synthesis gas, mainly CO and H_2. During World War II, a substantial part of Germany's aviation gasoline and other fuel products was produced from coal by this process (F-T synthesis) and the now-obsolete Bergius process.

The South African Coal, Oil, and Gas Corp, Ltd. (SASOL) plant in So. Africa produces gas, liquid hydrocarbons, and wax from coal-derived synthesis gas, via two versions of the Fischer–Tropsch synthesis. The synthesis gas is produced in a battery of Lurgi high-pressure, steam–oxygen gasifiers, indicated in Fig. 3.7 (see Chapter 4, Lurgi).

For the initial plant, brought into operation in 1955, part of the synthesis gas after purification is converted into hydrocarbons by the Arge synthesis (Arge-Arbeit Gemeinschaft Lurgi and Ruhrchemie) process, a fixed-bed catalytic reactor using iron–cobalt catalyst and operating at 450°F, 360 psi. The products are gasoline, LPG, oil, wax, gas, and alcohol.

Part is converted by Kellogg (M. W. Kellogg Co., U.S.A.) fluidized-bed process, using an iron catalyst and operating at 620°F, 330 psi. It produces more of the gasoline and alcohol products and less oil, wax, and gas.

The newer plant coming on-stream in 1979–1981 will use only a refined Kellogg process (fluidized bed), tuned to produce mainly gasoline and fuel oil, but also substantial amounts of ethylene, ammonia, sulfur, and tar products. It is about nine times the capacity of the earlier plant. Together, the SASOL plants will supply about 30% of the automotive fuels in the Union of South Africa in 1980. Feedstocks (syncrudes) are supplied to various refineries.

Char Oil Energy Development (COED) (F.M.C. Corp.)[4]

The Char Oil Energy Development (COED) process was designed to produce a variety of solid, liquid, and gaseous fuels by multistage fluidized pyrolysis of coal. In each stage the temperature is selected to be just short of the agglomerating point, and a fraction of the volatile matter in the coal is released. Typical temperatures in a four-stage process are: 600°, 850°, 1000°, and 1500°F, but the number of stages and the temperatures vary with the agglomerating characteristics of the coal. A coal drier at 375°F is used before the first stage.

Heat is generated by burning char in the last stage, and then using the hot recycle char and gases to heat the other stages.

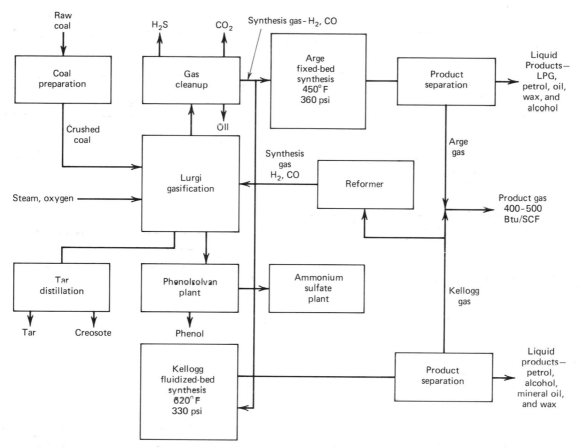

Figure 3.7. Fischer–Tropsch synthesis at SASOL 1954 plant. Later plant, 1980, uses only refined Kellogg process. *Source.* Noyes Data.[4]

Typical products from this four-stage process are char 59.5%, oil 19.3%, gas 15.1%, and liquor 6.1%, using Illinois No. 6 coal having a proximate analysis on a dry basis of: volatile matter 37.6%, fixed carbon 49.3%, and ash 13.1%. The moisture content is 9.1%.

The process envisions using each of these products. The oil, 0.4 to 1.5 bbls/ton of coal, would be hydrogenated into a syncrude refinery input, or into a fuel oil. The offgas, 8000 to 10,000 SCF/ton of coal, has a heating value of about 500 Btu/SCF, and would be purified and sold to industry. The char, typically 59.5% of the coal, would be used as a power plant fuel, or gasified in a commercial gasifier such as a Koppers-Totzek. The char contains about the same % sulfur as the coal.

SRC-II (Gulf Oil Co./Others)[14]

The solvent-refined coal process (SRC) has been further developed in a pilot plant at Ft. Lewis to produce mainly liquid and gaseous products instead of a solid, as with SRC-I. Based on favorable results of the pilot plant work, an intermediate-scale demonstration plant using commercial-size equipment had been planned. This 6000 t/D plant was to be located near Morgantown, West Virginia, and used to confirm pilot-plant results on a semicommercial scale, leading to full-scale plants of typically 33,500 t/D in the late 1980s, if the demonstration plant met expectations. At a rough equivalent of 3 barrels/ton, the full-scale plant output would be 100,000 barrels/day of oil equivalent.

Plans were to operate the 6000 t/D demonstration plant during the period 1983 to 1986. However in 1981 this 2-billion dollar project, a joint venture of U.S., German, and Japanese companies and governments, was cancelled due to a change in U.S. policy placing the responsibility for full scale commercial developments on industry. DOE subsidy, which had largely supported the development to date was discontinued for the specific plant, but

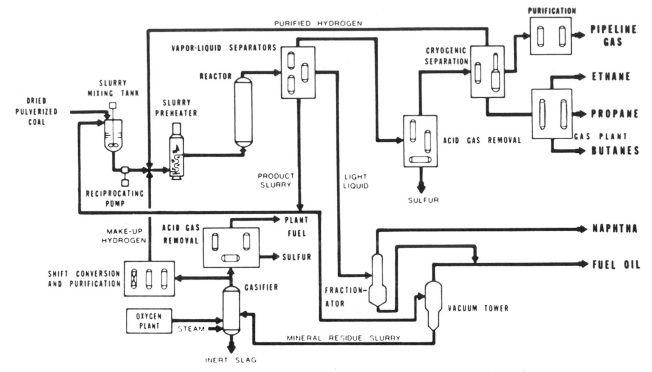

Figure 3.8. SRC-II process. *Source.* B. K. Schmid, J. C. Koenig, and D. M. Jackson, Reference 14.

continues for fundamental research. The development will no doubt continue as dictated by commercial requirements.

The major product will be a distillate fuel oil of less then 0.3% sulfur, with substantial amounts also of gases and naphtha. An overall selling price of $3.50 to $4.00/$10^6$ Btu, 1978 basis, is estimated from a future full-scale plant. This is expected to make it economic compared with petroleum-based fuels and SNG in the mid- or late 1980s.

A simplified flow diagram of a full-scale plant is shown in Fig. 3.8, and a typical slate of products is given in Table 3.9. The latter will vary considerably depending on the coal feed and the market needs. However, the main product is fuel oil, as indicated.

The main product stream can be traced (Fig. 3.8) starting with the dried, pulverized coal. After slurrying with recycle oil, it is mixed with hydrogen and enters the slurry preheater and then the reactor. There, it is catalytically cracked and dissolved. The sulfur and nitrogen are largely converted to H_2S and NH_3 gases. From here, the flow stream enters the vapor–liquid separator. The main product slurry passes downward and divides. Part enters the vacuum tower where the mineral residue slurry is separated out, and the product fuel oil feeds out.

A substantial part of the catalytically active product slurry from the vapor–liquid separator is recirculated back to the slurry mixing tank, as shown. This results in a high degree of catalytic cracking and conversion to products, the heart of the process.

There are a number of secondary loops. Light gases from the vapor–liquid separators pass through a purifying unit where the acid gases are removed and the H_2S is converted to elemental sulfur. Then, they undergo cryogenic separation. The purified hydrogen returns to the slurry stream. Pipeline gas (methane) is purified and feeds out. The heavier gas mixture of ethane, propane, and butane goes to the gas plant and thence to market.

Mineral residue slurry from the bottom of the vacuum tower is gasified with steam–oxygen to a syngas. Part of this is purified for plant fuel; part is shifted to hydrogen and purified for makeup hydrogen. There is substantial flexibility in proportions to accommodate various coals, but the process is sufficiently flexible to process substantially all U.S. coals.

An important feature of the process is the direct hydrogenation of carbon to methane. This is considerably more efficient than the methanation of

Table 3.9. SCR-II Products From Typical Commercial Plant[a]

Product	Yield
Major Products	
Methane	50 MM SCF/day
Ethane/propane	3,000 tons/day
Butane	300 tons/day
Naphtha	17,000 barrels/day
Fuel oil	56,000 barrels/day.
By-products	
Sulfur	1200 tons/day
Ammonia	180 tons/day
Tar acids	240 barrels/day
Thermal efficiency	72%

Source. Courtesy B. K. Schmid, J. C. Koenig, and D. M. Jackson.[14]
[a]West Virginia Coal, 33,500 tons/day.

syngas, in which one third of the hydrogen is simply converted to water (Eq. 4.6).

Use of SRC-II Products. Clean petroleum-type fuel from coal will be needed particularly as a replacement for petroleum oil in steam plants in large metropolitan centers in the East. Even though it appears more expensive than the use of high-sulfur coal with flue gas desulfurization (FGD), the plants are much simpler and occupy less space. There is no sludge pond. The penalty is, of course, less with intermediate-load plants and still less if the liquid-fueled plant is dual cycle. Petroleum-type fuel in an advanced dual-cycle plant (under development) may well be cheaper than the use of high-sulfur coal and FGD. SRC-II fuel oil is expected to become competitive with petroleum fuels sometime during the 1980s.[14]

The light liquid fractions, such as naphtha, produced by the SRC-II process, can be upgraded to a high-octane, unleaded gasoline blending stock. The methane produced should be fully competitive with SNG from direct coal gasification. Light hydrocarbons can be used directly or converted to ethylene in conventional cracking plants.

Alternatively SRC-II products may replace the corresponding refinery products on the market, making available a larger supply of premium products which can best be made in refineries: diesel fuels, jet fuels, home heating oil, and gasoline.

PEAT

The book *Peat,*[15] published in 1866, shows that the subject is not new. The industry was still young at that time and subject to the same prejudices against the use of peat as had plagued bituminous coal somewhat earlier, and plagues nuclear energy today.

While peat is generally viewed as the younger member of the coal series, a considerable part has grown from sphagnum moss. The new growth above leaves the underportion dead and buried. The plants soak up and hold moisture and build up to great depths, burying everything underneath, even fallen forests, and sometimes building to several feet above the surrounding land.

A large tract of forest cut down by the Romans in England for safety reasons was thus submerged in what became a peat bog of sphagnum moss. Seventeen centuries later, vast quantities of excellent timber, pine, oak, birch, and so on, were extracted from beneath the morass, some retaining the original axe marks and wooden wedges and, along with them Roman axe heads, bits of chain and even some Vespasian coins.

However, in general, peat is the product of partial decomposition and disintegration of sedges, trees, and other plant life, as well as mosses. It is made up of the same elements as coal, namely, carbon, oxygen, hydrogen, nitrogen, sulfur and lesser amounts of other elements.

In the United States, about three fourths of the peat reserves are in Minnesota, Wisconsin, and Michigan, a considerable amount in Florida, and the rest scattered in 26 other states.

In many parts of the world, peat is used as a fuel and was in fact so used in the United States until replaced by the abundant fuels of higher rank. It is now (1980) used primarily for soil conditioning, litter for animals, packaging, and in some chemical uses such as filtering and tanning.

Regarding its use as a fuel, the following facts are significant. Starting with the moisture content of an undrained peat bog, 92 to 95%, peat used as a commercial fuel, as in Ireland, is dried to 10 to 55% moisture, depending on the use and method of production. Usually, it is harvested from a drained bog and dried naturally by sun and wind. For production of briquettes, the peat is dried further using heat from fuel. The characteristics of peat are given in Table 3.10.

The relatively low sulfur and ash contents are a

Table 3.10. Characteristics of Peat

Form	Moisture (%)	Density (lb/ft³)
Air-dried, machine-cut, sod peat	25–40	20–25
Air-dried, hand-cut, sod peat	25–50	15–20
Milled peat	40–60	
Briquettes stacked	10–12	50–60
Briquettes loose	10–12	30–40

Proximate Analysis of Peat   *Ultimate Analysis of Peat*

Dry Basis		Dry, Ash-Free Basis	
Fixed carbon	30–40%	Carbon	53–63%
Volatile matter	55–70%	Hydrogen	5.5–7%
Ash	2–10%	Oxygen	30–40%
		Sulfur	0.3–0.5%
		Nitrogen	1.2–1.5%

Calorific Value

Form	Btu/lb
Dry, ash-free	9500–11,000
Air-dried, machine-cut, sod peat, 30% moisture	Approx. 6200
Milled peat, 40–60% moisture	3700–5300
Briquettes, 12% moisture	Approx. 8000

clue to the popularity of peat as a "pleasant" domestic fuel in countries where it is used. In spite of its volatile content it is not generally "smoky."

In areas of Ireland, peat is burned in 5000 kW electric generating stations supplying the surrounding territory.

REFERENCES

1. H. J. Rose, "Classification of Coal," in H. H. Lowry, Ed., *Chemistry of Coal Utilization*, Vol. 1, New York: Wiley, 1945, Chap. 2.

2. National Association of Purchasing Agents, National Committee on Coal, *Factors in Selecting Coal*, New York: Natl. Assn. Purchasing Agents, 1936.

3. H. F. Yancey and M. R. Geer, "Hardness, Strength, and Grindability of Coal," in H. H. Lowry, Ed., *Chemistry of Coal Utilization*, Vol. 1, New York: Wiley, 1945, Chap. 5.

4. I. Howard-Smith and G. L. Werner, *Coal Conversion Technology*, Chemical Technical Review No. 66, Park Ridge, N.J.: Noyes Data Corp., 1976.

5. "New Source Performance Standards, NSPS," *Federal Register*, June 11, 1979, p. 33580.

6. R. A. Loth, O. D. Gildersleeve, Jr., S. A. Vejtasa, and R. W. Zeren, "Comparative Evaluation of New Electric Generating Technologies," *Proc. 7th Energy Technol. Conf.*, Washington, 1980.

7. J. H. Field, L. W. Brunn, W. P. Haynes, and H. E. Benson, *Cost Estimates of Liquid Scrubbing Processes for Removing Sulfur Dioxide from Flue Gases*, Rept. of Invest. 5469, Pittsburgh: BuMines, 1959.

8. W. C. Patterson and R. Griffin, *Fluidized-Bed Energy Technology: Coming to a Boil*, New York: Inform, Inc., 1978.

9. A. V. Slack, "Flue Gas Desulfurization: An Overview," *Chem. Eng. Prog.*, August 1976, p. 94.

10. Robert Cairns, "A New Role for Coal: Clean Energy," *The Orange Disc*, September–October 1974, p. 2.

11. J. C. Tao, C. L. Yeh, and S. M. Morris, "Solid SRC A Building Block," *Proc. 6th Coal Gasification Conf.*, University of Pittsburgh, 1979.

12. L. E. Furlong E. Effron, L. W. Vernon, and E. L. Wilson, "The Exxon Donor Solvent Process," *Chem. Eng. Prog.*, August 1976, p. 69.

13. P. M. Yavorsky, S. Akhtar, J. J. Lacey, M. Weintraub, and M. A. Reznik, "The Synthoil Process," *Chem. Eng. Prog.*, April 1975, p. 79.

14. B. K. Schmid, J. C. Koenig, and D. M. Jackson, "Economic and Market Potential for SRC-II Products," *Proc. 6th Annu. Int. Conf. on Coal Gasification*, University of Pittsburgh, 1979.

15. J. H. Benham, *Peat*, New Haven: J. H. Benham, 1866.

16. C. C. Kang and P. H. Kydd, "H-Coal," 3rd *Annu International Conf on Coal Gasification*, University of Pittsburgh, 1976.

17. A. C. Fieldner, W. E. Rice, and H. E. Moran, *Typical Analyses of Coals of the United States*, Bull. 446, Pittsburgh: BuMines, 1942.

18. M. E. Kelly and S. A. Shareef, *Second Survey of Dry SO₂ Control Systems*, EPA-600/7-81-018, Research Triangle Park, N.C.: Industrial Environmental Research Labs, 1981.

19. "U.S., Germany, Japan to Discontinue SCR-II Project in West Virginia," DOE *Energy Insider*, July 6, 1981.

20. J. W. Patterson, "Heavy Medium Separation of Sulfur from Coal," *Fourth Gasification Conference*, University of Pittsburgh, 1977.

21. R. C. Forsyth, "Experiences with Flue Gas Desulfurization at the Bruce Mansfield Plant," *Fourth Gasification Conference*, University of Pittsburgh, 1977. See also: *The Bruce Mansfield Plant and the Environment*, New Castle: Pennsylvania Power Co.

22. W. J. Raymond, A. G. Sliger, and J. J. O'Donnell, "Kellogg-Weir Scrubbing System for Bruce Mansfield Unit No. 3," *Fourth Gasification Conference*, University of Pittsburgh, 1977.

4

Gaseous Fuels—Natural and Synthetic Gas, Natural Gas Liquids, and Hydrogen

INTRODUCTION

In this chapter, we seek to answer the questions, "What are these fuels and what are the above-ground processes used in preparing the natural fuels or manufacturing the synthetic fuels?"

No attempt is made to outline the still hypothetical geologic history of formation of the present oil and gas deposits. Nor is the prospecting, drilling, oil field development, and production covered, except for a few topics necessary to understand the nature of these fuels.

We have separated petroleum and natural gas into two chapters. However, nature has mixed them. There is generally gas in the oil and frequently oil in the gas. This relationship is covered in the present chapter.

Two new forms of natural gas have become important, namely, LPG and LNG. Liquefied Petroleum Gas (LPG), or bottled gas, mostly propane and butane derived from natural gas liquids (some from refining petroleum), has become important, both as a chemical feed stock and as a convenient heat source, particularly where natural gas is not available or not readily applicable.

Liquefied Natural Gas (LNG), liquefied abroad for shipment in tankers, is beginning to fill the shortage of natural gas in the United States. LNG is also stored extensively for "peak shaving." No essentially new characteristics are involved, but the processes and economics will receive attention.

Synthetic or manufactured gas was in widespread use long before the discovery of natural gas. Generally, this was a gas of about half the heating value of natural gas. The first gas company was chartered in London in 1812. The first oil well (69 ft deep), drilled by Col. Edwin Drake in Titusville, Pennsyl-

vania, was completed in August 1859. Natural gas was first used by the Chinese in ancient times to make salt from brine, and then forgotten. It was first used in modern times in Fredonia, New York, in 1820, but the real start of natural gas use was during the oil boom following 1859.

More recently, with dwindling supplies of natural gas and the emphasis on clean fuels, attention is again focused on synthetic gas, both low- and medium-Btu gases similar to the earlier manufactured gases and also high-Btu gas equivalent to natural gas and interchangeable with it.

Numerous lower-Btu gases occur as by-products or are manufactured for special purposes. Most synthetic gases are manufactured from coal, but since large quantities are now required, processes are sought that can use all the abundant coals. Earlier methods were restricted to noncaking grades.

Low- or medium-Btu gas may be used, for example, as a power plant fuel. The goal is a clean fuel, providing maximum heat energy per dollar, and using all abundant coals rather than compatibility with pipeline gas.

Hydrogen is important as a secondary form of energy. Also the production of synthetic fuels from coal is largely a matter of adding hydrogen. The essential facts about hydrogen are given. The "hydrogen economy" concept is discussed.

The reserves, rates of use, and general cost structure for natural gas have already been given in Chapter 1; the facts regarding transmission and storage are given in Chapter 14; and the chemical composition of the hydrocarbons involved are given in Chapter 11. Many of the processes of treating natural gas liquids are very similar to those for petroleum, which are treated in Chapter 2. A general introduction to coal conversion technology is given in Chapter 3, as well as a description of processes resulting in solid, liquid, and gaseous fuels. All processes primarily for the gasification of coal or hydrocarbons are covered in this chapter.

NATURAL GAS

Most oil and gas occurs in natural reservoirs from 500 to 25,000 ft below the surface of the earth. The reservoir is filled with porous rock or sand, which physically supports the overburden. The pore surfaces are generally lined with capillary or adsorbed water, usually saline, called interstitial or connate water, and the oil or gas occupies part of the remaining space. The water lining of the pores usually stays in place when the oil or gas is "produced" that is taken out. The bed is also characterized by a permeability or ability of the gas or oil to migrate toward a lower-pressure region. The reservoir has an anticlinal (concave downward) or equivalent impervious top,[1] as shown in Fig. 4.1, which traps the oil or gas as it is pressed upward by the water

below. Where both oil and gas are present, they separate into a gas cap above an oil layer, as shown at the left of the figure. However, gas may be present alone as shown at the right.

Temperature Gradient in the Earth

The mean earth temperature gradient is about 1.5°F/100 ft (2.75°C/100 m) below the surface. There is considerable variation, and gradients up to 10 times average occur in rare instances (see Chapter 9, Geothermal Energy). At 100 ft down, the temperature is the average yearly temperature at the surface, with a delay of many centuries. It is 2 to 10°F above the mean annual temperature. If it is taken as 50°F, then a reservoir at a depth of 25,000 ft with average gradient would be at 425°F. However, the temperature of practically all oil and gas reservoirs is below 212°F, with a few instances where oil has been produced up to 340°F. At the reservoir pressures, it is doubtful if petroleum could exist as a liquid above 350°F.

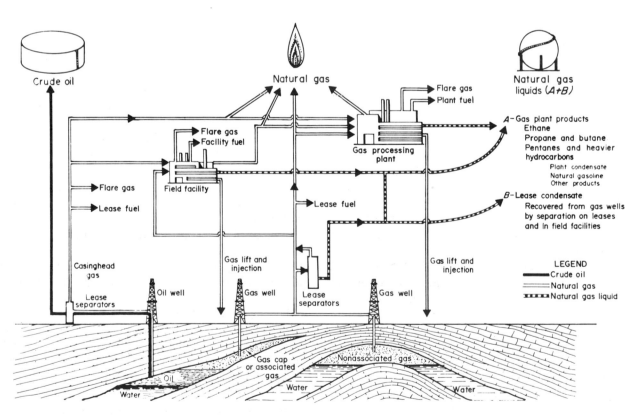

Figure 4.1.   Schematic flowchart of crude oil, natural gas, and natural gas liquids. Courtesy American Petroleum Institute.

Pressure Gradient

The overburden produces a pressure of about 1.0 psi per foot of depth, that is 25,000 psi at 25,000 ft depth. However, this is the pressure in the rock or sand. The pressure on the oil or gas within the pores of rock or sand is limited generally to that which would be created by a column of salt water up to the surface, that is, about 0.44 psi per foot of depth. The pressure may be anything up to this, occasionally somewhat more. It drops off as the oil or gas is produced, unless some method of pressure maintenance is employed. Typical reservoir pressures are several thousand psi.

Gas in the Oil[2]

If a barrel of oil and 1200 SCF of natural gas are compressed together the entire 1200 SCF (214 barrels) of gas will just be dissolved in the barrel of oil at a typical reservoir pressure and temperature of 3000 psi and 160°F. The oil with its dissolved gas will then occupy 1.61 barrel. This is a typical condition in the reservoir at the left, designated "oil" in Fig. 4.1.

The specific gravity, viscosity, and surface tension of this reservoir fluid are all less than the corresponding properties of the surface or "stock tank" oil. This enables it to flow to the well bore more readily than if the light solution gas were removed.

When this 1.61 barrel of oil and dissolved gas is produced (brought to the surface), it separates into 1200 SCF of "casinghead" gas and 1 barrel of crude oil, when admitted at atmospheric pressure and temperature into the "lease separator" shown at the left in Fig. 4.1. This is a "rich" gas containing large amounts of natural gas liquids, to be described later.

The heating value of oil is about 5.8 million Btu/barrel, whereas 1200 SCF of natural gas at 1031 Btu/SCF contains 1,240,000 Btu, or 21.4% of that in the oil.

Oil in the Gas (Natural Gas Liquids)

In the left-hand reservoir of Fig. 4.1, the gas in the "gas cap" above the oil is in intimate contact with the oil. It is called "associated gas" and has dissolved in it, or included in it, a considerable amount of propane and butane and the more volatile gasoline components such as pentane and heptane.

If the pressure of this gas were reduced, a sur-

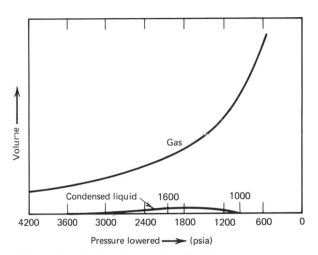

Figure 4.2. Condensed natural gas liquids versus pressure.

prising but very important phenomenon would take place. These valuable products, which are gaseous at the reservoir temperature and pressure, would liquefy and be lost forever, so dilute would they be in the gas reservoir.

This is illustrated in Fig. 4.2, which shows typical though qualitative variations of gas and condensed liquids as the pressure is lowered. From a bubble point at about 3600 psi, the liquid starts to condense, reaching a maximum at about 1600 psi. Then it starts to reevaporate until, at about 1000 psi, the liquid has completely disappeared. At atmospheric pressure, those components that are liquid under this condition separate out as "condensate."

As a result of this phenomenon, the only way these "natural gas liquids" can be brought to the surface and recovered is by producing the gas at or near the reservoir pressure. This is accomplished by reinjecting the gas into the reservoir after the valuable natural gas liquids have been removed, that is, by "cycling." Additional gas from other wells may also be injected to make up for the loss of volume from removing the liquids. This is continued until the production of natural gas liquids has reached its economic limit, after which the dry gas is produced by declining pressure.

When the high-pressure gas, rich in gasified liquids, is admitted to the lease separator (center of Fig. 4.1), most of the components that are liquid under normal atmospheric conditions separate out. They are known as "lease condensate" or simply as "condensate." The propane and butane, which are gaseous under normal conditions, remain in the gas and are separated out later.

Gas is frequently found quite separate from the oil, as in the right-hand reservoir of Fig. 4.1, and is termed "nonassociated gas." This is generally a "lean" or "dry" gas containing very little natural gas liquids.

Gas Processing Plant

All gas containing products that can be economically recovered or requiring purification before entering the pipelines then goes to the gas processing plant (or gasoline plant). Here, the remainder of the recoverable natural gas liquids is removed and the gas is purified for supply to consumers or recycled into a gas reservoir to maintain pressure.

As shown in Fig. 4.1, the gas plant products other than natural gas to the consumer include ethane, propane, and butane, as well as pentanes and heavier hydrocarbons. These may be in the form of bottled gas (propane and butane), ethane, plant condensate, natural gasoline, crude oil, and other products.

Propane and butane, as removed from natural gas, are bottled as liquids under slight pressure. The same is true of ethane. Thus, although these are gases in the reservoir and gases at atmospheric temperature and pressure, they are included in the term "natural gas liquids," an inclusive term for all of the products recovered from natural gas.

About one half of the natural gas liquids go direct to the consumer as bottled gas, and about half go to refineries as "condensate" or "natural gasoline," where they are further refined and blended with petroleum gasoline to raise its volatility and its octane rating. As shown in Table 2.4, natural gas liquids constituted about 6.5% of the input to U.S. refineries in 1969.

Liquefied petroleum gas (LPG), or bottled gas, was originally produced as a by-product of petroleum refining. The term "petroleum" has been retained, although most LPG is now recovered from natural gas liquids in the United States. That produced in refineries is now sometimes designated as "liquefied refinery gas" (LRG) to differentiate it, since propylene and butylene may be present because of the chemical reactions within the refinery.

Natural gasoline consists of further hydrocarbons in the gasoline range that are extracted from the gas in the processing plant, the lease condensate that separated naturally having been previously removed. The processes used will be described.

Processing Natural Gas

Natural gas liquids are recovered from the gas by the following physical processes, alone or in combinations:

Phase Separation	Absorption
Cooling	Adsorption
Compression	Refrigeration

Products in liquid or gaseous form at atmospheric temperature and pressure are separated in the lease separator or field facility, as shown in Fig. 4.1, into condensate and wet gas, that is, gas containing recoverable products. Cooling and refrigeration are used to increase the condensation and production of liquids from the gas.

Absorption is used to strip propane and butane from the gas as described in Chapter 2 for the gases from the top of a fractionating column.

As shown in Fig. 4.2, maximum condensation of liquids occurs at a high pressure. This phenomenon is the basis of separation by compression.

Natural Gasoline and Cycling Plants[3]

The majority of gas processing plants in the United States are of the following types:

Absorption	435
Compression	79
Charcoal	2
Refrigeration	16
Combination	16
Cycling	52
Total	600 (1956)

Purification of Natural Gas

Many natural gases can be used from the well directly. However, some must be processed to remove carbon dioxide, hydrogen sulfide, and other sulfur components such as the mercaptans, as well as water. When treatment is necessary, collecting pipelines from the wells transport the gas to a central processing plant where some or all of the following processes are carried out. These are simply typical, there being many variations depending on the type and amount of impurities to be removed:

1. Expansion to the processing pressure, which

due to cooling condenses propane, butane, pentane, and heavier hydrocarbons.

2. Dehydration by contact with glycol, which is recovered.

3. Sweetening to remove the H_2S and CO_2 by diethanolamine treatment and fixation in an absorber.

4. Elimination of the mercaptans by soda fixation.

5. Stripping (by cooling or by absorption in an oil) to extract longer chain molecules than methane and ethane and produce a gas conforming to the desired specifications.

The absorption oil and the condensate, after purification, undergo fractional distillation to produce the bottled gases—propane and butane—as well as stabilized natural gasoline and residues. (Stabilized gasoline has had volatiles removed which might cause dangerous pressures in containing vessels.)

These operations are all performed on a continuous basis.

Composition and Heating Value of Natural Gas

The average heating value of natural gas in the United States in 1969–1971 was 1031 Btu/SCF. This value applies closely to the gas flowing in main pipelines or delivered to consumers during this period.

The composition of natural gas varies widely depending on the particular wells or fields where it is produced.[4] Table 4.1 shows the composition of the gas and some of the well characteristics for a number of U.S. wells. These were selected to show the large variation, rather than being representative. However, those samples having heating values within ±10% of the national average are of course most typical (wells 1, 3, 8, and 9).

As discussed earlier, gas from the wells is processed to remove the valuable natural gas liquids and is purified and blended to form the much more uniform product supplied to consumers. Even so, there has been considerable variation in heating value in different parts of the country. For example, in Pittsburgh natural gas to consumers was typically 1129 Btu/SCF in 1958, 1208 Btu/SCF in 1974, and 1030 Btu/SCF in September 1980, compared with the national average of 1020 Btu/SCF for 1975–1980.

Natural gas is generally over 75% methane (CH_4), which has a heating value of 978 Btu/SCF (Table 11.2). Thus, it must contain some gas richer than methane to bring its heating value up to the values mentioned above. From Table 4.2, the heating values of the next higher alkanes are, in Btu/SCF, ethane 1713, propane 2438, and butane 3162. Obviously, a little of these will do the trick.

Table 4.1. Composition, Heating Values, and Well Characteristics of Selected Natural Gas Wells

Characteristic	Well 1	2	3	4	5	6	7	8	9	10
Location	Alaska[a]	Arix.	Cal.	Colo.	Fla.	Miss.	Wyo.	Tex.[a]	La.[a]	Colo.
Date	1970	1970	1970	1970	1970	1970	1970	1939	1939	—
Well depth (ft)	2040	1196	8107	6360	15,140	13,800	6034			
Wellhead press. (psig)	—	124	—	1600	3775	2000				
Open flow (MCFD)	166	610	10,100	1800	2145	760,000[a]				
Component (vol %)										
Methane, CH_4	98.4	00.0	99.0[a]	75.1	69.2	61.0	59.5	92.3	84.7	00.5
Ethane, C_2H_6	0.6	0.0	0.2	11.5	10.9	16.1[a]	9.5	3.1	6.1	4.0
Propane, C_3H_8	T[b]	0.0	T	6.3	4.3	9.7	16.6[a]	1.8	2.3	0.0
Other alkanes	0.0	0.0	0.0	3.8	2.3	7.2	12.0	1.5	2.4	0.0
Nitrogen	0.9	89.5[a]	0.7	2.3	2.7	2.3	2.3	1.6	4.0	3.2
O_2, Ar, H_2	0.1	0.8	T	0.1	0.1	T	0.1	0.2	0.0	0.1
Hydrogen sulfide	0.0	0.0	0.0	0.0	7.2[a]	1.8	0.0	0.0	0.0	0.0
Carbon dioxide	0.0	0.4	0.1	0.9	3.3	1.8	0.0	0.0	0.5	92.1[a]
Helium	0.03	9.23	T	0.04	0.02	0.02	0.01	0.00	0.00	0.00
HHV (Btu/SCF)	1008	0[a]	1006	1272[a]	1139	1436	1642[a]	1085	1101	75

Source. Courtesy BuMines IC 8518,[4] others.
[a]Reason for selecting that well.
[b]T = Trace.

Table 4.2. Composition and Heating Value of a Typical Pennsylvania Pipeline Natural Gas

Hydrocarbon	Volume Proportion	Heating Value Btu/SCF	Contribution (Btu/SCF)
Methane, CH_4	0.776	978	759
Ethane, C_2H_6	0.129	1713	221
Propane, C_3H_8	0.065	2438	160
Higher alkanes	0.030	3162[a]	95
Total			1235

[a]Used value for butane, C_4H_{10}.

For example a sample of natural gas taken from a pipeline supplied by a large number of Pennsylvania wells contained methane, ethane, and propane about in the volume proportion of 12:2:1, with higher alkanes making up only 3% of the total. As calculated in Table 4.2, such a gas would have a heating value of about 1235 Btu/SCF.

LIQUEFIED NATURAL GAS[5]

Liquefied natural gas (LNG) occupies about $\frac{1}{600}$ of its gaseous volume at atmospheric pressure, and hence is advantageous for transportation by ship, barge, truck, or rail, as well as for storage in tanks. It is generally stored or shipped in cryogenic insulated tanks at about $-260°F$ ($-127°C$) at nearly atmospheric pressure.

The density of liquid methane, the principal constituent of natural gas, is 0.424 relative to water, or 26.5 lb/ft³. Its molecular weight is 16, and its density as a gas is therefore 16/359 = 0.0446 lb/ft³ at 32°F and 1 atm. Thus, the liquid/gas ratio is 26.5/0.0446 = 594, or about 600 as mentioned. Its boiling point is $-263.2°F$ ($-164°C$) at 1 atm. It is stored about at its boiling point. The slight heat leakage in through the cryogenic tank causes a corresponding boil-off.

Ethane and propane, present in smaller amounts, have boiling points of -127.5 and $-43.7°F$, respectively, and hence do not boil off at the storage temperature.

LNG is a relatively new form of energy. Prior to 1964, it was practically nonexistent. It was proposed in 1930, and the first plant for "peak shaving" began operating in 1940. There were further installations until 1944, when an accident and disastrous fire at the Cleveland LNG plant effectively stopped further developments until the early 1960s; 131 people were killed and 29 acres gutted.

The first major overseas transport used a modified oil tanker, the Methane Pioneer, to transport LNG from Louisiana to England in 1959. With improvements in tanks, ships, liquefaction, and safety methods, the real start of LNG was about 1964. In the 16 years to the present (1980), it has achieved widespread use throughout the world, both for "peak shaving" and for overseas shipment of major gas supplies.

For example, in 1972 LNG installations for peak shaving in natural gas systems were either in use or under construction at 76 locations in the United States. For overseas transport in 1972, 13 LNG carriers were in operation with capacities up to 71,500 m³. Such a ship carries the equivalent of 1580 million SCF natural gas, worth about $1.58 million at $1.00/MCF. Twenty-four more ships were under construction, up to 125,000 m³ capacity, for delivery by 1976, and 63 more on order, up to 165,000 m³ capacity, for delivery within a few years.

The Alaska to Japan shipments of 48 billion SCF natural gas in LNG form in 1973, or the Algeria to El Paso scheduled movement of 365 billion SCF annually, amount to 0.2 amd 1.6%, respectively, of the total U.S. consumption. The world consumption is about twice the U.S. consumption. One hundred of the 125,000 m³ ships, making 10 trips a year, could transport about 7% of the world consumption, or 14% of U.S. consumption (1978). Thus, this development is already of large-scale import. In 1978, 84 billion SCF gas was imported from Algeria, up from 11 billion SCF in 1977. While still but 0.4% of the supply, the importation of LNG is growing rapidly.

In this brief treatment, we propose to give a few of the salient facts about the LNG development, to present some of the principal characteristics, processes, and costs. The latter are intended to show relative costs within a system, as actual costs have escalated greatly in a few years. The cost indices in Chapter 1 will *aid* in updating these costs to the date of interest. However, a large part of the LNG system is located abroad and depends on arbitrary costs, as with oil.

Why LNG?

The reasons for the rapid rise in the use of LNG are many. The principal reasons may be summarized as follows:

1. To avoid the waste of natural gas produced along with oil. Much of this was previously flared in fields remote from the markets. Some was reinjected to maintain pressure. The governments of oil-producing countries brought strong pressure to convert this gas to petrochemicals or later, when it became feasible, to LNG.
2. Natural gas available in remote regions had been considered valueless and the wells simply plugged.
3. The growing concern about pollution and the environment created a strong demand for clean fuels and, with it, a premium price.
4. The exhaustion of supplies in many areas of the world created an ever-increasing demand for supplementary supplies of fuel, particularly of clean fuel.
5. There were developments and improvements in every step of the LNG system that helped make it safer and more economic: gas purification, separation, treatment, liquefaction, transport, storage, and regasification.
6. Compared with pipelines, which are limited to land areas, LNG transports provided great flexibility in linking all major sources to all major markets.
7. There was a great need to store gas for peak periods. "Line pack" provided only limited storage. Underground storage was limited to areas of suitable geology. LNG could be stored anywhere.

LNG Applications

LNG has developed for two principal applications:

Peak Shaving. By storing gas the production and transmission facilities are more fully used dur-

ing periods of lower demand. The excess gas is liquefied and stored. During periods of heavy demand it is regasified and distributed. With the best insulated tanks the "boil off" may be as low as 0.05%/day, although 0.1 to 0.3% is more normal. The gas boiled off may be used or reliquefied. However where the entire storage is needed in a matter of weeks or a few months, large scale regasification facilities are employed, using heat exchangers.

Main or Supplementary Gas Supplies. This generally involves overseas transport by LNG tankers, with associated loading and unloading facilities and storage. In this case, the LNG system includes most of the purification, separation, and treatment prior to liquefaction, a minimum amount necessary for the gas collection pipelines being done near the wells. The liquefaction and regasification processes are similar in the two applications.

Characteristics of LNG

The principal characteristics of methane of significance in its liquefaction and storage are shown in Table 4.3. A few properties of the other constituents in LNG are shown in Table 4.4. Note that the other hydrocarbons all liquefy before the methane, a fact that will be made use of later. Butanes and pentanes solidify before the liquefaction temperature of methane is reached, a fact of importance in certain processes.

Purification and Liquefaction

The purification prior to liquefaction is similar to that for other natural gas already described. In addition, depending on the refrigeration process used, it may be necessary to separate out certain components that might liquefy or solidify in the gas compressors or engines.

A gas is liquefied by lowering its temperature to the boiling point and then removing further heat energy equivalent to its latent heat of vaporization. As shown in Table 4.3, the temperature of methane must be lowered from a normal temperature, say, 60°F, to −263°F (16 to −164°C), at 1 atm, a total of 323°F (179°C).

The refrigeration process involves compression, cooling, and expansion of a refrigerant, just as in a home freezer. However, in the liquefaction of natural gas, the temperature range is so great that

Table 4.3. Characteristics of Methane, CH$_4$[6]

Characteristic	English	Metric
Molecular weight	16.04	16.04
Boiling point at 1.0 atm	−263.2°F	−164.0°C
Melting point at 1.0 atm	−296.5°F	−182.5°C
Specific heat C_p	0.528 Btu/lb °F	0.528 kcal/kg °C
at	59°F, 1 atm	15°C, 1 atm
C_p/C_v	1.31	1.31
Critical temperature	−116.5°F	−82.5°C
Critical pressure	45.8 atm	45.8 atm
Density of gas	0.0448 lb/ft³	0.718 kg/m³
at	32°F, 1 atm	0°C, 1 atm
Density of gas at boiling point	0.1124 lb/ft³	1.8004 kg/m³
Density of liquid at boiling point	26.47 lb/ft³	424 kg/m³
Heat of vaporization at boiling point	248.4 Btu/lb	577,600 J/kg
Higher heating value	23,890 Btu/lb	55.55×10^6 J/kg
	978 Btu/ft³	36.43×10^6 J/m³
at	77°F, 1 atm	25°C, 1 atm

several steps of refrigeration are necessary. It is not generally practical to cool more than 60 to 90°C (108 to 173°F) in a single step.

A simplified diagram of a typical refrigeration cascade using propane, ethylene, and methane as refrigerants is shown in Fig. 4.3. Each refrigerant cools to near its dew point (a few degrees above its boiling point). Thus, the natural gas is cooled to ambient temperature by means of cooling water. After booster compression, it is further cooled to near −42°C by a propane refrigerant, then to near −105°C by ethylene, and to near −162°C by methane. It is then expanded to near atmospheric pressure to further cool and liquefy it.

As with all liquids, the boiling point and the dew point rise with pressure. Thus, at moderate pressure, propane can be cooled to near its dew point with cooling water before expanding it. Ethylene is additionally cooled in the propane vaporizer, and methane in the two preceding vaporizers, in order to reach temperatures near their dew points before expanding them. There are further loops omitted from Fig. 4.3 for simplicity. Each refrigerant is heat exchanged with its own expanded vapor.

There are many variations of this basic cascade. Other refrigerant combinations are used, such as ammonia/ethylene/methane, or Freon 22/Freon 15/methane.

Table 4.4. Characteristics of Hydrocarbons Important in LNG

Hydrocarbon		Mol. Wt.	Boiling Point at 1 atm		Melting Point at 1 atm	
			°F	°C	°F	°C
Methane	CH$_4$	16.04	−263.2	−164.0	−296.5	−182.5
Ethane	C$_2$H$_6$	30.07	−127.5	−88.61	−297.9	−183.3
Propane	C$_3$H$_8$	44.09	−43.7	−42.06	−305.8	−187.7
Ethylene	C$_2$H$_4$	28.05	−154.7	−103.7	−272.5	−169.2
Isobutane	C$_4$H$_{10}$	58.12	10.9	−11.7	−255.3	−159.6
n-Butane	C$_4$H$_{10}$	58.12	31.1	−0.5	−217.0	−138.3
Isopentane	C$_5$H$_{12}$	72.15	82.1	27.8	−255.8	−159.9
n-Pentane	C$_5$H$_{12}$	72.15	96.9	36.1	−201.5	−129.7

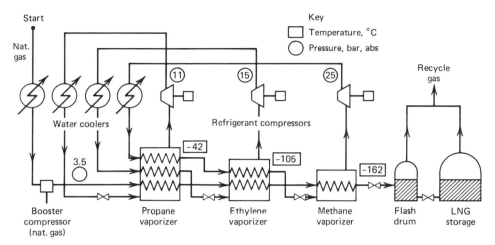

Figure 4.3. Simplified flow diagram of a conventional refrigeration cascade. Reproduced from Lum, "Liquefied Natural Gas,"[5] Applied Science Publisher, London, 1974.

The gases may be expanded isentropically in gas engines or turbines, instead of the isenthalpic expansion by throttling, as shown in Fig. 4.3. The external work thus saved helps to power the compressors. This is known as the Clausius cycle.

In a perfect gas, an isenthalpic expansion does not result in cooling. The enthalpy is a function of temperature only. However, in actual gases, in the range used for refrigerants, enthalpy varies also with pressure, and throttling does result in substantial cooling. This is known as the Thompson–Joule effect, and the throttling cycle of Fig. 4.3 is known as the Thompson–Joule cycle.

Both the Thompson–Joule cycle with throttling expansion and the Clausius cycle using isentropic expansion are in widespread use. In the latter, throttling is necessary for the final expansion, since moving parts are not practical at such low temperatures and with the phase change to a liquid. Theoretically, the Clausius cycle is more efficient since the external work of expansion is utilized. However, the overall economy of well-designed plants of either type appears to be nearly the same.

As natural gas is cooled, its components liquefy starting with the heaviest in the order butane, propane, ethane, and methane. These components can be separated out in successive flash drums or separators and used as the refrigerants in modified cascades of either closed- or open-cycle arrangement. In the open cycle, the refrigerants are separated from the main stream of natural gas, whereas in the closed cycle (Fig. 4.3), they are separated from each other in a separate refrigerant system.

Storage of LNG

The world's largest storage tank in 1973 was the 600,000 barrel tank on Staten Island which was destroyed in February of that year by a violent gas blast and fire. The tank was empty and was being repaired. Forty employees were killed. The 600,000 barrels amounts to 95,400 m³ LNG and is equivalent to 2.10 billion SCF natural gas. It weighs 44,600 tons; and at $1.00/MCF, it is worth $2.1 million.

By comparison, the largest LNG tankers of that day were of 71,500 m³ capacity (450,000 barrels, or 33,500 tons deadweight). Subsequent ships will have capacities up to those shown in Table 4.5.

Note that the Algeria to El Paso scheduled movement mentioned earlier is 1 billion SCFD, which can be compared with the tank and ship sizes of Table 4.5.

The ideal storage for LNG would be a Dewar flask, in which the double walls of the container are separated by vacuum, thus preventing heat transfer by conduction or convection. The surfaces are highly reflective and turn away radiation. However, such a vessel is practical only in small sizes. Instead, a double wall vessel, with ample heat insulation between, is generally used. There are several arrangements, and the development is continuing.

Both double-skinned metal tanks and prestressed concrete tanks with metal shells and liners are used. Most expensive and of lowest "boil off" (0.03 to 0.04%/day) is a double-skinned concrete tank with gas-impermeable inner lining. The insulation may

Table 4.5.   LNG Ship and Storage Tank Capacities

Container	m³	Barrels	Short Tons of Cargo	Billions[a] of SCF Gas Equiv.	Worth at $1/MCF ($ Millions)
1973 Tank	95,400	600,000	44,600	2.10	2.10
1973 LNG tanker	71,500	450,000	33,500	1.58	1.58
Ship for 1976 delivery	125,000	786,000	58,600	2.75	2.75
Ship soon after 1976	165,000	1,038,000	77,300	3.64	3.64
Ship planned	200,000	1,258,000	93,700	4.41	4.41

[a]Calculated as methane, SCF at 60°F, 1 atm.

be between the liner and the concrete or between the concrete and the outer shell.

A metal inner membrane exposed to the low temperature must be of aluminum, stainless steel, or steel with at least 5% nickel. The concrete must be prestressed or the equivalent. Tanks can be above ground or below ground. If above ground, dykes are required to contain the contents in event of rupture. Tanks above ground with earth sloped up around the sides at least 10 ft thick radially at the top are considered "underground."

Sidewall insulation may be of insulating powders made by firing minerals such as silica and magnesia, or may be evacuated powders such as expanded Pearlite or silica aerogel. Load bearing insulation requires solid blocks such as expanded polystyrene, foamed concrete, or balsa wood. To keep water from entering the insulation space, it is filled with dry gas such as nitrogen or dry natural gas.

Underground caverns and tank-like excavations, which depend on the frozen earth or rock walls for containment, are also used with mixed success. The boil-off frequently exceeds calculations and is subject to the vagaries of geology.

The design of any LNG storage facility involves a compromise between the cost of the containment and the cost of reliquefying the boil-off, the latter being a significant factor.

It has been found that the prestressed concrete can be exposed to the LNG without harmful effect. The outer metal shell is of normal steel. One of the more promising designs of the mid-1970s uses this principle. A prestressed concrete container is used without inner liner, with insulation and metal shell outside.

Revaporization of LNG

Natural vaporization from storage tanks is usually in the range of 0.1 to 0.3%/day. Thus, to obtain large quantities of gas on short notice, massive vaporizers are required. In peak-shaving installations,

liquefaction usually takes place over 200 to 220 days per year in temperate climate. The stored gas is used during some part of the remaining 165 to 145 days, reaching a peak for relatively few weeks.

Vaporizers are essentially heat exchangers using cooling water if available (actually heating water). Otherwise, combustion gases are used in direct-fired heat exchangers, or steam or hot water in indirect systems. The natural gas is generally released at 0°C and up to 75 atm pressure.

Many different heat exchange arrangements are used. In one typical arrangement, the LNG passes vertically upward through pipes embedded in aluminum panels, in countercurrent to water running down the panels from troughs at the top. The water exit temperature is limited to 0°C (−2°C for seawater) to prevent ice formation on the tubes.

Since regasification is intended for distribution, advantage can sometimes be taken of the easy transportation of LNG, shipping it to the area needed before vaporizing it. Truck, rail, or barge transportation to outlying areas have been used, but LNG pipelines are also under consideration.

By-product uses of the cheap refrigeration available in the vaporization of LNG are also considered such, for example, as air separation or freeze drying.

Cost of LNG

A cost estimate[5] has been made in 1972 of an LNG system for major overseas supply of 10⁹ SCFD (1 billion SCF/day) of natural gas. This is equivalent to the Algeria to El Paso scheduled supply mentioned earlier, although the conditions may of course differ. This estimate has been altered in form to agree with the other cost estimates in this book, and some changes have been made. Needless to say, this estimate is intended to show the general proportions only, since energy prices have escalated nearly fourfold from 1972 and 1980 (Fig. 1.2). See Chapter 1 for updating indices.

The system considered consists of a liquefaction plant, LNG storage at both supply and receiving ends, LNG tankers, port facilities, regasification plant, and connections to the distribution system.

The following assumptions are made:

1. Gas is available at the liquefaction plant at 15¢/million Btu.
2. Overall processing loss of gas, 10%.
3. Supply end storage 100,000 tons, about four days.
4. Receiving end storage 300,000 tons, about 12 days.
5. Ships of 125,000 tons dwt, 17 knots average speed.
6. 10^9 SCFD natural gas delivered.
7. LNG processed in four streams.
8. Distance by sea 3000 miles.
9. Prices mid-1972.
10. Gas of 1200 Btu/SCF.
11. Return on investment   15%
12. Taxes and insurance   3%
13. Depreciation, S. L.   5%
14. Total fixed charges   23%

Capital Costs

	$ Millions
Liquefaction plant and storage	500
6 Ships of 75,000 tons dwt, 125,000 tons displacement, at $75 million each	450
Port facilities, jetty, etc.	20
LNG storage, revaporization, and connections	20
Contingency, miscellaneous, and escalations	100
Total investment	1090

Production Costs

	¢/Million Btu
Gas supply into liquefaction plant	15
Fixed charges, 23%, 1090 × 10^8 ¢, 438 × 10^6 million Btu/yr	57
Operation and maintenance Labor and materials Fuel and utilities Port and other charges	18
Total production cost of delivered natural gas	90

Thus, with these assumptions, natural gas could be purified, liquefied, transported by ship, regasified, and delivered into pipelines 3000 miles away for 90¢/million Btu, including all necessary storage at the two ends. While all costs have escalated (see Chapter 1 for indices), the wellhead price of natural gas in the United States increased from 17.1¢ to $1.14/MCF, and about the same per million Btu, from 1970 to 1979 (Fig. 1.6). Thus, the economy of LNG is improving and its use is expanding rapidly in 1980.

Cost of Peak Shaving

Noting that 1 billion SCF of natural gas is about 25,000 tons, the storage in the foregoing system is roughly 4 billion SCF, or four days' supply at the supply end and 12 days' supply at the receiving end, a total of 16 days' supply. For peak shaving, the storage is filled in about 200 days. Thus, the ratio of average liquefaction capacity to total storage in a peak shaving setup is about 8% of that in a major supply. However, since the supply may not be steady, assume that 10% is required.

The daily revaporization capacity in the particular major supply described must be at least $\frac{1}{16}$ of the total storage, or 1 billion SCFD. This may not be too different from a peak shaving system. Thus, a peak shaving system with the same total storage as the major supply described above and having 10% as much liquefaction capacity would have *roughly* the following costs:

Summary of Assumptions—Peak Shaving System

1. Storage capacity 400,000 tons, 16 billion SCF
2. Liquefaction capacity 10^8 SCFD
3. Vaporization capacity 1 billion SCFD
4. Prices mid-1972
5. Gas of 1200 Btu/SCF
6. Return on investment   15%
7. Taxes, insurance, and depreciation   8%
8. Total fixed charges   23%

(In the United States, 18% is more average for maintenance, depreciation, taxes, insurance, and return.)

Capital Costs

	$ Millions
Liquefaction plant	80
LNG storage, revaporization, and connections	26
Contingency, miscellaneous, and escalations	11
Total investment	117

Production Costs

(To be added to the cost
of supply gas)

	$/Million Btu
Fixed charges, 23%, 117×10^6, 19.2×10^6 million Btu	1.40
Operation and maintenance	0.44
Total cost to be added to supply gas	1.84

A plant $\frac{1}{10}$ full size is taken as having a cost 40% of the cost of one of the four streams. (Cost varies about as the square root of size.) The operation and maintenance are taken the same percent of total investment as for the larger plant.

In 1972, the average price of gas to utilities was 29¢/million Btu. Adding this to the storage cost of $1.84/million Btu, gas for peaking is available at $2.13/million Btu.

Most of the cost of the normal supply is in wells, gas plants, and transmission facility. If additional facilities were built to supply a one-month load, they would cost $12 \times 0.29 = \$3.48$/million Btu, or about 1.63 times that obtained through storage.

While energy prices have escalated nearly four-fold from 1972 to 1980, not all equally, the principles illustrated by this example still apply. LNG storage for peak shaving is economic in many locations.

LIQUEFIED PETROLEUM GAS[7]

Liquefied petroleum gas (LPG), LP gas, or bottled gas, as it is variously called, consists of propane, C_3H_8, or butane, C_4H_{10}, or a mixture of the two. Small amounts of propylene, C_3H_6, and butene, C_4H_8, may also be present, as explained later. LPG is usually supplied as a liquid in tanks or steel "bottles." The vapor above the liquid is used as a gas and is continuously replaced by boiling of the liquid until the normal vapor pressure at the existing temperature is restored. Propane boils at $-44°F$ ($-42°C$) at 1 atm, and at 100°F (38°C) its vapor pressure is 190 psia (13 atm). It can thus be used down to quite low temperatures.

The two isomers of butane boil at 31 and 11°F (about 0 and $-12°C$). At 100°F, their vapor pressures are 3.5 and 4.0 atm, respectively.

In cold climates where the ambient temperature may be near or below the boiling point of butane, propane alone is used, except in industrial applications where the fuel is heated before use.

In 1922, the first year of record, 223,000 gal (approx. 500 tons) of LPG were marketed, used chiefly for cooking. This may be compared with the estimated 1975 consumption of 134 million tons, with hundreds of varied uses. The pattern of growth in the United States and in the world is shown in Table 4.6. Whereas the growth rate in the United States has stabilized at around 4%/yr, in the rest of the world it is much higher.

Reasons for Use of LPG

The reasons for the very extensive use of LPG are as follows:

Table 4.6. Growth in Consumption of LPG (in Millions of Tons/Yr)

	1922	1950	1960[a]	1970	1975—Est.
United States	(500 tons)	7.0	27.50	43.5	52.3
Rest of the world			7.47	44.3	82.0
Total			34.97	87.8	134.3

Source. Reprinted with permission from *Liquefied Petroleum Gases* by A. F. Williams and W. Lom, 1974, Ellis Horwood Ltd., Chichester, England.
[a]Not including Eastern Europe.

1. It exists both as a gas and a liquid at moderate temperatures and pressures. It can be easily liquefied for storage or shipment at slight pressure but can be used as a gas under normal atmospheric conditions.

2. It has a high heating value as a gas, propane 2438 Btu/SCF, n-butane 3162 Btu/SCF, compared with natural gas, about 1020 Btu/SCF.

3. It is clean. All undesirable impurities have been removed in the process of producing it.

4. As a liquid, it is a compact and convenient form of energy for storage or shipment. Propane contains 21,670 Btu/lb, n-butane 21,316 Btu/lb, approaching methane, 23,890 Btu/lb. The density of liquid propane is 31.6 lb/ft^3, and thus 1 ft^3 contains 685,000 Btu. This is more than liquid methane, 633,000 Btu/ft^3. LNG is intermediate, depending on its composition.

5. It is a hydrocarbon, highly suitable as a feedstock for many chemical processes.

These inherent properties have led to a very wide diversity of uses throughout the world. Where natural gas is plentiful, the largest use is for domestic and commercial use and for chemical feedstocks. Where most gas is manufactured, the LPG is used largely by the gas industry to upgrade manufactured gas or directly as an industrial fuel, as shown in Table 4.7.

Originally, propane and butane were removed

Table 4.7. Uses of LPG in North America and the United Kingdom, 1969

	Percent of Total	
	North America	U.K.
Domestic and commercial use	42.7	6.0
Industrial fuel	5.3	27.6
Gas industry consumption	1.0	66.4
Automotive sales	8.2	Small
Chemical industry use	41.7	Small
Agricultural, aerosols, etc.	1.1	Small
	100.0	100.0
Total sales, thousands of tons	40,000	1,300

Source. Reprinted with permission from *Liquefied Petroleum Gases* by A. F. Williams and W. Lom, 1974, Ellis Horwood, Ltd., Chichester, England.

from crude oil in order to stabilize it, that is, to lower its vapor pressure to a value permissible in the tanks or ships that were to hold it. Frequently, the propane and butane were simply flared. However, as the market developed owing to the inherent advantages cited above, this practice ceased.

Recently (1976), additional factors have greatly increased the importance of LPG and the demand for it:

1. The impending shortage of natural gas in the United States.

2. The rapidly growing demand for clean fuels to control air pollution, particularly by the large fuel importers, Europe and Japan.

3. Technologic developments in the long-distance transport of LPG.

Hydrocarbons Involved in LPG

The principal hydrocarbons involved in LPG are the so-called C_3 and C_4 hydrocarbons, that is, those having three or four carbon atoms. These include the alkanes or saturated hydrocarbons propane, n-butane, and isobutane; and the alkenes or unsaturated hydrocarbons propylene (propene) and the butenes (four isomers) (see Chapter 11).

In general, LPG from oil and gas fields contains only the saturated hydrocarbons propane C_3H_8 and butane C_4H_{10}, whereas LPG from oil refineries contains varying amounts of propylene, C_3H_6, and the butenes, C_4H_8, as well.

The unsaturated hydrocarbons do not occur to any extent in crude oil or natural gas but result from the cracking or other operations in the refinery. The amounts depend on which of several processes in the refinery is the source of the LPG. One sample of Swedish trade LPG contained 64% propylene.

In oil- and gas-producing countries, most of the LPG is produced in oil and gas fields, 66% in the United States and Canada and 98% in the Middle East and Africa (1975). In other countries, it is produced mainly in refineries, Europe 73% and Japan 100%. Altogether in 1975, about 55 million tons of LPG were produced in refineries, 80 million tons in oil and gas fields, a total of 135 million tons (52 million tons in the United States.)

Applications of LPG

While the total use of LPG in the United States in 1975, 52 million tons, or 2.2 quads, represents a small part of the total gas and oil used, 52.7 quads, it is nonetheless an extremely important fuel because of the possibilities it has opened up. Typical examples are:

1. Domestic fuel for water heating, cooking, refrigeration, waste disposal, and for space heating and air conditioning, particularly for mobile, vacation, and farm homes.
2. Industrial fuel for steam boilers, heat-treating furnaces, space heating, waste disposal, drying and firing of clay products, glass and plastics processing, and the generation of inert atmospheres.
3. Automotive propulsion, tractors, buses, and cars.
4. Commercial cooking and baking.
5. Feedstocks for innumerable chemical processes, particularly steam reforming for town gas, SNG, or hydrogen or feedstocks for the vast petrochemical industry.

Characteristics of the LP Gases

Very complete characteristics of the LP gases, including their thermodynamic properties, are given in the references.[7-9] Characteristics of the unsaturated LP gases, propylene and the butenes, will be found in Ref. 9.

Specifications for commercial LP gases have been fixed by the Natural Gas Processing Association. Commercial propane has not over 2.5% C_4 and higher, commercial butane not over 2% C_5 and higher. The maximum vapor pressures specified practically preclude any lighter components.

Production, Purification, and Liquefaction of LPG

The processes involved in the production, purification, and liquefaction of the LP gases have been described under the headings of petroleum, natural gas, and LNG. As shown in Fig. 4.3, in the liquefaction of natural gas, in which propane is used as a refrigerant, the propane can be liquefied by compression to about 11 atm and cooling to near ambient temperature by means of cooling water.

SYNTHETIC OR MANUFACTURED GAS

At the time of this writing (1980), a number of new methods are under development for producing synthetic gas from coal. The synthetic gases fall generally into three categories: low-Btu gas of about 150 Btu/SCF (producer gas), medium-Btu gas of about 300 Btu/SCF (syngas), and a high-Btu or pipeline-quality gas of about 1000 Btu/SCF (SNG). Some of these are already in use in industrial plants in various parts of the world. However, most of the development in the United States is directed to alternate energy sources for the future.

A number of by-product or manufactured gases, mostly under 600 Btu/SCF, have been in common use for many years, some since the beginning of the gas industry about 1800. These include producer gas, blast furnace gas, blue water gas (or water gas), carbureted water gas, and coke oven gas.

In the period 1920 to 1940, about 11,000 coal gasifiers were in operation in the United States.[10] But with the general availability of cheap natural gas, these had nearly all disappeared by the 1970s. In 1979, except for by-product gas production, only six small coal gasification installations were operating in the United States, and 15 others were in various stages of design, construction, or start-up. Elsewhere in the world, where natural gas was not available, the gasification of coal and heavy hydrocarbons has continued. In 1979, there were at least 95 coal gasification plants, some quite large, operating in at least 27 different countries, producing low- and medium-Btu gas.

In the following sections, the basic chemical and physical principles of coal gasification are first presented. The general types of coal gasifiers are then identified. Examples are given of gasification systems past and present and of those under development in 1980.

Basic Processes

The earliest coal gasification involved simply the incomplete combustion of coal in air, resulting ideally in a 1:2 volume mixture of CO and N_2 called "producer gas":

$$\tfrac{1}{2} O_2 + C + 2 N_2 \rightarrow CO + 2 N_2 \qquad (4.1)$$

In later units, steam was added, Eq. 4.4, and the resulting CO and H_2 increased the heating value to about 150 Btu/SCF. This was the forerunner of the low-Btu gas now being produced by much improved methods but still blowing coal with air and steam.

Under suitable conditions of temperature and pressure and with proper catalysts, all to be described later, the following chemical reactions can be made to "go":

	Heat Output		
Combustion	Btu/lb mole	MJ/kg mole	
$C + O_2 \rightarrow CO_2$	170,000	395	(4.2)
Boudouard Reaction			
$C + CO_2 \rightarrow 2\ CO$	$-72,190$	-167.9	(4.3)
Gasification (Carbon-Steam Reaction)			
$C + H_2O \rightarrow CO + H_2$	$-58,350$	-135.7	(4.4)
Water Gas Shift			
$CO + H_2O \rightarrow H_2 + CO_2$	13,830	32.16	(4.5)
Methanation*			
$CO + 3\ H_2 \rightarrow CH_4 + H_2O$	108,000	251.2	(4.6)
Hydrogasification			
$C + 2\ H_2 \rightarrow CH_4$	39,380	91.6	(4.7)

Coal is mainly carbon, C, and natural gas is mainly methane, CH_4. Thus, the three steps, Eqs. 4.4, 4.5, and 4.6, can be used to convert coal to methane, a substitute natural gas (SNG) interchangeable with natural gas. However, this requires high temperatures and pressures, complex equipment, and substantial energy to convert essentially all of the carbon to methane. It is very costly. Manufactured gases through the 1970s have used mainly the steps of Eqs. 4.1 through 4.5, with pressures up to about 300 psi (20 atm), resulting in heating values less than half that of natural gas.

A common terminology has grown up around the three equations, Eqs. 4.4 through 4.6, and although it is used rather loosely, a knowledge of it is necessary in order to understand the literature.

The first step, Eq. 4.4, is often referred to as the "gasification" step, even though the whole process is broadly gasification. The resulting gas, a mixture of carbon monoxide, CO, and hydrogen, H_2, is called "water gas" because it is obtained by passing steam through incandescent coal. In actual practice, it contains various other gases as well, depending on the particular process used. It is also called "synthesis gas," or "syngas," since it is frequently used for the synthesis of hydrogen, ammonia, methanol, and a whole range of hydrocarbons as well as for heating gas.

The second step, Eq. 4.5, is known as the "water gas shift" reaction. It uses some of the CO and makes additional H_2. The CO_2 is then removed. Looking ahead to the methanation step, Eq. 4.6, H_2 and CO are needed in a 3:1 ratio. Equation 4.4 has produced them in a 1:1 ratio. Actual synthesis gas may have H_2 and CO in nearly 2:1 ratio, some H_2 having been given off directly by the coal. In any

*At 77°F (25°C), 1 atm.

event, some of the CO must be shifted to H_2. This reaction, usually carried out in a separate catalytic reactor, is controllable by the amount of steam admitted, so that the desired 3:1 ratio of H_2 to CO can be obtained.

The third step, Eq. 4.6, is known as the "methanation" step for obvious reasons. It is generally carried out in a separate reactor, although in processes under development (1976) as much as 60% of the methanation takes place in the gasifier unit.[11,16]

The approximate chemical composition of coal is $CH_{0.8}$, and its conversion to methane, CH_4, is a matter of adding hydrogen. In Eqs. 4.4 and 4.5, the hydrogen of water is used. However, carbon can be reacted with hydrogen from any source to produce methane. This is known as "hydrogasification," Eq. 4.7. A steam moderator is used, and a gas very rich in methane results. The term "methanation" however usually refers to the process of Eq. 4.6.

The methanation reaction of Eq. 4.6 requires a nickel catalyst and must be followed by dehydration to remove the water. This reaction is generally preferred. However, an iron catalyst can be used, leading to a CO_2 by-product which must be removed:

$$2\ CO + 2\ H_2 \rightarrow CH_4 + CO_2 \qquad (4.8)$$

Overall, the three equations, 4.4 through 4.6, result in the conversion (add 2 of Eq. 4.4 to Eqs. 4.5 and 4.6):

$$2\ C + 2\ H_2O \rightarrow CH_4 + CO_2 \qquad (4.9)$$

$$24 + 36 \rightarrow 16 + 44$$

Carbon plus water, in the relative weights shown, plus heat energy are converted to methane and carbon dioxide.

Methanol (Methyl Alcohol, Wood Alcohol). Syngas (H_2 and CO) can be converted to methanol after shifting to a 2:1 ratio of H_2 to CO:

$$CO + 2\,H_2 \rightarrow CH_3OH \text{ (methanol)} \qquad (4.10)$$

A zinc oxide/chromium oxide catalyst is needed in a reactor operating at 3000 to 4500 psi (200 to 300 atm), or a highly reactive copper catalyst in a reactor at 750 to 1200 psi (50 to 80 atm). A thermal efficiency, coal to methanol, of 40 to 50% is obtainable in large plants (7500 tons/day methanol from 15,000 tons/day coal.

Methanol was originally derived from wood (wood alcohol), but for many years it has been produced primarily by reforming natural gas or naptha. However, it can be made from coal in the future.

Direct conversion of methanol to high-octane gasoline, using a zeolite catalyst, has been demonstrated[12] and is under development:

$$x\,CH_3OH \rightarrow (CH_2)_x + x\,H_2O \qquad (4.11)$$

Nearly 90% of the hydrocarbons formed are in the gasoline range and the 10% remainder is LPG.

Hydrogen. Syngas can also be converted to hydrogen by shifting all the CO to H_2, Eq. 4.5, and removing the CO_2 in an absorber. To date (1980), most hydrogen has been produced by steam reforming of natural gas or naptha, or by partial oxidation of hydrocarbons (see Hydrogen, later in this chapter).

Ammonia. Hydrogen can in turn be converted to ammonia (NH_3), the basis of many industrial products and fertilizers. The Haber process is usually used:

$$N_2 + 3\,H_2 \rightarrow 2\,NH_3 + \text{heat} \qquad (4.12)$$

This requires very high pressures and high temperatures in the presence of a catalyst. Typical conditions are 200 to 1000 atm and 750 to 1050°F (400 to 600°C).

Hydrocarbons. Syngas can be converted to a wide range of hydrocarbons by the Arge and Kellogg versions of the Fischer–Tropsch synthesis, using iron/cobalt and iron catalysts, respectively. Syngas is converted to a syncrude which is supplied to refineries for the production of gasoline, LPG,

gas, oil, alcohol, and wax (see Fischer–Tropsch synthesis, Chapter 3).

Possible Conversions. Thus, there is a two-way street. Syngas from coal can be used as a gaseous fuel or be converted to methane, methanol, ammonia, and a whole range of hydrocarbons. Natural gas or liquid hydrocarbons can in turn be reformed into methane or into hydrogen, which can be converted to methanol or ammonia. Methanol can be converted directly to gasoline. The choice depends on what materials are available, what are needed, and the economics.

Energy Considerations

The amounts of heat energy generated by reactions 4.2 through 4.7 are shown alongside, the negative sign indicating an endothermic process, requiring the addition of heat. Significant gasification by the processes 4.1 through 4.7 proceed mainly at temperatures above 1500°F (815°C). The energies shown are for reactions taking place at 1832°F (1000°C).

The gasification step, Eq. 4.4, is endothermic, and heat must be provided by combustion or some other source. From Eq. 4.9, each pound of carbon, having a HHV of 14,096 Btu, is converted to $\frac{2}{3}$ lb methane having a HHV of 23,890 Btu/lb, or 15,927 Btu. At 100% efficiency and equal input and output temperatures, 1831 Btu must be added for each pound of carbon converted.

In a practical process, there are losses in every stage. The gasification must be at high temperatures and pressures. Coal is used, not pure carbon. Energy is expended in preparing the coal and in removing impurities from the gas. Some is lost with the ash. Heat energy is lost in the various reactors. Some is lost in the cooling or quenching water, some in the output gas. Carbon conversion is not complete.

If the heat is provided by burning part of the coal in oxygen, a high-purity oxygen plant is required, which in turn uses coal. A steam plant may be required using further coal. Thus, overall thermal efficiency from coal to high-Btu SNG may be of the order of 50%. The efficiency of conversion to low- or medium-Btu gas is considerably higher since some of the steps are not required.

In addition, the high capital cost of all the equipment must be charged to the output gas, with the result that the cost per Btu of SNG is expected to be six to eight times the cost per Btu of the coal input when full-scale gasification plants are in operation. For medium-Btu gas, the cost may be three and a half to five times that of the coal. For pilot plants, the ratios may be somewhat higher. Note that in the period 1960 to 1970, coal and natural gas energy was supplied to utilities at the *same* price per Btu.

Temperatures, Pressures, and Atmospheres

In the proximate analysis of coal, the temperature is raised to 1740°F (950°C) in the absence of air, as indicated in Fig. 4.4. All of the volatile materials are driven off, but none of the fixed carbon is gasified. The coal passes through a plastic stage between 750 and 930°F (400 and 500°C) and may cake or be noncaking.

In a gasifier, air or oxygen and steam are used at temperatures generally above those to which coal is carried in the proximate analysis. However, certain fundamentals must be observed:

1. Significant gasification occurs only above 1500°F (815°C).

2. The temperature, pressure, and time required for the gasification reactions to occur varies with the rank of the coal. Highly reactive lignites and sub-bituminous coals can be gasified at reasonable rates at temperatures of 1700 to 1900°F (927 to 1038°C) and at pressures of 30 to 35 atm. The dry-ash Lurgi gasifier operates in this range. Higher-rank coals require much higher temperatures and pressures for complete gasification in reasonable times.

3. Below about 1900°F (1040°C), coal burns to a dry ash, but at about 1900°F the ash starts to fuse and form clinkers, which may adhere to the walls and clog the grates. Dry-ash removal thus limits the gasification temperature to below about 1900°F. To operate above this temperature, the temperature must be sufficiently high that the ash can be removed as a liquid slag or flyash.

4. If the devolatilization is at temperatures appreciably below 2000°F (1090°C), tars and condensables may form that must later be removed for most gas applications. However, this depends on the type of gasifier. In the Texaco or Koppers–Totzek entrained gasifiers, shown as examples, operation is at 2000 to 3000°F and 3300 to 3500°F, respectively, and no tars are formed. The fluidized-bed Westinghouse gasifier with longer dwell time, operating in the 1600 to 1900°F range, also produces a gas free of tars. In the Lurgi gasifier, devolatilization starts at about 1000°F as indicated, and condensable materials including oxygenated compounds and tars are formed. See Lurgi later.

5. Significant methane is not formed by processes 4.6 and 4.7 above 2000°F (1090°C). This is an important factor in processes intended for SNG, since the formation of appreciable methane in the gasifier reduces the later methanation requirements. Both of these processes are exothermic, and the resulting heat can be used efficiently in the gasifier. Thus, many processes are two-stage, carrying out initial gasification in the 1700 to 1900°F range and final complete gasification at a higher temperature. As indicated in Fig. 4.4 the BiGas process operates at 1700 and 2700°F.

6. High pressure is favorable to the formation of methane.

7. Combustion in an oxidizing atmosphere above 1900°F results in the formation of nitrous oxides. Since gasifiers produce hydrogen in the gasifier proper, the atmosphere is reducing and nitrous oxides are not formed. The sulfur in the coal is converted to H_2S and a small amount of COS. Organic nitrogen is changed either to elemental nitrogen or ammonia gas.

8. High pressures of 300 to 1200 psi (20 to 80 atm) are needed for the complete gasification of the higher-rank coals. Since the new gasifier developments are considered necessary to gasify *all* coals, they are almost invariably pressurized. The two-stage Combustion Engineering atmospheric entrained gasifier is an exception.

9. Typical pipeline pressures are about 1000 psi. Atmospheric or low-pressure gasification leads to considerable expense in pressurizing the product gas. Some gasifiers therefore operate at near pipeline pressure for this reason.

10. In all cases, the coal must pass through a plastic stage between 750 and 930°F. Except

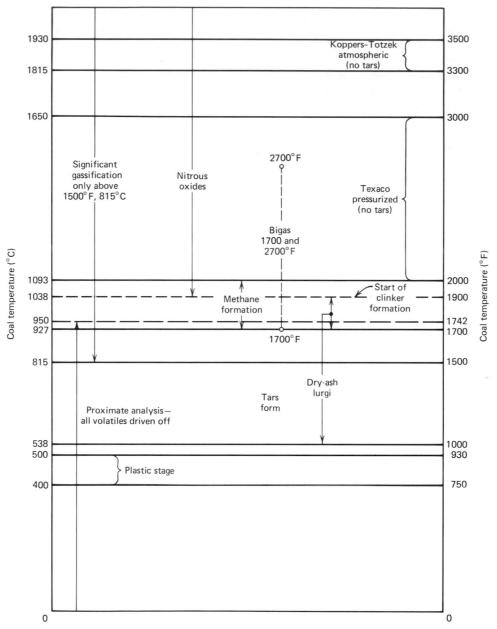

Figure 4.4. Gasification temperature considerations.

for noncaking coals or coke, some measure must be taken to keep it from adhering in a solid mass and clogging the gasifier. The methods used are described in a later section, Types of Gasifiers.

11. Operation at low temperatures requires excess moderating steam, especially when oxygen-blown, and results in large quantities of foul water to be treated.

12. Large high-pressure vessels, over 10 atm, are extremely expensive.

13. Temperatures above 1000°F (540°C), the metallurgical limit, require refractory lining, water cooling, or other device for keeping the metal below 1000°F.

Early Gasification Schemes

Table 4.8 shows the composition and heating values of a number of commercial gases. The heating values of these manufactured or by-product gases vary from 92 Btu/SCF for blast furnace gas to 574 Btu/

Table 4.8.   Characteristics of Typical Gaseous Fuels (Composition, vol%; HHV, Btu/SCF)

	Pittsburgh Nat. Gas 1958	Producer Gas[a]	Blast Furnace Gas	Blue Water Gas[b]	Carbureted Water Gas	Coke Oven Gas	Lurgi Process Gas[c,d]
Hydrogen, H_2		14.0	1.0	48	37	46.5	57.8
Carbon monoxide, CO		27.0	27.5	38	33.5	6.3	25.6
Methane, CH_4	83.4	3.0		1.2	12.5	32.1	13.9
Ethane, C_2H_6	15.8						
Ethylene, C_2H_4[e]					6.0	3.5	
Benzene, C_6H_6[e]					2.8	0.5	
Oxygen, O_2		0.6		0.6	0.3	0.8	
Nitrogen, N_2	0.8	50.9	60.0	7.2	3.7	8.1	
Carbon dioxide, CO_2		4.5	11.5	5.0	4.2	2.2	
Higher heating value	1129	163	92	290	530	574	440
Specific gravity	0.61	0.86	1.02	0.56	0.64	0.44	

[a]From bituminous coal.
[b]From coke.
[c]CO_2 and H_2S free.
[d]Illinois No. 6 coal in Westfield, Scotland, tests.
[e]Illuminants.

SCF for coke oven gas, compared with about 1000 Btu/SCF for natural gas.

Blast Furnace Gas.   The combustible is CO from the incomplete combustion of coke in the air blast. However, this is diluted with CO_2 from complete combustion as well as the full nitrogen content of the air, so that the CO constitutes only 27.5% by volume. This accounts for the low heating value.

Producer Gas.   Producer gas is made by partial combustion of coal with air and steam, and produces a much higher proportion of H_2 than in the blast furnace. It is still diluted with CO_2 and the full nitrogen content of the air and results in the comparatively low heating value of 163 Btu/SCF. However, the process is straightforward and simple. Pure oxygen is not required. If bituminous coal is used, it is restricted to a noncaking variety, so that it will feed down through the firebed by simple gravity action as it burns. The gas composition varies with the amount of steam used. Steam lowers the temperature and hinders the formation of clinkers, but the rate of gasification increases rapidly with rise in temperature. As with blast furnace gas, its sensible heat may also be used if it feeds directly into the process using it.

Water Gas (or Blue Water Gas).   In the production of water gas, steam is passed through a bed of coke or coal which has been heated to incandes-

cence. H_2 and CO are produced by the reaction of Eq. 4.4, which is endothermic. Some further H_2 and CO_2 are produced by the shift reaction, Eq. 4.5 (see Table 4.8).

The process is frequently alternating. First, air is fed through the firebed in one direction to bring the bed up to incandescence by normal combustion. Then, steam is passed through in the reverse direction, producing the water gas, a mixture of combustibles, CO, H_2, and a small amount of CH_4 diluted with nitrogen and CO_2. A representative heating value is 290 Btu/SCF. As with producer gas, noncaking varieties of coal or coke are required to secure good gravity feed. Different cycles, with and without reversal, are used.

Carbureted Water Gas.   The heating value of water gas can be more than doubled by mixing it with pure fuel gases obtained by cracking oil or other hydrocarbon, using the sensible heat of the water gas. There is obviously a large variation in heating value possible, depending on the amount of carbureted, but the value shown, 530 Btu/SCF, is representative of much of the "city gas" used before the advent of natural gas, or where natural gas is not readily available. It was used extensively in Europe before the North Sea and Russian discoveries of the 1970s.

Coke Oven Gas.   In a coke oven, the volatiles are driven off in the absence of air. Much of the hydrogen in the coal is driven off directly, while a considerable amount combines with carbon to form methane. These principal combustibles, with lesser amounts of CO and C_2H_4, ethylene, provide the heating value of 574 Btu/SCF. This is a relatively high-quality gas and is frequently sold at a premium, with the cheaper producer gas being used to heat the ovens.

The Lurgi Gas Generator[13]

The Lurgi pressurized gas generator for continuous complete gasification of coal to synthesis gas, CO and H_2, and some methane, CH_4, was developed during the 1930s. High-pressure steam and oxygen are used. It is in successful operation in various parts of the world. Since 1959, it has been used at the Sasol plant in South Africa to provide syngas for conversion to synthetic fuels by the Fischer–Tropsch process (see Chapter 3). In Pakistan, it is used to produce ammonia. In Australia, it is used for heating gas. In Scotland, prior to the North Sea discoveries of natural gas in the 1970s, it was used in the manufacture of "Towns Gas."

The Lurgi was a major improvement over the earlier atmospheric-pressure gasifiers. It requires on the order of $\frac{1}{2}$ lb oxygen and $2\frac{1}{2}$ lb steam with each pound of coal, and so requires an oxygen plant and a high-pressure steam plant as part of an installation. It operates at a pressure of 25 to 35 atm. It may also be blown with air to form producer gas.

The reactions of Eqs. 4.2 through 4.7 are involved, although the water gas shift and the methanation take place only to a limited extent. Separate shift and methanation units are not used as in processes to be described later. A vertical cylindrical combustion chamber is used, with coal fed from the top through a lock hopper and steam and oxygen fed from the bottom (see Fig. 4.5). The effluent gases are passed through purifiers to remove the CO_2, H_2S, moisture, tars, and so on.

This unit is most suitable for noncaking or lightly caking coals. However, a mechanical stirrer can be provided in the upper part to continuously break up the softening mass of coal and so prevent the formation of a closed gas-tight layer or cake. Coals with moderately high swelling/caking indices require specially designed stirrers. Char, or coal specially pretreated to eliminate agglomeration, may alternatively be used.

The cleaned, CO_2-free gas from an oxygen- and steam-blown reactor typically contains about 50% H_2, 35% CO, and 15% methane.

A typical vessel of 19 ft overall height and 10 ft outer diameter will gasify about 170 tons of coal per day. An overall gasification efficiency of 48.8 to 59.6% was obtained in tests of certain U.S. coals.[14] The 12 ft diameter Lurgi can process up to 600 tons/day. A 16 ft diameter unit under development (1980) is expected to reach 1000 tons/day.

The counterflow of gas and coal results in a relatively low-temperature effluent gas, typically 1000°F (538°C), which improves the thermal efficiency. However, the relatively low-temperature carbonization and devolatilization of the coal in the upper part of the bed results in condensable tars that must be removed later.

The inability to accept fines (-28 mesh) is a considerable disadvantage. Mechanically mined coal may contain as much as 20% fines.

The operation below 1900°F for dry ash removal requires excess steam flow and results in large volumes of foul water to be treated. The more recent development of the slagging Lurgi at higher temperatures will lessen this problem. However, the treatment of large amounts of foul water, the sensitivity to coal rank and size, and the production of tars are still disadvantages.

In spite of these obvious drawbacks, the Lurgi is today (1980) a principal workhorse of commercial gasifiers throughout the world. The 16 ft diameter Lurgi, under development, is expected to process all ranks of coal and to accept fines as well.

Lurgi with U.S. Coals

In 1972–1974, a Lurgi gas generator in Westfield, Scotland, was modified to gasify certain U.S. coals[14] (Rosebud-Montana, Illinois, No. 5 and No. 6; Pittsburgh No. 8). These had higher swelling and caking indices than the Scottish coal for which the installation was designed. The results for the Illinois No. 6 coal are shown in Table 4.8.

Types of Gasification Reactors

Gasification reactors differ in the way the reactants are brought together and in the pressures and temperatures used. The three most common types of flow are shown in Fig. 4.6.[15] In the fixed bed, the oxidant and moderator, such as oxygen and steam, are forced through the bed. To date (1980), this is limited to noncaking or lightly caking varieties of coal or to coke.

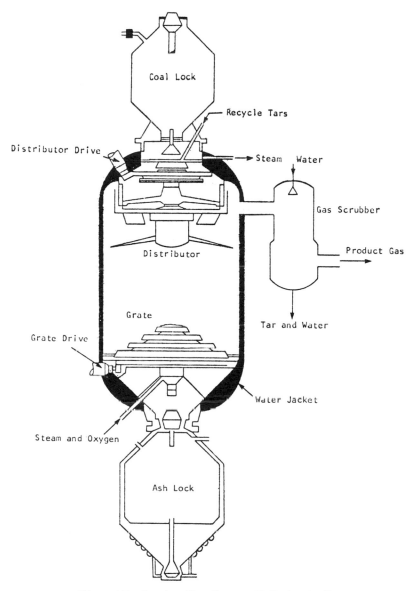

Figure 4.5. Lurgi gasifier. Courtesy N. Berkowitz.[16]

The *fluidized bed* was described in Chapter 3 as used for the direct combustion of coal. The good temperature control and high heat transfer rates are also advantageous for coal gasification. Operation is limited to temperatures below 1900°F (1073°C) to avoid formation of clinkers.

In early gasifiers of this type complete gasification was difficult to achieve in one stage due to carryover of fines and coal removed with the ash.[15] However, in second generation gasifiers[30] these problems have been solved. Essentially all U.S. coals available for gasification have been gasified in a single stage with very high carbon utilization, essentially no tars, and with no pretreatment required other than crushing the coal.

In the *entrained bed,* the ground coal is entrained in the flow of steam and oxygen. High temperature is generally required for complete gasification because of the short residence time of the fuel in the gasifier and the relatively low carbon inventory. Entrained gasifiers may operate upflow or downflow and may use one or more stages.

In *other systems* under development, the gasification takes place in a molten bath of iron or

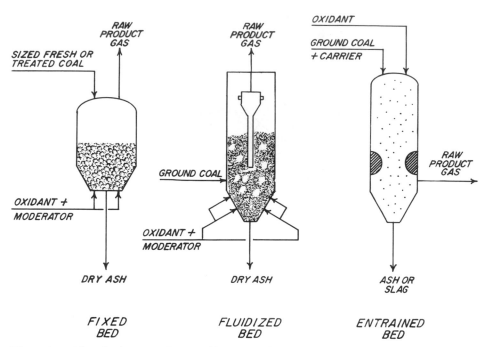

Figure 4.6. Three basic types of coal gasifiers. Reproduced from *International Journal of Energy Research*, No. 2, 1980.[15]

salt. A "dilute phase" is also used in some systems, particularly for the initial treatment of the coal particles to destroy their agglomerating tendency.

Commercial Gasifiers, "First Generation Systems"

Examples of the three main coal gasification types in current (1980) commercial operation are: the Lurgi pressurized, fixed-bed, dry ash or slagging,[13] the Winkler atmospheric, fluidized-bed,[16] and the Koppers–Totzek[16] atmospheric entrained gasifier. The smaller Wellman or Wellman–Galusha[16] fixed-bed gasifiers, up to 72 tons/day, are used for the production of producer gas, about 170 Btu/SCF. In 1980 parlance, these are "first generation gasifiers."

The *fluidized-bed Winkler* is operated at atmospheric pressure and at temperatures below 1800°F (980°C). This usually restricts it to lignites and subbituminous coals, since coals of higher rank are not sufficiently reactive at such low temperatures.

The *Koppers–Totzek* (K-T) gasifier, which has been used commercially since 1938, is a squat, cylindrical, refractory-lined vessel into which pulverized coal entrained in oxygen is injected. The coal–oxygen stream is shrouded by steam flow to protect the vessel walls from excessive heat, the

reaction taking place at 3300 to 3500°F (1815 to 1930°C). The K-T gasifier accepts all ranks of coal without any prior treatment except pulverization. At the high temperature, gasification is very rapid, even at atmospheric pressure. Units with four burner heads can gasify up to 850 tons/day. Raw gas exits at 2300 to 2700°F (1260 to 1480°C). The loss of heat is serious, and new developments will incorporate heat recovery. The compression of downstream gas also requires considerable energy. K-T gasifiers can also be blown with air for producer gas.

In addition to coal gasifiers, a large number of units for gasifying heavy hydrocarbons, such as high-sulfur residual petroleum fuels and tars, are in commercial use. For example, over 75 plants in 22 countries have been built employing the Texaco[15] pressurized, entrained, heavy hydrocarbon gasifier, primarily producing syngas for conversion to ammonia, methanol, and oxo-chemicals.

In general, commercial gas manufacturing plants are classified as:

1. Light hydrocarbon reforming processes. See hydrogen production later.

2. Heavy oil processes, as just described.

3. Coal gasification processes.

New Requirements—1980

A large number of new developments in coal gasification and liquefaction are taking place, particularly in the United States, based on the concept that the requirements as natural gas and oil dwindle will be at least an order of magnitude greater than in the past. Correspondingly larger capacity conversion units will be mandatory. Methods of much higher efficiency and capable of converting all types of coal will be required. Conversion beyond syngas to SNG will be needed. Operation at near pipeline pressure, 1000 psi, will be required to avoid large compression losses. High pressures would also increase methane formation in gasifier stages operating below 2000°F and reduce the amount of downstream methanation needed for SNG.

Nearly a hundred different coal conversion schemes have been under development.[17] Gasifiers under development since about 1974, which offer potential advantages over the first generation systems and which have generally progressed to development units of 100 tons/day or more, are termed "second generation systems" in 1980 parlance.

Size Considerations

Commercial "first generation" gasifiers convert typically 1000 tons/day of coal; 12 ft Lurgi, 600 tons/day; four-burner Koppers–Totzek, 850 tons/day; Winkler fluidized bed, 1100 tons/day; and Wellman or Wellman–Galusha, 72 tons/day.

Future requirements are conceived as 5,000 to 20,000 tons/day. This would be comparable to other full-scale energy facilities as follows:

1. A 50,000 barrels/day oil-from-coal plant, at 3 barrels/ton, would use 17,000 tons/day coal.

2. The two tar-sand oil plants of Chapter 2 were of 50,000 and 120,000 barrels/day capacity.

3. A single fractionating tower in a refinery may process 100,000 to 200,000 barrels/day crude oil.

4. A 50,000 barrels/day oil plant (275 billion Btu/day) is comparable to a 250 million SCF/day SNG plant (250 billion Btu/day).

5. At 50% thermal efficiency, the 250 million SCF/day SNG plant would require 20,800 tons/day coal (12,000 Btu/lb). This would require 30 12 ft Lurgi gasifiers plus five on standby.

6. A modern generating station with two 1.25 GW units, operating base load at 40% efficiency, would need 21,300 tons/day 12,000 Btu/lb coal. With stack gas scrubbers and 34% efficiency, it would need 25,100 tons/day. And if operated as a dual-cycle plant at 50% thermal efficiency, using gas produced from coal at 60% T.E., it would need 28,400 tons/day coal.

Thus, these size assumptions (5,000 to 20,000 tons coal/day) are low if anything.

On a more global scale, and further into the future, we used 19.8 quads of natural gas energy and 14.1 quads of coal energy in 1978 (Table 1.2). Had the gas been SNG from coal at 50% T.E., it would have required 39.6 quads of *additional* coal energy. This is 2.8 times the coal we actually mined. Had we used 250 million SCF/day plants, it would have required 217 plants.

Second Generation Gasifiers

Principal conversion technologies under development in 1980, so-called "second generation systems," can be divided into four groups: (1) fluidized or entrained gasifiers producing syngas that may contain some methane but which requires massive shifting and methanation to convert it to SNG; (2) hydrogasification processes that gasify fluidized or entrained coal with hydrogen and steam, yielding a methane-rich gas that needs little or no shifting and methanation; (3) gasifiers producing syngas similar to (1) but in a liquid medium, which may also catalyze the process; and (4) processes that yield a whole spectrum of liquid and gaseous products.

BiGas.[17]   An example of group (1) is the BiGas process of Bituminous Coal Research, which employs two stages at 1700 and 2700°F (927 and 1480°C) (Fig. 4.4) and operates at about 1100 psi (75 atm). A 120 tons/day pilot plant at Homer City, Pa., has been operating since 1976. Pulverized coal and steam, or a slurry, are injected into the upper (dense phase) at 1700°F (927°C). There it is contacted by hot syngas moving up from the lower (dilute phase) to form CO, H_2, and CH_4. Residual unreacted char is then carried out with the product gas, separated in a cyclone, and fed into the lower stage, where it is almost completely gasified at 2700°F (1480°C).

Synthane, CO₂ Acceptor. Other examples are the BuMines Synthane process[16,17] and the CO$_2$ Acceptor process.[16,17] The Synthane process being developed at Bruceton, Pa., in a 75 tons/day pilot plant since 1976, used 1000 psi (68 atm) pressure but lower temperature, 1800°F (980°C), which resulted in a syngas containing 60% of the final methane. The pilot plant was closed in 1979.

The CO$_2$ acceptor process of Consolidated Coal Co., being developed in a 40 tons/day pilot plant at Rapid City, N.D., starting in 1972, is unusual in requiring no oxygen. Dolomite is calcined and heated by burning spent char in air. The hot dolomite is fed through a fluidized bed of crushed coal and steam and provides heat by combining with CO$_2$ from the shift reaction, Eq. 4.5. The spent dolomite feeds out the bottom and is recalcined. At the relatively low operating temperature, below 1600°F (870°C), gasification proceeds rather slowly, and feedstocks may be limited to lower-rank coals and lignite.

Texaco Process. Syngas from any source can of course be purified, shifted, and methanated into SNG. The "second generation" gasifiers producing syngas include a number of fluidized or entrained gasifiers which were undergoing tests in 1979 in pilot plants of 100 to 200 tons/day capacity. Among these are the single-stage pressurized gasifier developed by Shell and Shell–Koppers at Shell's Hamburg refinery in West Germany and the two-stage Combustion Engineering atmospheric pressure gasifier in Connecticut.

The Texaco[15,16] single-stage entrained, pressurized, downflow coal gasification reactor is operated at 2000 to 3000°F (1090 to 1650°C) (Fig. 4.4). It has been under test at the Texaco Montebello Laboratories in California with a wide range of coals and operating conditions and is now being applied in a variety of demonstration, semicommercial, and design projects (1980).

These include a 150 tons/day coal gasification unit at the Ruhrchemie Chemical plant in West Germany, in operation since 1977. At TVA, a natural gas ammonia plant at Muscle Shoals was being converted (1979) to coal, using a Texaco coal gasifier. A 1200 tons/day ammonia-from-coal plant was being designed with DOE funding (1979).

Texaco and Southern California Edison, with EPRI support, have announced[15,18,19] a 1000 tons/day, 92 MW dual-cycle demonstration coal-to-gas-to-electric power plant, to be described later.

HYGAS, HYDRANE. The second group, hydrogasification, is represented by the IGT (Inst. of Gas Technology) HYGAS[16,17] or the BuMines HYDRANE[16,17] process. Hydrogen-rich gases for these processes can be produced by a variety of processes (see Hydrogen later). Final methanation is required for only 2 to 5% residual CO in the HYDRANE process. HYGAS is being tested in an 80 tons/day pilot plant in Chicago. HYDRANE is at the bench scale in 1979.

ATGAS, Kellogg Molten Salt. The third group is represented by the ATGAS[16,17] molten iron process of Applied Technology Corp. and M. W. Kellogg's Molten Salt process.[16,17]

In the PATGAS and ATGAS processes, crushed coal is injected with steam into a molten iron bath. Any coal can be used without pretreatment, regardless of caking tendency, fines, or sulfur content. Operation is at about 50 psi (3 atm) and 2500°F (1370°C). A limestone slag at the surface combines with the H$_2$S, which is thus removed with the ash. The raw gas exits at the top at atmospheric pressure. After heat recovery and dust removal, it is pressurized to 600 psi (40 atm), then shifted and CO$_2$ removed to become 315 Btu/SCF PATGAS. If further methanated and dehydrated, it becomes ATGAS, an SNG of about 940 Btu/SCF.

In the Kellogg Molten Salt process, dried coal is entrained in steam–oxygen and fed directly into the molten salt gasifier. In addition to catalyzing the coal–steam gasification at a low enough temperature for considerable methane formation, the molten-salt medium makes pretreatment of agglomerating coals unnecessary.

General Conversion. The fourth group, producing a variety of liquid and gaseous fuels, is represented by the COED–COGAS system, Synthoil, Solvent Refined Coal, SRC and SRC-II, and the Fischer–Tropsch Synthesis. The latter starts with syngas. These are described in Chapter 3 under Coal.

Gasification–Combined Cycle (GCC) Power Plant

During the period 1960 to 1969, natural gas and coal were being supplied to utilities in the United States at the same price, about 26¢/million Btu. In the interim 1967 to 1978, the costs of both fuels to utilities have escalated more than fourfold (coal 4.66, gas 5.27, Table 1.3). The use of natural gas or oil is no longer permitted in new base load power plants.

In mid-1976 dollars, coal energy was estimated at $1.00/million Btu, SNG from coal $6.00 to $8.00/million Btu[18] for new plants coming on line within a few years. Intermediate-Btu gas-from-coal was $3.50 to $5.00/million Btu. Thus, the effective ratio of gas to coal fuel has increased 3.5 to 8 times since 1967. This would appear to rule out gas as a power plant fuel, certainly SNG. However, several factors have entered the picture to make synthetic gas a viable fuel today (1981):

1. The dual-cycle plant[20] had been shown to be economic even with the low fuel prices of the late 1960s. With higher fuel prices, the savings are much greater with the more efficient dual-cycle plant. Its complexity is more justified. The gas turbine/steam turbine dual-cycle plant *requires* a liquid or gaseous fuel at present. In the EPRI study of mid-1975 (*EPRI Journal* November 1976), all of the combined cycle plants resulted in costs per kW about 85% of single-cycle plants. For base loads, a medium-Btu gas combined-cycle plant resulted in as low or lower busbar cost/kWh as any other option. A number of oil- or gas-fired dual-cycle plants are either in operation or nearly so (1980).[19] The dual-cycle plant typically had a thermal efficiency of 50% compared with 40% for a conventional plant (before sulfur controls).

2. Stricter environmental controls have required the elimination of most of the sulfur and nitrogen compounds from the flue gases of power plants. If this sulfur is removed by wet flue-gas scrubbers in coal-fired plants, the plant cost is increased over 30% and the typical efficiency of a base-load plant is reduced from about 40% to 34%.[18] In comparison, the addition to a coal gasifier plant to remove the sulfur is relatively minor, and the sulfur is a salable by-product instead of a huge sludge pond. If the gas turbine is fired at not over 2000°F (1090°C), nitrogen oxides can easily be kept within EPA limits. This greatly reduces the differential between coal and medium-Btu gas from coal.

3. There have been important improvements in coal gasification systems suitable for power plants, as described earlier in this chapter.[18,19]

4. In an integrated gasifier–combined cycle (GCC) plant, full advantage can be taken of heat recovery that is not possible in a decoupled plant with gas supplied from a distance.

The net result is that a gasification–combined cycle plant appears to be economic in 1980, using the state-of-the-art turbine inlet temperature of about 2000°F. Southern California Edison Co. and Texaco, with EPRI support, have jointly announced plans[15] to proceed with a 1000 tons/day, 92 MW demonstration plant to establish the technical and economic viability of this system. This plant is described in the following paragraphs.

Demonstration Plant

Principal features of the demonstration plant are shown in Fig. 4.7, a Texaco oxygen-blown coal gasifier, a heat recovery steam generator (HRSG) or "waste heat boiler," gas turbine, and steam turbine. Also integrated with the plant are the coal grinding and slurrying system, sulfur removal system, and oxygen plant for the gasifier.

The main gas flow is shown in heavy lines from the grind–slurry system to gasifier, sulfur removal, fuel gas reheat in the gasifier, to the gas turbine. The exhaust from the gas turbine flows through the heat recovery steam generator (HRSG) to the atmosphere.

The water–steam flow is shown in light lines, from the condenser through heaters and boiler feed pump, and through the "economizer" coils of the HRSG. Here, for heat balance reasons, the flow splits, part being evaporated in the HRSG and part in the coal gasifier, taking up the available heat there. The superheater is in the hottest part of the HRSG. From it, the superheated steam flows through the steam turbine which exhausts to the condenser.

Many flow loops have been omitted for simplicity. A substantial part of the superheated steam (about 37%) is supplied to the oxygen plant as indicated. The first heater is supplied from the HRSG, and the second one is a heat coil in the gasifier (not shown). Circulating water for the condenser and the oxygen plant is cooled in a forced-draft cooling tower.

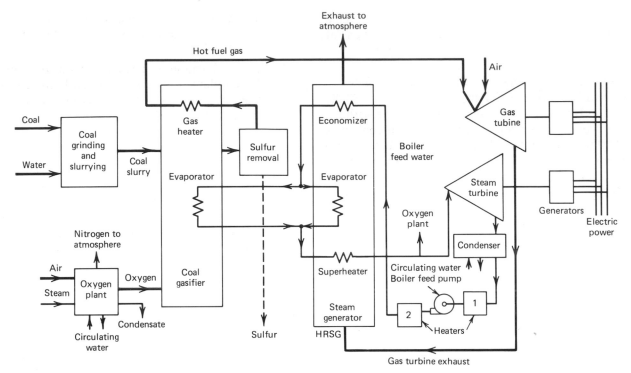

Figure 4.7. Gasification–combined cycle electric power plant.

Sulfur removal is actually two units, a Selexol unit, 97% effective in removing the H_2S and COS from the raw gas, and a Claus-type sulfur removal unit with Beavon-type tail-gas treating unit that removes sulfur from these gases as well as from flash gas and gas from the effluent water treatment (not shown). An alternate fuel system is provided for the gas turbine.

For the comparatively small 1000 tons/day, 92 MW demonstration plant, many of the refinements of a 1000 to 1500 MW full-scale plant are not justified. The full-scale plant would no doubt incorporate a reheat steam turbine and fuel-gas expansion turbines, with resulting increase in thermal efficiency.

A state-of-the-art gas-turbine inlet temperature of 2000°F (1090°C) is used. Efficiency will be improved substantially in the future if developments underway permitting 2400°F (1320°C) inlet temperatures are successfully completed.

Texaco Coal Gasifier.[15] In the Texaco pressurized, entrained, downflow gasifier, ground coal in a slurry of 61% solids in water is introduced through a specially designed burner at the top, where it is mixed with oxygen. The reaction takes place in a refractory-lined chamber to produce a medium-Btu gas consisting mainly of CO, H_2, and CO_2 and having a HHV of 283 Btu/SCF.

The ash in the gasifier, which operates at 2000 to 3000°F, is melted to a slag, which is water quenched and withdrawn through a lock hopper at the bottom as gravel-like particles. This slag is completely innocuous[15] and easy to handle, and will be disposed of on site.

As shown in Fig. 4.7, the gas is initially cooled by the cool gas stream after sulfur removal. It is further cooled in shell-and-tube-type gas coolers (syngas coolers or evaporators) by producing 1200 psig saturated steam, which after superheating to 950°F in the HRSG is fed to the steam turbine and the oxygen plant.

The demonstration plant will be located at the So. Cal. Edison Coolwater generating station at Daggett, Cal., 12 miles from Barstow, in the Mojave Desert. This site includes 2700 acres and presently (1980) contains four generating units, 65 and 81 MW oil-fired steam units and 472 MW total in two combined-cycle units. In each of the latter units, two gas turbines exhaust to a supplementary-fired HRSG that supplies steam to a single steam turbine. Thus, ample space and all facilities for the new demonstration plant are available on site, as well as experience in dual cycle and HRSG. While out of the Los Angeles Basin, it is still conveniently located for demonstration.

A heat rate of 11,060 Btu/kWh or thermal efficiency of 30.9% is expected in the demonstration plant. The cost in 1978 dollars, escalated per schedule, will be $234 million, or $253/kW. The gross output of the plant is 100,700 kW, the auxiliary power 8295 kW, resulting in a net output of 92,405 kW.

The sulfur removal system will remove 97% of the sulfur from the 0.7% sulfur western coal normally used in this area. It will remove 93% of the sulfur from high-sulfur coal (3.5% S) that may be used in limited tests during the three-year demonstration period.

Successful conclusion of this demonstration project will pave the way for full-scale GCC plants in the mid-1980s.

HYDROGEN

Two aspects of hydrogen will be treated here. First is the manufacture of hydrogen and its characteristics, uses, and costs. Second is the "hydrogen economy" concept,[21,22] which has been advanced proposing that most of U.S. energy in the future be converted to hydrogen for transmission, storage, and use.

Use of Hydrogen. The U.S. and world production of hydrogen[23] is given in Table 4.9 for the years 1965 and 1971, together with the U.S. BuMines projection for the year 2000. The U.S. usage of hydrogen is also given in Table 4.10, again with the BuMines projection for the year 2000. Note that 2580 billion SCF of hydrogen represents 0.81 quads of energy. This may be compared with 22.5 quads of natural gas energy used in the United

Table 4.9. Hydrogen Production (Billions of SCF)

	1965	1971	2000
United States	1,155	2,580	10,500
Rest of the world	1,576	4,170	31,100
Total	2,731	6,754	41,600

Source. Ref. 23.

Table 4.10. Hydrogen Uses in the United States (Billions SCF)

	1965	1971	2000
Chemicals	597	1,131	3,400
Petroleum refining	375	1,334	3,400
Other	183	133	1,500
Shale Oil			200
Gasification of coal			2,000
Total	1,155	2,580	10,500

Source. Ref. 23.

States in 1971. Most of the hydrogen was made by reforming natural gas.

The "chemicals" category includes principally ammonia, methanol, and oxo-chemicals. By the year 2000, hydrogen in substantial amounts is expected to be needed for upgrading shale oil and for the gasification of coal.

Manufacture of Hydrogen

Typical steps involved in the production of hydrogen by different processes are outlined below. For simplicity, the purification steps and the catalysts have been omitted. When hydrocarbons "run out," the first two processes, accounting for most of the present production, can no longer be used. In the other processes, the hydrogen derives from water.

Steam–Methane Reforming (or Naphtha). Naphtha, if used, is first vaporized. Steam and heat are then added with the steam/carbon mole ratio of 6:4 (excess steam). The reforming reaction is

$$C_nH_m + n\,H_2O \rightarrow n\,CO + \frac{(2n + m)\,H_2}{2} \quad (4.13)$$

The CO is then shifted to H_2 in two stages by the reaction of Eq. 4.5.

Partial Oxidation of Hydrocarbons. This has been referred to earlier as gasification of heavy hydrocarbons to a syngas, with the CO all subsequently shifted to H_2. The process may be indicated schematically as

$$C_nH_m, S, O_2, \text{ and } H_2O \rightarrow CO, H_2, \text{ and } H_2S \quad (4.14)$$

After shifting and purification, the pure hydrogen results.

Coal–Steam–Oxygen. Coal gasification to a syngas, followed by shifting to hydrogen, has been described earlier.

Steam–Iron Process. The main unit of this process is an oxidizer-reducer operating continuously at elevated temperatures. When steam is introduced into the oxidizer section, the lower section, a mixture of 5% Fe and 95% FeO, is converted to a mixture of 80% Fe_3O_4 and 20% FeO, according to Eqs. 4.15 and 4.16, producing hydrogen, which is drawn off:

$$4 H_2O + 3 Fe \rightarrow 4 H_2 + Fe_3O_4 \quad (4.15)$$

$$H_2O + 3 FeO \rightarrow H_2 + Fe_3O_4 \quad (4.16)$$

The oxides are then fed up into the reducer section where they are reacted with producer gas to form the hot Fe and FeO mixture which feeds down into the oxidizer. The reactions in the reducer section are

$$Fe_3O_4 + 4 CO \rightarrow 3 Fe + 4 CO_2 \quad (4.17)$$

$$Fe_3O_4 + CO \rightarrow 3 FeO + CO_2 \quad (4.18)$$

Nuclear Heat–Steam–Coal. The process using nuclear heat is essentially the same as the coal–steam–oxygen process, except that the heat is supplied by a loop from a nuclear reactor into the gasifier unit instead of oxygen feed to burn some of the coal.

Electrolysis of Water.[24,25] The cells are similar to the alkaline fuel cell described in Chapter 13, except in reverse. Electric energy is supplied and hydrogen and oxygen evolved at the two electrodes. The cathode supplies electrons, and hydrogen is generated:

$$2 NA^+ + 2 H_2O + 2\epsilon \rightarrow 2 NaOH + H_2 \quad (4.19)$$

At the anode oxygen is generated and electrons are given up:

$$2 OH^- \rightarrow H_2O + \tfrac{1}{2} O_2 + 2 \epsilon \quad (4.20)$$

The cells typically produce 6.5 to 7.5 SCF of H_2 and 3.25 to 3.75 SCF of O_2 per kWh, corresponding to 60 to 69% energy efficiency. With 1973 technology, 75 to 80% efficiency was feasible.[21,23] With improvements, 85% is expected in the hydrogen economy. At a large hydro project,[26] 60% is obtained (1972).

Cost of Producing Hydrogen[27]

The estimated cost for new plants built in 1973 was \$.45/MSCF by steam–methane reforming, with about half the cost being natural gas feed stock at \$.48/MSCF. The cost by steam–naptha reforming was about \$.74/MSCF, the higher cost being mainly the cost of feedstock.[23] With partial oxidation of fuel oil, the cost was \$.68/MSCF, the higher cost being primarily the higher plant cost, including an oxygen plant.

The cost by electrolysis was roughly four times that by steam–methane reforming.[28]

By 1979, hydrogen by steam-methane reforming was estimated at \$1.88/MSCF, again about equal to the cost of natural gas in large quantities (to utilities, *Monthly Energy Review,* Chapter 1, Ref. 4), \$1.80/MSCF average.

Hydrogen by conventional electrolysis (70% efficiency) was estimated at about \$7.00/MSCF, at 3¢/kWh, the average electric rate to industrials. It was still about four times the cost by steam–methane reforming. However, solid polymer electrolyte cells (SPE), at 85% efficiency, and lower plant cost, were stated[29] to have the potential of about \$4.60/MSCF at 3¢/kWh, or about 2.5 times the cost of steam–methane reforming.

Energy Properties of Hydrogen

Hydrogen is the lightest gas, relative densities (molecular weights) being: H_2 2, He 4, CH_4 16, O_2 32, N_2 28, and air 29.

Its HHV in Btu/lb is the highest: H_2 61,031, methane 23,890, natural gas about the same, gasoline 20,300, and jet fuel 18,700.

Its HHV in Btu/ft³ is less than that of other fuels: H_2 314, CH_4 978, and natural gas about 1020.

The boiling points of a few gases are:

	H_2	He	CH_4	O_2	N_2	Absolute Zero
°F	−423.17	−452.07	−263.2	−297.33	−320.4	−460
°C	−252.87	−268.73	−164.0	−182.96	−195.8	−273

The boiling point of hydrogen is close to absolute zero. Some 30% of its HHV is consumed in liquefying it. The in–out efficiency of hydrogen storage is about 60%. This compares with the overall processing loss in an LNG system of about 10%.[5]

Liquid Hydrogen. The specific gravity of liquid hydrogen at boiling point and 1 atm is 0.0708, compared with LNG 0.424, gasoline 0.719, and water 1.0, or, respectively, in lb/gal, liquid H_2 0.591, LNG 3.54, gasoline 6.0, and water 8.35.

The HHV in Btu/gal is thus: liquid H_2 36,069, LNG about 84,600, gasoline 121,800. LNG and gasoline contain more energy per gallon than liquid hydrogen. Overall, for the same energy content, liquid hydrogen has about 33% of the weight and 3.38 times the volume of gasoline.

Specific Heat. The specific heat C_p of gaseous hydrogen is the same as for any diatomic gas, N_2, O_2, or air (see Chapter 13), namely, 29 J/g-mole °K, or 6.90 Btu/lb-mole °R. This gives it a tremendous advantage in cooling electrical machines, since it holds as much heat per unit volume as air, whereas the power to circulate it is proportional to its density, which is $\frac{2}{29}$ that of air.

The "Hydrogen Economy" Concept[21,26]

The setting of the proposed hydrogen economy is as follows. As fossil fuels become exhausted and we turn to nuclear or solar sources, what is the "energy form" that is to be transmitted, stored, delivered, and used? It is predicated that nuclear siting problems become so serious that nuclear plants must be built at great distances from the load, further that for environmental reasons overhead transmission lines are practically ruled out. Only clean fuels can be burned at the load.

In this setting, hydrogen is offered as the solution. At large nuclear plants remote from cities, electricity is generated, converted to dc, and used to produce hydrogen by the electrolysis of water. The hydrogen is piped underground to convenient storage areas and to use areas, where it is used as follows:

1. To generate electric power locally.
2. In industry for heat, feed stocks, and reducing gas.
3. In transportation for use in heat engines, probably liquefied.
4. In commercial and residential areas for space and water heating.

This is illustrated in Fig. 4.8.

It is granted that energy in hydrogen made by electrolysis will cost more than the electric energy that produced it. The envisioned savings and advantages result from the transmission, storage, and use. Some of the pertinent facts considered are:

1. Hydrogen burns to water and is therefore nonpolluting.
2. The largest users of energy are transportation, large industry, and space heating.
3. Underground electric cables cost 9 to 20 times the cost of overhead lines and are not suitable technically for long distances.
4. Sites for pumped-storage hydro (the only viable large-scale "electric" storage) are limited.

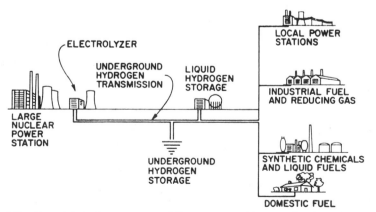

Figure 4.8. Complete hydrogen energy delivery system. Courtesy Derek P. Gregory.[21,26]

5. Gas by pipeline is one of the cheapest methods of energy transmission. In 1970, some 252,000 miles of natural gas trunklines transported 22.4 trillion ft^3 of gas in the United States, about 30% of the nation's energy.

6. While hydrogen has about one third the heating value of natural gas per ft^3, it is very light and can be transmitted at three times the velocity over the same pipelines (with larger compressors) at comparable overall costs per Btu-mi (Actually 1.4 times, see Chapter 14).

7. Pipelines are up to 48 in. in diameter. A 36 in. line will transmit about the same amount of energy as 10 500 kV transmission lines (11,000 MW).

8. A total of 5.681 trillion SCF of natural gas was stored underground in 1972 in the United States. This was about 25% of all gas used and 7% of the nation's total energy input. Hydrogen can probably be stored similarly, but at one third of the energy density. Helium, the next lightest gas, is stored underground at Amarillo, Tex.

9. LNG was in use or under construction for "peak shaving" at 76 locations in the United States in 1972. Liquid hydrogen can be used similarly but entails a 30% energy loss in the process of liquefaction. (40% overall, Table 4.11)

10. The liquid hydrogen tank at the John F. Kennedy Space Center stores 900,000 gal, 37.7 billion Btu, or 11 million kWh. This is 73% of the world's largest pumped-storage hydro project at Luddington, Mich.

11. Liquid hydrogen storage, like LNG, is applicable anywhere. It is not limited by geography, as is underground storage or pumped-storage hydro.

12. Hydrogen is now produced electrolytically (1973) at remote hydro plants in British Columbia, Norway, Egypt, and elsewhere. The efficiency from electricity to hydrogen is 60% at the British Columbia plant.[21] With current (1976) technology, 75% is attainable; and with developments, 85% is anticipated.

13. Gas burners would have to be changed to burn hydrogen in place of natural gas.

14. A hydrogen–air mixture ignites easier than a natural gas–air or gasolene–air mixture (one tenth the energy required).

15. Of the 97 persons aboard the Hindenburg, 62 survived the hydrogen explosion.

16. To store the same energy, liquid hydrogen has about 30% of the weight and 3.39 times the volume of gasoline or kerosene (see earlier section). Liquid hydrogen is attractive for SSTs.

In addition to the "facts" given above, a number of developments are cited, which if successful would improve the "hydrogen economy." Among them are:

1. Improved electrolytic cells.
2. Chemical splitting of water molecules at temperatures within range of nuclear reactor heat.[21]
3. Flameless catalytic heating.
4. Improved and large-scale fuel cell electric generation.
5. Hydrogen–oxygen closed-cycle steam turbine.
6. Large-scale metal hydride energy storage[21] (see Chapter 14).

Since the hydrogen economy concept includes nuclear electric plants and local generating stations, it is more properly considered as a supplementary technology, handling part of the transmission and storage functions. It is an alternate "form" of energy in *part* of the system.

The "Electric Economy"[22]

The "electric economy" has been proposed as an alternate to the "hydrogen economy." It assumes that the necessary transmission lines can be built. It is a solution of the near-term problem (1975 to 2000), substituting coal and nuclear energy for the diminishing supply of natural gas and oil. It is also a solution to the long-term problem, adequate energy for the world, since regardless of the source, be it nuclear fission or fusion or solar energy, electricity is a suitable link from source to user.

The end use of energy in the United States in 1972 is shown in Fig. 4.9. The four largest uses, transportation, space heating, process steam, and direct heat, account for 80% of the end use of energy. Consequently, it is in these uses that substitution of electricity for gas and oil must be made to be effec-

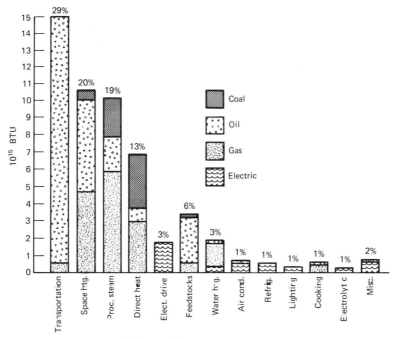

Figure 4.9. End use of energy in the United States, 1972.

tive. These uses have been examined to see to what extent electricity could be substituted.[22]

Not all energy uses could be converted to electric supply. Examples of those that could not be with near-term technology are aircraft, heavy mobile units, and feedstocks.

However, substantial substitutions are technically and economically feasible by the year 2000 in many applications. For example, in transportation a substantial part of the cars, trucks, and intercity trains could be electric drive. In the residential sector, space heating is a prime candidate, using the heat pump. In the industrial and commercial areas, much of the process steam and direct heat could be produced from electricity. Altogether, if all feasible substitutions were made, the use of gas and oil could be reduced to 30 to 40% of that which would

be required if the 1972 pattern were continued into the year 2000.[22]

Hydrogen Versus Electric Economy

As mentioned, there are some areas where conversion to electric supply is not feasible. However, in those cases where it is, it may be compared with conversion to a hydrogen economy. This will be illustrated by several examples.

Let the system components involved in the two concepts and their typical efficiencies and designations be as shown in Table 4.11. Assume that fossil fuels have run out and all energy is from nuclear plants, via an electrical or hydrogen link.

Automotive Load. Consider first the supply to a vehicle, battery powered if the system is electric, and battery or engine powered if the system is hydrogen. The steps are as follows:

Input/Output											Efficiency (%)
				Electric							
	Fuel →	Nuc →	Etr →	Bat →	Trp						
5.3		35	90	60							18.9
				Hydrogen							
	Fuel →	Nuc →	Elzr →	Gtr →	FC →	Bat →	Trp				
13.2		35	80	90	50	60					7.6
	Fuel →	Nuc →	Elzr →	Gtr →	Liq →	Eng →	Trp				
66.1		35	80	90	60	10					1.51

Table 4.11. Efficiencies and Designations of System Components Involved in the Electric and Hydrogen Economies

Component	Efficiency (%)	Symbol	Component	Efficiency (%)	Symbol
Nuclear electric plant	35	Nuc	Gas furnace	60	GF
Electrolyzer (electric to hydrogen)	80	Elzr	Gas fired steam boiler	75	GFB
Electrical transmission	90	Etr	Clean-fuel-fired steam-electric plant	40	StPl
Gas transmission (or hydrogen)	90	Gtr	Fuel cell	50	FC
Battery charging and use	60	Bat	Pumped-storage hydro	75	PSH
Automotive engine, liquid fuel, overall	10	Eng	General load other than heat		Load
			Nuclear fuel		Fuel
Hydrogen liquifier (or hydrogen storage)	60	Liq	Transportation		Trp
			Electric Steam Generator	100	ESG
Electric heat pump, annualized	170	HtP			

The nuclear fuel required is 5.3 times the useful energy on the vehicle with the electric system, 13.2 or 66.1 times with the hydrogen system depending whether a battery or liquid hydrogen is used on the vehicle.

Transportation is about 26% of total energy use (Table 2.2). This illustrates the much larger capacity in nuclear plants that would be required if the relatively inefficient hydrogen link is used.

Commercial–Residential Electric Load. As a second example, consider loads requiring electricity, motors, lights, TVs, and microwave ovens, the top line on Fig. 4.8:

Input/Output								Efficiency (%)
			Electric					
	Fuel	→	Nuc	→	Etr	→	Load	
3.2			35		90			31.5
			Hydrogen					
	Fuel →	Nuc →	Elzr →	Gtr →	StPl →	Load		
9.9		35	80	90	40			10.1

Three times as much nuclear plant capacity would be required with the hydrogen system.

Direct Heat and Process Steam. As a third example, consider the electric steam boiler on an electric system versus a gas-fired boiler on a hydrogen system:

Input/Output								Efficiency (%)
			Electric					
	Fuel	→	Nuc	→	Etr	→	ESG	
3.2			35		90		100	31.5
			Hydrogen					
	Fuel →	Nuc →	Elzr →	Gtr →	GFB			
5.79		35	80	90	75			18.9

Even in the most favorable situation, where hydrogen can be used directly, the nuclear capacity needed is nearly double.

Storage. Gas storage on an annual basis cannot be compared with electric storage for daily peaking. For the load that must be supplied by electric plants in any case, top line of Fig. 4.8, storing hydrogen is comparable to storing nuclear fuel in a present nuclear plant. A year's supply can be stored in the reactor.

Conclusions

Thus, it is evident that if the country were forced into a largely hydrogen economy by environmental considerations, the economic penalty would be very great indeed. The greatest penalty would be in transportation, the least in space heating and process steam. Hydrogen for chemicals, feedstocks, and reducing gas would be needed in any event if fossil fuels were exhausted.

REFERENCES

1. R. F. Zaffarono, *Natural Gas Liquids*, I.C. 8441, Pittsburgh: BuMines, 1969.

2. *Oil and Gas Production,* Compiled by Engineering Committee, Interstate Oil Comp. Comm., Norman: University of Oklahoma Press, 1951.

3. I. F. Avery and L. V. Harvey, *Natural Gasolene Plants in the United States,* I.C. 7790, Pittsburgh: BuMines, 1957.

4. L. E. Cardwell and L. F. Benton, *Analyses of Natural Gases,* I.C. 8518, Pittsburgh: BuMines, 1970.

5. W. L. Lom, *Liquefied Natural Gas,* New York: Wiley, 1974.

6. R. C. Weast and M. J. Astle, Eds., *Handbook of Chemistry and Physics,* 61st ed., Boca Raton, Fla.: CRC Press, 1980.

7. A. F. Williams and W. L. Lom, *Liquefied Petroleum Gases,* Chichester, England: Ellis Horwood Ltd., 1974.

8. L. C. Denny, L. L. Lakon, and B. E. Hall, Eds., *Handbook of Propane—Butane Gases,* 4th ed., Los Angeles: Chilton Publ. Co., 1962.

9. *Physical Constants of Hydrocarbons Boiling Below 350°F,* ASTM, Special Tech. Publ. No. 109, 1963.

10. Russell Bardos, "Low and Medium Btu Coal Gasification—A Status Report," *Proc. 6th Annu. Int. Conf. on Coal Gasification,* University of Pittsburgh, 1979.

11. R. Lewis, W. P. Haynes, J. P. Strakey, and R. R. Santore, *Synthane Process Update, Mid-'78,* Proc. Fifth Int. Conf. on Coal Gasification, Univ. of Pittsburgh, 1978.

12. S. L. Meisel, J. P. McCullough, C. H. Lechthaler, and R. B. Weisz, "Gasolene From Methanol in One Step," *Chemtech,* February 1976, p. 86.

13. J. Cooperman, J. D. Davis, W. Seymore, and W. L. Ruckes, *Lurgi Process—Use for Complete Gasification of Coals with Steam and Oxygen under Pressure,* Bull. 498, Pittsburgh: BuMines, 1951.

14. *Trials of American Coals in a Lurgi Gasifier at Westfield, Scotland, 1972–1974,* R&D Report No. 105, Washington: ERDA, 1974.

15. W. G. Schlinger, "Coal Gasification—Development and Commercialization of the Texaco Coal Gasification Process," *Proc. 6th Ann. Conf. on Coal Gasification,* University of Pittsburgh, 1979.

16. N. Berkowitz, *An Introduction to Coal Technology,* New York: Academic Press, 1979.

17. I. Howard-Smith and G. J. Werner, *Coal Conversion Technology,* Chem. Tech. Rev. No. 66, Park Ridge, N.J.: Noyes Data Corp., 1976.

18. M. J. Gluckman, N. A. Holt, S. B. Alpert, and D. F. Spencer, *The Near Term Potential for Gasification–Combined Cycle Electric Power Generation,* Energy Tech. VI Conf., Wash, D.C., 1979.

19. *Preliminary Design Study for an Integrated Coal Gasification-Combined Cycle Power Plant,* Final Rept., AF-880, Palo Alto, Cal.: EPRI, 1978.

20. P. A. Berman and F. A. Lebonette, "Combined Cycle Power Plant Serves Intermediate System Loads Economically," *Westinghouse Engineer,* November 1970, p. 168.

21. S. Linke, Ed., *Proceedings of the Cornell International Symposium and Workshop on the Hydrogen Economy,* Ithaca, N.Y.: Cornell University, 1973.

22. P. N. Ross, "Hydrogen Economy Rebuttal," Ref. 21, p. 27.

23. H. Goff, "Status of Conventional Hydrogen Production," Ref. 21, p. 49.

24. W. C. Kincaide and C. F. Williams, "Storage of Electrical Energy Through Electrolysis," Ref. 21, p. 170.

25. D. C. Fink, Ed., *Standard Handbook for Electrical Engineers,* 10th ed., New York: McGraw-Hill, 1968.

26. D. P. Gregory, "The Hydrogen Economy," *Scientific American,* January 1973, p. 13.

27. S. Katell, "Hydrogen Cost Analysis," Ref. 21, p. 240.

28. J. D. Balcomb, "Hydrogen Production Economics," Ref. 21, p. 70.

29. L. J. Nuttall, "Production and Application of Electrolytic Hydrogen," in W. N. Smith and J. G. Santangelo, Eds., *Hydrogen Production and Marketing,* ACS Symposium Ser. 116, 1979, Washington, D.C.: ACS, 1980.

30. J. D. Holmgren, C. E. Seglem, L. A. Salvador, and M. W. Dyos, "Coal Gasification/Combined Cycle System is Ready for Commercialization," *Modern Power Systems,* Crawley, Sussex, England, March, 1981.

5

Water Power: Hydro, Pumped Storage, and Tidal

INTRODUCTION

To understand the place that water power occupies in our energy picture, one needs the answers to the same questions as for coal, oil, or gas. What is it? What are the principal facts of its importance as an energy source? And what are the processes involved in harnessing it for human use?

But here the similarity ends. Who would go to the site of a coal-fired plant on his honeymoon? What Indian maiden was sacrificed each year to the gods of a gas well? And what crowds gather generation after generation to gaze in awe at a coal mine? The thundering roar of falling water in the Cave of the Winds behind Niagara Falls spells *power* as does no other energy source. It can be seen and felt as well as heard. It appeals to three of our five senses.

To most of us, the fact that water releases just as much energy in falling 1°F as in falling 778 feet down a waterfall is totally unbelievable, though true. It is of course much easier to harness the falling water, and man learned how to do this over 5000 years ago, about the same time he learned to use the wind. True, the process has been refined. The water falling on the "old mill wheel" no longer turns the millstone directly. An electric generator and transmission line has been interposed to distribute the energy hundreds or even thousands of miles from the mighty falls. But "the hydro" will always hold a nostalgic place in our list of energies—one not likely to fade with the passing years, even though its proportional share becomes less and less.

Water power has always been understood by everyone, and there is no point in complicating it in this chapter. There are just some "facts of life" to be stated that will help us understand it in a more quantitive way. To begin with, what is it?

First came the natural falls. Part of the water could be diverted to a water wheel, which could turn a millstone. The mighty Niagara Falls are essentially just this. At first, only part of the 160 ft fall was used for direct mechanical power, before the days of electricity. Then with electricity, the full 160 ft fall was used; and later canals were built to include the drop in the upper and lower rapids, 312 ft in all. Still later, pumped storage was added to pump the extra water available at night (after tourist hours) into a pool for use the next day.

International treaties fixed the amount of water each country, Canada and the United States, could divert and still keep one of the "seven wonders of the world" functioning as one of natures greatest show places, day and night. Today (1980), on the American side 1950 MW conventional hydro and 240 MW pumped storage are in use, generating about 1950 MW-yr electricity annually, about 17.1 billion kWh. In the United States as a whole, some 2000 billion kWh are used annually by about 200 million people, or about 10,000 kWh per capita. At this rate, the American plants at the Falls supply about 1.7 million people.

After man's use of the waterfalls came the millpond, with its little dam, forerunner of the great multipurpose dams built by The Bureau of Reclamation, The Corps of Engineers, The TVA, and so on, in later years. The millpond brought storage and greater flexibility. The miller could close the gates at night and go to bed. The energy was all there in the morning, perhaps with even greater drop (head) than had he used it continuously. More and more, the large dams are being used in this fashion.

At Niagara, before the advent of pumped storage, the water had to be used as it arrived or not at all. This is called "run of river," a term also applied to hydro plants with storage for only one or two days. Similarly, when a large dam is built with just a few hydro units, there may be enough water to run them steadily all year round. Then why shut down any of them and waste the precious energy? By simply adding more turbine generators, with no more "dam" expense or very little, the available energy most of the year can be used for "peaking." During high-water periods, the units would be kept running night and day. But during most of the year, while the river flow is relatively low, and there are enough turbine generators to use all the allowable water in several hours a day, it would be used during the peak load periods. It is much cheaper to add hydro generator units to carry this peak load than to build fossil fuel plants of the same capacity. Furthermore, the hydro generators can be started or shut down quickly (2 to 5 minutes) and thus provide excellent backup when they are not fully loaded in peaking service.

The words "allowable water" have now been used twice, first at Niagara for international division of rights and then for scenic beauty. At the multipurpose dams, water is needed for irrigation, for municipal water supplies, for navigation and maintenance of downriver flow, for recreation, for fish and wild life, and for power. The water level in the reservoirs must also be regulated to provide flood control in season. The formula is different at every project, but generally, except for flood periods, the generating plant may use just so much water a day (or in a given period), with the restriction that the downriver flow not be reduced below some prescribed minimum. This daily or periodic allowance, within the minimum flow restrictions, if the storage is adequate, is used during the peak load periods to displace the most expensive energy that might otherwise be used. With an economic dispatch computer, the "hydro" is assigned a fixed incremental cost such that the allowed amount of water is used to minimize the overall fuel and operating cost of the system.

Multiple Dams. With just one dam on a great river and limited storage, tremendous quantities of water may be spilled during flood season. Yet most of the year, the flow will support only limited hydro generation. As more dams are built upriver, each one impounds a part of those flood waters and releases them more gradually throughout the year. Hence, the downriver plants now have enough flow spread over the year to support additional generation.

Construction of the Grand Coulee Dam on the Columbia River in Washington impounded the flood waters for 151 miles, clear to the Canadian border.

This so regulated the flow of the Columbia River that the economic generation in downriver sites was increased by 50 to 100% (100% to the Snake River and 50% beyond). This was an important factor in the justification of the Grand Coulee project.

The initial generating capacity at the "high dam" at Grand Coulee was *only* 2.0 GW, which still made it the largest hydro plant in America at that time (Hoover Dam had 1.1 GW installed). The excess water pouring over the dam in high-water periods was enough to supply every person in the United States with several gallons of fresh water in less than an hour. It was a sight to behold! Later, as the need for peaking energy grew in the Northwest, the generating capacity was increased to 6 GW (by 1979), with an ultimate 9 GW planned. The water is used mainly for peaking.

Pumped-Storage Hydro. Starting in the 1930s, the available hydro became insufficient for peaking in many parts of the world, and the pumped-storage hydro plant came into existence. At first it developed slowly, but then in the 1950s, 1960s, and 1970s, there was a rapid expansion as the need grew and the economies and emergency power capabilities of these plants became apparent.

The idea was simply to pump water to the upper reservoir during the off-peak hours, using the most efficient generation, and then use it to provide additional capacity during the peak load periods. The upper reservoir could be above the dam of a hydro project or a pool at high elevation built specifically for this purpose.

In some cases, a hydro plant can be arranged for pumping back upstream, storing energy in the pool above the dam during light-load periods (See Salt River Valley later). However, an adequate pool below the dam is not usually available. Furthermore, the construction of both the turbine and the electrical machine, as well as some of the hydraulic gear, is different when used both for generating and pumping. These features will be covered later.

Waves and Tides. Harnessing the waves[15] or tides to run machines or generate electricity has always been a challenge to man's ingenuity, today no less than in the past. Four different wave-harnessing machines are currently (1980) being tested in England.[1] Environmental problems are receiving attention at the same time.

In regions of high tidal range, the conversion of the enormous energy into electricity has engaged the attention of many groups of engineers. There have been many studies and a few successful installations. Best known is the installation at the mouth of La Rance River on the northwest coast of France.[2,3] Average tides over the earth are but a few feet. But in many coastal areas, the tidal flow from a considerable expanse of ocean is funneled into a constricted or V-shaped pocket and reaches great heights. Ten to fifteen foot tides are not uncommon; and in the Bay of Fundy, tides reach 70 ft in height. At the proposed power sites in the Bay of Fundy, an average head of 18.1 ft could be developed at Passamaquoddy, on the United States–Canada border, and 33.0 ft at Hopewell in Canada.

The Bay of Fundy projects have been studied since the 1930s. In 1956–1961, the Passamaquoddy site was studied exhaustively by a Joint United States–Canada Commission. The arrangement favored was a two-pool project integrated with hydro developments in Maine that would provide storage of the variable tidal energy. The project was found to be "not economically feasible under present conditions" (1961) by the International Commission. It was recommended that it be deferred until "other less costly energy resources available to the area are exhausted."

The principal facts about these various water power developments are given in the following sections of the chapter, together with an outline of the various arrangements used or considered.

HYDRO POWER

Hydro Capacity and Generation

While the installed hydro-generating capacity in the United States has been doubling every 15 yr on the average (4.4%/yr) from 1920 to 1980, the total generation has been doubling every 12 yr (5.8%/yr). Hence, as shown in Fig. 5.1, hydro is a steadily diminishing fraction of the total generating capacity. At the turn of the century, it was about 38%. It held at about 25% from 1920 to 1945, but since then has been steadily diminishing, reaching 12% about 1978 (68 GW). This includes about 10 GW of reversible pumped-storage capacity.

The proportion of hydro energy generated, Fig. 5.2, shows a similar decline, with irregularities due to rainfall and economic conditions. Hydro-generated electric energy in the United States amounted to 36.3% of the total in 1902, and was about that in 1920. It declined to 15% in 1976, and was about 10% in 1977, a year of exceptionally low river flow in the west, where over 60% of the U.S. hydro is concentrated.

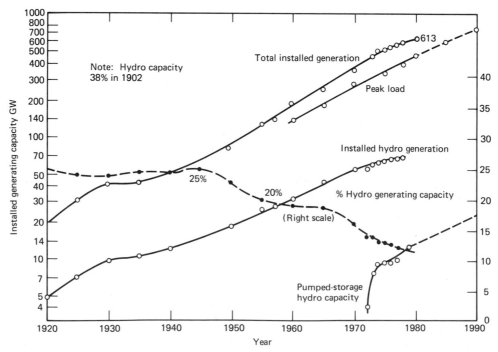

Figure 5.1.   Installed hydro and total generation in the United States.

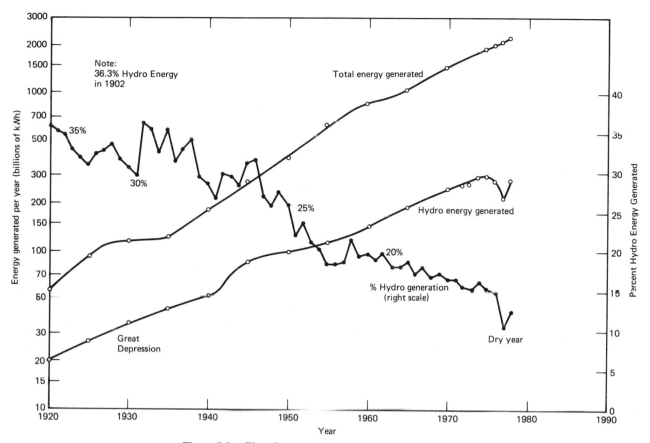

Figure 5.2. Electric energy generation in the United States.

To summarize, hydro in 1978 accounted for 4.1% of the total energy production in the United States and about 13% of the electric energy generated. It is a steadily diminishing proportion, even though the amounts of prime hydro capacity (total hydro less pure pumped storage) and the net hydro generation are still increasing.

It may be noted that the electric energy generated by hydro in 1978, about 300 billion kWh, if converted to quads, is

$$300 \times 10^9 \times 3413 \times 10^{-15} = 1.024 \text{ quads}$$
$$(1.024 \times 10^{15} \text{ Btu})$$

whereas hydro is credited with 3.2 quads in 1978 (Table 1.2). Hydro is credited with the same energy production as would be required in average coal-fired plants to produce the same amount of electricity. Otherwise, it would be meaningless to add hydro and fossil fuel energy to obtain the combined energy produced. Actual hydro efficiencies are of the order of 80 to 90%.

Developed and Undeveloped Hydro

In 1976, the installed prime hydro capacity in the United States was about 58.6 GW, of which 35.6 GW, or 61%, was in the Mountain and Pacific states. Some 110 GW remained to be developed, of which 19 GW was in the Mountain states and 67.5 GW was in the Pacific states, including about 32 GW in Alaska. Altogether, 86.5 GW, or 79% of the remaining undeveloped prime hydro, was in the Mountain and Pacific States, including Alaska. Thus, while the prime hydro-generating capacity has doubled 3.7 times in the last 60 yr, only 1.5 doublings remain. Since this involves the less desirable sites and additions to present sites, the growth rate will probably decrease. Quite the reverse is true of pumped-storage hydro, to be described later.

Federal Hydro Projects

The total hydro capacity in the United States in 1976 to 1978 was 68 GW. Of this, some 27.7 GW,

Table 5.1.   Federal Power Projects 1976

Project	Installed Capacity (GW)		Allocated Cost of Hydro ($/kW)	Hydro Installation Date
	Total	Hydro		
Central Valley	1.141	1.14	209	1940s
Columbia Basin	13.455	13.44	395	1950s
Hoover and Parker-Davis	1.699	1.70	137	1930s
Missouri Basin	2.671	2.67	505	1960s
Southeast Power Administration	2.401	2.40	380	1950s
Southwest Power Administration	1.847	1.85	365	1950s
Tennessee Valley Authority (TVA)	25.795	3.20	299	1930s–1940s
Other	1.344	1.34	601	
Total or average	50.343	27.7	343	

or 40%, was concentrated in the seven large Federal power projects listed in Table 5.1. Most of these involve multipurpose dams, the cost allocated to electric power generation varying from $137/kW (Hoover Dam, 1936) to $500 to 600/kW for the later projects. The average cost is $343/kW.

The corresponding costs if built, for example, in 1980 can be estimated from the actual construction dates (Table 5.1 or 5.3) and the producer energy–price index given in Chapter 1. For the Columbia Basin installations, typically contracted in the early 1950s, the increase through 1980 is more than fourfold. Such comparisons are necessarily very rough, merely orders of magnitude.

Energy and Power Relations

The simple energy relations of still and moving water are as follows:

Potential Energy. The potential energy of a pound of water at a height of H ft above datum is H ft-lb.

Kinetic Energy. A mass, in falling H ft under the influence of gravity g, acquires a velocity V such that

$$V^2 = 2gH \qquad (5.1)$$

where $g = 32.2$ ft/sec^2.

Thus, 1 lb of water, in falling a height H, acquires a kinetic energy, $\frac{1}{2} mv^2$, of

$$\text{K.E.} = \frac{1}{2} \times \frac{1}{g} \times V^2 = H \text{ ft-lb} \qquad (5.2)$$

That is, its potential energy, or static head, H, is just converted to an equal kinetic energy, or "velocity head," H, as required by the law of conservation of energy. The "velocity head" is then the height from which the water (or other fluid) would have to fall to attain the velocity at which it is flowing.

At intermediate positions in the fall, the head is split between static head and velocity head; but neglecting friction losses, the sum of the two is always equal to the initial static head.

Power. If a quantity Q ft^3/sec of water falls through a height of H ft, the power P expended is

$$P = 62.5QH \text{ ft-lb/sec} = 0.1135QH \text{ Hp} \quad (5.3)$$

where the density of water is taken as 62.5 lb/ft^3.

This power can be imparted to a waterwheel by:

1. Riding it down slowly—"weight of water"—wheel diameter same as head.
2. Flow times pressure at the bottom elevation—pressure (reaction turbine).
3. Converting the head to velocity head, usually in a nozzle—"velocity head" (Pelton wheel).
4. A combination of velocity head and pressure.

Water wheels have been made to function in each of these ways.

Waterwheels

Early methods of harnessing water power were the "overshot" wheel, the "undershot" wheel, and the simple paddle wheel dipped in a rapidly flowing stream.

The diameter of the overshot wheel was approximately the height of the falls. Water flowed into the buckets at the top and was discharged at the bottom. Thus, the full head was effective in turning the wheel. Such a wheel operated simply by the weight of the water, the velocity being negligible, and the pressure being atmospheric throughout. The power developed is $0.1135QH$ Hp, as given by Eq. 5.3, multiplied by the efficiency of the wheel. High efficiencies of the order of 80 to 90% can be obtained by careful design.

In an undershot wheel, the water was conducted by the equivalent of a penstock and sluiced through the bottom of the wheel. This operated primarily by the pressure of the water against the paddles, with some recovery of velocity head. The design was ill suited to this type of operation, and the efficiency was usually quite low. It remained for a later generation to design efficient reaction turbines operating in this mode.

A unique design of a waterwheel (Mitchell or Banki turbine) for very small installations, as in developing countries, is described in a VITA manual.[14] It can be easily constructed from simple materials. The jet of falling water passes twice through the runner before discharge into the tailrace, resulting in an efficiency of about 80%.

Some waterwheels were operated similarly to a paddlewheel ship in reverse. The "ship" or barge was anchored, and the rapidly flowing water turned the wheel. In this case, the energy was in the kinetic energy or velocity head of the water. The power in the water approaching the wheel is given by Eq. 5.3, where Q is $V \times A$, the velocity times the projected area of the wheel in the water; H is the velocity head, $V^2/2g$. If the conversion efficiency, or power coefficient, is P_c, the output power is

$$P_{out} = 0.1135 P_c VA \frac{V^2}{2g} = 17.6 \times 10^{-4} P_c A V^3 \quad (5.4)$$

Similarly to a wind turbine, the power coefficient is below 0.5 practically.

Hydraulic Turbines

Hydraulic Turbine Relationships

From the seemingly innocuous observation that a dimensionless coefficient such as efficiency, pres-

sure ratio, or flow ratio of a hydraulic turbine could not possibly depend on the arbitrary units in which man measured things, some of our smart forebears reached the valuable conclusion that they must therefore depend only on dimensionless ratios of the various physical quantities involved. For water turbines, there are two such ratios: (1) ratios of linear dimensions and (2) Reynolds number, $vd\rho/\mu$.

That the ratio of any two linear dimensions is dimensionless is obvious. The quantities in Reynolds number, velocity $\times$ linear dimension $\times$ density/viscosity, have the dimensions in length L, mass M, and time T:

$$LT^{-1} \times L \times \frac{ML^{-3}}{ML^{-1}T^{-1}} = 1$$

Reynolds number is also dimensionless. It is found by experience that with water as the fluid and with the ranges of dimensions and velocities involved in hydraulic turbines, from models to full scale, the variation of coefficients with Reynolds number is not great. Turbulent flow is not involved. Thus, the variation with Reynolds number is at first neglected in model testing of hydraulic turbines, and later an empirical correction is made to account for the small difference in efficiency in going from model to full scale.

Thus, homologous model turbines have the same ratios of all linear dimensions as the full-scale machines they simulate, but they can have diameters and heads as appropriate for the model. The "laws of proportionality" are given in Table 5.2 and show the variation of power output P_{out}, angular velocity n, and flow Q with head H and diameter D.

Constant Runner Diameter. Since all heads are changed proportionally and since velocity head is $V^2/2g$, the fluid velocities and the flow Q must vary as $H^{1/2}$. Power is proportional to $H \times Q$, hence to $H^{3/2}$, as given in Table 5.2. With a constant runner diameter, the rotational speed n varies with linear velocity, hence with the flow Q and with $H^{1/2}$.

Thus, the power output, rotational speed, and flow for a given turbine can all be expressed at a 1 ft head, if desired, and easily ratioed to any other head.

Constant Head. Similarly, for constant head, the power output is proportional to the flow Q. With fixed velocity heads, the velocities are all fixed, and

Table 5.2. Laws of Proportionality for Homologous Turbines

For Constant Runner Diameter D		For Constant Head H		For Variable Diameter and Head	
Quantity	Proportional to	Quantity	Proportional to	Quantity	Proportional to
P_{out}	$H^{3/2}$	P_{out}	D^2	P_{out}	$D^2 H^{3/2}$
n	$H^{1/2}$	n	D^{-1}	n	$D^{-1} H^{1/2}$
Q	$H^{1/2}$	Q	D^2	Q	$D^2 H^{1/2}$

the flow Q is therefore proportional to D^2. Hence, P_{out} is proportional to D^2.

In order to have the same linear velocities at corresponding locations, the rotational velocity n must be inversely proportional to D, as given in the table.

Variable Diameter and Head. If both diameter and head are varied, the output, rotational speed, and flow must vary as shown in the third column of Table 5.2, which combines the two effects.

These relations provide a sound basis for model tests. The diameter can be reduced, with all proportions maintained, and the head can be reduced to a convenient value for scale model testing. Then at the speed, output, and flow given in the third column of Table 5.2, the phenomenon taking place in the model is nearly a small-scale replica of that in the full-scale machine.

This equivalence has now been shown to be so close with accurately constructed models that scale model tests are being accepted to demonstrate compliance of the full-scale installation, in many cases. The entire hydraulic flow path leading up to and following the turbine, as well as the turbine itself, must of course be modeled accurately.

Types of Hydraulic Turbines

While there have been many different types of hydraulic turbines since the days of the "old mill wheel," modern machines fall mainly into four classes: the Pelton, or impulse, wheel for the highest heads of 500 to 5500 ft; the Francis reaction turbine for intermediate heads of 5 to 2000 ft; and the Kaplan, or propeller reaction-type, wheel for low heads up to about 40 ft. The Kaplan turbine may be of fixed- or variable-pitch design. A "mixed flow" or Deriaz-type turbine, having variable-pitch blades, is used successfully for heads up to 90 ft and higher.

For some applications involving backup and peaking use, with highly varying flow, maximum energy may be obtained by using the variable-pitch Kaplan turbine up to 150 ft head (Duke Power, Cowans Ford).

As indicated, the proper fields of application overlap considerably in terms of head, since there are many other factors that determine, to a lesser extent, the best type for any particular application. These are, for example, size, presence of entrained material, simplicity of hydraulic arrangement, character of flow, and so on. The impulse wheel is not reversible, so that combination pump–turbines, to be discussed later under Pumped-Storage Hydro, are limited to the Francis and propeller types and the mixed-flow turbine. There is then a strong incentive to push the reversible Francis turbine to higher and higher heads, rather than use the more expensive arrangement of impulse wheels and separate multistage pumps.

All types of hydraulic turbines have a stationary reducing passage, or nozzle in the case of the impulse turbine, in which the static head is partially or wholly converted to velocity head before the water enters the turbine runner.

Francis Reaction Turbine. In the reaction turbine, this takes the form of a converging duct and scrollcase that smoothly accelerate the water from the bottom of the penstock, or dam, into the runner. The stay vanes and wicket gates, shown in Fig. 5.3, are shaped to form a part of this converging passage. There is still a substantial positive pressure in the runner accelerating the water, with the outlet opening smaller than the inlet.

Below the turbine, if vertical shaft type, is the draft tube which diverges gradually to the tailrace, converting the remaining velocity head into static head, with an efficiency of about 85%. Thus, there is a continuous column of water from the top of the

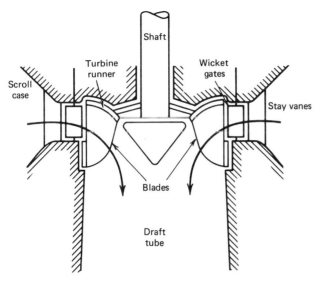

Figure 5.3. Francis reaction turbine.

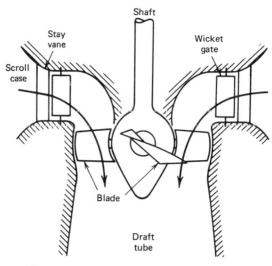

Figure 5.4. Kaplan or propeller-type turbine.

dam to the tailrace, and the turbine can theoretically be located up to 34 ft above the tailrace, at which point the water would separate. Practically elevations up to 25 ft above the tailrace can be used in turbine applications. For pumping, the unit must be set much lower.

As long as the water enters the tailrace at low velocity (usually less than 1 ft velocity head) and there is a smooth channel with little friction loss from forebay to tailrace, the turbine converts most of the energy of the falling water, about 90% at design load, into useful output.

Wicket Gates. Wicket gates (Figs. 5.3 and 5.4), which provide part of the whirl or entering velocity and also serve to control the flow from shutoff to full load, have proved to be the most practical form of valve, being little subject to sticking or being blocked by foreign objects. The number of gates ranges from 12 for small units to 28 for large units. The connection from each gate to the operating ring includes a "breaking element" to protect the rest of the gates in event of an obstruction.

Kaplan or Propeller-Type Turbine. The conformation of the Kaplan-type turbine is shown in Fig. 5.4. This arrangement is ideally adapted to low heads and has the highest specific speed of all four turbines. The Kaplan turbine is frequently supplied with variable-pitch blades, which are adjusted by a cam in accordance with the gate opening to provide the maximum efficiency at each load. This results in

a practically flat efficiency curve from about 40 to 100% load. Without this feature, a unit designed for maximum efficiency at full load would have about 50% efficiency at 40% load. Thus, it is particularly valuable for units that must run at part load for considerable periods.

Several variations of the propeller-type turbine are mounted with shaft horizontal, or nearly so. These are:

1. The "straight flow" or "rim generator" type, in which the generator windings are located around the rim of the propeller. This is being applied for the first time in N. America at a tidal plant in Nova Scotia. See "Potential Tidal Sites" later.
2. The "slant axis" or "tube turbine" shown in Fig. 5.18 (see Tidal Power later).
3. The "bulb type" turbine, Fig. 5.17 (see Tidal Power).

These units are particularly well adapted to the very low heads, under 50 ft, encountered at all tidal power sites. However, their principal application is at low-head river plants.

Pelton or Impulse Turbine. In the Pelton wheel (Fig. 5.5), applicable to the highest heads, up to 5500 ft, the static head is converted completely to velocity head in a "needle nozzle," which also serves as the shut-off valve. The diameter of the upstream portion of the nozzle pipe is kept such that

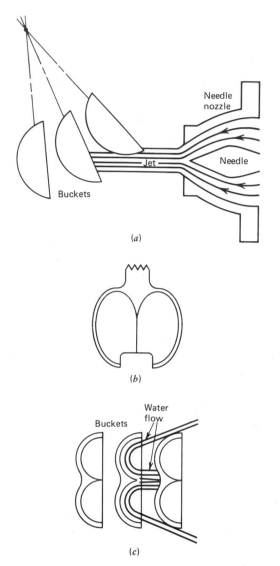

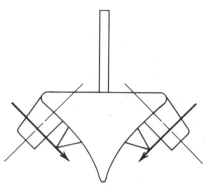

Figure 5.6. Runner of the Deriaz-type or mixed-flow turbine with adjustable blades.

Figure 5.5. Impulse-type hydraulic turbine or Pelton wheel: (*a*) jet impinging on buckets; (*b*) front face of a divided bucket; (*c*) bucket spacing showing water flow path.

the velocity head is not over 1% of the total. Thus, over 99% of the conversion from static to velocity head occurs in the nozzle.

The bucket wheel runs in air at atmospheric pressure. The entering water jet is split by the sharp edge of the double bucket, one half flowing around the curvature of each bucket. The water velocity is nearly reversed with respect to the bucket, but with sufficient exit angle to just miss the next bucket, as shown in Fig. 5.5. If the buckets are moving with half of the jet velocity, the exit velocity is nearly zero with respect to ground, and practically all of the energy of the water has been converted to mechanical energy.

Pelton wheels may be set with the shaft horizontal or vertical. In the latter arrangement, as many as six jets can be used around the circumference of the wheel, without interference. The maximum efficiency of the horizontal-shaft arrangement is about 90%, but efficiencies as high as 91.5% have been measured with vertical-shaft multijet units.

Mixed-Flow Turbines, Deriaz Type. In a vertical-shaft Francis turbine, the water enters horizontally and turns downward in the turbine, exiting vertically, as shown in Fig. 5.3. In a propeller-type turbine, the water flow through the blades is parallel to the shaft as shown in Fig. 5.4. In a mixed-flow turbine, the water flow is slantwise through the blades, as shown in Fig. 5.6. The adjustable blades are larger than in the propeller type and pivot on slant axes, as shown. In the extreme position, they completely shut off the water passage, overlapping each other.

Specific Speeds of Hydraulic Turbines

A given turbine, operating at a given head, will obviously have a maximum efficiency at some speed n between zero and infinity. At this speed of maximum efficiency, it has a certain output, P_{out}.

For homologous machines operating at the same head, $n \propto 1/D$ and $P_{out} \propto D^2$, and $n\sqrt{P_{out}}$ is therefore a constant. If the head is allowed to vary among homologous machines, $n\sqrt{P_{out}}$ varies. But if the head is held at 1 ft, then $n\sqrt{P_{out}}$ is the same for all homologous machines, regardless of their diameters. It is called the "specific speed" n_s and is measured in units of rpm $\times \sqrt{\text{Hp}}$. That is,

$$n_s = n\sqrt{P_{out}} \qquad \text{at 1 ft head} \qquad (5.5)$$

where n is the speed of maximum efficiency and P_{out} is the power output at that speed. The power at 1 ft head is related to the actual power and head by the laws of homologous machines (Table 5.2). The specific speed is also the speed in rpm at best efficiency that the runner would have if sized to develop 1 Hp at 1 ft head.

A typical propeller-type turbine may have a specific speed in the range of 110 to 160 rpm, whereas Francis turbines have lower specific speeds, of 20 to 100 rpm. The mixed-flow turbine is intermediate between the Francis and Kaplan turbines. The pelton wheel has the lowest specific speed, 4 to 5.5 rpm. This is the n_s per jet. A vertical-axis multijet machine may have six jets, which raises the specific speed by $\sqrt{6}$, to 9.9 to 13.5.

Thus, the specific speed is a measure of the relative turbine speed for any given turbine output, the propeller type being the highest, the Francis type next, and the impulse wheel the lowest. Considering that a high-speed machine is generally smaller and of lower cost for the same power output, there is an advantage in using the highest specific speed practical for a given application. Thus, reaction-type turbines are being pushed to higher and higher heads. However, the impulse turbine alone is practical for the highest heads. The propeller-type unit of highest specific speed is normally applicable only up to about 40 ft head but has been used to 150 ft as mentioned earlier.

Hydro Plants

While there was a 68 GW total hydro capacity in the United States in 1976, in some 1156 plants, 91% of this capacity (95% of the net hydro generation) was concentrated in 434 plants having capacities of 10 MW or more. Principal features of the 33 larger prime hydro plants, approximately 400 MW and over, are shown in Table 5.3. These plants include 46% of the total hydro capacity.

These are all prime hydro plants, with the exception of Smith Mountain, No. 20, in which reversible pump-generators are used at a prime hydro site, a trend discussed under Pumped Storage. By 1979, there were 13 pumped-storage plants in the United States of over 400 MW. These are included in Table 5.5.

The oldest of the large prime hydro plants is Wilson Dam, at Muscle Shoals in Alabama, built in 1925 by the Corps of Engineers and now a part of the TVA. While the Niagara Falls development dates from 1895, the machines in use in the 1920s on the American side used only the drop in the falls and the upper rapids and have since been closed down. All of the allowable water is now used in the higher-head Robert Moses Plant, No. 2, at Lewiston, which uses the full 310 ft head of the falls and the upper and lower rapids. Associated with it, as with its Canadian partner, the Sir Adam Beck station in Queenston, Ontario, is pumped-storage hydro, to be described later.

The largest of the plants shown in Table 5.3 are in the 2 GW range, except the Grand Coulee plant, which was increased to 6 GW in 1978–1979, with an ultimate planned development of 9 GW.

At these large plants, the heads range from 49 ft at Rock Island to 632 ft at Dwarshak. Nearly all use Francis reaction turbines. The low-head Rock Island installation uses a bulb-type Kaplan turbine, shown in Fig. 5.17.

A majority of these plants would be classed as "run of river," with pondage from a few hours to a few days of full load. Notable exceptions are Hoover Dam (15 months) and upriver dams such as Glen Canyon in Arizona (22 months), Oahe in South Dakota (8 months), Shasta in California (7 months), the Garrison plant in North Dakota (10 months), and a few others. At Grand Coulee, there is storage for about 0.51 months of full load (4.76 GW) above the main dam. The other plants are essentially run of river, with the important difference, compared with a falls, that they have enough storage to use the water for daily, and in some cases weekly, peaking over a large part of the year.

Plants with high plant factor, such as Dalles, McNary, and Bonneville, have obviously operated practically base load. Plant factors are based on rating; and as capability frequently exceeds rating, a plant factor over 100% is possible, as at Bonneville. The plant factors are for a single year and can vary considerably from year to year and with added capacity. Some 81% of the capacity of these larger plants is in the Mountain and Pacific States, the Columbia Basin alone accounting for 72%.

As mentioned, the 33 plants of Table 5.3 account for 46% of the U.S. hydro in 1979. The remaining 54%, some 39 GW in 1131 smaller prime hydro and in pumped-storage plants, is much more evenly distributed over the country. Pumped storage, which constituted about 13 GW of the 72 GW of total hydro in 1979, is a rapidly growing component and can be located near the load center. To date (1980), pumped-storage capacity has been predomi-

Table 5.3. U.S. Prime Hydro Plants 400 MW and Over, 1978–1979. Department of Energy Reports

No.	Plant	Year in Service	MW Nameplate	Head (ft)	Pondage full-load (hr)	Plant Fact. 1978	Cost ($/kW)	State	River	Data from Inventory of Power Plants, April 1979[5,a] MW Nameplate	No. of Units	MW per Unit
1	Grant Coulee	1941	4763	333	375	46	86	Wash.	Columbia	6133	b	b
2	Robert Moses, N.F.	1961	2196	310	—	—	203	N.Y.	Niagara	2190	13/12	150/20PS
3	John Day	1968	2160	100	7.3	59	178	Ore./Wash.	Columbia	2160	16	135
4	Hoover	1936	1345	530	10,964	29	—	Ariz./Nev.	Colorado	1347	c	c
5	Rocky Reach	1961	1213	92	2.7	58	197	Wash.	Columbia	1242	7/4	102/132
6	Dalles	1957	1125[d]	86	4.2	89	153	Ore./Wash.	Columbia	1814	d	d
7	Chief Joseph	1955	1689	171	4.8	66	204	Wash.	Columbia	1981	e	e
8	McNary	1953	986	88	16.9	82	270	Ore./Wash.	Columbia	986	2/14	3/70
9	Wanapum	1962	974	79	13.4	66	188	Wash.	Columbia	831	10	83.1
10	Glen Canyon	1964	950	558	16,244	45	199	Ariz.	Colorado	952	8	119
11	Robert Moses, S.L.	1958	912	81.3	—	—	245	N.Y.	St. Lawrence	912	16	57
12	Priest Rapids	1960	789	76	4.4	78	—	Wash.	Columbia	788	10	78.8
13	Wells	1967	774	67	6.2	62	245	Wash.	Columbia	770	10	77
14	Wilson	1925	630	88	7.5	54	83	Ala.	Tennessee	628	f	f
15	Oahe	1961	595	193	5,641	56	333	S.D.	Missouri	595	7	85
16	Boundry	1967	551	261	20.9	78	145	Wash.	Penc Oreille	634	4	158.6
17	Bonneville	1938	518	57	9.1	102	173	Ore./Wash.	Columbia	522	g	g
18	Conowingo	1928	477	87	13.3	41	157	Md.	Susquehanna	512	7/4	36/65

No.		Year						State	River			
19	Big Bend	1965	468	67	38.2	25	233	S.D.	Missouri	464	8	58
20	Smith Mountain	1965	432	192	71.5	14	119	Va.	Roanoke	547	[h]	[h]
21	Shasta	1944	452	480	4,782	45	44	Calif.	Sacremento	453	2/3	84/95
22	Little Goose	1970	810	98	6.1	45	214	Wash.	Snake	810	6	135
23	Lower Granite	1975	810	—	—	40	376	Wash.	Snake	675[j]	5	135
24	L. Monumental	1969	405	100	5.1	83	371	Wash.	Snake	810	6	135
25	Dwarshak	1973	400	632	—	59	706	Ida.	N. F. Clearwater	433	2/1	90/253
26	Garrison	1956	430	172	7,421	73	390	N.D.	Missouri	400	5	80
27	Ice Harbor	1961	603	998	4.1	60	213	Wash.	Snake	603	3/3	90/111
28	Libby Dam	1975	420	300	—	43	979	Mont.	Kootenai	420	4	105
29	Brownlee	1959	360	277	776	84	191	Ore.	Snake	585	4/1	90/225
30	Rock Island	1933	213	49	1.1	84	261	Wash.	Columbia	261	[i]	[i]
31	Noxon Rapids	1959	397	152	131	59	260	Mont.	Clarks Fork	398	4/1	71/114
32	Hells Canyon	1967	392	210	6.5	70	168	Ore.	Snake	392	3	130.5
33	Ross	1952	360	395	1,150	38	297	Wash.	Skagit	360	4	90

[a] Units projected in 1979 are included in these columns.

[b] Grand Coulee: 17 × 125 MW + 1 × 108 MW + 3 × 600 MW + 3 × 700 MW. Pumped storage units are included in Table 5.5.

[c] Hoover: 14 × 83 MW + 1 × 40 MW + 1 × 50 MW + 1 × 95 MW.

[d] Dalles: 2 × 3 MW + 2 × 14 MW + 14 × 78 MW + 8 × 86 MW (8 × 86 MW not included in column 4).

[e] Chief Joseph: 2 × 2.4 MW + 19 × 64 MW + 8 × 95 MW.

[f] Wilson: 4 × 23 MW + 4 × 31 MW + 10 × 25 MW + 3 × 54 MW.

[g] Bonneville: 1 × 4 MW + 2 × 43 + 8 × 54 MW.

[h] Smith Mt.: All pump generators; 2 × 66 MW + 1 × 115 MW + 2 × 150 MW. Pondage given is upper, as straight hydro plant.

[i] Rock Island: 1 × 1.2 MW + 1 × 15 MW + 3 × 20.7 MW + 6 × 22.5 MW + 8 × 51 MW.

[j] 810 MW is correct. Sixth 135 MW unit was added in May 1978.

nantly in the East, where hydro for peaking capacity was lacking.

In many hydro plants, later units are of much larger capacity than originally planned, as manufacturers have been able to achieve larger units, transportation facilities have improved, and needs for peaking energy have skyrocketed. The units being built for Grand Coulee in 1971 were of 700 MW, compared with the original units (largest in 1941) of 108 MW. The later units were assembled at the site.

While a natural falls can be tapped economically for a small or large part of its available energy, a large river, such as the Columbia or Susquehanna, cannot be economically harnessed until the load can absorb the output of a very large power plant. Early studies of power in the Pacific Northwest included consideration of transmission lines as far east as Chicao to absorb the tremendous amounts of power that would be available from an economic installation. This was never necessary, as the load built rapidly with the availability of low-cost electric energy, and steam-generating plants are now necessary to supplement the hydro power. These considerations and the many other factors of importance in hydro projects can best be illustrated by two examples, Conowingo and Grand Coulee.

Conowingo was a private-enterprise, single-purpose power project having relatively small interaction with other plants or with the surrounding country. Grand Coulee was a multipurpose, public enterprise, primarily power and irrigation, having profound interaction with downriver plants and involving irrigation and settlement, over the ensuing 60 yr of over 2000 square miles of land in south-central Washington. Thus, these two examples are at opposite ends of the spectrum of hydro development. The Columbia Basin development, of which Grand Coulee is a part, like the TVA, is a Federal project. However, unlike the single authority of the TVA, the dams and major features of the Columbia Basin were built by the Bureau of Reclamation and the Corps of Engineers and the transmission system by the Bonneville Power Administration. Both TVA and Grand Coulee were started during the Great Depression of the 1930s to aid in the recovery and to conserve great natural resources.

Conowingo Hydro Project

The Conowingo Dam on the lower Susquehanna River in Maryland and the original power plant,

Figure 5.7. Aerial view of Conowingo power plant. Courtesy Philadelphia Electric Co.

both built in 1926–1928, are shown in Fig. 5.7. The dam, nearly a mile long, with 89 ft head, is traversed by U.S. Route 1, the Old Baltimore Pike. This plant held seven 36 MW units for a capacity of 252 MW, exceeded only by Niagara Falls, 285 MW, at that time. Figure 5.8 shows the plant with the addition of four 65 MW outdoor units in 1964, bringing the capacity to 512 MW as it stands today (1980). Some of its statistics are given in Table 5.3.

The Susquehanna is a mighty river, draining 47% of the state of Pennsylvania, 13% of New York State, and a little of Maryland. Its flow in the 1972 flood (1,000,000 cfs) was equal to the design overflow of the Grand Coulee Dam and nearly equal to the flow in the lower Columbia during the 1894 flood (1,170,000 cfs). However, its normal flow is much less, and its minimum flow drops to 15,000 cfs. At 89 ft head, 15,000 cfs will supply (Eq. 5.3) $0.1135 \times 15,000 \times 89 = 152,000$ Hp. At 80% efficiency, this is $152,000 \times 0.8 \times 0.746 = 91,000$ kW, or 91 MW, only 36% of the *initial* installed capacity.

Thus, it was not until the load in the Philadelphia area had grown to a point that it could efficiently utilize the output of a large hydro plant that the Conowingo project became attractive. Three conditions had to be fulfilled: (1) The system must be able to use the full capacity of 252 MW during substantial parts of the year, when this flow was available; (2) it must also have steam or other capacity adequate for base load so that the full 252 MW could also be used for several hours of peaking in low water periods; and (3) the system must have a peak load of such short duration that even in low-flow

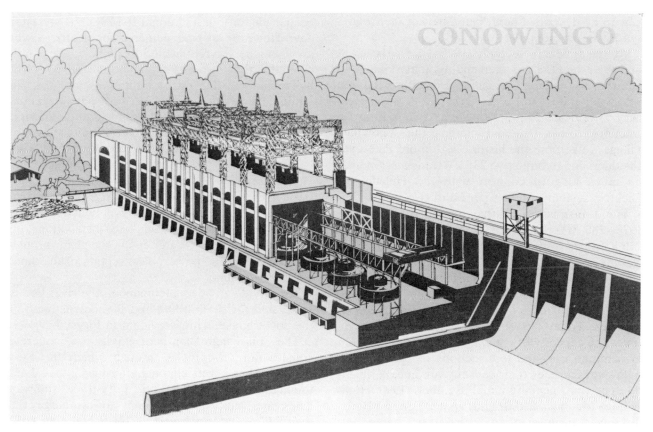

Figure 5.8. Conowingo power plant showing the four 65 MW outdoor units installed in 1964. Courtesy Philadelphia Electric Co.

periods it could be supplied by the hydro generation, without requiring the installation of added steam generators. From Table 5.3, the pondage is adequate for 13 hr of present full capacity, or about 25 hr of the initial 252 MW capacity.

It was not until 1926 that these three conditions were met and it became economic to dam the lower Susquehanna. After elaborate plans, the Federal license was granted in 1926, and in less than two years from that date the dam and powerhouse had been built and the first units were running and supplying power to Philadelphia over newly built 230 kV transmission lines. This record is not likely to be bettered in present times (1980).

In an average year, the initial plant supplied 1.25 billion kWh, corresponding to an average load of 142.7 MW, or a plant factor of 56.6%. As noted in Table 5.3, the present factor with 512 MW installed is substantially lower (41%), although this varies from year to year with river flow.

At the time of building Conowingo, the smaller Holtwood plant, 51 ft head, 112 MW, belonging to a neighboring power company, had been in operation since 1910, some 14 miles upstream. The development of 89 ft head at Conowingo would back water right up to the Holtwood tailrace. While this would not interfere with the operation of the existing Holtwood plant, it would eliminate part of the fall below the dam that Holtwood was licensed to develop. Both companies recognized that this energy could be harnessed far more economically at Conowingo than by further development at Holtwood. An equitable arrangement was made to share the financial benefits at Conowingo to compensate for the use of these rights.

Thus, multiple dams on a river generally have interactions of various kinds. Between Conowingo and Holtwood, the interaction was relatively small. In other cases, the interactions are very extensive. Many of the great river systems have been developed as coordinated public-power enterprises, in which the locations of dams and their multiple uses are optimized as one vast system. Examples of these are given in Table 5.1, the Grand Coulee project, to be described next, being part of the Columbia Basin development.

At the time it was built (1926–1928), the Conowingo project was a tremendous undertaking. In addition to building the dam and installing the first seven generators in the powerhouse, it was necessary to relocate 16 miles of Pennsylvania Railroad trackage, to evacuate and inundate the Conowingo village, to reroute the historic Baltimore Pike over the dam, and to build two 230 kV transmission lines 58 miles long to connect with the Philadelphia Electric system—all in a period of two years.

The transmission of that amount of power in 1928, 280 MW max., for 70 miles altogether over 230 kV lines from Conowingo to Philadelphia, represented something of a stability problem. However, by 1936 twin 287 kV lines were being built to carry much of the Hoover Dam generation (900 MW) some 270 miles over the Mojave Desert to Los Angeles. Today (1980), 765 kV lines with over 12 times the capacity of the early Conowingo lines are in common use, and test sections of 1100 kV line with over 25 times their capacity are in experimental operation, pending initial use about 1990. High-voltage dc has also found application for long-distance transmission, as discussed in Chapter 14.

By 1964, the Philadelphia Electric Co. was operating as part of the Pennsylvania–New Jersey–Maryland (PJM) interconnected power pool. Such pools operate the entire interconnected system for greatest economy, regardless of which system owns the facilities in use at any particular time. The resulting savings are equitably distributed.

In the original Conowingo plant, provision had been made for four additional 36 MW units at a later date. However, as peak demands for electric power grew on the interconnected system, studies indicated that the new units should be as large as possible within existing space limitations and that an outdoor installation would be most advantageous. This technology had been developed in the interim. This led to the four 65 MW outdoor units shown in Fig. 5.8, bringing the installed capacity to 512 MW.

Hydraulic Arrangements. The flow path through the Francis-type hydraulic turbine is shown in Fig. 5.9. Note the butterfly valve, which has been superseded by wicket gates in later installations, up to about 400 ft head. Water enters the plant through 11 openings in the powerhouse section of the dam, one for each unit.

To provide an adequate channel, the river bed is excavated for several hundred feet downstream of the powerhouse. This is evident in Fig. 5.7.

The Conowingo Dam is of gravity-type concrete construction, the rocks through which the Susquehanna has cut its channel affording an excellent foundation at that location. Fifty-three spillway gates are built into the dam to discharge the overflow water. Each gate is $22\frac{1}{2}$ ft high and 40 ft wide. They are operated by huge cranes which can be seen at the highway level. In cold weather, the formation of heavy ice, which would interfere with operation of the gates, is prevented by the use of electric heaters and the use of compressed air blown into the water.

Years Later. By 1978 the Philadelphia Electric System had grown to over 6000 MW of total

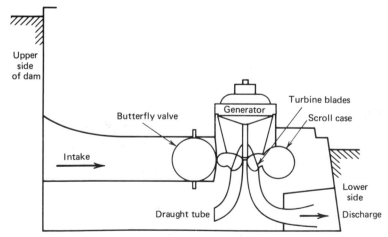

Figure 5.9. Flow path through the Conowingo turbines.

generation, with an additional 8000 MW authorized. It was operating as a part of the PJM Pool of 37,000 MW, with three large pumped-storage hydro plants—Muddy Run, 900 MW, to be described later; Yards Creek, 330 MW; and Seneca (Kinzua), 400 MW—in addition to Conowingo, Safe Harbor, Holtwood, and several smaller run-of-river hydro plants. Over 2200 MW (2.2 GW) of hydro was thus available for peaking in the interconnected system, 58% of it being pumped storage. The operation is described under Pumped Storage Hydro later.

Grand Coulee

The Conowingo project was completed in 1928. One year later came the stock market crash of 1929, followed by the Great Depression of the early 1930s. Industry in America was at a low ebb. Many were out of work. The Grand Coulee project was one of the great public enterprises that helped to start the wheels of industry, to provide employment for those that were idle in all parts of the country, and, as economic conditions became normal, to supply electric power at low rates for new and expanding industries and for homes. It was a part of the national program of industrial recovery.

The setting for the Grand Coulee Project is shown in the Columbia Basin map of Fig. 5.10 and the aerial view of Fig. 5.11. Millions of years ago, the Columbia River as it flowed westward south of the Canadian border had cut a channel 1500 ft deep through solid granite and basalt. In the Ice Age, a glacier advancing southward crossed the canyon of the Columbia east of the Okanogan River, raising the water surface of the Columbia River some 1500 ft, causing it to cut a new channel to the southwest, now called the Grand Coulee. This channel, 50 miles long, up to 900 ft deep, and 5 miles wide, includes at its lower end a cataract now known as Dry Falls, some 400 ft high and 6 miles long. From there, the river meandered over the land until it reached the former channel of the river.

When the glacier receded, the Columbia River

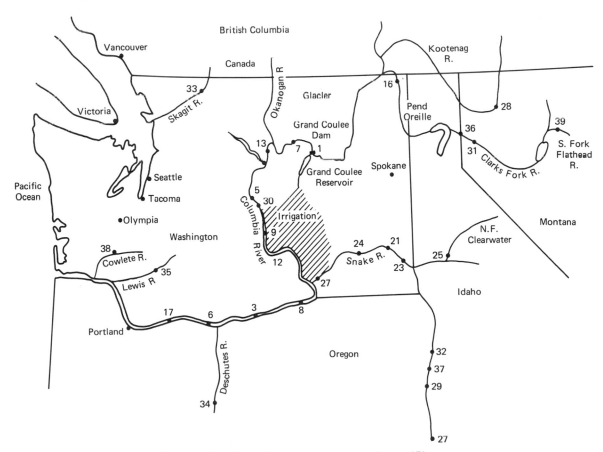

Figure 5.10. Major Columbia Basin power plants, 1979 ■■■.

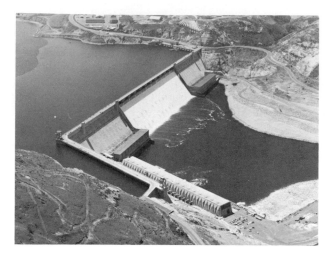

Figure 5.11. Aerial view of Grand Coulee, Courtesy U.S. Department of the Interior—Grand Coulee Project.

returned to its original channel and left the Grand Coulee as one of the marvels of nature.

South of the Grand Coulee, between it and the Snake and Columbia Rivers, as shown in Fig. 5.10, lay thousands of square miles of fertile but arid land, where the water of the Columbia had once flowed, probably the finest expanse of arid land awaiting reclamation in the west.

The Grand Coulee project envisioned use of the Grand Coulee above Dry Falls as a storage reservoir for irrigation of the dry land to the south, using water pumped from the Columbia River. Power would be supplied by a hydro project on the Columbia River, close to the north end of the Grand Coulee. The dam itself would elevate the water 350 ft above low water, and pumps would raise it an additional 280 to 360 ft to the reservoir.

In 1934, the Grand Coulee Low Dam was being constructed exclusively for power purposes, as a part of the national program for industrial recovery. It constituted about one third of the work involved on the ultimate Grand Coulee High Dam. The overall project was conceived as involving four additional steps which had not yet been authorized by Congress:

1. The Grand Coulee High Dam and powerplant.
2. The Grand Coulee Pumping Plant and pipeline to the Grand Coulee.
3. The Grand Coulee north and south dykes forming the reservoir.

4. The main canals and water distribution system for irrigation.

Altogether, the construction would require about 11 yr. At an estimated rate of settlement of 20,000 acres per year, the irrigation project would extend over 60 yr.

The fall of the Columbia River from the Canadian Border to the sea is about 1700 ft. Hydro power from much of this fall could be developed by a series of dams, most of which have now been built (see Table 5.3 and Fig. 5.10). Additional power was available from the tributaries. However, the key project was the Grand Coulee High Dam. It would create a lake (Lake Franklin Roosevelt) 151 miles long to the Canadian Border and utilize the full 341 ft drop in that stretch of river. The irrigation reservoir, 23 miles long, with a capacity of over a million acre ft, would regulate the irrigation flow and permit maximum use of off-peak and secondary energy for pumping. From the Grand Coulee Reservoir, 1,000,000 acres of land could be irrigated by gravity flow, and an additional 200,000 acres by auxiliary pumping not exceeding 100 ft head.

The Grand Coulee High Dam would so regulate the flow of the Columbia River as to increase the firm power available at downriver sites by 100% down to the Snake River and by 50% below the Snake River. Thus, the Bonneville Dam, near Portland, Oregon, at tide level, which had been started earlier, and the Grand Coulee Dam were the first two great dams to be built in the Columbia Basin. Today (1980), there are 26 dams and hydro plants of 200 MW or over in operation. Nineteen of these exceed 400 MW in capacity, as shown in Fig. 5.10, and are included in the 33 largest hydro plants in the United States listed in Table 5.3.

The Grand Coulee High Dam project included 18 108 MW generators, 17 of them later increased to 125 MW, located nine in each of two powerhouses, on either side of the central overflow, as shown in Fig. 5.11. These were installed between 1941 and 1951.

Later, a third powerhouse was built on the right bank, below the dam, as shown in the foreground in Fig. 5.11. Three 600 MW and three 700 MW units were installed in 1978–1979, bringing the capacity to 6133 MW, as indicated in Table 5.3. An ultimate capacity of 9000 MW (9 GW) is planned.

The pumping station, designed for 20 30,000 Hp pumps, is located just upstream of the west abutment of the dam (left bank). It pumps irrigation water from above Grand Coulee Dam to the reservoir, a height of 280 to 360 ft. Pipes to the canal leading to the reservoir can be seen in the upper center of Fig. 5.11. Later pumps were of 65,000 Hp. Two could be supplied from each 125 MW generator.

In 1973, two 47 MW reversible pumped-storage units were installed, but difficulties have kept these shut down through 1979. Additional pumped-storage capacity of 214 MW is planned in 1980–1981, as shown in Table 5.5, altogether 308 MW pumped-storage capacity. When not needed for irrigation pumping, these units will provide 308 MW of additional peaking capacity, at very little added expense, since both reservoirs and the powerhouse are already existing.

While the High Dam and its 18 generators were in full operation by 1951, the irrigation and land settlement project, as mentioned, would extend over a much longer period of about 60 yr. Water costs were charged half to revenues from the sale of electric power and half to landowners. This resulted in a charge to the settlers of $5/acre per year for water.

The Grand Coulee Dam was constructed about a mile below the head of the Grand Coulee, near the point where the water had been diverted during the ice age. It is of straight gravity construction, 4000 ft long and 500 ft high above the lowest foundation (333 ft head). It contained over 10,500,000 cubic yards of concrete before the third powerhouse. Its spillway, controlled by drum gates, is 1800 ft long, with a capacity of 1 million cfs. From 1913 until 1934, the discharge of the Columbia River at this point has varied from 17,000 to 492,000 cfs, and the annual runoff has varied from 56,830,000 acre-ft in

360,000 cfs, respectively. The 2 GW plant of the 1950s and 1960s used less than the average flow at full load, with a little over a month of full-load storage. The ultimate 9 GW plant requires 3.3 times the average flow at full load and will have about eight days of full-load storage. The water will be used primarily for peaking.

At the minimum flow of 17,000 cfs, the ultimate plant of 9 GW, requiring 360,000 cfs at full load, could be operated only 1.1 hr a day. However, the storage of 74 GW-days (Table 5.3) could be used for example to provide three additional hours of 9 GW for 66 days of the lowest flow. This simply illustrates the flexibility afforded by the storage. However, optimum use of the plant requires estimates of the future flow and an optimization of the overall system operation.

PUMPED-STORAGE HYDRO ENERGY

It comes as a surprise to many that it should be economical to use fossil fuels or nuclear energy, or even hydro energy, to pump water uphill during the night in order to generate additional power during the daytime peaks. Since part of this energy is lost in the storage cycle, why not generate all energy from fuel when it is needed?

The answer lies in the nature of the average utility load cycle and the options that are available to supply this load. A composite load pattern of all U.S. electric utilities is shown in Fig. 5.12. This may be taken as typical. To supply these loads the utility may choose a mix of the following options (some no longer permitted):

Type of Plant	Capital Cost	Fuel Cost
1. Nuclear plant	Very high	Very low
2. Efficient coal-fired plant	High	Low
3. Gas- or oil-fired (no longer)	Moderate	Higher
4. Gas turbine or diesel plant	Low	High
5. Pumped-storage hydro plant	Moderate	Low
6. Hydro plant	Moderate or high	Zero

1929 to 98,800,000 acre-ft in 1927, the mean being 79,000,000 acre-ft, corresponding to 109,000 cfs average.

At the rated head H of 333 ft, from Eq. 5.3, at 90% efficiency, $P = 0.0762QH$ kW = 25 kW/cfs. Thus, the plant ratings of 2, 6, and 9 GW (millions of kW), correspond to flows of 80,000, 240,000, and

Pumped-Storage Economics

The objective is to determine what combination of these plants will carry the system load at the lowest overall cost. The problem may be illustrated by a simple example.

Suppose there were only three choices as shown

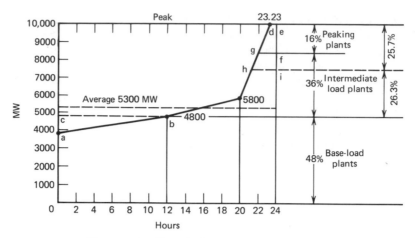

Figure 5.12. Hypothetical load–duration curve.

in Table 5.4, using 1970 costs. Base-load units are of high capital cost and low production cost. Peaking plants are of low capital cost and very high production cost. The intermediate-load plants are intermediate. Let the capital charge be 15%/yr. What combination of these plants would be selected to carry the load of Fig. 5.12?

By trial and error, it is readily found that with the costs of Table 5.4, the load of Fig. 5.12 can be supplied most economically by (1) 16% of peaking plants supplying 1.7% of the energy, (2) 36% of intermediates supplying 12.5% of the energy, (3) 48% of base-load plants supplying 85.8% of the energy.

Note that 16% capacity in peaking plants with a production cost 10 times that of efficient base-load plants is economic if they can be procured at 42% of the capital cost per kW. These plants supply only 1.7% of the energy.

This mix of plants represents the actual proportions required to be added in the United States in the 20 years 1970 to 1990, as determined by an extensive analysis.[6] The single load–duration curve of Fig. 5.12 and the costs of Table 5.4 were selected to be "in the ballpark" and to yield the same mix of

plants, using a 15% capital charge. Thus, they are purely designed to illustrate the principles involved and are only roughly representative of actual cases.

As an example, consider a load of 2 hr duration, point g on Fig. 5.12. For a peaking plant, the capital cost, 15% of $97.5/kW, must be spread over 365 × 2 hr during a year. This results in a capital cost of 2.00¢/kWh. Adding the production cost of 2.00¢/kWh, the total is 4.00¢/kWh.

Similarly, to supply the 2 hr load with an intermediate plant, the capital cost is 3.60¢/kWh, the production cost is 0.40¢/kWh, and the total is 4.00¢/kWh. Two hours is the break-even point. For shorter-duration loads, the peaking plant is cheaper. For longer-duration loads, the intermediate plant is best.

Similarly, it can be established that a base-load plant can supply loads over 12 hr duration cheaper than an intermediate-load plant. With points b and g determined, the optimum mix of plants is fixed, as given earlier.

Note that part of the base-load capacity will be idle for 12 hr a day, as shown on Fig. 5.12, area a–b–c. Note also that the 6000 MWh that could be generated at very low incremental cost by the base-load units, area a–b–c, is more than the 2216 MWh generated by the peaking units, d–e–f–g. Pumped-storage units could use this low-cost energy for peaking and for some intermediate load.

Use of Pumped Storage

The minimim pumped-storage production cost is the base-load plant production cost divided by the

Table 5.4. Typical Generating Station Costs of 1970

	Base Load	Intermediate Load	Peaking Load
Capital cost ($/kW)	234	175	97.5
Production cost (¢/kWh)	0.200	0.400	2.000

turn-around efficiency, that is, 0.200/0.75 = 0.267 ¢/kWh, in this case. There may be transmission costs and some operating cost of the pumped-storage plant. Assume a production cost of 0.300¢/kWh.

With reversible pump turbines and a favorable location, the capital cost of a pumped-storage plant may be less than an intermediate-load plant. Assume for illustration that it is $125/kW, 1970 basis (See Table 5.6). Consider pumped storage as an alternate to the gas turbine peaking plants in the problem.

With these costs, pumped storage is more economic than gas turbines. For example, a 1 hr load would cost 6.00¢/kWh with gas turbines and 5.44¢/kWh with pumped storage.

With both capital and production costs of pumped storage less than for intermediate-load plants, the pumped-storage plant is more economical than the intermediate-load plant for *all* loads. However, there is a limit to the amount of pumped-storage energy available at this low cost. Without increasing the base-load capacity, no more than 6000 MWh × 0.75 = 4500 MWh can be supplied at this low cost. This would dictate 25.7% pumped-storage capacity, supplying the 4500 MWh area d–e–i–h on Fig. 5.12, thus reducing the intermediate load capacity to 26.3% supplying 10.7% of the energy.

While it is not possible to pursue this further, with such a simplified model, additional low-cost base-load energy from weekends and from spare capacity needed for emergencies can supply much of the remaining 10.7% of intermediate-load energy most economically by using the hydro generation over a longer period.

Thus, while early pumped-storage plants for daily cycling in the 1960s were designed with 4.5 to 6 hr duty cycles, most of the later plants in the 1970s have been built with 8.5 to 10 hours of full-load pondage. Some have still larger pondage to store energy over each weekend for use during the week. For example, Muddy Run, 800 MW, largest in the world when completed in 1968, has 14.8 hr of full-load pondage. Twelve hours of pumping and 12 hr of generation is not uncommon. The hydro generation is increased with the rapid increase of morning load, and vice versa, thereby lessening the shock on the thermal machines. Thus, pumped storage may supply a substantial part of what has traditionally been termed "intermediate load" as well as peaking load.

Advantages of Pumped Storage

A number of the advantages of pumped storage have already been mentioned. However, the remarkable growth in this form of power supply from 1960 to 1971, from 78 to 5000 MW in 16 installations in the United States and from 180 to 6200 MW in 31 installations in Europe, So. America, and Japan, can only be explained by a variety of reasons. These include the following:

1. Most economic generation of peaking energy and part of the intermediate-load energy.
2. Most acceptable environmentally. No objectionable emissions during peak hours (or any other time).
3. High reliability of hydro units.
4. Low operating expense; simple operation.
5. Low maintenance expense; long life.
6. Keeps efficient fossil fuel-fired units, particularly supercritical units, and nuclear units operating at highest efficiency, near full load, around the clock.
7. Reduces maintenance and forced outages on these machines by steady full-load operation.
8. Upon loss of a large unit, the pumping load can be dropped immediately, providing the equivalent of spinning reserve. In fact, the pump-generators can be reversed and generate full load, thus providing double their capacity in (almost) spinning reserve until other units can take over.
9. They can be started quickly from standstill and thereby improve the system reliability.
10. They provide a new dimension in flexibility afforded to the system operator in rapid load pickup and reduction. The rapid changes can be taken first on the hydro units and then shifted at a more moderate rate to the preferred units.
11. They can provide a substantial block of power from cold start in the event of a major system outage.
12. A falls cannot generally be dammed to provide storage. But pumped storage to a nearby pool can save some of the water for later use as with a dam. At Niagara Falls,

this is the most economic way of using the extra water available at night, after tourist hours.

Each of these benefits provided by pumped storage has been an important factor in one or more of the applications. Thus, knowledgeable system planners view pumped storage as providing intangible benefits of great value in system operation and reliability, in addition to their direct economic benefits in displacing higher cost energy.

For all these reasons, pumped-storage hydro experienced a remarkable growth in the 1960s and 1970s. In the late 1970s, the rapidly escalating cost of fuel oil made pumped storage more attractive compared with combustion turbines for peaking. By 1979, there were 13.3 GW of pumped storage in the United States, a 166% increase in eight years, and a rapid continuing growth can be expected as the technology develops.

Pumped-Storage Hydro Installations in the United States

The rapid growth of pumped-storage hydro capacity in the United States is shown in Fig. 5.1. The installations accounting for this growth are detailed in Table 5.5, which shows the 31 U.S. installations through 1979, 10 projected installations, and the controversial Cornwall project on the Hudson River. The MW values given are all nameplate ratings, and many have greater capability than shown. For example, the Muddy Run capacity, It.11, is given as 880 MW by Philadelphia Electric. For this reason, small differences are found in various tables and references.

The MW ratings shown are those installed initially plus any additions up to 1979. As noted, a few of the stations included in pumped-storage tables are listed as "hydro" in the DOE inventory of generating capacity.[5] There is no sharp dividing line, since even if a unit is reversible, it might be used mainly as a conventional hydro unit if the river flow is sufficient. In Fig. 5.1 the total hydro capacity is given, and the capacity of reversible pumped-storage units included in the figure is shown, regardless of how they are used.

Thus, the installed capacity of pumped-storage hydro in the United States through 1979 (nameplate rating), as given in Table 5.5, was 13,316 MW, with projected increases shown in the table that would bring this to 23,625 MW by 1990. Numerous other projects are in the early stages of license application and study.

Note the increase in unit ratings to 383 MW, It.27, Raccoon Mountain, with 400 MW, It.37, Oak Creek, projected in 1985. Several installations use single-stage Francis-type pump turbines at heads above 1000 ft, for example, 1226 ft, It.9, Cabin Creek, and over 1600 ft at Helms and Montezuma, Its.33 and 35, projected in 1981 and 1982. Reversible, modified Francis-type pump turbines can be developed for at least 2000 ft head.[4]

Typical Installation—Muddy Run

Pumped storage is still in its infancy, most of the development throughout the world having occurred in the last two decades. New arrangements are continually being devised, and new milestones in machine capability occur every few years. To give a picture of the pumped-storage development as of 1980, we will describe a typical large installation, Muddy Run, cite the many variations in configuration that have proved best for various installations, and summarize the environmental issues that have been raised and the measures taken to make these plants environmentally acceptable.

An overview of the Muddy Run pumped-storage hydro project is given in Fig. 5.13. It is located 12 miles upstream from the Conowingo hydro plant described earlier. The Conowingo pool extends 14 miles up the Susquehanna River to the Holtwood plant. Thus, the lower reservoir for Muddy Run is the Conowingo pool, shown in the foreground.

At this point, the river level lies 400 ft below the surrounding Piedmont Plateau. The upper reservoir, about 1000 acres, shown in the center, was formed by damming Muddy Run, which ran in a ravine parallel to the river and about ½ mile inshore from the plant. The dam (not visible) has a maximum height of 250 ft from the bottom of the ravine and a length of 4400 ft. A canal about a half-mile long connects the upper pool to the intake structures as shown. When full, the water level is 410 ft above the lower pool. A drawdown of 50 ft to a 360 ft head constitutes the working storage of the installation.

From the intake structures, four shafts 24½ ft in diameter extend vertically downward and then horizontally toward the plant. The shafts subsequently divide into eight tunnels, each 14 ft in diameter, which connect with the eight 100 MW reversible-pump turbines in the station.

Figure 5.13. Aerial view of the Muddy Run pumped-storage hydro development. Courtesy Philadelphia Electric Co.

The usable volume of the reservoir, corresponding to 50 ft drawdown, is 35,500 acre-ft (total volume 60,500 acre ft). At a nominal head of 400 ft, 35,500 acre-ft of water has a potential energy of 14,560 MWh. However, due to the variable head and various hydraulic and electrical losses, the energy available in the 50 ft drawdown is 11,800 MWh, or enough to run the plant at rated load, 800 MW, for 14.75 hr. This is expressed as 118 unit-hr and is so used by the dispatchers. The use of this capacity for daily and weekly storage is described later.

The Pump Turbine Units. These are 180 rpm modified Francis turbines, as shown in Fig. 5.3, with fewer blades and vanes than would be used for turbining only. Each machine is rated 138,000 Hp at 353 ft head on generator mode and 150,000 Hp at 427 ft head on pump mode, discharging 2610 cfs. They are set 20 ft below minimum tail water. The runner and shaft weigh 80 tons, and the runner is 215 in. in diameter.

The Generator–Motors. These are rated 100,000 kW (nameplate), 13.8 kV. Each rotor weighs 230 tons. The machines are arranged for reduced-voltage starting at 6600 V. Each two machines connect to a 13.8 kV to 231 kV transformer, thence through 220 kV transmission lines to a large nuclear plant (Peach Bottom, over 2000 MW) on the Philadelphia Electric system, 4 miles distant.

Wicket Gates. These, as shown in Fig. 5.3, control the flow of water.

Recreation Lake. The 50 ft weekly fluctuation of the upper pool level precludes its use for recreation. However, a 100 acre, constant-level recreational lake with associated park facilities was provided by an additional dam at the eastern end of the reservoir. Also, fisherman parking and access were provided along the river at the plant. These features augment the boating, fishing, and picnicking facilities of Conowingo Lake, which since 1928 has been a fine recreation spot, accessible to Philadelphia, Baltimore, and Washington as well as to nearby residents.

Cost. The capital costs of the Muddy Run plant are given in Table 5.6, along with four other pumped-storage plants for comparison. These were all installed between 1965 and 1973. The capital costs vary from $41/kW at Yards Creek (1965) to $156/kW at Luddington (1973), with Muddy Run (1967) being intermediate, at $102/kW. Note that for three of the installations, including Muddy Run, the two largest and nearly equal cost items are (1) reservoirs, dams, and waterways; (2) equipment.

For Seneca (Kinzua), the reservoir item is 57%, and for Luddington, 66% of the total project, and accounts for their higher costs.

Operation. The Muddy Run plant is designed for automatic, unattended operation, although nor-

Table 5.5. Pumped-Storage Hydro Installations in the United States, 1979[4,5,7,a]

No.	Plant	Year in Service	MW Nameplate	No of Units	MW per Unit	Head ft	State	River	Remarks
1.	Rocky River	1929	32	1	32		Conn.	Housatonic	7 MW in 1919
2.	Flat Iron	1954	9	1	9	290	Colo.	Big Thompson	[b]
3.	Hiwassee	1956	60	2		246	N.C.	Hiwassee	[b], pumped[7]
4.	Lewiston	1961	240	12	20	75	N.Y.	Niagara	
5.	Taum Sauk	1963	408	2	204	790	Mo.	E. Fork Black	
6.	Smith Mt.	1965	547	5	[c]	185	Va.	Roanoke	
7.	Yards Creek	1965	387	3	[d]	700	N.J.	Yards Creek	
8.	Senator Wash	1966	6	1	6		Calif.	Colorado	[b]
9.	Cabin Creek	1967	300	2	150	1226	Colo.	Clear Creek	
10.	O'Neill	1967	25	6	4	50	Calif.	San Luis Creek	[b]
11.	Muddy Run	1967	800	8	100	400	Pa.	Susquehanna	
12.	San Luis	1968	424	8	53	300	Calif.	San Luis Creek	[b]
13.	Salina	1968	258	6	43	246	Okla.	Salina Creek	
14.	Edward Hyatt (Oroville)	1968	196	2	98	675	Calif.	Feather	+ 4 × 117 MW[b]
15.	Thermolito	1968	84	3	28	102	Calif.	Feather	
16.	Seneca (Kinzua)	1970	422	2 / 1	198 / 26[f]	798[e]	Pa.	Allegheny	+ 33 MW[b]
17.	Mormon Flat	1971	49	1	49	133	Ariz.	Salt	
18.	Northfield Mountain	1972	846	4	212	744	Mass	Connecticut	1000MW[7]
19.	Castiac	1972	1275	6	212	1018	Calif.	L.A. Aqueduct	
20.	Horse Mesa	1972	87	1	87	266	Ariz.	Salt	
21.	Blenheim-Gilboa	1973	1000	4	250	1085	N.Y.	Scholarie Creek	
22.	Jocassee	1973	610	4	152	293	S.C.	Keowee	
23.	Luddington	1973	1980	6	330	361	Mich.	Lake Michigan	

104

No.	Name	Year	MW	No. of units	MW/unit	ft head	State	River	Remarks
24.	Grand Coulee	1973	94	2	47	272	Wash.	Columbia	Shut down + 214 MW projected 1980, 1981
25.	Bear Swamp	1974	600	2	300	770	Mass.	Deerfield	
26.	Carters	1977	250	2	125	344	Ga.	Coosawattee	
27.	Raccoon Mt.	1978	1530	4	383	1040	Tenn.	Tennessee	
28.	Fairfield County	1978	512	8	64	169	S.C.	Broad	
29.	Truman	1979	81	3	27		Mo.		+ 81 MW proj. 1980
30.	Wallace	1979	104	2	52	983	Ga.		+ 104 MW proj. 1980
31.	Mt. Elbert	1979	100	1	100	406	Colo.		+ 100 MW proj. 1980
	Total 1979[a]		13,322						
32.	Clarence Cannon		31	1	31	57	Mo.		proj. 1981
33.	Helms		1053	3	351	1624	Calif.		proj. 1981
34.	Bath County[g]		2100	6	350	1079	Va.		proj. 1982
35.	Montezuma		500	2	250	over 1600	Ariz.		proj. 1982
36.	Rocky Mountain		675	3	225		Ga.		proj. 1983
37.	Oak Creek		800	2	400		Colo.		proj. 1985
38.	Davis		1000	4	250		W.Va.		proj. 1986
39.	Prattsville		1000				N.Y.		proj. 1988
40.	Stony Creek		1655	2	800,855		Pa.		proj. 1989
41.	Boyd County		1000	6	167		Neb.		proj. 1990
	Projected 1980–1990		10,309						
42.	Cornwall		1800		1050		N.Y.		Hudson	Proj. indefinite.

[a] Including units projected for 1979.[5]

[b] Listed as "Hydro" in DOE Inventory of Power Plants 1979.[5]

[c] 2 × 66 MW + 1 × 115 MW + 2 × 150 MW.

[d] 2 × 138 MW + 1 × 112 MW.

[e] 665 ft when discharging into the Allegheny Reservoir, 798 ft when discharging into the Allegheny River below the dam (Ref. 7, p. 231).

[f] Conventional hydro unit.

[g] From Ref. 7: 457 MW/unit, 1260 ft head (p. 291); 357 MW/unit, 1079 ft head (p. 298).

mally there is an operator on duty and some work going on in the plant. The plant is actually operated remotely from Conowingo, in accordance with instructions from the power pool dispatcher.

Pumped Storage Operation in a Power Pool. Muddy Run is one of three pumped-storage plants operated as part of the Pennsylvania–New Jersey–Maryland (PJM) power pool which includes 11 operating companies (six members). Together, these companies had an installed generating capacity of 37,323 MW (1974) and supplied 21 million people in the Middle Atlantic States, representing about one tenth of the population of the United States. The entire interconnection is operated for maximum economy as a single system, and economies from this operation, as compared with separate operation, are shared equitably. All units are dispatched by the power pool dispatcher, who determines which of the available units to run and how much each unit should generate or pump at any given time.

The three pumped-storage installations on this system, Yards Creek, Muddy Run, and Seneca (Kinzua) (20% to PJM), are described in Tables 5.5 and 5.6. Altogether, the mix of units on the PJM system in 1974 was:

	MW	Percent
Fossil steam	25,271	67.7
Run-of-river hydro	941	2.5
Pumped-storage hydro	1,286	3.5
Nuclear	1,711	4.6
Combustion turbine and diesel	8,114	21.7
Total	37,327	100.0

For pumped-storage units, the incremental production cost is mainly the cost of pumping water to the upper reservoir. The pump/generation ratios are shown in Table 5.6. For Muddy Run, 1.6 times as many MWh are expended in pumping the water up as are realized later in generation. This energy therefore has a cost 1.6 times the incremental cost of the energy actually used for pumping.

In this respect, pumped-storage differs from run-of-river hydro, for which the water is free. Run-of-river hydro, with daily and some weekly storage, is dispatched very simply by assigning an arbitrary cost that results in just using up the available water while displacing the most expensive alternate energy. If pumped storage is limited, it is dispatched in the same way, using an arbitrary cost above the actual incremental cost. This is the "limited energy mode."

However, if during high river flow, for example, energy can be supplied cheaper by other sources (unlimited energy mode), then the actual incremental production cost of pumped storage is used, and it comes on only when it has the lowest incremental cost. In this mode, it does not use the full storage capacity.

For simplicity, the dispatch has been described neglecting transmission losses. Actually, these losses are usually taken into account by a "transmission loss formula" or an "economic dispatch computer."[8]

At Muddy Run, the storage is normally fully used each week (limited energy mode), and this results in a reservoir level pattern as shown in Fig. 5.14. Starting Monday morning with a full pool, it is drawn down further each day, then partially refilled at night, until it reaches minimum on Friday evening. It is then refilled over the weekend to repeat the cycle.

This description has been highly oversimplified in order to explain the general principles involved. Daily, weekly, and seasonal forecasts are necessary to determine the amount best to pump and use each day. Some variations from strict economy considerations are necessary to secure the many other advantages described earlier. However, the actual resulting pattern is usually close to that shown in Fig. 5.14.

Other Pumped-Storage Arrangements

The Muddy Run plant is typical of the majority of installations up to the present (1980). An existing body of water forms the lower pool. A specially constructed reservoir at a higher elevation forms the upper pool. All pumping and generation is between these two pools. The chief benefits are the supply of low-cost peaking and intermediate energy and keeping the base load units running efficiently near full load.

However, a number of different arrangements have been used or proposed, and the prime purposes and benefits are different. Several of these alternate situations and arrangements will be outlined briefly. There is room for much innovation in the use of pumped-storage in the future.

Table 5.6. Typical Pumped-Storage Costs[4]

Installation	Yards Creek	Muddy Run	Luddington	Northfield Mountain	Seneca (Kinzua)
Capacity (MW)	387	800	1980	846	422[5]
Year installed	1965	1967	1973	1972	1970
Head (ft)	700	400	330	212	813
Pump/generation ratio	1.5	1.6			1.4
	Capital Costs (Thousands of Dollars)				
Land and land rights	448	1,401	5,294	1,321	—
Structures and improvements	2,008	14,270	34,700	19,735	7,489
Reservoirs, dams, and waterways	6,753	34,279	204,928	49,989	29,647
Equipment costs	6,475	30,666	59,537	46,002	13,191
Roads, railroads, and bridges	215	998	3,398	4,825	—
Total cost	15,889	81,614	307,857	121,872	51,887
Cost (in $/kW)	41	102	156	153	123

Salt River Dams—Mormon Flat and Horse Mesa

The Salt River valley above Phoenix, Arizona, has a long history of irrigation. From 300 B.C. until about A.D. 1400, the Hohokum Indians irrigated the land on either side of the river by a system of canals. For 1700 years they converted about 140,000 acres of desert into cropland that supported a population of about 2 million people. The Hohokums ("the people who went away") then disappeared. The first successful irrigation in recent times came with the Federal Reclamation Act of 1902, which made possible the Roosevelt Dam shown in Fig. 5.15. This was followed by five other dams as shown, which now constitute the irrigation and water supply system for the Salt River Valley. Hydro units were installed at Roosevelt Dam and at the three downstream power dams, Horse Mesa, Mormon Flat, and Stewart Mountain.

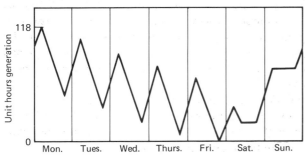

Figure 5.14. Average weekly pumping/generating pattern for Muddy Run.

As the population and load grew, water supply requirements took precedence over all other considerations. This seriously restricted the use of existing hydrogeneration and would do so even more with new larger units. In winter, when very little or no water release was needed, the hydrogenerating capacity was effectively "locked up."

The introduction of reversible-pump turbines at Mormon Flat and Horse Mesa largely removed these restrictions, since water could be used for peaking or spinning reserve without affecting its storage for irrigation and water supply. All hydro generation at these two stations then became spinning reserve when not running on needed irrigation water, and the pumped storage supplied some peaking energy. While the amount of highly efficient base-load capacity for conversion to peaking was limited, it would increase as the generation grew and become more valuable with increasing oil costs. The chief advantages of pumped storage at Mormon Flat and Horse Mesa are cited as:

1. Levelize thermal unit daily load pattern.
2. Reduce production cost through oil displacement.
3. Firm up summer generating capacity during low water-release periods.
4. Unlock winter generating capacity.

However, the principal year-round benefit has been spinning reserve.

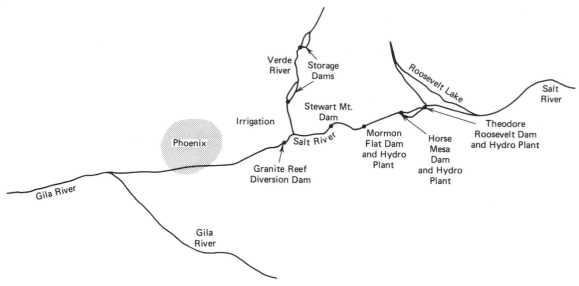

Figure 5.15. Dams in the Salt River valley.

At both Mormon Flat and Horse Mesa, the upper and lower pools for pumped storage were already in existence, the lower pool being that of the downriver dam. In both cases, a new powerhouse for the pumped-storage unit was placed a short distance below the dam, where the pump could draw water from the lower pool. The penstock penetrates the dam, the shut-off control being a wheeled gate on the upstream face of the existing dam, operated by a hydraulic hoist.

Lewiston (Tuscarora) Pumped Storage

The Lewiston pumped storage has the prime purpose of preserving the scenic beauty of Niagara Falls. The extra water available at night after tourist hours is saved until the next day. However, the 310 ft head of the Robert Moses plant is four times that of the pumped storage (75 ft). Thus, the energy stored for the next day is roughly five times that required to store it. Other than this unusual feature, the arrangement is typical, with the upper Niagara River forming the lower pool and a specially constructed upper reservoir.

Seneca (Kinzua) Pumped Storage

The Seneca pumped storage is tailored to match the operation of the Allegheny Flood Control Reservoir on the upper Allegheny River that forms its lower pool. The 106 acre upper reservoir, 665 ft above the Allegheny Reservoir, is atop Seneca Mountain. As

shown in Table 5.5, No. 16, this plant has two 198 MW pump turbines and one 26 MW conventional turbine. The smaller unit operates at a head of 798 ft between the upper reservoir and the river below the dam. During much of the year, this provides the prescribed minimum river flow and takes advantage of the larger head. The two large units are used to fill the upper pool from the Allegheny Reservoir. Unit No. 1 discharges to the Allegheny Reservoir when generating, with a head of 665 ft. However, unit No. 2 has a divided draft tube and may discharge either into the Allegheny Reservoir, head 665 ft, or into the lower river, at a head of 798 ft. Thus, during parts of the year when large amounts of water are being discharged down the river, advantage can be taken of the higher head to generate additional power.

Underground Pumped Storage (UGPS)

Newest and most promising environmentally of the pumped-storage alternatives is underground pumped storage. The lower reservoir(s) and the plant(s) are placed underground. This concept has received considerable study[7] and appears competitive with other sources of peaking energy, $242 to 300/kW (1974 through 1980), though higher than conventional pumped storage, $41 to 156/kW (1965 through 1973) (Table 5.6). One such installation was reportedly underway in Finland in 1972.

The conclusions of a conference devoted to pumped storage in 1974 were as follows[7]:

1. Present-day know-how in engineering and construction is adequate to design and build successfully a pumped-storage project with an underground reservoir.

2. Equipment for such plants is currently available.

3. Project costs may be competitive with alternative means of peaking capacity.

4. Site environmental problems would generally be less than those encountered in building (above ground) projects.

5. This type of development allows considerable flexibility in plant location and operation.

As shown in Table 3.6 and Fig. 3.2, EPRI expects the underground pumped hydro development to be far enough along to be a viable commercial option by 1991.

High heads and small reservoir volumes are needed to minimize the costs of both excavations and equipment. However, the lower cost of reversible-pump turbines generally dictates that these ends be obtained by the use of two stages, of about 2000 ft each, or 4000 ft total. Each stage would have reversible Francis-type pump turbines, with an intermediate-level balancing reservoir of small capacity at the 2000 ft depth. Three or more such stages could of course be used, but economic studies to date favor two stages.[7]

This basic arrangement is shown in Fig. 5.16. The intermediate balancing reservoir may be sufficient to allow for 1 or 2 unit-hr mismatch between the operation of the upper and lower power plants. The pumps must have adequate submergence at low reservoir and so must be located somewhat below their respective reservoirs, as shown in Fig. 5.16.

The power plants of several recent pumped-storage installations have been underground, for example, Northfield Mountain, No. 18, and Raccoon Mountain, No. 27, of Table 5.5. Location of one of the reservoirs underground as well has obvious environmental advantages:

1. One less reservoir above ground. The more objectionable of the two above-ground reservoirs can be eliminated. When one above-ground reservoir already exists, placing the second one underground reduces environmental problems by an order of magnitude.

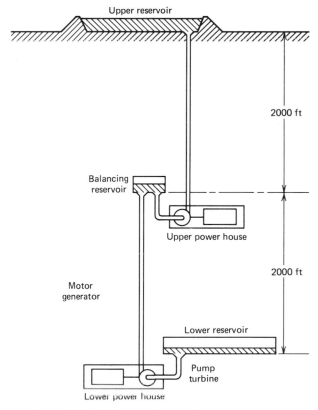

Figure 5.16. Underground pumped-storage concept.

2. With the high head possible, regardless of topography, both reservoirs can be smaller.

3. Eliminating the topography factor greatly increases the flexibility of site selection.

Power Parks. Large power concentrations serving groups of utilities have been proposed for protection, environmental isolation, and other reasons. Certain combinations involving underground pumped storage fit in well with these ideas. Two of these are the nuclear/UGPS combination and the hydro/UGPS combination. The UGPS provides load leveling for the nuclear plant. Its upper reservoir provides the cooling pond, and both share the transmission facilities. Similarly, for a hydro plant with limited storage, an associated UGPS saves or reuses the water and greatly expands the peaking capacity. Examples can be found at Niagara Falls and Grand Coulee. A nuclear or highly efficient coal-fired base-load plant added to this complex has the same desirable attributes as described earlier.

Siting Factors. While siting is much more flexible with UGPS, there are some new factors to be considered. There must be rock strata of adequate "competency" at the required depths so that tunnel labyrinths can be mined for the reservoir without supports. A competent stratum of some 250 ft in thickness would normally be required to house the lower reservoir and powerhouse. Its structure must be such that there is little leakage. The intermediate level has similar requirements, and the intervening layers must not present undue problems in the sinking of shafts. Several shafts are required for an installation.

Environmental Coordination

A key ingredient in any large hydro project or pumped-storage installation today is its environmental coordination. Involving as it does two sizable reservoirs—a power plant and transmission lines—the usual above-ground pumped-storage installation is highly visable and almost certain to raise environmental objections. Just as certainly it offers benefits of lower cost and more reliable electric service to the community. Like all industry, it provides employment for some and, if properly directed, may enhance the recreational facilities for the whole community.

The features of a pumped storage which may be environmentally objectionable include:

1. *The Change in Land Use and Appearance.* This includes the placing of a reservoir where once was dry land. It includes the appearance of a power plant, of reservoirs, dams, dikes, hydraulic facilities, electric switching stations, and transmission lines where once was an undisturbed "natural" countryside.
2. *The Effect on the Ecology of These Facilities.* This involves the removal of several square miles of habitat from former nature dwellers (habitat loss). It includes the warming or cooling of certain waters, the changing level of formerly fixed-level pools, the pumping of fish and microorganisms through turbines, back and forth. It includes the mixing of upper and lower strata of previously unmixed waters and changing the downstream flow of rivers (collectively, habitat alteration).
3. *Unknowns and Fears.* Many human activities in the past have had unpredicted, and sometimes disastrous, effects the nature of which became known only after it was too late.

Thus, some fear the effect of any large-scale change in an area. The possibility of reservoir leakage, of fracture of dams, of groundwater alteration, or of seismic effects fall in this category.

These factors are all considered in any new pumped-storage project,[7,9] and detailed studies are required as part of the licensing procedure. The environmental measures taken include extensive concealment to lower visual impact. Underground power plants, reservoirs, and transmission facilities are hidden by natural topography and trees and by plantings. At Raccoon Mountain in the scenic "Grand Canyon of the South," the lower reservoir is the natural Tennessee River, and the other facilities are almost completely concealed.[7] Rather than generalize, the measures taken at one major installation will be described.

Luddington Pumped Storage (Table 5.5, No. 23).[9] The upper reservoir, 330 ft above the shore of Lake Michigan, displaced about 1400 acres of orchards and pasture land and reduced the habitat for rodents, deer, wildfowl, and so on, by that amount. The reservoir is encircled by an earth embankment 6 miles long. Environmental measures include the following:

1. The visual impact is softened by judicious reforestation and landscaping and by extensive grass planting. Also, selected roadside plantings beautify all highway approaches to the project. This results in a pleasing appearance to passing motorists and those visiting the project.
2. Transmission lines are designed and routed for aesthetic reasons, taking advantage of natural screening as much as possible. All wooded areas along the route are preserved and augmented where necessary by roadside plantings. In general, lines are below the crests and on the far sides of hills from major roads.
3. Where lines cross rivers, towers are set well back to minimize visual impact for fishermen and boaters.
4. As part of Michigan's autumn color tour area, reservoir embankments and penstock slopes are landscaped to harmonize with the surrounding terrain in key areas.

5. The beach area is narrow, rock-strewn, and inaccessible due to cliffs. It is not well suited to public use. This is unchanged.

6. The fluctuating upper pool level results in an uniced surface most of the winter, making it attractive to migrant ducks and geese and to local waterfowl.

7. While the area was farmland and not particularly recreational, some facilities are provided which augment rather than compete with existing recreational facilities in the area. These include: (a) Vista Point with views of construction and of the shoreline of Lake Michigan and a Visitors Center with facilities and an auditorium for audio-visual presentations; (b) a scenic overlook atop the dike, including a well-equipped picnic area, overnight camping/trailer area, and playgrounds; the Overlook, 370 ft above Lake Michigan, affords a 30 mile view of the lake shore: (c) reserved natural areas for future conversion to recreational use.

8. The effect on Lake Michigan is, of course, minimal. Some small fish in the immediate area will pass through the pump turbines in both modes, a small percentage being injured or killed. Jetties, breakwaters, and changes in water velocity, temperature, and dissolved oxygen will change the habitat and consequently fish and fish food behavior in the area of the plant. Studies, including regular sampling of fish and water properties, started three years before plant operation. These studies continue in order to quantize the effects and determine whether any corrective measures are necessary to protect the fish in Lake Michigan.

TIDAL ENERGY

Tidal mills were in common use as far back as the tenth and eleventh centuries in England, France, and other countries. They operated the same as other water mills, except that the water for the millpond was caught in a small tidal inlet at high tide. The gates were then closed, and at low tide the mill could be operated for several hours. Then the process was repeated. Since tides occur 50 minutes later each day, the miller had to adjust his hours accordingly. However, since there are two tides each day, there was always one period suitable for mill operation during the daylight hours.

Some of the later tidal mills were quite large. A mill in Rhode Island in the eighteenth century had a 20 ton wheel 11 ft in diameter and 26 ft wide. A large tidal mill in Hamburg, Germany, in 1880 was used for pumping sewage. However, with the advent of more convenient power sources, most such mills disappeared by the end of the nineteenth century. A few remain. Slade's Mill in Chelsea, Mass., 100 Hp, founded in 1734, was still operable up to recent years. It is now a historic site. A mill at the Deben estuary in England has been operating since 1170.

In modern times, a great deal of consideration has been given to tidal power on a large scale for the generation of electricity. Instead of small inlets, large tidal estuaries would be dammed with entrance gates that could be closed at high tide. The resulting head at low tide would provide several hours of hydroelectric generation.

Tidal Pool Arrangements

Several different schemes have been employed or proposed for extracting energy from the tides:

1. *Single-Pool Ebb Tide.* This is the scheme used in the tidal mills of the past and is also the basis of some modern designs for electric power generation.

2. *Single-Pool Flood Tide.* Here, the gates are closed at low tide, and power is generated at high tide while water is flowing into the pool. This plan is not generally considered alone since the average head is greater with ebb-tide generation. The pool generally has a greater area when full, and hence more water is available at the highest head with ebb-tide generation.

3. *Single-Pool Two-Way Operation.* With the adjustable-blade, reversible-pump turbine, shown in Fig. 5.17, power can be generated by the flow of water in either direction. The small-diameter generator is completely enclosed and immersed, leading to the name "bulb-type" turbine generator. The slant-axis turbine generator shown in Fig. 5.18 can be used in the same way. Some greater energy, though not double, can be obtained by the two-way operation. In addition, when the pool is nearly full, additional water can be

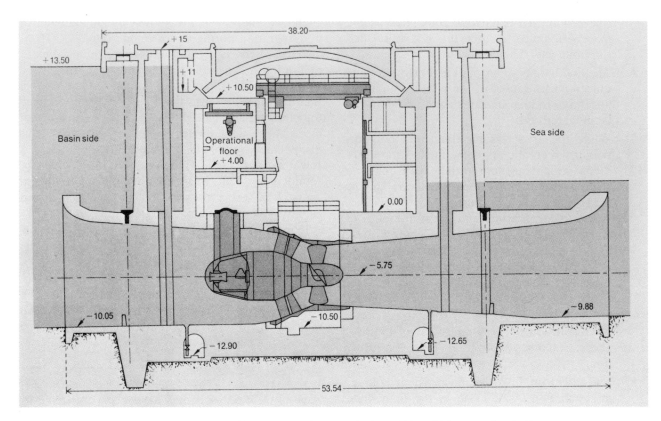

Figure 5.17. Bulb-type turbine generator in use at the Rance River tidal power plant in France. Courtesy *IEEE Spectrum.*[2]

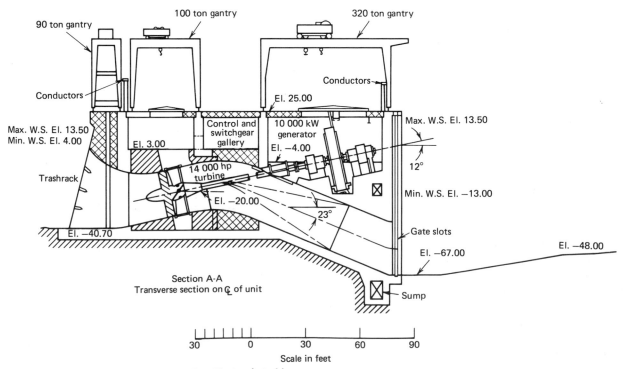

Figure 5.18. Slant-axis turbine generator. Courtesy *IEEE Spectrum.*[10]

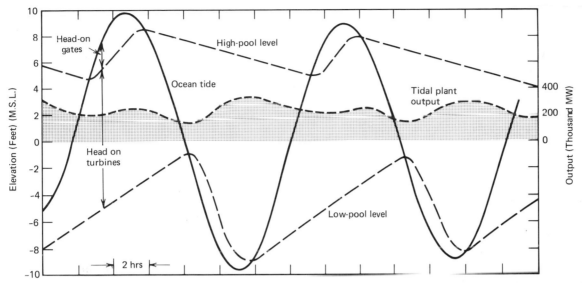

Figure 5.19. Output curves for a two-pool (high–low) tidal plant operation. Courtesy *IEEE Spectrum*.[10]

pumped into it using an external source of power. This water pumped in at a few feet of head later falls through a much larger head in generating energy and returns several times the energy required for pumping. The blades can also be feathered to use the duct as a sluiceway or gate. The considerable flexibility afforded by this arrangement makes it possible to shift much of the generation to peak periods, when energy is more valuable. Thus, the system can be operated either for maximum economy or for maximum energy.

4. *Two-Pool Ebb- and Flood-Tide System.* The two ebb tides in a day each last for about 6 hr, and the two generating periods are of this order. For a single-pool ebb-tide arrangement, no energy is generated for about half of the day. However, if a nearby estuary is operated as a flood-tide system, it produces energy while the other is idle. The energy supply is still not continuous but is more nearly so. It would take less storage to "firm it."

5. *Two-Pool One-Way System (High and Low Pools).* In order to have some continuous energy, the two-pool system was devised. It requires two pools physically adjacent, a "high pool" and a "low pool." The high pool is filled each high tide. The low pool is emptied each low tide. Water flows from the high pool to the low pool through the tur-

bines, generating some power continuously, as shown in Fig. 5.19. The power still fluctuates as shown, and there is also a large variation from neap to spring tides, to be described later.

Low-Head Hydro Generators

Even the highest-head tidal plants are less than 50 ft head, which is low compared with most hydro plants on rivers. As mentioned, the bulb-type and slant axis configurations shown in Figs. 5.17 and 5.18 are suitable for these low heads. They are also used in low-head river hydro plants. For example, the bulb-type machine is used at the Rock Island hydro plant on the Columbia River (No. 30 of Table 5.3), 49 ft head. However, most low-head hydro plants use the vertical-axis Kaplan turbine (Fig. 5.4). A straight-flow turbine, to be described later, will be used at the Annapolis River tidal plant in Nova Scotia.

The Tides

The maximum tidal ranges do not occur every day. Tides are caused by the attraction of the moon and the sun on the waters of the earth and on the earth itself. The moon has the greater differential pull. It draws the water away from the earth, piling it up (about 3 ft) on the side toward the moon. It also draws the earth away from the water, piling water up on the other side, almost the same amount. Thus, there are two low tides and two high tides each lunar day.

The sun acts in the same way but has a lesser tide. When the sun and moon are pulling together, or opposite, (new moon or full moon), their tides are in phase. This results in a high tidal range or "spring tide." About seven days later, when the sun and moon are pulling at right angles (half moon), their tides tend to cancel each other, and the tidal range is small (neap tide).

At the spring and fall equinoxes, when sun, moon, and earth can be exactly in line, the spring tides are a little higher. Neap tides are about one third of the tidal range of average spring tides. At the Rance River tidal power plant in France, the energy generated during neap tides is only 80,000 kWh/day, compared with 1,450,000 kWh/day during an equinoctal spring tide, a ratio of 18:1.

Potential Tidal Sites

The required conditions for economic tidal power generation include a high tidal variation, an estuary pool of narrow enough mouth so that it can be dammed and deep enough so that turbines can be set below the low tide. The problems of silting and adverse effects on the ecology (fishing) and on navigation must all be within acceptable bounds, and the measures required thereby must be included in the costs.

There are at most a dozen locations on the earth where all of these requirements *may* be met. The question of economics cannot be resolved until each site is studied. The principal potential or currently used sites are:

1. The mouth of the Rance River near Saint-Malo, on the northwest coast of Brittany, France. Here, a 240 MW plant with 24/10 MW, bulb-type units, as shown in Fig. 5.17, has been in operation since 1966. It produces electricity cheaper than oil, coal, or nuclear plants in France.[2,10] The average head is 28 ft.

2. Passamaquoddy[10,11] on the lower Bay of Fundy, between Maine, United States, and New Brunswick, Canada (see Fig. 5.20). Here, the average tidal range is 18.1 ft (5.5 m). Both one-pool and two-pool arrangements have been studied. An International Commission conducted extensive studies in

1956 through 1961 and found the project not viable under existing economic conditions. Independent U.S. studies were made in 1962–1963, but as yet no arrangement has proved economic. The project is deferred until lower cost sources available to the area are exhausted.

3. Hopewell, Canada, on the upper Bay of Fundy, with an average tidal range of 33 ft (10 m) (see Fig. 5.20). While believed by Canadian Engineers to be a better site than Passamaquoddy, energy costs would still be higher than other sources available to that area. Thus, this site is not being developed at present (1980). Several other sites in the Bay of Fundy have been studied, as indicated in Fig. 5.20 and Table 5.7.

4. Annapolis Royal, at the mouth of the Annapolis River in Nova Scotia (Fig. 5.20). Construction started in 1980 on the first tidal electric plant in North America.[12] The demonstration plant, which is expected to start operation in 1983, will utilize a 20,000 kW, straight-flow horizontal axis machine in which the generator surrounds the 23 ft diameter turbine. Smaller straight-flow machines have been used for some years in West Germany. The tidal ranges cited, 29 ft maximum, 15 ft minimum, are approximately the same as listed at site 7, Table 5.7. The dike and two 33 ft wide gates already exist, having been built in 1963 to protect 4300 acres of farmland along the Annapolis River from flooding by high tides. The powerhouse will be constructed in the existing dike. The project is being carried out by the provincially owned Tidal Power Corp.; $25 million of the estimated cost, $43 to $47 million, is being provided by the Canadian government because of its interest in the turbine's potential for hydroelectric projects elsewhere in Canada.

5. An experimental site at Kislaya Guba, north of Murmansk, on the Barents Sea in the Soviet Union. Here, a French 400 kW unit has been in operation since 1968. A completely assembled plant, with all hydraulic facilities required at the site, was floated into place and sunk, and dikes were added on either side to close the narrow inlet.[3]

6. On the Sea of Okhotsk, in the Soviet Union, a major site is under study now (1980).

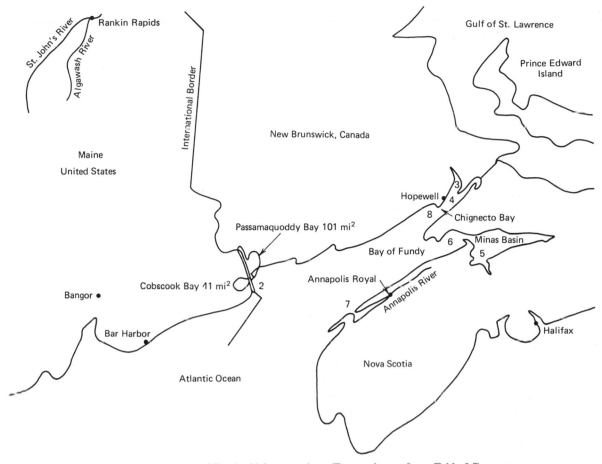

Figure 5.20. Bay of Fundy tidal power sites. (For numbers refer to Table 5.7).

7. On the White Sea, in the Soviet Union, a major site is under study now (1980).

8. On the Severn River in Great Britain, a site with a tidal range of 47 ft (14.5 m) has a calculated output of 2400 million kWh annually. Studies to date have not resulted in an economic plan.

9. At the Chausey Islands, about 20 miles off Saint-Malo, France, a site with a potential of 34 billion kWh annually has been the subject of numerous studies. This compares with 500 to 550 million kWh from the Rance River site. However, due to problems of environment, economics, and implementation time, this project has been shelved (1980).

10. At San Jose, Argentina, a site has an energy potential of 75 billion kWh/yr, but unfortunately has a tidal range of only 20 ft (6 m).

11. There are potential sites in Korea.

Together, these sites represent a potential of the order of 400 billion kWh/yr, of which roughly 200 billion kWh are currently under consideration (1980), and 0.5 billion kWh/yr are actually being produced (Rance River). Tidal ranges for some of these sites are given in Table 5.7.

Passamaquoddy

The two-pool plan judged best by the International Commission (1961) for Passamaquoddy is shown in Fig. 5.21. The high pool includes all of the 101 square mile Passamaquoddy Bay. It is bounded by Deer and Moose Islands and several smaller islands with connecting dams. The low pool includes Cobscook Bay and the waters inside of Campobello Island, altogether 41 square miles. It is formed by the series of interisland dams shown. Altogether some 7 miles of rock-filled dams are required. There are 90 filling gates, marked FG, for the high pool, and 70 emptying gates, marked EG, for the low pool. The 30 10 MW, vertical-axis, fixed-blade propeller-type turbines, 300 MW in all, are located at Carrying Place Cove on Moose Island, between

Table 5.7.   Tidal Ranges and Energy Potential at Tidal Power Sites

Site[a]	Location	Energy Potential (Millions kWh/yr)[b]	Tidal Range (ft)			
			Max.	Min.	Avg.	RMS
1.	Rance River Estuary, France	500–550	44		27.8	
2.	Passamaquoddy, Univted States/Canada	1840	25.7	11.3	18.1	
3.	Mary's Point to Grindstone Island to Cape Maringouin, Bay of Fundy	2590	43.7	19.1	31.4	32.4
4.	Ward Point to Joggins Head, Bay of Fundy	1630	44.4	21.6	33	32
5.	Economy Point to Cape Tenney, Bay of Fundy	5140	52.9	23.9	38.4	37.3
6.	Clark Head to Cape Blomidon, Bay of Fundy	18,190	46.4	19.8	33.1	32.1
7.	Entrance to Digby Gut, Bay of Fundy	800	29.5	11.7	20.6	20.3
8.	Chignecto Bay, Bay of Fundy	920				
9.	Severn River, England	2400	48	22	35	

[a]Sites 2–8 shown on Fig. 5.20.
[b]Energy at sites in Bay of Fundy is average natural energy, Bernstein formula, except Passamaquoddy, 1961 study.

the two pools. These are conventional hydro units as shown in Fig. 5.4.

The calculated project cost, 1961 basis, was $532 million. The project would provide a minimum generation of 95 MW, a maximum generation of 345 MW (short-time 15% overload on 300 MW rating) and would produce an estimated 1843 million kWh/yr.

This tidal plan was considered primarily in conjunction with additions to a Rankin Rapids hydro plant on the St. John's River in northern Maine or with a pumped-storage hydro project on the nearby Digdequash River, to firm it and produce an energy pattern matching the Maine–New Brunswick load pattern (61% load factor). The two alternate firming arrangements would increase the total cost to $567.5 millions (additions to Rankin Rapids hydro) or to $568.9 millions (pumped storage). Rankin Rapids would be used purely to borrow and return energy, the net delivery remaining 1843 million kWh/yr, as produced by the tidal plant. The pumped storage would have some losses, partly compensated by flow in the Digdequash River, resulting in a net delivery of 1759 million kWh/yr.

Neither arrangement proved economic compared with other sources available, and the project was deferred.

Shortly after this 1956–1961 study, the U.S. Department of Interior studied the tidal project combined with hydro on the St. John's River as a 500 to 1000 MW peaking energy source. Full details are given in Ref. 10. The two-pool arrangement was the same as in Fig. 5.21, except with two 50 unit plants,

1000 MW total, at Carrying Place Cove, with units of the type shown in Fig. 5.18. Again viability was not proved, and the project has been deferred.

Rance River Tidal Power Plant

The main structures of the Rance River plant are shown in Figs. 5.22 and 5.17.[2,10] Navigation lock, power plant, dike, and gates are all incorporated in one structure. The 10 MW turbine generators, 24 in number, are of the reversible, adjustable-blade bulb type shown in Fig. 5.17. This provides complete flexibility to operate on both ebb and flood tides and to obtain additional energy by pumping. The plant could produce about 547 million kWh/yr but has been operated for maximum economy at "nearly 500 million kWh/yr." Some 130 million kWh/yr of this is due to pumping.

Compared with Passamaquoddy, the investment at the Rance River plant is much less. With the higher head, the two-way operation plus the pumping feature, and the much smaller extent of civil works, the Rance River project was economically justified. It produces electricity cheaper than oil, coal, or nuclear power-generating plants in France.[2] A comparison of key factors is given in Table 5.8.

The delivery of 3.5 times as much energy for 7 times the capital cost at Passamaquoddy fully accounts for the difference in viability at the two sites. Even with low-interest government financing, the capital costs are predominant, especially since the fuel cost is zero.

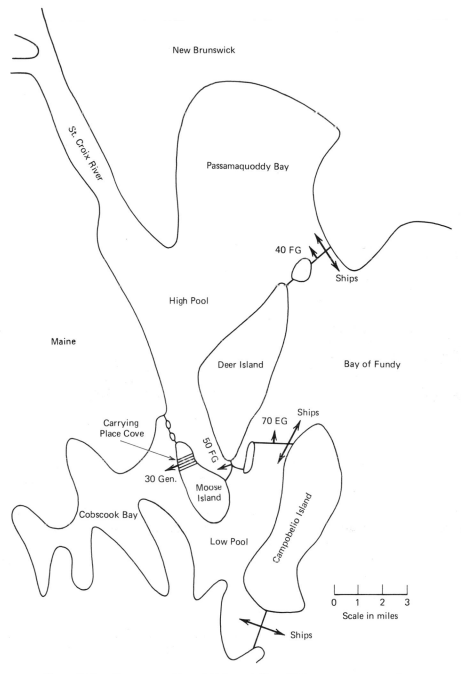

New Brunswick

Passamaquoddy Bay

St. Croix River

40 FG

Ships

High Pool

Maine

Deer Island

Bay of Fundy

Carrying
Place Cove

70 EG

Ships

50 FG

30 Gen.

Moose
Island

Campobello Island

Cobscook Bay

Low Pool

0 1 2 3
Scale in miles

Ships

Figure 5.21. Passamaquoddy and Cobscook Bays tidal power arrangement.

Economics of Tidal Energy

The characteristic pattern of load of an electric utility system consists of high daytime loads five days a week, including high peaks of a few hours' duration each day. Lesser daytime loads occur over the weekends. At night, the load is relatively light. An overall load factor of 50 to 60% is typical. To pro-vide for this load pattern, the power company has installed a mix of highly efficient base-load units, intermediate-load units capable of "load follow" or considerable variation, and peaking capacity that can most efficiently supply the peak loads lasting only a few hours a day. The annual cost of all these facilities can be combined and divided by the total energy generated to get a single "cost per kWh."

Figure 5.22.   Rance River tidal power plant. Courtesy *IEEE Spectrum.*[2]

For a large tidal installation such as Passamaquoddy and the rather limited "systems" of Maine and New Brunswick, the United States–Canadian Commission took as a basis that the tidal project, including auxiliary steam and hydro or storage plants, must be able to supply a load of exactly the same weekly load pattern, at a lower overall cost per kWh, to be economic. This is certainly one simple, clear-cut basis of economic evaluation. However, it is not the only basis and not *necessarily* the best. For example, the U.S. Department of the Interior later proposed justifying Passamaquoddy, combined with a hydro plant, as a source of peaking energy. The Rance River plant,

Table 5.8.   Some Comparisons of Passamaquoddy and La Rance

	Passamaquoddy		La Rance
	1956–61 Study	1962–63 Study	
Installed capacity (MW)	300	1000	240
Auxiliary works	Addition to hydro plant or pumped storage	Hydro plant	None
Approximate cost ($ millions)	570	1200	80
Output (millions kWh/yr)	1843	Peaking	500–550
Total length of dams and power plant	7 miles	7 miles	2500 ft
Emptying and filling gates	160	160	6
Total pool area (sq. mi.)	142	142	8.5
Maximum water depth for dams or dikes (ft)	125–300	125–300	70

simply operated for maximum economy on the much larger French system without any firming auxiliary, was justified.

Maximum Energy from the Tides

The tidal range varies about 3:1 from spring tide to neap tide, and the actual energy available from a tidal power installation varies considerably more (18:1 at the French Rance River installation[2]). However, with some simplifying assumptions, the maximum energy available with a given tidal range H ft and a given tidal flow Q lb of seawater, into and out of the estuary each tide, is $2HQ$ ft lb per lunar day. At 353 lunar days per year

$$\text{Max. tidal energy} = \frac{2HQ \times 353}{778 \times 3413} = 266 \times 10^{-6}HQ \text{ kWh/yr} \quad (5.6)$$

Or, in terms of the volume of seawater, V ft³, and a seawater density of 64 lb/ft³, it may be expressed as

$$\text{Max. tidal energy} = 416 \times 10^{-4}HV \text{ kWh/yr} \quad (5.7)$$

The assumptions made are these: First, consider the tidal basin to have vertical sides over the tidal range. Usually, the pool area is slightly greater at high tide. Next, assume that the gates are closed at low tide. At high tide, the pool is allowed to fill quickly, in an hour or so, through a substantially infinite battery of reversible hydro units, having 100% efficiency at all heads. The average head is $H/2$ ft, the flow is Q lb, and the energy generated is $HQ/2$ ft-lb.

The gates are now closed until low tide. The pool is then emptied through the same hydro units, and again $HQ/2$ ft-lb is generated, or a total of HQ ft-lb/tide. With two tides per lunar day, $2HQ$ ft-lb is generated as stated above.

The tidal basin does not have vertical sides but sloping sides. For this reason, the ebb-tide generation is slightly greater than $HQ/2$, and the flood-tide generation is less than $HQ/2$. However, $HQ/2$ is a good average.

For a two-pool arrangement, with the smaller pool a fraction s of the total and turbines only between the two pools, the maximum energy is $HQ(1 - s)s$ per lunar day, where Q is the normal tidal flow of both pools and H is the tidal range. This is $(1 - s)s/2$ times the maximum potentiality of the site, $2HQ$. If $s = 0.29$, the maximum theoretical energy

from the two-pool arrangement is $0.71 \times 0.29/2 = 10\%$ of the maximum for the site.

Pumping

The maximum energy has been determined without any use of external energy. If, just after filling the pool through the generators, while the sea and pool levels are equal, 10% additional water is pumped into the pool to a level of $1.1H$, the average head involved is $0.05H$ and the energy is $0.05H \times 0.1Q = 0.005HQ$. Now the gates are closed until low tide, and $1.1Q$ is discharged through the turbines at an average head of $0.55H$. The energy generated is $0.605HQ$. Deduct the $0.005HQ$ used for pumping, and the net energy generated is $0.600HQ$, or 20% greater than without pumping (all at 100% efficiency).

Thus, pumping can ride piggyback on the tides. The water pumped is raised from low tide to high tide by the ocean. It produces additional energy, a theoretical 20% more with 10% pumping. If this calculation is repeated with 20% pumping, it is found that there is an additional 40% energy. The pumped supplement varies linearly with the fraction pumped. The limit is the height of the dam and pool sides. This would appear to be an excellent way of increasing the output during neap tides, when the dam and pool have unused capacity.

Passamaquoddy Example

It is stated that the tidal flow into the Passamaquoddy and Cobscook Bays during each tidal cycle is about 70 billion ft³ and that the average tide is 18.1 ft. The combined area is 142 square miles. The latter two figures give $18.1 \times 142 \times 5280^2 = 71.7$ billion ft³ and confirms that the tidal flow given is average. From Eq. 5.7, the maximum tidal energy is

$$416 \times 10^{-4} \times 18.1 \times 70 \times 10^9 = 52,700 \text{ million kWh/yr}$$

With the United States–Canada plan of 1961, 1843 million kWh/yr would be derived from the tidal project, or 3.5% of the maximum (without pumping). Presumably, this system was optimized under the established ground rules (see Economics of Tidal Energy earlier). Apparently, a very small fraction of the theoretical maximum energy can be extracted by a practical system.

La Rance Example

The Rance River estuary above the dam is 8.5 square miles, the average head is 28 ft, and the approximate tidal flow is

$$V = 28 \times 8.5 \times 5280^2 = 6.64 \text{ billion ft}^3$$

Thus, the maximum tidal energy is

$$416 \times 10^{-4} \times 28 \times 6.64 \times 10^9 = 7734 \text{ million kWh/yr}$$

This plant, as operated, delivers "nearly 500 million kWh/yr" to the French grid,[2] that is, about 6% of maximum. Of this, pumping contributes about 130 million kWh/yr, or 1.7%. Had the plant been operated for maximum energy instead of maximum economy, it would have delivered 550 million kWh/yr, or 7.1% of maximum theoretical energy (without pumping).

The two-way operation and higher head are factors in the somewhat greater percentage of maximum than at Passamaquoddy, and the pumping is a very important factor. Use of the two-pool arrangement in the Passamaquoddy study, while providing continuous energy, greatly lowered the maximum theoretical energy.

Balancing the Design

Any tidal project involves a large fixed cost for civil works. Maximum energy is obtained with a very large power plant and low load factor. As the capacity is reduced, the load factor is increased, and the derived energy is reduced. At some point, the minimum cost per kWh or the lowest overall cost of the system with which the plant is integrated is obtained.

For the 1961 Passamaquoddy ground rules, this resulted in a powerplant cost 31% of the total project ($151 million of $484 million). For the later 1963 plan, the 500 MW plant was 31% and the 1000 MW plant was 46% of the total.

REFERENCES

1. K. Probert and R. Mitchell, "Wave Energy and the Environment," *New Scientist*, Aug. 2, 1979, p. 371.
2. H. Andre, "Cheap Electricity from the French Tides," *Spectrum*, February 1980, p. 54.
3. F. L. Lawton, Ed., "Tidal Power," *Proc. International Congress on the Utilization of Tidal Power, Halifax, 1970*, New York: Plenum Press, 1971.
4. "Hydroelectric Plant Costs—1978," 20th Ann. Suppl., Washington, D.C.: Supt. of Documents, 1979.
5. "Inventory of Power Plants in the United States," April 1979, Washington, D.C.: Supt. of Documents, 1979.
6. P. N. Ross and L. G. Hauser, "Some Future Dimensions of Electric Power Generation," *Westinghouse Engineer*, January 1971, pp. 2–7.
7. *Pumped Storage*, Engineering Foundation Conference, New York: ASCE, 1974.
8. W. H. Osterle and E. L. Harder, "Loss Evaluation IV—Economic Dispatch Computer—Principles and Application," *Trans. AIEE (IEEE)*, 1956, pp. 387–394.
9. *Pumped Storage Development and its Environmental Effects*, Int. Conf., Univ. of Wisconsin, American Water Resources Assn., Urbana, Ill., 1971.
10. G. D. Friedlander, "The Quoddy Question—Time and Tide," *Spectrum*, September 1964, p. 96.
11. *Investigation of the International Passamaquoddy Tidal Power Project*, Rept. of the Int. Jt. Commission, 1961.
12. L. Anderson, *"Turbine Will Tap Bay of Fundy's Tides off Nova Scotia in Demonstration Project,"* Wall Street Journal, February 27, 1980.
13. E. J. Lerner, "Low Head Hydro Power," *Spectrum*, November 1980, p. 37.
14. H. W. Hamm, *Low Cost Development of Small Water Power Sites*, 3706 Rhode Is. Ave., Mt. Ranier, MD. 20822: VITA, 1967.
15. M. E. McCormick, *Ocean Wave Energy Conversion*, New York: Wiley, 1981.

6

Nuclear Energy

INTRODUCTION

In this chapter and the companion chapter on nuclear physics of energy, we shall endeavor to answer the questions, What is nuclear energy? What are the processes of converting it from the ore in the ground to useful electric energy? And how safe is it? This chapter covers primarily the question of nuclear plant safety, the various reactor systems used for power generation, the arrangement of a typical nuclear plant, and fuel processing and reprocessing.

The reserves and rates of use of nuclear energy have already been given in Chapter 1. The facts about transmission (electrical) and storage are given in Chapter 14, and the physics of nuclear energy (selected topics) is covered in Chapter 12.

The costs of fossil fuels are given in Chapter 1. Nuclear-electric generation, at first competitive only in high coal-cost regions such as New England, had become the lowest cost option for base load in nearly all areas by 1970. In this sense, the nuclear energy cost structure is covered in Chapter 1, under electric energy costs.

What Is Nuclear Energy? While covered in more detail in Chapter 12, the simplest answer is as follows: When heavy elements like uranium are broken apart (fissioned), the mass of the fragments is less than the original mass. The energy corresponding to the loss of mass is given off and is called *fission energy.* It is 9×10^{16} J/kg of mass converted, or 931 meV/amu (millions of electron volts per atomic mass unit), according to the Einstein relation $e = mc^2$. That is,

joules = kilograms × (velocity of light in meters/sec)^2

where $c = 3 \times 10^8$ m/sec. A 1 kg mass converted to heat energy is enough to run a large turbine generator (a 1 GW unit) for a year at 35% thermal efficiency. This would require about 2.5 million tons of coal, 400 6000-ton train loads.

Likewise, if two or more light elements like hydrogen are joined together (fused) to form a heavier one, the resulting mass is less than the sum of the components, the difference being converted to energy, as above. The energy corresponding to the loss of mass is given off and is called *fusion energy.*

Fission energy and *fusion energy* together are called *nuclear energy,* both arising from changes in nuclear mass.

Growth of Nuclear Power. Nuclear generation of electricity in the United States has been steadily growing ever since the first power-size nuclear plant at Shippingport, Pennsylvania, 60,000 kW (90,000 kW later) in 1957.[1] By 1973, 16 yrs later, only 1.2% of all energy, about 3.6% of electric energy, was being produced by nuclear plants (see Table 1.2); but in 1975, more than half of all new units on order were nuclear.

In 1978, nuclear plants in the United States had a capacity of 54 GW (54,000 MW)* and generated about 12.6% of the electricity used. In some parts of the country, 40% of the electricity used was from nuclear plants.

Considering only reactors in operation, under construction, or on order in late 1976, there would be 208 GW of nuclear generating capacity in the late 1980s, about 25% of the total U.S. capacity expected at that time. If the further projected expansion of nuclear capacity takes place, over 400 GW of nuclear power should be in operation by the end of the century, generating around 50% of the U.S. electricity.[2] Energy Secretary Schlesinger estimated that by the year 2000, some 300 new nuclear plants would be needed in addition to the 63 currently operating in 1977 (*Wall Street Journal,* June

*1965, 1 GW; 1970, 6 GW; 1973, 21 GW; 1975, 40 GW.

8, 1977, p. 1). With the change in federal policy starting in 1976, this program has considerably slowed.

THE CASE FOR NUCLEAR ENERGY

Since many of the citizens of the United States have fears, great or small, regarding the safety of nuclear plants, and some seek to curtail or eliminate their use, this chapter would be remiss if it did not enumerate these worries and the facts regarding them as presented by some of the best informed.[2] This treatment of a controversial subject, involving ideologies, is a departure from the goals of the book as stated in the preface. However, these answers, insofar as they are factual, are such an important part of the nuclear development that it was judged best to include them.

The following information is largely from a presentation[2] to the National Academy of Engineers by Manson Benedict in 1976, with his permission. The questions that have been raised are as follows:

1. Critics Claim That Nuclear Plants Are Unreliable

As shown in Table 6.1, nuclear plant reliability is comparable with that of coal and oil.

2. Critics Are Concerned About Occupational and Public Health Hazards

As shown in Table 6.2, nuclear systems have far less occupational hazard than coal. Statistics on injuries and lost-time accidents are similar. Note that about 2200 billion kWh were used in the United States in 1978 (Fig. 5.2).

Public Safety. No member of the U.S. public has ever been injured, much less killed, as a result of accidents in nuclear plants (1977). This amazing safety record is a tribute to the precautions that

Table 6.1. 1975 Statistics—Reliability of Generating Plants

Type of Plant	Availability % of Time	Generation % of Rated
Nuclear	74	64
Oil	70	42
Coal	80	55

Table 6.2. Occupational Hazards. 1973 Report—Council on Environmental Quality

Type of Generation	Occupational Deaths from Generation of 200 Billion kWh of Electricity
By nuclear	4.6 (mostly uranium mining)
By surface mined coal	79
By deep mined coal	120

have been taken in the design, operation, and regulation of the potentially hazardous nuclear facilities.

In regard to the *probability* of accidents in nuclear power plants, as a result of which people *might* be harmed, the Rasmussen study and report for the Nuclear Regulatory Commission places the probability associated with the production of 200 billion kWh of electricity at 0.8 death and 8 injuries. A study and critique by the American Physical Society would triple these figures, mostly because of exposure of people to down-wind and time-persistent effects. The public health effects are still small, however, compared with those of electric generation by coal.

Both Public and Occupational Effects. Taking into account *all* industrial and civilian consequences of electrical generation, 1976 studies place the relative health effects of generating 200 billion kWh of electricity at:

Coal	600 deaths
Nuclear	11 deaths

Thus, the probability of deaths from nuclear generation is far less than that from the use of coal.

3. Critics Feared Long-Time Health and Genetic Effect of Low-Level Radiation From Normal Operation of Nuclear Facilities

Fears are now largely dispelled (1977). As a result of strict control of effluents, exposure to the public from nuclear plants has been far less than 1% of the natural radiation to which everyone is exposed (Table 15.6).

4. Fear of Catastrophic Consequences of Sabotage by Irrational or Malevolent Groups

Nuclear plants are more readily protected against sabotage than fossil fuel plants. They have no vulnerable coal pile or tank farm. A year's fuel supply is within the reactor, protected by its radiation and the radiation shield. Physical protection to primary piping is also afforded by massive shielding. An emergency cooling system provides redundant protection. An elaborate security system of barriers, alarms, and armed guards is a part of every nuclear plant.

5. Fear of Radiation From Nuclear Wastes

The National Research Council's Committee on Radioactive Waste Management is convinced[2] that technical measures are available for the safe packaging, transportation, and long-term storage of these wastes, with minuscule risk to the public.

The preferred procedure involves shipment of spent fuel from nuclear plants to a reprocessing plant where uranium and plutonium would be separated and freed of contaminating radioactivity. The radioactive wastes would be converted to a water-insoluble glass (about 80 ft^3/yr for a 1000 MW plant). This would be packaged in seal-welded stainless-steel containers and shipped in shielded carriers to a federal waste repository. The repository would be a geologic formation at least 500 m below ground known not to be subject to permeation by flowing water.

The principal problem is public acceptance, not technical feasibility.

6. Fear of Diversion of Fissile Material for the Fabrication of a Bomb by Terrorists

Uranium enriched to about 3% U-235, as used in current (1977) light-water reactors, is too dilute for use as a bomb. Plutonium separated in reprocessing plants could be used. Diversion is not feasible while the plutonium is diluted with millions of curies of radioactivity as it leaves a reactor and is in spent-fuel storage or in shipment to a reprocessing plant. After decontamination and concentration in the reprocessing plant, it is most vulnerable to theft and must be carefully guarded. After mixing with uranium oxide in the right proportion for reactor fuel, about 15% plutonium, it is no longer a feasible bomb material. Thus, it must be carefully guarded during shipment in pure form, possibly including irradiation to make it radioactive. Alternatively, it can be processed into fuel elements in the same protected area where it is separated.

If the entire operation of plutonium decontamination, separation, and mixing to fuel proportions is carried out in one heavily guarded facility, a single facility could serve 60 GW or more of nuclear plants, more than the 1977 total (42 GW), and about one third that projected in the late 1980s. Actually, of course, plants through the late 1980s will use mainly uranium fuel. The above comparisons are given to visualize the guarding problem later when reprocessed fuel may be used in present or breeder reactors. The guarding problem is not great.

7. Fear of Spread of Nuclear Weapons to Countries That Do Not Have Them

This fear is well founded if these countries build reprocessing plants. However, it is not a valid argument against expanded use of nuclear energy within the United States. The placing of reprocessing plants under international control is the best possibility of preventing the diversion of civilian plutonium to military uses.

8. Fear That the Amount of Uranium in the United States Is Insufficient to Make the Development Worthwhile

ERDA estimates (1976) 640,000 tons of uranium oxide in reasonably assured reserves, and about 3,600,000 tons are estimated from these deposits and others to be discovered. It requires about 150 tons of uranium oxide per 1 GW plant per year. Thus, the 640,000 tons of assured reserves would fuel the 208 GW of capacity projected in the late 1980s for 20 yr. The estimated 3,600,000 tons, if found, would fuel the 400 GW of nuclear capacity projected for the year 2000 for 60 yr. However, it would be fully committed before 2000, since each new plant commits a 40 yr supply (its lifetime). Breeder reactors would extend the energy available from this uranium about 70-fold.

Use of Breeder Reactors

Most major nations of the world consider the breeder reactor and the use of plutonium for energy essential to their economy and are pursuing its development and use vigorously (1977). In the United States, the breeder development has suffered delays, which become more serious as time is lost. The world schedule of breeder reactor installations in 1979 is shown in Fig. 6.1.[4] The U.S. schedule is already several years behind the dates originally planned.

By using the breeder reactor, the 150,000 tons of

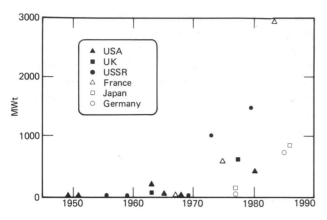

Figure 6.1. Schedule of breeder reactors. International development: Five other nations—France, United Kingdom, West Germany, Soviet Union, and Japan—have made significant commitments to the LMFBR concept. France, United Kingdom, and the Soviet Union already have completed demonstration plants (1979). Courtesy DOE.

depleted uranium in storage in ERDA's uranium enrichment plants in 1976 could generate as much electricity as 400 billion tons of coal, nearly all the commercially minable coal in the United States.

The breeder reactor would also make efficient use of the plutonium produced in the present light-water reactors. Thse have a breeding ratio of about 0.6. They produce about 60% as much plutonium fuel as the uranium fuel burned.

Opposition to the vigorous and immediate development of the breeder reactor in the United States (1977) appears to stem from (1) the toxicity of the plutonium fuel, (2) the risk of theft by would-be terrorists, and (3) the possibility of a nuclear excursion from maloperation of the reactors. (4) In strange contrast to item 8, above, it is also argued that the breeder development should be delayed because we have plenty of uranium for a long time. (5) Solar energy and nuclear fusion are also offered as substitutes. Considering these five points in order:

1. Fear of Health Hazards From the Toxic Plutonium Fuel

The techniques for safe handling of plutonium and the far greater number of curies of accompanying fission products have already been worked out by ERDA in its fuel reprocessing and fabrication plants. There has been no case of serious plutonium poisoning in over 20 yr (to 1976) of handling large amounts of this toxic material.

2. Fear of Theft of Plutonium by Terrorists

Theft of plutonium can be prevented by the same methods discussed for light-water reactors, item 6.

3. Fear of a Nuclear Excursion Due to Maloperation

The 1000 MWt Clinch River Breeder, originally scheduled in the 1980s, will have two independent and redundant reactor shut-down systems. It has nevertheless been postulated that both might fail, the sodium coolant boil away, and the core melt down and reassemble quickly enough for a supercritical configuration to occur and release substantial energy. The amount postulated is 1200 MJ.

Three measures for preventing or mitigating the effects of such an unlikely accident are:

1. Confinement in a structure strong enough to contain the energy release. This measure is being required by the Nuclear Regulatory Commission for Clinch River.
2. Provision of an internal blanket of U-238 to reduce the energy release from such an event.
3. An inherent built-in shut-down, or reactor fuse, which would respond directly to increase in power and insert a neutron absorber without any external control system.

Each of these measures should effectively eliminate offsite effects from a nuclear excursion in a fast reactor.

4. Proposal to Delay for Economic Reasons, Based on Ample Uranium Supply

Two grave uncertainties are involved in this proposal: (1) the length of time for full development and deployment of the breeder reactor, and (2) the actual reserves of uranium other than assured reserves. The consequences of delay vary from a substantial economic penalty to a huge penalty and danger of a major power shortage in the United States.

The breeder reactor is considerably more expensive, as will be explained later. However, its fuel cost is much less since it is spared the cost of natural uranium and uranium enrichment.

Studies by ERDA and others[2] show that even if

the breeder cost 25% more than a light-water reactor, the fuel cost saving would result in a favorable return on the higher investment in the breeder and the estimated $10 billion needed to complete its development and commercialization.

The breeder is the *only proved way* to provide the energy needed in the future. Vigorous pursual of the development currently will save costly blunders in the future when "time has run out." The U.S. world leadership and export business is being lost to other nations as they take the lead in the breeder development. According to the *EPRI Journal*, the cost to the nation of deferring the fast breeder (FBR) will range from $100 billion to $3.5 trillion over the next 75 yr.[3] "If our other known major energy resources, coal, oil, gas, and uranium, prove to be adequate, the benefits of the breeder are still significant. But if some of these energy resources fall short of expectations, deferment of the breeder could cost us dearly."

5. Why Not Solar Energy or Nuclear Fusion?

Solar energy, though very valuable for low-temperature comfort and process heating, is too diffuse, too variable with time of day or season, and too uncertain because of weather to be competitive economically with the breeder for major electric power generation.

Controlled fusion energy is a highly desirable goal, well worth the major development effort it is receiving. However, there are major unsolved engineering problems that leave it far from certain ever to generate a single kWh of economic electric energy.[19]

1981 View

While these statements were written in 1977, no changes in any of the foregoing are required as a result of the Three Mile Island nuclear plant accident of March 28, 1979.

In the words of Peter Sandman and Mary Paden, writing in the *Columbia Journalism Review,* "In the week following the accident, a hypothetical resident standing naked at Three Mile Island north gate absorbed roughly 85 millirems of radioactivity, less than the annual difference between what residents of Harrisburg and Denver receive."

While still not certain, the prospects for nuclear fusion are certainly much brighter in 1981. The Fu-

sion Act of 1980 commits the United States to the development of a demonstration fusion reactor by the year 2000.[18]

This concludes the treatment of the controversial aspects of nuclear energy. The development itself will now be outlined.

NUCLEAR ELECTRIC POWER GENERATION

The conversion of nuclear energy into electric energy can be conveniently broken apart into two steps, as shown in Fig. 6.2. The nuclear reactor and steam generator in the nuclear plant perform the same function as the boiler in a conventional fossil fuel-fired plant. They convert the energy in the fuel into heat energy which is transferred to water to produce high-pressure steam. The balance of the plant uses the high-pressure steam to drive a steam turbine, which turns the electric generator, producing electricity. The turbine generator part of the system is the same, diagrammatically, whether the steam comes from a fossil fuel-fired boiler or from a nuclear steam-generating system.

There are important differences in the steam temperatures and pressures involved and the fuel costs, which result in substantial differences in the turbine and auxiliary equipment design. While these do not appear in the simplified flow diagram of Fig. 6.2, they will be covered in discussing the different systems. However, it will be convenient to separate the nuclear steam-generating system from the turbine generator system. Thus, the various nuclear reactor types, and the resulting steam-generating systems, can be discussed without going into the turbine generator plant to any great extent.

In general, we shall be discussing (1) light-water and heavy-water reactors, which produce relatively low steam temperature and pressure and result in a thermal efficiency of about 33% for the overall system, and (2) high-temperature gas-cooled reactors and liquid metal-cooled fast breeder reactors, which produce high steam temperatures and pressures, comparable to modern fossil fuel-fired plants, with a resulting thermal efficiency of about 40%.

Following a brief discussion of the various reactor–steam generator systems, one particular system, the pressurized-water reactor, will be described in more detail as an example. This section gives some idea of the sizes, shapes, and features

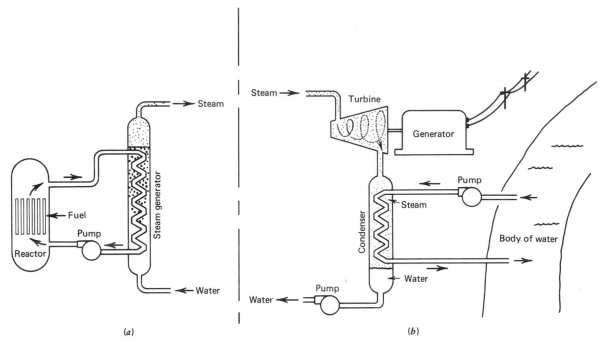

Figure 6.2. Sections of a Nuclear Power Plant. (*a*) nuclear-stem system; (*b*) steam-electric system. Courtesy DOE.

involved and the interaction of the nuclear steam and the steam electric parts of the overall system.

The nuclear fuel handling and processing is described in a final section of the chapter.

Types of Reactor–Steam Generator Systems[4,5]

Since practically all nuclear power in the United States currently (1981) is from light-water reactors, these will be treated first. They are fueled with slightly enriched uranium, about 3% U-235 compared with 0.7% in natural uranium. In Canada and a few other countries, natural uranium reactors with heavy-water moderator are used. While 20 to 25% more expensive to build, they avoid the need for uranium enrichment plants. These will be covered next. The high-temperature gas-cooled reactor is of considerable importance and will also be covered. It uses a graphite moderator and helium coolant. The liquid metal-cooled fast breeder reactor, now under development and beginning use, will be outlined. It is the principal hope of the future. Two other breeders, the gas-cooled fast breeder reactor and the light-water breeder, will also be described. Finally, the various other schemes, proposed or experimental, are summarized briefly.

Light-Water Reactors[5]—PWR, BWR

Two alternative arrangements of the light-water reactor are shown in Fig. 6.3. In the *pressurized-water reactor (PWR),* Fig. 6.3*a*, the water in the reactor is kept just below the boiling temperature at the reactor pressure of 2250 psi. Steam is generated in a separate heat exchanger. A pressurizer, not shown, is connected to the primary loop to maintain the desired pressure. The light water (ordinary water) serves both as moderator and coolant. The primary loop is shielded, but the steam line does not require shielding. Typical temperatures and pressures are shown in the figure.

Control rods containing boron or other neutron-absorbing materials are used to control the reactivity during normal operation or in start-up and shutdown. These may be supplemented by a boron "shim," a variable amount of boron dissolved in the water of the primary loop. The shim provides for the longer-term variations and reduces the number of control rods required. A typical neutron balance of a PWR is given in Table 12.5.

In the *boiling-water reactor (BWR),* Fig. 6.3*b*, the steam is generated directly in the reactor. The core is somewhat larger and more complex, with a greater number of control rods. However, the ex-

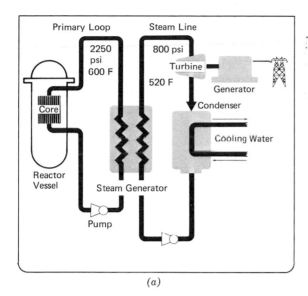
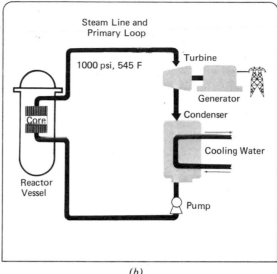

(a) (b)

Figure 6.3. Light-water reactors: (a) Pressurized water reactor. In a pressurized water reactor (PWR), the water heated by the core is circulated through a closed system, called a "loop". This first loop carries the heat from the core to a steam generator where the heat is transferred to a second loop. It is in this second loop that the steam is generated to produce electricity. (b) Boiling water reactor. There are two distinct types of light-water reactors. In both, the heat extracted from the core is used to make steam. In a boiling water reactor (BWR), the steam is generated directly by the heat from the core. This steam runs a turbine to generate electricity. Thus it is a "direct'cycle" system. Courtesy DOE

ternal steam generator and pressurizer are not needed. Also, the pressure is much lower. For example, an 800 MWe nuclear power plant requires a boiling-water reactor vessel about 70 ft high by 20 ft in diameter. A pressurized-water reactor vessel for the same capacity is about 40 ft high by 16 ft in diameter, or about 37% as much volume.

The boron shim cannot be used in the BWR arrangement. The shielding includes the turbine and condenser. The overall costs of these two arrangements are similar so that they continue to be sold competitively on an equal basis (1977).

In both cases, the steam temperature and pressure are relatively low, and a thermal efficiency of about 33% is realized. This compares with about 40% in a modern fossil fuel-fired steam plant of 1977, without scrubbers, or in a breeder reactor plant (LMFBR), to be described later. Flue gas scrubbing reduces the efficiency of a coal-fired plant to about 34%.

Heavy-Water (Natural-Uranium) Reactor[6]

A heavy-water moderator in which the hydrogen atoms are deuterium is required in order to use natural uranium as a fuel in a practical power reactor. The loss of neutrons by capture is too great in

ordinary water. However, it requires about three times as much heavy-water moderator to slow neutrons down to thermal energies, and the reactor is consequently much larger for the same power.

The usual arrangement is that shown in Fig. 6.4, a tube-type reactor. The fuel is placed in tubes in the reactor core, through which the cooling fluid flows and carries the heat generated to the separate steam generator. The cooling fluid may be organic compounds, gas, water, or heavy water. The heavy-water moderator fills the reactor vessel around the fuel tubes as shown.

Coolant temperatures in heavy-water reactors may be up to 700°F, depending on the cooling fluid used.

The natural-uranium reactor uses a relatively small part of the available fuel, and is considered as an interim design until economical breeder reactors are available.

High-Temperature, Gas-Cooled Reactor (HTGR) (Fig. 6.5)[5,7]

The use of helium in this design permits achieving high temperatures at modest pressures and minimizes corrosion problems, offsetting the fact that gas is not as good a heat-transfer medium as liquid.

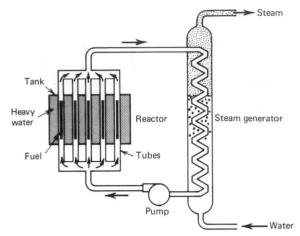

Figure 6.4. Nuclear steam supply components in a heavy-water reactor. Courtesy DOE.

An external steam generator is used as with the PWR.

At the higher temperature, superheated steam, comparable to that in a fossil fuel-fired plant, is available for the turbine, with corresponding thermal efficiency of about 40%. The primary loop is shielded as with the PWR.

Graphite is used both as the moderator and as the structural material of the reactor core. Fuel rods and control rods (of boron carbide and graphite) are passed through vertical openings in the graphite blocks. The helium coolant passes through other vertical openings.

While a number of HTGR systems were ordered

following its development in the early 1970s, serious difficulties developed in the design and only two were built, one of which is operating in 1981. Nevertheless, because of its high-temperature capabilities and high thermal efficiency, limited development on it is being continued[5] (1976).

Development work is also being carried out (1976) on two advanced concepts: (1) the direct-cycle gas-cooled reactor, where the heated gas is used directly in a gas turbine to generate electricity at a higher thermal efficiency; and (2) the very-high-temperature reactor, which could operate at temperatures up to 2000°F (1093°C) for industrial processes requiring such temperatures.

Liquid-Metal Fast Breeder Reactor (LMFBR)[5,9,10]

The general arrangement of the liquid-metal fast breeder reactor (LMFBR) is shown in Fig. 6.6. The use of sodium permits achieving high temperatures at normal pressure. Also, sodium is a very efficient heat transfer medium. This permits a very high power density and a relatively small-sized reactor. However, the handling of sodium introduces some design and operating complications. These, including the additional sodium loop (the first is highly radioactive), increase the capital cost. The expected lower overall cost results from the lower fuel cost, as has been mentioned.

In the initial installations, enriched uranium will be used as the fuel. However, the normal operation

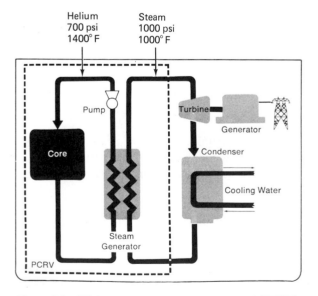

Figure 6.5. High-temperature, gas-cooled reactor (HTGR). Courtesy DOE.

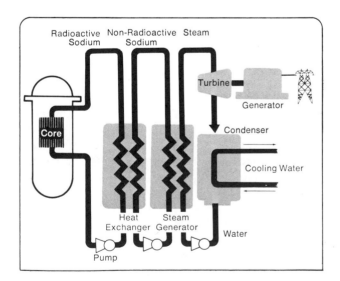

Figure 6.6. Liquid-metal fast breeder reactor (LMFBR). Courtesy DOE.

of the breeder, as plutonium becomes available, will be the plutonium–uranium fuel cycle. The fuel loaded will be U-238 mixed with about 17% Pu-239. This fuel is formed into pellets of the uranium–plutonium oxide mixture for insertion into the reactor. In operation, this will breed more plutonium than it burns, the breeding ratio being 1.3 or 1.4. The supply of U-238 is very large; the stockpiles of depleted uranium at DOE's enrichment plants, some 150,000 tons in 1976, are adequate for all U.S. electrical generation for about a century.

A typical neutron balance of a LMFBR is given in Table 12.5.

While a number of experimental breeder reactors have been operated in the United States and elsewhere in the world and commercial breeders are operating in some countries in 1981, the U.S. plans of 1976 called for a three-step program before widespread utilization of the breeder:

1. FFTF[11]—a Fast Flux Test Facility, at Hanford, Washington, for fuel and component testing for fast breeder reactors, a 400 MWt test reactor.
2. CRBRP—the Clinch River Breeder Reactor Plant, a 350 MWe, 1000 MWt demonstration plant planned for operation in the early 1980s.
3. PLBR—a Prototype Large Breeder Reactor which would make general deployment of this technology possible in the 1990s.

The latter two steps have been indefinitely delayed (mid 1981). However, a policy statement (Wall St. Jour. Oct. 9, 1981) removed the ban on commercial reprocessing and directed government agencies to proceed with the Clinch River breeder reactor.

Light-Water Breeder Reactor (LWBR)[5]

A principal incentive for a light-water breeder reactor (LWBR) is the use of the light-water technology, highly developed in 1977, and the possibility of converting present light-water reactors to breeders at a future time, without major modifications. This would significantly improve the fuel utilization of pressurized light-water reactors.

The diagram of the LWBR is the same as that of the PWR, Fig. 6.3*a*, and the conditions are nearly the same. However, a uranium–thorium fuel cycle is used, breeding the fissile material, U-233, in a thermal neutron flux.

A demonstration LWBR has gone into operation at Shippingport, Pennsylvania, in 1977, having a breeding ratio slightly over 1.0.

Successful development of the LWBR can ultimately make available for power production about 50% of the potential energy of thorium resources, an energy resource much greater than known fossil fuel reserves.

Gas-Cooled Fast (Breeder) Reactor (GCFR)[5,12]

Although the GCFR development is not as far advanced (1977) as the LMFBR, it is being pursued because of its potentially high breeding ratio and low capital cost.

Diagrammatically, it would appear the same as the HTGR, Fig. 6.5, although it would use somewhat higher helium pressure (about 1300 psi). It operates on a uranium–plutonium fuel cycle. That is, plutonium is burned, providing energy and also breeding U-238 into more plutonium. The physical construction of the GCFR would be the same as described for the HTGR.

Other Reactors

A large number of other reactor designs have been used experimentally, or in small numbers. Examples are the following:

Molten-Salt Reactor.[13] A molten-salt reactor, using U-233 fuel, was placed in operation at Oak Ridge in 1968. A mixture of lithium, beryllium, zirconium, and uranium fluorides is melted to form a liquid that is pumped through pipes like water. When the molten salt is pumped through hundreds of parallel channels in a graphite core, the mass and geometry of a solid reactor core is reproduced, and the reactor operates accordingly. The molten salt is the coolant and is pumped through the heat exchanger. This of course increases the inventory of fuel required, which increases the cost considerably.

Molten-Salt Breeder Reactor.[13] A molten-salt breeder reactor was also designed, with molten thorium fluoride surrounding the core and in passages through it, but separate from the fuel.

Homogeneous Reactor.[14] In the aqueous homogeneous reactor, the fuel, the moderator, and the coolant are all intimately mixed and circulated. The moderator, either H_2O or D_2O, that is, either ordinary water or heavy water, carries the fissionable material, either in solution or suspension, and the whole is circulated through the reactor vessel, heat exchanger, and pump.

An experimental 150 MWe plant operating on this principle was designed in 1956, the PAR (Pennsylvania Advanced Reactor) Homogeneous Reactor. At the recommendation of the design team it was not built.

Pebble Bed Reactor.[13] The West German high-temperature or "pebble bed" reactor is another arrangement of a gas-cooled reactor for process heat or power generation. The fuel and moderator in this case are encased in marble-sized pellets forming a porous bed through which the inert helium cooling gas flows.

Hanford "N" Reactor.[15,4] This reactor, also called the "New Production Reactor," at Hanford, Washington, for the production of plutonium for defense stockpiles, provided for the recovery of waste heat. This had been dissipated in the cooling water heretofore. However, since 1966 a nearby steam-electric plant has used the hot coolant water from this reactor to generate 790 MW of electric power. The production reactor uses U-235 fuel to breed plutonium from U-238 in a huge graphite moderator structure.

Alternatives Being Considered in 1977.[8] As an alternative to the LMFBR and use of plutonium fuel, a number of other arrangements were being considered in 1977. While the primary aim was reduced likelihood of weapons proliferation, this takes the form of increased fuel burnup with the once-through cycle, greater use of thorium, and techniques in which fissile material of a type suitable for bombs is not separated outside of the reactor. It may be produced and burned in place.

The LMFBR had already been selected by the United States and other major nations as most economic, and these other schemes are generally more costly. They include the following:

1. Increased enrichment and more frequent refueling of light-water reactors to improve the fuel burnup—about 10%.

2. The use of thorium in the fuel to reduce the uranium—about 25%. This would extend the period before breeders or some alternative energy source would be a necessity.

3. The "spectral shift" reactor in which light water is gradually replaced by heavy water over the life of the fuel, to improve the burnup—about 25%.

4. A thorium–uranium fuel cycle, with uranium recycle, to reduce the uranium requirements to about one half that with LWRs.

5. The "tandem fuel cycle." Spent fuel from LWRs would be used in heavy-water reactors to add about 10,000 MWd/tonne to the fuel burnup and thus extend the resources about 25%.

6. The use of heavy-water reactors, the Canadian CANDU, in the United States with a denatured thorium cycle, requiring about one half the uranium of LWRs.

7. The use of HTGR with 10% instead of 90% enrichment.

8. An accelerator-driven subcritical power generation scheme. A beam of protons on natural or depleted uranium produces neutrons at fission energies by spallation reactions and thus fissions the uranium. The resulting heat energy would generate electric power to drive the accelerator, and for normal use. A once-through system would be expected to require no enrichment, have less spent fuel, and require only 5 to 10% of the initial fertile material.

None of these proposals for extending the uranium supply has been adopted to date (1981).

A Typical Nuclear Power Plant of the 1970s and 1980s[6]

During the 1970s, the maximum unit sizes of turbine generators, with fossil-fuel-fired boilers or nuclear reactors to match, increased from below 1 GW to well over 1 GW. One or two such units might be used in the largest plants. The unit selected for this description is an 800 MWe (0.8 GWe) nuclear plant with one unit. A diagram of the principal components is given in Fig. 6.7.

The reactor is a pressurized-water reactor (PWR), supplying hot water at 2250 psi and 600°F, to three parallel steam generators, all of which feed

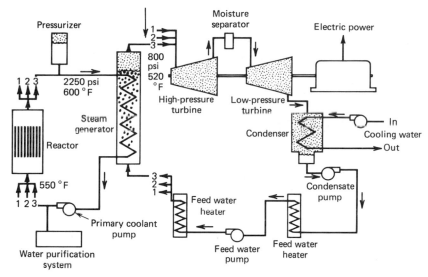

Figure 6.7. A complete nuclear power plant based on a pressurized water reactor. Courtesy DOE.

steam at 800 psi and 520°F to the single high-pressure turbine. This modular arrangement, in which a single reactor supplies several coolant "loops," each comprising a steam generator, coolant pump, and piping, provides flexibility in design. More loops can be used for larger sizes without exceeding the size limitations of steam generators or pumps. Thus, for the 0.8 GWe plant shown, there are three coolant loops (one shown in detail), but only one reactor and one turbine generator. The pressurizer and water purification system are connected to one of the three loops.

At the comparatively low temperature of 520°F at 800 psi, the steam becomes quite wet as it passes through the high-pressure turbine. A moisture separator is used before it enters the low-pressure turbine. No reheat stages are used. With the comparatively low fuel cost, this arrangement produces electricity at the lowest cost. Superheating and reheating are not justified economically. However, the thermal efficiency, as a result, is relatively low, about 33%.

From the low-pressure turbine, the steam flows into a condenser where the remaining heat is transferred to cooling water from some source, either a natural body of water or a cooling tower. The steam is condensed to water (condensate) at a pressure well below atmospheric, that is, at a vacuum.

From here, its pressure must be raised to 800 psi before it enters the steam generators. This is accomplished by the condensate pump and feed-water pumps, of which one is illustrated. Usually, two or

more are required in series. Several stages of feed-water heating are used between the condensate pump and the steam generators to raise the temperature to near that in the steam generator. Two heaters are illustrated, although more are generally used to accomplish this function. More detail is given in Chapter 13.

Arrangement of Principal Components. The physical arrangement and relative sizes of the nuclear-steam components are shown in Fig. 6.8. Note the size of the man entering the door of the containment. The several steam generator-coolant loops surround the reactor, and all are enclosed in the containment structure. Only the steam pipes to the turbine, containing nonradioactive steam and the returning feed water, pass through the walls of this protecting structure.

The arrangement of principal components in the complete plant is shown in Fig. 6.9. The steam–electric portion of the plant is the same as for a fossil fuel-fired plant, except for some of the auxiliaries as mentioned, and will not be described.

The Reactor. A cross-sectional view of the pressurized-water reactor is given in Fig. 6.10. The vessel is 40 ft high and about 16 ft in outside diameter and weighs about 500 tons empty and over 750 tons when filled with fuel and coolant. The walls are about 1 ft thick. This vessel contains the core, which is the heart of the system, control rods and space for their withdrawal, and the coolant—light

Figure 6.8. Cutaway view of a nuclear steam supply system based on a pressurized-water reactor. Note size of man entering door at right. Courtesy DOE.

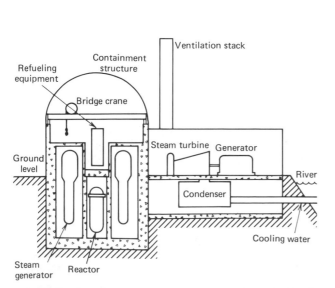

Figure 6.9. Location of principal components in a pressurized-water nuclear plant. Courtesy DOE.

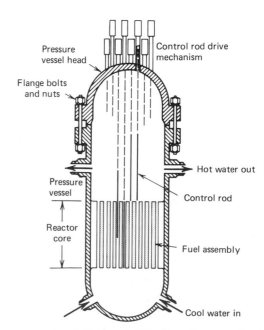

Figure 6.10. Vertical cross section view of a pressurized-water reactor vessel. Courtesy DOE.

water in this case. The top or "pressure vessel head" is removable for refueling or maintenance and is held down by a complete circle of large bolts.

The "cool" water (550°F) enters near the bottom, flows up through the core, where it is heated, and exits as "hot" water (600°F). The power density in such a reactor is very high, compared with the power density in fossil fuel-fired boilers; a tremendous quantity of water must be circulated to carry this heat to the steam generators, with the temperature differential of only 50°F. Some 330,000 gal/min flow through the reactor and one third of this through each coolant loop.

Alternatively, the "cool" water may enter at the same level as the "hot" water leaves and be directed to the bottom by baffles, as in Fig. 6.8.

The Core. The core is fueled with uranium dioxide, UO_2, a ceramic. It is enriched to about 3% of the U-235 isotope. The UO_2 is formed into cylindrical pellets which are enclosed in sealed zircaloy tubes about 12 ft long and a little less than $\frac{1}{2}$ in. in diameter. About 200 of these tubes, or fuel rods, are arranged in a square pattern to form a fuel assembly, as shown in Fig. 6.11. They are spaced apart with egg crate-like spacers to allow the coolant to flow freely up through the assembly.

Each fuel assembly is 8 in. square by 12 ft long and weighs 1300 lb; UO_2 is almost as heavy as lead. Some 200 fuel assemblies form the core of the reactor. They are arranged in a pattern so that from the top the core looks circular. The core is nearly 12 ft across.

Figure 6.11. A fuel assembly for a nuclear power plant. Courtesy DOE.

Control rods containing neutron-absorbing material pass through vertical passages in some of the fuel assemblies. There may be as many as 90 control rods. These, together with a boron shim dissolved in the coolant water, control the reactivity, or heat-generating rate, of the reactor.

Pressurizer and Purifier. The purpose of the pressurizer is to maintain the water pressure at 2250 psi in the reactor and the steam generator loops. It is the unit at the left in Fig. 6.8. A small amount of water is vaporized to steam near the top of the pressurizer, by electric heat, to control the pressure in the system. The purification system continuously removes impurities from the coolant water.

Steam Generator. The steam generator, shown in Fig. 6.8, is 10 to 15 ft in diameter and about 60 ft high. It weighs about 250 tons. Hot water enters the tubes at 600°F and leaves at 550°F, generating steam at 800 psi, 520°F, from the water surrounding the tubes. Some 10 million lb ($5\frac{1}{2}$ million ft³) of steam per hour is supplied to the turbine by the three steam generators working in tandem, for the 800 MWe plant output.

Coolant Pumps. The coolant pumps of several thousand horsepower in each of the three loops raise the water pressure about 100 psi to compensate for the pressure drop in the reactor, the steam generator, and the piping.

Containment Building. The containment building of steel or concrete houses the nuclear-steam components, as shown in Figs. 6.8 and 6.9. While some 99.99% of the fission products and most of the associated radiation are confined within the fuel elements, the comparatively low level of radiation associated with the main nuclear-steam components is held within the containment building, sealed off from the rest of the plant and the outside.

In many nuclear plants, this containment building gives the characteristic architecture—a domed structure as in Fig. 6.9—whereas in others, while containment is used, it does not show in the plant architecture.

Ventilation Stack. The ventilation stack has the appearance of a smokestack, as indicated in Fig. 6.9. However, its function is to discharge high in the atmosphere small amounts of highly diluted, filtered, gaseous wastes. The amounts are strictly controlled within safe limits by rigid standards. As mentioned earlier, the exposure to the public from nuclear plants is far less than 1% of the natural radiation to which everyone is exposed.

Nuclear Fuel[16,17]

Uranium, thorium, and plutonium are nuclear fuels. However, the relatively small amount of thorium used to date (1981) has been available on the open market. Plutonium is produced by irradiation of uranium. Thus, this section is primarily an account of uranium production and processing. Most uranium produced currently (1981) is from ore bearing 2 to 20 lb uranium per ton, average about 5 lb per ton. In the United States, the majority of such ores are found in the western part of the country, notably northwestern New Mexico, central Wyoming, and the Colorado–Utah border region. While some uranium is mined in open pits, the largest part is mined underground. After mining, the uranium goes through the processes of milling to "yellow cake" (70 to 90% U_3O_8), refining to UO_3, "orange oxide," then to UO_2, then to UF_4, "green salt," and finally enriching from 0.7% to about 3% U-235, in a gaseous diffusion plant (currently) or centrifuge (future).

It requires about 150 tons uranium oxide, U_3O_8, to run a 1000 MWe nuclear plant for a year, about 2.5 million tons coal for a corresponding coal-fired plant, about 10,000 times as much coal as yellow cake. Thus, after milling, the transportation cost of uranium is negligible. It is shipped to the most suitable locations for refining, enriching, and fabricating.

Fuel Fabrication for Reactors.   Following enrichment, the uranium is converted chemically to an oxide, UO_2, for light-water reactors or to a carbide for the higher-temperature reactors. The UO_2 is formed into cylindrical pellets and loaded into fuel rods, about 200 rods forming a fuel element, Fig. 6.11. Each fuel rod is filled with helium for thermal bonding and sealed. The cladding also contains some 99.99% of the fission products within these tubes.

Reprocessing of Spent Fuel.   After three or four years of operation in a reactor, the useful life of a fuel rod is terminated by radiation damage, build-up of poisons (fission products), and depletion. After a period of "delay cooling" at the reactor site,

the valuable uranium and plutonium in the spent fuel can be recovered at a reprocessing plant. By remote control behind massive concrete shielding walls, the structural supports are cut away, the remainder dissolved, and the uranium and plutonium recovered and purified by successive extractions. The waste is concentrated for disposal.

The plutonium is a fuel for breeder reactors or, as is well known, has military uses. The uranium reenters the fuel cycle, either being recycled through "enrichment" or blended directly with enriched uranium for reactor fuel.

Waste Disposal.   A preferred method of disposing of nuclear waste was described at the beginning of this chapter. However, since a final decision on waste disposal and on reprocessing of civilian fuel has not yet been reached (1981), the treatment in this chapter has been limited to a few brief paragraphs. Note: The ban on commercial reprocessing was removed in Oct. 1981.

REFERENCES

1. "Shippingport Issue," *Westinghouse Engineer,* March 1958.

2. M. Benedict, "The Case for Nuclear Power," Founders Award Lecture, National Academy of Engineering, November 18, 1976, Washington, D.C.: NAE.

3. R. H. Malès, "Fast Breeder Called Energy Insurance Policy," Excerpt from Address to Nuclear Reg. Subcomm. of Senate Comm. on Environment and Public Works, 4-28-77, *EPRI Journal,* August 1977, p. 35.

4. *Nuclear Power from Fission Reactors—An Introduction,* DOE, November 1979.

5. *Advanced Nuclear Reactors—An Introduction,* ERDA–76–107, May 1976, Oak Ridge, Tenn.: ORNL Technical Information Center.

6. R. L. Lyerly and W. Mitchell, III, *Nuclear Power Plants,* AEC/DOE, Oak Ridge, Tenn.: ORNL Technical Information Center, 1973.

7. "HTGR—The High-Temperature Gas-Cooled Reactor," *Orange Disc,* May–June 1974, p. 22.

8. R. P. Smith, "Thorium Won't Solve Proliferation Problems," *New Engineer,* July–August 1977, p. 25.

9. "Breeder Reactors," *Westinghouse Engineer,* January 1968, entire issue (J. W. Simpson, "Industry/Government Approach"; J. C. Rengel, "Fuel Considerations"; R. J. Creagan, "Basic Concepts"; J. H. Wright, "Core Design"; J. C. R. Kelly, Jr., and P. G. DeHuff, "Prototype Plant.")

10. W. M. Jacobi, "A Demonstration Power Plant Design for the Liquid-Metal Fast Breeder Reactor," *Westinghouse Engineer,* November 1971, p. 180.

11. "Fast Flux Test Facility Progressing," *Westinghouse News,* August 1977, p. 1.

12. D. B. Coburn, "Gas-Cooled Fast Breeders," *Power Engineering,* February 1969, p. 30.

13. J. M. Dukert, *Thorium and the Third Fuel,* AEC/DOE, Oak Ridge, Tenn.: ORNL Technical Information Center, 1970.

14. W. E. Johnson, D. H. Fax, and S. C. Townsend, "PAR—Homogeneous Reactor," *Westinghouse Engineer,* March 1957, p. 34.

15. J. F. Hogerton, *Nuclear Reactors,* AEC/DOE Oak Ridge, Tenn.: ORNL Technical Information Center, 1965.

16. J. F. Hogerton, *Atomic Fuel,* AEC/DOE, Oak Ridge, Tenn.: ORNL Technical Information Center, 1964.

17. A. L. Singleton, Jr., *Sources of Nuclear Fuel,* AEC/DOE, Oak Ridge, Tenn.: ORNL Technical Information Center, 1968.

18. E. J. Lerner, "Magnetic Fusion Power," *Spectrum,* December 1980, p. 44.

19. S. McGrady, "Is Fusion a Falling Star?" *Reader's Digest,* July 1981, p. 54.

7

Solar Energy

INTRODUCTION

In a sense, all energy on earth, except the faint radiation from the stars, is solar energy, since the earth itself with its store of nuclear, geothermal, and rotational energy was thrown off from the sun. The sun in ages past supplied the energy now contained in our fossil fuels—coal, oil, and gas. It continues to supply the hydrocycle of evaporation, rain, and waterfalls and causes the wind to blow and move windmills and sailboats. Tidal energy derives from the rotational energy stored in the sun–earth–moon system. Energy from the sun also causes the ocean temperature differences, from which we can derive energy. And most important, the sun provides direct sunlight that warms the earth, grows our food and all our organic material, and now supplies some of the heat and electricity we have traditionally obtained from fuels.

Obviously, an arbitrary dividing line is necessary for the material to be included here under "solar energy." Nuclear, geothermal, and tidal energy and the fossil fuels are not generally classed as solar. These are treated in separate chapters in this book. Wind power, hydro, and current growth (biomass), which some class as solar, are also treated in separate chapters. This chapter covers solar collectors and concentrators for water heating and for space heating and cooling, and for the generation of electricity, including the ocean thermal system. Passive solar systems, though at least equally important, are not covered in this book.

The basic principles of solar thermoelectric and solar voltaic (photovoltaic) systems are treated in Chapter 13. However, the application of photovoltaic conversion to buildings is covered here. Similarly, the basic principles of absorption-type cooling and of the Rankin heat-to-work cycle are given in Chapter 13. But their applications to solar cooling and to solar power plants and ocean thermal energy conversion (OTEC) are covered in this chapter.

THE TOOLS OF SOLAR ENERGY

The tools available are of two types: information and devices. The functioning of the devices is dependent on the characteristics of solar energy: How much? When? Where? At what frequencies? Direct or diffuse? At what angles? It is also dependent on the characteristics of materials: their light transmission, reflection or refraction; their thermal conductivity (insulation) and heat storage capacity; the flow characteristics of fluids; and their cost.

Solar Information

The sun puts out 3.92×10^{26} W continuously. Of this, the earth intercepts 1.783×10^{17} W, or 129.6 W/ft² projected area, based on the mean sun–earth distance of 92.9 million miles. This latter figure is called the "solar constant," and the proposed value by NASA,[1] following the space flights, is 1353 W/m², or 125.7 W/ft², which is equivalent to 429 Btu/ft² hr. The value 442 Btu/ft² hr was commonly used in the past[2] and corresponds to the earlier calculation of 129.6 W/ft². In any event, the value fluctuates ±3% from this mean as the earth's distance from the sun varies ±1.5% annually in its elliptic orbit. The exact value is of theoretical interest only, except in space, since there are many losses between outer space and the collector on a roof.

Wavelength

The sun's energy, as it reaches our outer atmosphere, is of very high frequency, or short wavelength, from 0.3 to 3.0 μ (1 μ is 10^{-6}m). It is distributed approximately as follows:

7.82% in the ultraviolet spectrum.
47.33% in the visible spectrum (0.4 to 0.76 μ).
44.85% in the infrared spectrum.

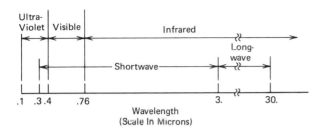

Figure 7.1. Electromagnetic spectrum. Diagram of the portion of the electromagnetic spectrum used in solar heating devices, showing the shortwave and longwave portions of the thermal spectrum. (1 micron = 10^{-6} meters.)

This distribution is characteristic of a body radiating at 6000°K.

As the radiation penetrates the atmosphere, the ultraviolet is selectively absorbed more than the infrared and changes this proportion. This has little effect on the efficiency of solar collectors but does matter in photovoltaic cells, where the materials used are frequency sensitive in their performance (Chapter 13). However, radiation reaching the earth is still "short wave" or high frequency, as defined by Fig. 7.1. Glass and certain plastics are transparent to most of this spectrum, and about 90% will pass through one pane of glass, 80% through double glazing.

When the radiation is absorbed by materials and reradiated, it is being radiated by a body at 300 to 400°K, not 6000°K. The frequencies reradiated are much lower, or the wavelength is much longer, and are shown as "long waves" in Fig. 7.1. Window glass is essentially opaque to these frequencies, and this energy is trapped within the collector. This is known as the "greenhouse effect." The only energy to escape is that *reflected* back, which can be kept quite low (about 5%) with good absorbent paints, materials, or construction. Heat also escapes by conduction, aided by convection, which will be treated later.

Global Radiation

From a global standpoint, of the total radiation reaching the earth's atmosphere (1514 × 10^{15} kWh/yr), approximately the following holds true:

50% is reflected back into space by the clouds.
15% is reflected back by the earth's surface.
5.3% is absorbed by bare soil.
1.7% is absorbed by marine vegetation.

0.2% is absorbed by land vegetation.
27.8% is absorbed in evaporating water.

This latter energy is largely reradiated into space when the water condenses in the clouds.

The ratio of the earth's surface to its projected area is 4. Thus, if 50% of the solar constant is spread over the whole earth, the average energy on a horizontal square foot is 429/8 = 53.6 Btu/ft² hr, or 1287 Btu/ft² day, average for the year. The measured values (Table 7.1), ranging from 1219 Btu/ft² day on the horizontal for Winnepeg, Canada, to 1916 for Phoenix, Arizona, annual mean, appear not inconsistent with this global picture. Values would be higher near the equator and lower near the poles. However, this is of academic interest since measured values have been recorded.

Clear-Day Radiation

More insight into the phenomenon taking place as the sun's radiation passes through the atmosphere can be obtained from Fig. 7.2. Passage through the atmosphere splits the radiation reaching the surface into a direct and a diffuse component and reduces the total energy through selective absorption by dry air molecules, dust, water molecules, and thin cloud layers. Heavy cloud coverage eliminates all but diffuse radiation.

The range of conditions is shown in Fig. 7.2. For a surface always perpendicular to the sun's rays, on a clear day at medium and low latitudes, within 4 hr either side of noon.[1] Starting with the solar constant, 1353 W/m², or 429 Btu/ft² hr, some 56 to 83% penetrates as direct sunlight, and 5 to 15% as diffuse sunlight, for a total of 71 to 88%. This is 961 to 1191 W/m², or 305 to 378 Btu/ft² hr. The diffusion is caused by dry air molecules and dust, the latter of course being highly variable.

The rest of the radiation is reflected or absorbed. The reflection, 1.1 to 11%, is highly variable because of the dust, which causes most of it. As mentioned earlier, clouds reflect 50% of all solar radiation, but Fig. 7.2 is for a clear sky. Absorption accounts for 11 to 23% of the solar radiation in dry air molecules, dust, and water vapor.

The maximum reception, 88% of the solar constant, or 378 Btu/ft² hr, if applied for 8 hr/day would yield 3024 Btu/ft² day on a surface normal to the radiation. The actual mean daily insolation (annual) at Phoenix, Ariz., is 1919 Btu/ft² day on a horizontal surface, close to the highest obtained in U.S. cities.

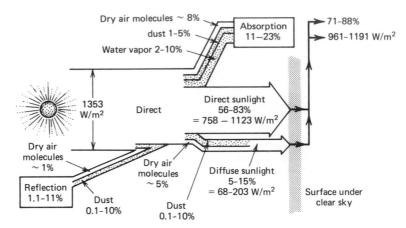

Figure 7.2. Sunlight penetration of atmosphere (clear sky). From *McGraw Hill Encyclopedia of Science and Technology,* Copyright 1977. Used with the permission of the McGraw Hill Book Co.

In June, the mean daily insolation in Phoenix is 2725 Btu/ft² day. The sun's rays at noon are about 10° from vertical. However, by 4 hr either side of noon, they are over 60° from vertical.

Degree Days (DD)

Empirically, it has been found that no central heat is required to maintain a house at 70°F with the outside temperature 65°F. Minimal sun on the house, the ground, greenhouse effect, lighting and cooking heat, and occupants accomplish this on the average. With 65°F as a base, each degree lower is counted as 1 degree day, or 1 DD. One day at 55°F average temperature or five days at 63°F each constitute 10 DD, base 65°F.

Climatological Data[2-4]

Climatological data have been collected for most major cities in the United States and throughout the world. Samples of the U.S. and Canadian statistics are given in Table 7.1, covering the range of degree days and solar insolation from Winnepeg, Canada, to New Orleans, Louisiana, and Los Angeles, California.

For solar heating and cooling of houses and buildings, the principal data needed are the monthly and annual degree days, the solar insolation on level ground, and the ambient temperatures. These data are given for the selected sample cities in Table 7.1. Some cities have more than one recording point, for example, airport and city stations. The values in Table 7.1 are from Ref. 3, for consistency in certain examples to follow. Complete tables will be found in ASHRAE Systems.[4]

Also, of supplementary use are the number of days of precipitation, the percentage and hours of sunshine, and the mean sky cover, which are a part of weather data.[2]

Heat Loss Estimating Factors

The heating, refrigerating, and air conditioning industry has developed empirical heat loss tables, such as Table 7.2, from which the heat loss of a house or building at any given design temperature can be estimated. Such tables have long been used for heating requirements and of course serve for solar heating as well. While much more refined calculations can be made, this empirical table serves ideally for this chapter to provide an insight into the heating system needed as a function of the house size and insulation and its location.

Each city or location has an accepted design temperature. Typical values are shown in Table 7.1 for a few cities.[2] For example, for Pittsburgh, it is 0°F. In general, it is 15°F above the lowest temperature on record at that location.

From Table 7.2, one can calculate the heat loss per hour (or day) through the insulation of the structure, at the design temperature. The loss is taken as zero at 65°F. Thus, if the design temperature is 0°F (Pittsburgh), the heat loss per degree is then the heat loss at the design temperature ÷ (65 − 0).

In addition, there is a heating loss due to infiltration. For n changes of air per hour, in a house of volume V ft³, the heat loss due to infiltration is

$$L = n \times V \text{ ft}^3 \times 0.075 \text{ lb/ft}^3 \times 0.24 \text{ Btu/lb} \times 24 \text{ hr}$$
$$= 0.432nV \text{ Btu/DD} \qquad (7.1)$$

Table 7.1. Typical Climatological Data

Location	Jan.	Feb.	Mar.	Apr.	May	June	July	Aug.	Sept.	Oct.	Nov.	Dec.	Annual
				(a) Total Heating Degree Days (Base 65°F)									DD$_{annual}$
Albuquerque, N.M.	913	700	596	283	58	0	0	0	7	218	616	893	4284
Greensboro, N.C.	815	682	544	203	59	0	0	0	23	209	500	787	3822
Lincoln, Neb.	1237	1015	833	401	171	31	0	5	76	301	729	1066	5865
Los Angeles, Cal.	331	270	266	194	113	70	20	14	23	77	158	279	1815
Madison, Wis.	1494	1253	1078	590	297	72	14	40	173	473	909	1336	7729
New Orleans, La.	364	257	193	4	0	0	0	0	0	20	193	322	1389
Phoenix, Ariz.	428	292	185	59	0	0	0	0	0	16	182	389	1551
Pittsburgh, Pa.	1066	923	763	382	160	11	0	5	58	299	626	983	5276
San Francisco, Cal.	331	270	266	194	113	70	20	14	25	77	158	279	1815
Shreveport, La.	553	427	304	81	0	0	0	0	0	47	297	477	2186
Winnipeg, Can.	2009	1719	1465	814	405	148	38	70	322	682	1251	1757	10680
				(b) Mean Daily Solar Radiation (Btu/ft² on Horizontal) (Insolation)									
Albuquerque, N.M.	1135	1438	1887	2323	2536	2724	2544	2345	2087	1648	1246	1036	1912
Greensboro, N.C.	755	1002	1305	1729	1965	1072	1994	1751	1519	1213	896	686	1407
Lincoln, Neb.	701	940	1279	1563	1828	2009	1980	1872	1519	1198	763	634	1912
Los Angeles, Cal.	948	1268	1692	1919	2124	2274	2393	2171	1858	1357	1080	907	1666
Madison, Wis.	565	813	1233	1457	1747	2034	2049	1742	1446	994	556	496	1261
New Orleans, La.	790	956	1237	1521	1657	1635	1539	1536	1413	1328	1026	731	1280
Phoenix, Ariz.	1095	1504	1921	2370	2669	2728	2404	2256	2094	1666	1250	1032	1916
Pittsburgh, Pa.	584	786	1188	1477	1797	2963	2019	1779	1499	1985	679	524	1289
San Francisco, Ca.	719	1045	1506	1890	2133	2203	1993	1764	1565	1225	849	646	1462
Shreveport, La.	833	1029	1394	1722	2020	2005	2072	1917	1530	1276	896	730	1452
Winnipeg, Can.	483	829	1338	1618	1880	1935	2094	1740	1176	755	439	339	1219
													Design Temp. (°F)
				(c) Mean Monthly Ambient Temperature (°F)									
Albuquerque, N.M.	34	39	45	54	63	72	77	73	68	55	43	34	0
Greensboro, N.C.	37	39	46	57	66	73	77	75	70	57	46	39	
Lincoln, Neb.	25	28	37	52	61	72	77	75	66	54	39	30	−10
Los Angeles, Cal.	54	55	55	59	61	64	68	68	68	64	59	55	
Madison, Wis.	19	21	32	45	55	56	70	68	59	50	34	23	
New Orleans, La.	52	55	59	68	73	79	81	81	77	68	59	54	20
Phoenix, Ariz.	50	55	59	66	75	84	90	88	82	72	59	52	25
Pittsburgh, Pa.	32	32	41	52	63	72	75	73	66	55	45	34	0
San Francisco, Cal.	50	54	54	55	57	59	59	59	63	61	57	52	
Shreveport, La.	46	50	57	64	72	79	82	82	77	66	55	48	
Winnipeg, Can.	1	5	19	37	52	59	63	63	54	43	25	9	

For example, consider a house having the following heat losses at a design temperature of 0°F:

1200 ft² of sidewall, 2 in. insulation	× 9 =	10,800 Btu/hr
150 ft² of windows, single glazed	× 79 =	11,850
60 ft² of doors, with storm doors	× 23 =	1,380
2000 ft² of ceiling, 6 in. insulation	× 4 =	8,000
and heated basement		0
Total		32,030 Btu/hr

Thus, there is 32,030 × 24/65 = 11,826 Btu/DD heat loss through the insulation of the structure. Suppose the house has a volume of 15,000 ft³ and

Table 7.2. Heat Loss Estimating Multipliers (Btu/ft² hr) at Design Outside Temperature[2]

	Design Outside Temperature (°F)						
	−30	−20	−10	0	+10	+20	+30
Sidewalls: frame or brick[a]							
No insulation	26	23	21	18	16	13	10
2 in. insulation	13	12	10	9	8	7	5
3⅝ in. insulation	7	6	6	5	4	4	3
Windows							
Single glazing	113	102	90	79	68	57	45
Double glazing	69	62	55	48	41	35	28
Doors							
1½ in. solid wood	49	44	39	34	29	25	20
1½ in. solid wood with metal storm door	33	30	26	23	20	17	13
Ceilings: Attic Above[a]							
No insulation	32	29	26	22	19	16	13
3⅝ in. insulation	8	7	7	6	5	4	3
6 in. insulation	5	5	4	4	3	3	2
10 in. insulation	3	3	2	2	2	2	1
14 in. insulation	2	2	2	2	1	1	1
Floors							
Concrete slab	10	9	8	7	6	5	4
Over crawl space (no insul.)	14	13	11	10	8	7	6
Over crawl space (4 in. insul.)	5	5	4	4	3	3	2
Over heated basement	0	0	0	0	0	0	0

[a]Based on Fiberglas batt insulation (k = 0.3 Btu/ft² hr °F in.).

there is one change of air per hour (an average value). The infiltration loss is then

$$0.432nV = 0.432 \times 1 \times 15,000 = 6480 \text{ Btu/DD}$$

The total heat loss is 11,826 + 6480 = 18,306 Btu/DD. On a 55°F day, this house will require 183,060 Btu to maintain it at 70°F. It is thought of as an 18,306 Btu/DD house. The application of this information will appear later under solar heating and cooling of buildings.

Typical well-insulated houses have Btu/DD values about as shown in Table 7.3.[2] To these values must be added the Btu/DD due to infiltration.

Devices Used In Solar Energy Systems

Some of the devices required specially for solar energy systems are the following. In general, many devices long standard in more conventional systems are used, and, everything is better insulated to conserve energy.

Solar Collectors

Solar collectors as described in the following paragraphs are primarily for residential service. Collectors for commercial, industrial, and institutional service are of course similar in principle. The flat-plate solar collector (Fig. 7.3) for air or water is essentially a large shallow box, well insulated on the bottom and sides, and with a heat-conducting plate painted with nonreflecting, heat-absorbing paint on the inside or coated with a selective coating. The top is sealed with one or two layers of glass or plastic with a thin dead-air space between. Approximately 90% of the solar radiation will penetrate one pane of glass, 80% through double glazing. With good nonreflecting surfaces, about 95% of the entering radiation is absorbed and 5% is reflected back through the glass.

For hot-air systems, the air flows above the conducting plate, absorbing heat. For hot-water systems, water pipes are bonded to the plate as shown in Fig. 7.3. Plastic collectors and piping are also used.

Table 7.3.   Typical Heat Losses of Well-Insulated Houses[a,b][2]

Size of House (ft²)	Heat Loss (Btu/DD)
750	4,000
1000	5,000
2000	7,500
3000	10,000
4000	12,500
5000	15,000

[a]To this must be added the Btu/DD due to infiltration.
[b]For *average* U.S. houses of 1980 the *total* heat loss is about 10 Btu/sq. ft. of floor. DD, eg for 2000 sq ft —20,000 Btu/DD.

Collectors are generally set facing south and elevated to a slope of (15° + latitude) ± 10° (not critical) above the horizontal. At this position, the sun's rays strike the collectors nearly perpendicularly at noon during the entire heating season, and the energy collected is maximum.

Collectors for domestic hot-water heating must function all year round. However, since the June insolation is 2.2 to 5.7 times the December insolation (Table 7.1) and the collector losses are greatly reduced in summer, it is still best to optimize the system for winter operation. Thus, the collector slope of (15° + latitude) above horizontal is usually best for these installations also.

When the solar energy is used for both heating and air conditioning, the relative summer and winter requirements and the insolation will determine whether one or the other should be favored. Optimum collector angles have been developed[5] for summer maximum, winter maximum, and all year-round maximum.

Cost of Collectors. A rough rule of thumb[2] for the cost of collectors is $7.50/ft², F.O.B. factory (1974); and for the complete solar heating system, including pipes, ducts, valves, vanes, storage, pumps or blowers, collectors, controls, and installation, approximately four times the cost of the collectors alone. Double-glazed collectors cost about 20% more than single-glazed ones. Overall system costs from $20 to $40 per ft² of collector are typical of 1974.

These costs are for an *auxiliary* solar heating system, supplying heat to a conventional heating system as illustrated in Fig. 7.11 later. They do not include the conventional furnace and associated

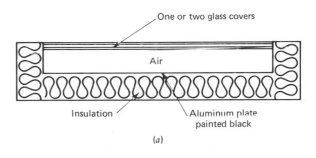

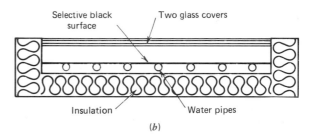

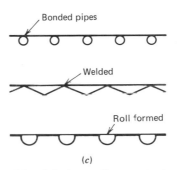

Figure 7.3. Solar heat collectors: (*a*) hot air; (*b*) hot water; (*c*) alternate bonding methods.

heat-distributing system. Price indices for updating these costs to the date of interest are given in Chapter 1.

Collector Theory. Collectors are generally operated "on–off," that is, with a fixed flow or none at all. A typical liquid flow rate is 0.022 gpm/ft² (0.015 l/sec m²) of collector area. It is not critical. Air flow rates from 1 to 4 cfm/ft² (5 to 20 l/sec m²) are used. The collector efficiency depends on the flow rate; 2 cfm/ft² is typical.

The collector output cannot exceed the entering insolation I_t multiplied by the cover transmittance τ and the plate absorptance α. If the entering fluid were at ambient temperature, the flow of heat from the plate to the fluid would raise the outlet temperature and cause some heat loss to ambient as well as useful heat to the fluid. The useful fraction is called F_R. Thus, with the entering fluid at ambient temperature, the useful energy transferred to the fluid is $F_R A I_t \tau \alpha$, where A is the area of the collectors.

In general, the entering fluid is at a temperature above ambient, the differential being $T_i - T_a$, the inlet temperature minus the ambient temperature. If the whole collector was at the inlet temperature T_i, the loss per unit area would be $U_L(T_i - T_a)$, where U_L is the collector loss coefficient per unit area. However, the gradient reduces this loss to $F_R A U_L (T_i - T_a)$. Thus, the useful energy Q_U from a collector is

$$Q_U = F_R A \left[I_t \tau \alpha - U_L(T_i - T_a) \right] \qquad (7.2)$$

Or, the efficiency of the collector is

$$n = \frac{Q_U}{A\, I_t} = F_R \tau \alpha - F_R U_L \frac{(T_i - T_a)}{I_t} \qquad (7.3)$$

Thus, with the inlet temperature at ambient ($T_i - T_a = 0$), the efficiency is simply $F_R \tau \alpha$, typically 68 or 77% with two or one covers, as shown in Fig. 7.4. However, it is reduced by a collector loss quantity that depends on the ratio of the temperature differential to the insolation. With twice as much temperature differential and twice as much insolation, the efficiency is the same.

If the efficiency n is measured and plotted against the quantity $(T_i - T_a)/I_t$, the n-axis intercept is obviously $F_R \tau \alpha$ and the slope is $-F_R U_L$. For example, at the highest possible insolation (barring reflectors), namely, 88% of the solar constant, $0.88 \times 125.7 = 111$ W/ft² (1191 W/m²); and with water en-

tering the collectors at 100°F (38°C) and an ambient temperature of 32°F (0°C), the quantity $(T_i - T_a)/I_t$ is $(100 - 32)/111 = 0.61$. The efficiency of a typical liquid heat collector would then be between 55 and 62%, as shown in Fig. 7.4.

Heat Storage

Heat storage is generally accomplished by a well-insulated bin of gravel ($1\frac{1}{2}$ in. washed gravel) for a hot-air system or a large tank of water for a hot-water system. The bin may be of an insulating material, such as rigid polyurethane foam, or it may be a well-insulated section of culvert set vertically. The specific heats and densities of water and gravel are shown in Table 7.4. Water holds five times as much heat per pound and 3.1 times as much per cubic foot as gravel for the same temperature rise.

It has been found empirically[3] that little is to be gained by storage above 0.82 ft³/ft² (0.25 m³/m²) of collector area for an air system (pebbles), or 1.84 gal/ft² (75 l/m²) of collector area for a water system. That is, 82 lb pebbles or 15.4 lb water per square foot of collector is a reasonable amount of storage, about 5 to 1 by weight. Note that at the same temperature rise, these provide the same heat storage. This applies to systems such as Fig. 7.11 to be discussed later. With less storage than this, the system efficiency drops off rapidly.

An idea of the size of storage may be given by an example. A 15,000 Btu/DD house located in Pittsburgh, Pa., would require about 360 ft² double-glazed, flat-black collectors for 50% solar heating with average storage and flows for a liquid system and 461 ft² of such collectors for an air system.

The pebble-bed storage of $0.82 \times 461 = 378$ ft³ might be a bin 6 ft by 10 ft by 6 ft 3 in. high, containing 37,800 lb (18.9 tons) gravel. It would normally be set on concrete blocks covered with wire mesh to let the air rise uniformly through it. The

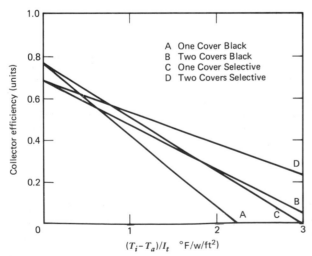

Figure 7.4. Typical efficiencies of liquid heating collectors (air heating collectors similar).

A   One Cover Black
B   Two Covers Black
C   One Cover Selective
D   Two Covers Selective

Table 7.4.   Heat Storage in Water or Rock (Gravel)

	Water	Gravel (Hot-Air System)
Specific heat (Btu/lb °F)	1.0	0.2
Density (lb/ft³)	62 at 100°F	100 (Avg. Value)
Specific heat capacity (Btu/ft³ °F)	62	20

flow is reversed for heat retrieval, with the air coming out of the hot end of the storage.

The hot-water storage tank of $1.84 \times 360 = 662$ gal (88.5 ft³) might be 4 ft in diameter and 7 ft long, or a 4.5 ft cubical tank.

Gravel for the above storage, 19 tons, even at $10 a ton, is only $190. Most of the cost of storage is in the structure and the installation.

Heat-of-Fusion Storage. The storage can be reduced in size and made more efficient through the use of a material that changes phase at the desired storage temperature. A eutectic salt is used, such as Glauber's salt, which has a heat of fusion of 108 Btu/lb at 100°F. It thus stores 108 Btu/lb when the salt has all liquefied, compared with 30 Btu/lb max. for gravel, even with a 150°F temperature rise. It has the additional advantage that all storage and retrieval is at approximately 100°F, so that both collector and storage efficiencies are substantially increased. Its cost is of course greater than gravel or water and a large heat exchange area is required.

Solar Cells

The theory of photovoltaic cells is covered in Chapter 13. These cells convert the incident sunlight directly into dc electricity, with an efficiency of 10 to 15%. They are usually made from single-crystal silicon and are relatively expensive. A 1966 book[6] gives "about $400/watt." At that time, they could be afforded only in satellite power applications where they were the least expensive alternative for small, long-duration supplies. In terrestrial applications, their use factor is much less due to fewer hours of sunlight, and other alternatives were cheaper.

However, in 1979, with the dendritic web process, a current price of $10/W is mentioned,[7] with an efficiency of 15.5%. The long-range goal is to achieve 22% efficiency. The Department of Energy has set a goal of about $1/W for solar cells by 1986. Figure 7.5 shows a battery of dendritic-web furnaces and the web before and after application of connections.

As a part of the DOE program, scheduled for mid-1979, the Mississippi Country Community College will be the first large building to draw electric power from photovoltaic (solar) cells. With a 250 kW peak output, this will be the largest photovoltaic system ever assembled and will satisfy all the power requirements of the 60,000 ft² facility.

(a)

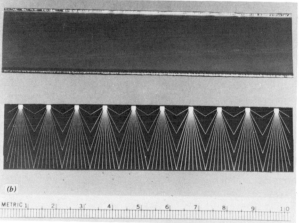

(b)

Figure 7.5. (a) Dendritic-web furnaces. (b) Dendritic-web solar cells, without and with connections. Courtesy Westinghouse Electric Corp.

The solar cells are arranged at the focus of parabolic mirror reflecting troughs that track the sun, and provide about 30 suns intensity on the cells. The cooling water duct below the solar cells provides building heat in cold weather.

Later installations use fixed solar cell panels without the expensive tracking mirrors. As the cost of solar cells is reduced this arrangement becomes more economic.

Solar Thermoelectric Devices

Solar energy concentrated onto a thermoelectric converter is another available tool (see Chapter 13). Its use is economic only in isolated locations or in space.

Solar Rankine-Cycle Conversion

If enough solar energy is concentrated on a boiler to produce high-pressure steam or vapor, a Rankine-cycle power plant can be used to convert the energy to electricity. (see Chapter 13, Rankine Cycle). Several pioneering applications of large-scale solar power are now underway (1981), as a part of the DOE Solar program, and are described later. Among these are (1) a large field of reflectors (heliostats) concentrating the solar energy on a boiler atop a tower (central receiver system); (2) a field of parabolic dishes, each focusing the sun's rays on a receiver at its focus, heating a fluid that is collected for power generation, or operating a small heat engine directly to generate electricity; (3) a field of parabolic troughs tracking the sun and heating the working fluid in a pipe at its focus; (4) large fixed solar bowls, each generating steam in a line focus receiver that tracks the sun; and (5) the ocean solar system in which a low-temperature Rankine cycle is operated between the sun-warmed surface waters of the ocean and the colder deeper waters.

Concentrators

Any terrestrial use of solar thermoelectric, solar Rankine cycle, or solar chemical requires concentrating the solar energy by mirrors or reflectors in order to use the high-cost converters more efficiently and, in the case of the Rankine cycle, to attain the temperature needed for its operation. Ocean solar uses low temperatures but tremendous volumes and is not applicable on land.* Reflectors or reflecting surfaces are also effective in increasing the output from the more conventional solar collectors.

One needs to be reminded that diffuse radiation comes from all directions and cannot be reflected or focused in the same manner as direct radiation. However, diffuse radiation striking a reflecting surface, such as a flat surface painted white or the snow-covered ground, is reflected as diffuse radiation. With this reminder, the principal tools of concentration used in solar energy systems are the following.

Reflecting Surface. A white or bright surface is used to reflect both direct and diffuse radiation in a desired general direction. Two examples are:

1. The cover of a "backyard solar furnace"[2] covers the collector glass for protection in summer but folds out on the ground in winter. Its white inside surface, or snow if present, increases the radiation entering the collector by about 32% on a winter solstice day at 35° latitude, due to direct radiation.

2. At the George A Towns Elementary School in Atlanta, Georgia (Fig. 7.6), some 10,360 ft² of solar collectors supply 60% of the total energy for heating, cooling, and hot water for the 32,000 ft² structure. A 100 ton absorption cooler is used for the air conditioning. The back slope of each row of collectors is a reflector for the next row, some 13,000 ft² of reflectors altogether. The reflectors were removed in 1981 and larger collectors installed.

Heliostat. A heliostat is a large flat or slightly parabolic mirror surface, with elevation and azimuth control, for reflecting the sun's rays continually on a particular spot. The heliostat is typically constructed of 25 4 ft by 4 ft mirrors arranged 5 × 5 (400 ft²) or 48 3 ft by 3 ft mirrors (432 ft²) arranged as shown in Fig. 7.7. If parabolic, the individual mirrors are flat, but are mounted to form a parabola with the focus at the distance of the spot to be heated. An example of their use is given later, under Solar Power Plant.

For a field of heliostats to keep the sun's rays directed on a spot as the sun's path from sunrise to

---

*Except see Solar Ponds later.

Figure 7.6. The George A Towns Elementary School in Atlanta, Georgia, has an area of 32,000 square ft, accommodates 500 students, and operates throughout the year. Courtesy Westinghouse Electric Corp.

sunset changes with the seasons, each heliostat has a different elevation and azimuth at every instant. A central computer determines these settings, and the individual heliostats follow its commands. As mentioned, only the beam radiation can be focused.

Parabolic Mirror, or "Dish." All light rays parallel to its axis converge at its focus. Thus, if the

Figure 7.7. Westinghouse prototype low-cost, mass reproducible heliostat, 432 square ft (12 ft × 36 ft). Courtesy Westinghouse Electric Corp.

"dish" is kept directed toward the sun, the direct radiation intercepted is concentrated at the focus and can be used for:

1. Heating a fluid for power generation or process heat at a central location.
2. Driving a small heat-engine generator.
3. Changing solar energy to chemical energy in a small catalytic reactor for collection and use in a central power plant.
4. Providing a high-temperature heat source for direct heating needs such as solar cooking. Originally, this arrangement was called a solar furnace and was used scientifically in the 1800s.

Since the sun's rays are parallel, all parabolic mirrors in a field, such as in Fig. 7.8, have the same elevation and azimuth, thus simplifying the control.

Parabolic Mirror Troughs. Long parabolic trough-shaped mirrors are used to concentrate the sun's rays on a pipe located at the focus, producing hot water for power generation or process heat as mentioned earlier. They are also used to concentrate energy on photovoltaic cells arranged along the focus, as mentioned earlier.

Figure 7.8. Parabolic mirrors–solar total energy system. Large-scale experiment No. 2, Shenandoah, Georgia. This system will supply electricity, industrial process heat, and space heating and cooling for a new knitwear factory. Courtesy Westinghouse Electric Corp.

Bowl Collector. The parabolic mirror and trough are considered "distributed receiver systems," as contrasted with the heliostat, which is part of a "central receiver system." The bowl collector is being developed as part of the distributed receiver program.

The bowl, a 60° section of sphere, is stationary and facing upward. As the sun moves across the sky, the line at which the rays converge moves, and with it the thermal receiver which is a steam generator. A 65 ft diameter bowl was in operation at Crosbyton, Texas, in 1981, producing high temperature steam to verify the performance and provide design data for a proposed 5 MWe plant using 10–200 ft diameter bowls.

Thermal Insulation

The most valuable tool for use or conservation of solar energy is insulation. As clean fossil fuels have become more expensive, it has become economic to use them more efficiently.

In 1935, with coal at $3 to $4 a ton, the cost of heating a $10,000 six-room house, with minimal insulation, was about $25/yr in Pittsburgh, Pennsylvania. To save $5/yr, or 20%, one could hardly afford to spend more than $80 (6% interest). Insulation and weather stripping were applied more for comfort than for savings.

While the cost of the house has increased about five times, to about $50,000 (1979), the cost of heating it has increased about 20 times, to about $500/yr, with gas. Even at 1979 mortgage rates of 9

to 10%, $1000 can be afforded to effect a 20% fuel saving. The extra comfort comes free.

The saving can be readily estimated from Tables 7.2 and 7.1. For example, going from no attic insulation to $3\frac{5}{8}$ in. insulation saves 16 Btu/ft² hr at the design temperature of 0°F (Pittsburgh, Pa.). For 2000 ft² of ceiling, this is 32,000 Btu/hr at design temperature, or 32,000 × (24/65) = 11,800 Btu/DD. In Pittsburgh, with 5276 DD/yr, the saving is 5.276 × 11.8 = 62.3 million Btu/yr. In a gas furnace of 60% efficiency, the saving in fuel is 62.3/0.6 = 103.8 million Btu/yr. At $2.50/million Btu (Pittsburgh, 1979), this is a saving of $260/yr and would justify an expenditure of $2600, or $1.30/ft², not counting the comfort. The $3\frac{5}{8}$ in. batts of insulation cost about 30¢/ft². The labor would depend on contractor or "do-it-yourself." With an unfloored attic, it is a simple job.

With solar heat, the total capital cost of an installation can generally be substantially reduced by ample insulation, using the best that can be accommodated in the structure.

Insulations are rated in R values, the reciprocal of the conductivity k (Btu/hr per square foot of area per inch of thickness per °F of temperature drop) for loose materials, or the reciprocal of the conductance C (Btu/hr per square foot of area per °F of temperature drop, for the thickness listed) for various "batts" of insulation.

For example, typical Fiberglas insulation is of a material having a heat conductivity of 0.3 Btu/ft² hr °F in., or an R value of 3.33, but a finished batt 6 in. thick would have an R value of 6 × 3.33 = 20, or a conductance of 0.05 Btu/ft² hr °F.

APPLICATIONS OF SOLAR ENERGY

Having described the energy from the sun and the various tools for using it in lieu of fuels, we now come to selected applications illustrating how these principles and devices are being applied. Some of these applications are quite old, some currently (1979) in widespread use, and some are pioneering applications, exploring the feasibility and economics of future applications.

Older Systems

In addition to wind power and hydro, which are secondary forms of solar energy, direct solar energy has been harnessed in the past for a variety of tasks,

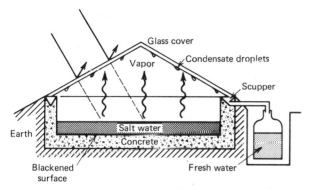

Figure 7.9. Roof-type solar still. From *McGraw Hill Encyclopedia of Science and Technology,* Copyright 1977. Used with the permission of the McGraw Hill Book Co.)

such as evaporating salt water to produce salt, drying fruits, vegetables, and tobacco, drying wood and peat, and distilling fresh water. These are all uses in which solar energy can, and frequently does, replace fuels. A roof-type solar still is shown in Fig. 7.9.

Solar Water Heaters

Solar water heaters are a simple and economic use of solar energy to replace fuels. A typical design is shown in Fig. 7.10. Generally operating on a ther-

mosyphon principle, with the hot-water tank above the collector, it continually feeds heated water to the upper part of the tank, and draws colder water from the bottom. These are available commercially and are widely used in many areas in which the probability of freezing is low. They must be drained at night if freezing is expected.

The water temperature typically reaches 180°F (82°C) in summer and 120°F (49°C) on sunny winter days. Roughly, 1 gal hot water per square foot collector per day is produced.

Thousands of these thermosyphon water heaters are now (1979) in use in Israel, N. Africa, Japan, and Australia. In freezing climates where air, or water with antifreeze, must be used in the collectors, a heat exchanger and indoor tank are required. These systems, frequently integral with solar house heating, are described in the next section.

Solar Home and Building Heating and Cooling

While limited solar home heating has been used for many years, the great impetus has come with the impending shortage of gas and oil and the associated steeply rising prices (1979). Strong government support has been given in the United States, particularly since the oil embargo of 1973–1974. Insu-

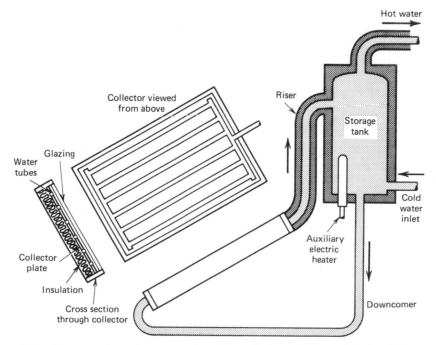

Figure 7.10. Thermosyphon solar water heater. (From *McGraw Hill Encyclopedia of Science and Technology,* Copyright 1977. Used with the permission of the McGraw Hill Book Co.)

lation and solar heating are encouraged by tax concessions. Environmental concerns over pollution by fuels has also stimulated the use of solar energy. There are literally thousands of designs and experimental installations being evaluated. These feature both conservation and use of solar energy.

Much can be accomplished, particularly in new buildings or housing, by judicious use of insulation and by efficient design and orientation to secure more solar exposure in winter, less in summer, and by the location of windows and the use of absorbing or reflecting colors. These techniques are termed "passive solar energy" techniques and are not covered in this book.

The devices already described, and the applications to follow, all fall in the category of "active solar energy." This section covers the use of solar collectors, usually to supplement the normal heating system of the house or building.

General Arrangement of a Solar Heating System

The size, cost, and fuel savings of a solar heating system can best be illustrated by considering a specific design. In general, we shall be speaking of a solar hot-air auxiliary system, as in Fig. 7.11a, or a solar hot-water auxiliary system, as in Fig. 7.11b. A "stand-alone" solar heating system is usually uneconomic.

In the air-based system, Fig. 7.11a, an optional air-to-water heat exchanger supplies heat to a preheat tank for the domestic hot-water heater. A conventional heating unit adds any additional space heat needed. For the house space heating, a fan and dampers, with suitable controls, provide the priority: (1) solar heat to house, (2) solar to storage, (3) storage to house, and (4) fuel or electric heat to the house.

In the water-based system (Fig. 7.11b), a heat exchanger is usually required between the collectors and storage. Thus, all solar heat passes through storage. Antifreeze in the collectors, if used, is not mixed with the comparatively large volume of storage water.

A second heat exchanger is needed between the storage tank and the air heating system of the house. A heat exchanger is also used between the storage tank and the hot-water preheater.

Some air-based systems arrange to bypass the pebble-bed storage in summer while heating domestic hot water only, whereas in the liquid-based system all heat passes through storage, summer and winter.

For domestic applications, an electric heat pump, if used, is in parallel with the solar heating system as shown in Figs. 7.11a and 7.11b. For the larger commercial, industrial, and institutional applications, the "solar augmented heat pump" may be in series. The collectors provide efficiently large quantities of low-temperature heat, which is then elevated to the required heating temperature by the heat pump. The combination is considerably more economical than a solar installation to provide the same heating.[8]

Collector, heat exchanger, and storage parameters assumed for the typical systems of Fig. 7.11 are shown in Table 7.5. The collector flows are as discussed earlier, and the symbols were defined under Collector Theory.

Economics of Solar House Heating—Specific Collector Areas

In order to give an idea of the economics of solar heating, the liquid-based system just described (Fig. 7.11b, Table 7.5) will be applied to a 15,000 Btu/DD house in each of the 11 sample cities for which climatological data are given in Table 7.1. The necessary area of collectors for 50% solar heating will be determined for each city; 50% solar is usually close to optimum, as will be shown shortly.

Then, based on the simplifying assumption that the capital cost of the complete installation is $30.00/ft² of collectors (1979), the capital cost of each installation will be determined. Since half of the conventional fuel is saved, its cost in dollars can then be expressed as a percent annual return on the solar installation cost. This will show, on a consistent basis, the *relative* advantage of solar heating in various locations and depending on which conventional fuel is displaced by solar.

Required Collector Areas. There are well established month-by-month calculation methods for determining the required collector area from the climatological data and the house heat loss, Btu/DD.[3,10] However, a much simpler method will be introduced here,[9] which gives collector areas within ±10% of the more accurate values and is completely consistent for general studies and estimates.

Intuitively, one knows that the collector area needed is proportional to the heat loss of the house, Btu/DD, and the number of DDs per year, DD_{annual},

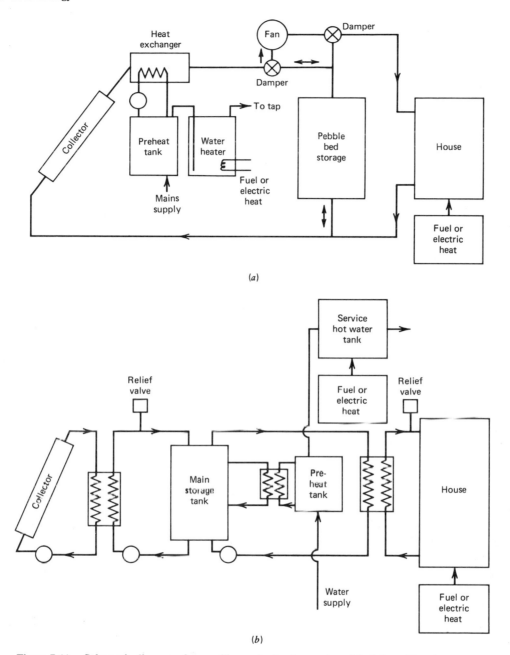

Figure 7.11. Schematic diagram of an auxiliary solar heating system: (*a*) air-based heating system; (*b*) liquid-based heating system.

in the location considered. The product of these two quantities gives the total heat energy needed by the house in a year.

It is also obvious that summer insolation is of no use in the winter. But if the amount of energy received by each square foot of collector during the cold winter months were known, the collector area required at 100% efficiency could be obtained by di-

viding the energy needed by that received per square foot of collector. The problems are: (1) what is winter? and (2) what is the system efficiency?

It was decided to define "winter" or "cold period" empirically as that minimum duration of the year in which 80% of the DDs occurred. The heat energy required in this "cold period" is then the Btu/DD of house × 0.8 DD$_{annual}$. From Table 7.1, it

Table 7.5.   Base Solar Heating Systems

	Air[a]	Liquid[a]
Collector Parameters		
Collector fluid	Air	50/50 Ethylene/glycol and water
Flow rate per unit collector area[b]	10 l/sec m², 2 cfm/ft²	0.0139 kg/sec m², 0.020 gpm/ft²
Collector slope	(Lat. + 15°) ± 10°	(Lat. + 15°) ± 10°
Azimuth	South	South
Covers	Double glazed	Double glazed
Absorber	Flat black	Flat black
Normal absorption $F_R (\tau\alpha)_n$	0.49	0.68
Loss $F_R U_L$ (W/°C m²)	2.84	3.75
Collector Heat Exchanger Parameters		
Effectiveness	1	0.7
Useful fraction reduction $F_{R'}/F_R$	1	0.97
Storage Parameters[b]		
Materials	Pebble bed	Water
Quantity per unit collector area	0.25 m³/m²	75 l/m²
	0.82 ft³/ft²	1.84 gal/ft²
Load Heat Exchanger Parameters		
Size	None	Sized for negligible loss of efficiency[c] $(\epsilon_L C_{min}/UA = 2)$

[a]Both air and liquid systems use hot-air heat in the house. Hot-water radiators require considerably higher temperatures and more or better collectors. They are not considered here.
[b]Hot-water preheat tank about 1.5 times service tank capacity.
[c]For meaning of symbols, see Ref. 3.

is not hard to find out which months are included in this "cold period." For example, in Greensboro, N.C., 0.8 DD$_{annual}$ = 3058 DD. The months of November to February plus 0.5 March, obviously the coldest months, include 3056 DD. Thus, the "cold period" for Greensboro, N.C., is Nov.–Feb. + 0.5Mar. Its duration, M_{cold}, is 4.5 months.

Now, the insolation received on level ground in Greensboro during the "cold period" can be accurately determined from Table 7.1 for these 4.5 months. It is

$$I_{cold} = [896 + 686 + 755 + 1002 + (0.5 \times 1305)] \, 30$$
$$= 11.97 \times 10^4 \text{ Btu/ft}^2 \text{ of level ground}$$

An average month duration of 30 days is used.

But how much energy is received on 1 ft² of collector elevated at the proper angle? During the "cold period" the sun is about 15° below the equator. True, the sun gets as far as 23.5° below and is at the equator on March 21 and September 23. But 15° is an excellent average for the "cold period".[9] Hence, at noon during the "cold period,"

the sun's rays strike the earth at an angle s = (lat. + 15°) from vertical. If the collector is elevated at approximately (lat. + 15°) above horizontal, it receives the sun's rays nearly perpendicularly all "winter" at noon. It is obviously not critical. However, the energy received per square foot of collector during the "cold period" is much more than falls on 1 ft² of level ground. It is $I_{cold}/\cos s$.

This quantity is so important that it deserves a name. It is the "winter insolation on collectors." It may surprise one to learn that the "winter insolation on collectors" is more in Winnipeg, Canada, than in Phoenix, Arizona. Not only is the latitude higher (and cos s smaller) but the duration of "winter," M_{cold}, is longer and more days of insolation are included. Even though the January insolation on level ground in Phoenix is 2.27 times that in Winnepeg, the energy received by 1 ft² of collector in Winnepeg during the "cold period" is 29% greater than in Phoenix. Contrary to popular belief, you may get "more for your money" from a solar installation in Winnepeg, although there are many other factors.

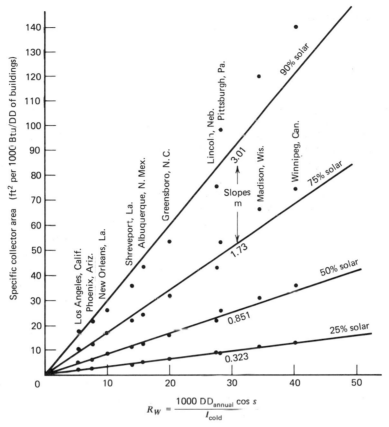

Figure 7.12. Specific collector areas for the solar heating system of Fig. 7.11*b* and Table 7.6. Curves apply to cities having M_{cold} from 3.5 to 6.0 months.

It has been found empirically[9] that the required specific collector area A_s (the collector area required per 1000 Btu/DD of house) is

$$A_s = m \, \frac{1000 DD_{annual} \cos s}{I_{cold}} \qquad (7.4)$$

The fraction is a weather ratio, R_w, depending only on the location; m is a constant depending only on the system and the percent solar desired.

For the water-based system of Fig. 7.11*b*, with double-glazed, flat-black collectors and average flows and storage:

$m =$	0.323	0.851	1.73
for solar	25%	50%	75%

This is shown in Fig. 7.12, in which the specific areas from Eq. 7.4 are compared with accurately calculated areas (dots) for 10 sample cities from Winnepeg to Los Angeles. The weather ratios R_w

for 11 sample cities are calculated in Table 7.6, column 8.

Correction Factor. In some sunny southern cities with relatively high winter ambient temperature, short winter, and high insolation, the percentage of collector losses is much reduced, and less collector area is required than given by Eq. 7.4. This "severity factor" is

$$SF = \frac{(212 - \overline{T}a_{cold}) M_{cold} \cos s}{I_{cold}} \qquad (7.5)$$

where $\overline{T}a_{cold}$ is the mean ambient temperature during the cold period, calculated much like I_{cold} but from Table 7.1(c). For 8 of the 11 sample cities, SF is in the range of 40 to 47, and no correction is needed. None was used in plotting Fig. 7.12. However, for three cities, Los Angeles, Phoenix, and Albuquerque, SF is in the range of 27 to 29, and a correction factor of 1.25 was needed to bring the accurately calculated areas up to the 50% solar curve, as

Table 7.6.   Computed Weather Ratios $R_w{}^a$

Location	Cold Months, Including 0.8DD$_{ann}$	M_{cold} (mo)	I_{cold} (Btu/ft² × 10⁴)	Lat. (deg.)	$s = 15°$ + lat.	cos s	R_w	$\overline{Ta}_{cold}$ (°F)	$212 - \overline{Ta}_{cold}$ (°F)	SF
1	2	3	4	5	6	7	8	9	10	11
Albuquerque, N.M.	Nov.–Feb. + 0.5Mar.	4.50	17.39	35.0	50.0	.643	15.8	38.3	174	29.0
Greensboro, N.C.	Nov.–Feb. + 0.5Mar.	4.50	11.97	36.0	51.0	.629	20.1	40.9	171	40.4
Lincoln, Neb.	Nov.–Feb. + 0.77Mar.	4.77	12.07	40.5	55.5	.566	27.5	31.5	181	40.5
Los Angeles, Cal.	Nov.–Mar. + 0.77Apr.	5.77	22.08	33.6	48.6	.661	5.4	56.1	156	26.9
Madison, Wis.	Nov.–Mar. + 0.19Apr.	5.19	11.82	43.1	58.1	.528	34.5	26.5	186	43.1
New Orleans, La.	Dec.–Feb. + 0.87Nov.	3.87	10.11	29.6	44.6	.712	9.8	54.9	157	42.8
Phoenix, Ariz.	Dec.–Feb. + 0.73Nov.	3.73	13.63	33.3	48.3	.665	7.6	53.6	158	28.8
Pittsburgh, Pa.	Nov.–Feb. + 0.82Mar.	4.82	10.64	40.3	55.5	.569	28.2	36.6	175	45.1
San Francisco, Ca.	Nov.–June	8.00	32.97	37.5	52.5	.609	5.7	54.8	157	23.2
Shreveport, La.	Nov.–Feb.	4.00	10.46	32.2	47.5	.676	14.1	49.8	162	41.9
Winnipeg, Can.	Nov.–Mar. + 0.50Oct.	5.50	11.42	49.5	64.5	.431	40.3	14.6	197	40.9

$^a R_w = (1000DD_{ann} \cos s)/I_{cold}$ (DD$_{ann}$ is from Table 7.1); SF = (3 × 7 × 10)/4 = 11.

shown in Fig. 7.12. Similarly, for any cities falling in this SF range, collector areas from Eq. 7.4 must be divided by 1.25. This is explained more fully in Ref. 9. SF is calculated in the last three columns of Table 7.6 for the 11 sample cities. The corresponding corrections for other solar percentages for cities with SF in the 27 to 29 range are:

25% solar 1.19   75% solar 1.30
50% solar 1.25   90% solar 1.46

Those familiar with the *f*-chart method[3] of determining the required collector area will recognize that this simplified method is equivalent to considering the "cold period" as one long month and that the two parameters $1000DD_{annual} \cos s/I_{cold}$ and SF are precisely Beckman's y and x, x being the minor parameter related to collector losses.

Unfortunately, if the "cold period," M_{cold}, is longer than six months, there is no sufficiently dis-

Jan	Feb	Mar	Apr	May	June	July
1606	1365	1190	702	409	184	126

tinct "cold period" and the method does not work.[9] For San Francisco, 80% of the DDs are spread over eight months, and Eq. 7.4 is inaccurate. The method is limited to M_{cold} not over six months.

Space and Water Heating.   Before proceeding to the promised economic comparison, the method of including water heating in the calculation will be illustrated by an example.

Consider a 21,253 Btu/DD house (°F), (467 W/°C), located in Madison, Wis. Suppose there are four occupants, and each one uses the average amount of domestic hot water, 26.4 gal/day (100 l/day) and that this must be heated from a mains temperature of 51°F (11°C) to the use temperature of 140°F (60°C). With a liquid-based system defined by Fig. 7.11*b* and Table 7.5, what area of collectors would be needed to realize 25, 50, and 75% solar fraction?

The annual heat load for domestic hot water is: 365 days/yr × 26.4 gal/person × 8.35 lb/gal × 4 persons × (140 − 51)°F = 28.64 × 10⁶ Btu.

In Madison, with 7729DD$_{annual}$ the 21,253 Btu/DD house requires for space heat 7729 × 21,253 = 164.26 × 10⁶ Btu. Thus, the water-heating load is 17.4% of the space heating load. It can be accounted for by increasing DD$_{annual}$ to 1.174 × 7729 = 9074DD, that is, by adding 1345 DD/yr, or 112 DD to each month. The revised Madison DDs are, then, from Table 7.1:

Aug	Sept	Oct	Nov	Dec	Annual
152	285	585	1021	1448	9073

and 0.8DD$_{annual}$ = 7258. Oct.–Mar. + 0.06Apr. include 7257DD; M_{cold} = 6.06 months. Thus, I_{cold} = 14.23 × 10⁴ Btu/ft²; $212 - Ta_{cold}$ = 182°F; cos s = 0.528. From these data,

$$R_w = \frac{1000\ DD_{ann} \cos s}{I_{cold}} = 33.7.$$

For 25, 50, and 75% solar, the specific collector areas are A_s = 33.7 (0.323, 0.851, 1.73) = 10.9, 28.7,

Table 7.7. Costs, Fuel Savings, and Return for a Liquid-Based Solar Space-Heating System Defined by Fig. 7.11*b* and Table 7.5, Supplying 50% of the Heating Requirements in a 15,000 Btu/DD House

City 1	R_w 2	Collector Area (ft²) 3	Cost at ($30/ft² (Dollars) 4	DD_{ann} 5	0.5DD_{ann} × 15,000 Fuel Sav. (× 10⁶ Btu) 6	Annual Fuel Saving (Dollars)				Annual Return (%)			
						Gas 7	Oil 8	Elec. Res. 9	Heat Pump 10	Gas 11	Oil 12	Elec. Res. 13	Heat Pump 14
Albuquerque, N.M.	15.8	162	4,860	4,284	32.1	137	251	437	257	2.8	5.2	9.0	5.3
Greensboro, N.C.	20.1	257	7,710	3,822	28.7	123	225	391	230	1.6	2.9	5.1	3.0
Lincoln, Neb.	27.5	351	10,530	5,865	44.0	188	345	599	352	1.8	3.3	5.7	3.3
Los Angeles, Calif.	5.4	55	1,650	1,815	13.6	58	106	185	109	3.5	6.4	11.2	6.6
Madison, Wis.	34.5	440	13,200	7,729	58.0	248	454	790	465	1.8	3.4	6.0	3.5
New Orleans, La.	9.8	125	3,750	1,389	10.4	44	81	142	83	1.2	2.2	3.8	2.2
Phoenix, Ariz.	7.6	78	2,340	1,551	11.6	50	91	158	93	2.1	3.9	6.8	4.0
Pittsburgh, Pa.	28.2	360	10,800	5,276	39.6	169	310	539	317	1.6	2.9	5.0	2.9
Shreveport, La.	14.1	180	5,400	2,186	16.4	70	128	223	131	1.3	2.4	4.1	2.4
Winnipeg, Can.	40.3	514	15,420	10,680	80.1	342	627	1091	642	2.2	4.1	7.1	4.2

and 58.3 ft² of collectors per 1000 Btu/DD of house. The collectors required are therefore 21.253 (10.9, 28.7, 58.3) = 232, 610, and 1239 ft² at 25, 50, and 75% solar, respectively. Since SF = 182 × 6.06 × 0.528/14.23 = 40.9, no correction is necessary.

Solar Costs, Fuel Savings, and Return. The economics of solar space heating varies widely with the severity of the winters and the amount of winter sunshine on the collectors. In general, there is a fixed cost for ductwork, dampers, controls, fans or pumps, heat exchangers, storage, and installation. The variable costs depend on the area of collectors required, and the sizes of storage and mechanical equipment.

In cities requiring very little heat, such as Key West, Fla., the fixed costs may not be justified. In cities of heavy heating load, poor winter sunshine on the collectors, and low fuel costs, a solar heating system is generally uneconomic. However, in between are many locations where moderate or high heat loads, coupled with good winter sunshine on collectors, make solar heating more attractive. This is particularly true where fuel costs are high, as with electric resistance heat or bottled gas or oil, and less so at present (1979) where natural gas supplies are still ample or low-cost electricity is available (parts of Canada and United States with large hydro supplies). These relationships may be illustrated by an example.

Consider the application of the liquid-based system of Fig. 7.11*b* and Table 7.5 to provide 50% solar

heating in a 15,000 Btu/DD house in each of the 10 sample cities. The required collector area is:

$$A = 15 \times 0.851 \times \frac{1000\ DD_{ann}\ \cos s}{I_{cold}}$$

= 12.77 × column 8 of Table 7.6 (divided by 1.25

if SF is in the range 27 to 29)

The resulting collector areas are shown in column 3 of Table 7.7. The cost at $30.00/ft² (1979) is shown in column 4. Except for very small systems, the fixed and variable costs should be represented in this figure, and the table thus provides a rough idea of the costs and return.

The fuel savings are shown in column 6. They are 0.5 DD_{annual} × 15,000 Btu/DD, expressed in Btu. The corresponding annual fuel savings in dollars are given in columns 7 to 10. based on the fuel costs and furnace (or heat pump) efficiencies shown in Table 7.8. The fuel prices are U.S. national averages of 1979 rather than for the individual cities. Residential natural gas, $2.56/MCF from Table 1.6, is taken as $2.51/10⁶ Btu (1020 Btu/SCF). Residential electricity is 4.63¢/kWh from Table 1.8. Residential fuel oil at 65.4¢/gal from Table 1.6 is considered to be No. 2 distillate, sp. gr. 0.85, HHV 19,600 Btu/lb from Table 2.3. Heat pump and furnace efficiencies are from Table 4.11.

As shown in Table 7.7, with the fuel costs of Table 7.8 applied uniformly, the return on a 50%

Table 7.8. Fuel Costs and Furnace Efficiencies—Typical 1979

Fuel	Cost ($/ million Btu)	Usual Units	Furance Eff. (%)	Cost ($/ million Btu) Delivered to Rooms
Natural gas	2.51	$ 2.81/MCF	60	4.27
Heating oil	4.70	65.4¢/gal	60	7.83
Electric resistance	13.62	4.63¢/kWh	100	13.62
Electric heat pump	13.62	4.63¢/kWh	170	8.01

solar installation varies from 1.2% when displacing natural gas in New Orleans to 11.2% when displacing electric space heat in Los Angeles, a spread of 10:1. The total system cost in Los Angeles or Phoenix may be low to absorb the fixed cost.

As shown, the return is more than three times larger when displacing electric resistance heat than when displacing natural gas.

Somewhat unexpectedly, the rates of return are as high in northern cities as in southern cities, based on equal fuel costs. Winnipeg, Canada, is quite comparable to Phoenix, Arizona, in percent return, and the actual savings are seven times as great.

The relatively poor return in New Orleans is due to the relatively large collector area required for the DDs (125/1389 = 0.09) compared with Albuquerque (162/4284 = 0.038). This is mainly due to the poorer winter sunshine in New Orleans.

These are real returns and do not consider tax incentives and outright grants being offered by the government currently to stimulate the solar heating development. Nor do they reflect the fact that these returns are tax free and are worth double the values shown, to a tax payer in a 50% bracket.

These costs and savings can be readily updated to current and local fuel costs and local solar system quotations. Also, the required collector area for a house of any Btu/DD in almost any part of the world, using this typical solar heating system, can be quickly determined from the basic climatological data by calculating I_{cold} as shown and applying Eq. 7.4.

Stand-Alone System. As mentioned, a stand-alone system is seldom economic. For example, a month-by-month calculation for Albuquerque. New Mexico, for the system just described showed that for a 52% solar fraction the portion of each month's average load that was carried by the solar equipment was as follows:

Jan	39%	May	100%	Sept	100%
Feb	46	June	100	Oct	95
Mar	60	July	100	Nov	55
Apr	88	Aug	100	Dec	38

Thus, the 52% system carries 38 and 39% of the average load in December and January. But the average ambient temperature is 34°F in these months (31° below the 65° base), and the design temperature is 0°F (Table 7.1). Thus, at times, more than twice the average heat would be needed. A double-size system would supply 38 to 39% of the requirements during these periods. Thus, a stand-alone system would have to be about five times as large as the 52% auxiliary system to tide over several days of design temperature, as would be expected of a conventional heating system.

Solar Cooling

For solar cooling (air conditioning), the collector system is almost identical with that for solar space heating. The collectors are set for optimum summer conditions,[5] at about lat. −15°. The absorption-type refrigeration system is used, as described in Chapter 13. The load heat exchanger supplies heat to evaporate ammonia from the water–ammonia solution, replacing the gas flame in Fig. 13.36. The cooling coils absorb heat at low temperature from the air being circulated through the building, as in a mechanical air-conditioning system.

If the solar facilities are to be used for both heating and cooling, a "best overall" collector slope must be selected. The load from the load heat exchanger is switched from house heating to absorption cooling as needed.

Solar Power Plant

A pioneering solar power plant of 10 MWe capacity is currently (1979) being built in the Mojave Desert,

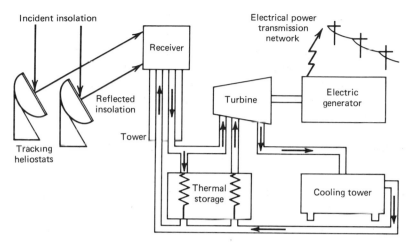

Figure 7.13. Central receiver solar thermal power system. *Source.* U.S. Department of Energy, "Environmental Development Plan, Solar Thermal Power Systems," DOE/EDP-0004, March 1978. Courtesy DOE.

near Barstow, California. It is shown in principle in Fig. 7.13. It is composed of 1818 large movable mirrors (heliostats) which keep the sun's rays focused atop a 283 ft high tower. The steam generated runs a nearly conventional turbine generator set. As shown in the diagram, the system includes heat storage to run the unit during at least part of the night. When completed in 1981 or 1982, it will be by far the largest solar power plant ever attempted. It is a part of DOE's solar program.

Ten megawatts is enough power for a city of 10,000 people. However, it is less than 1% of the size of the current largest fossil-fuel or nuclear units, which exceed 1000 MW. Two such units usually comprise a large station.

The 10 MWe solar plant at Barstow is viewed as the forerunner of 100 to 300 MWe plants in about 10 years, and possibly a 40,000 MWe (40 GWe) complex by the end of the century. This system is also seen as a retrofit for existing gas-fired stations in the Southwest, extending their useful lives as the natural gas runs out.

The 1818 heliostat plant at Barstow is based on experience with a 300 heliostat plant at DOE's Sandia Laboratory, near Albuquerque, New Mexico, where the components of the system are undergoing test. DOE is also involved in a United States–French experimental station in the Pyrenees Mountains in southern France. In this case, the heliostats are aimed at a huge 10 stories high parabolic mirror which concentrates the radiation on the boiler. Beyond the boiler, the turbine

generator arrangement is similar to that at Sandia and Barstow.

The Sandia experimental unit is rated 6.5 MWe, compared with 10 MWe for Barstow, even though it has only one sixth the number of heliostats. In both cases, the heliostats are composed of 25 4 ft by 4 ft silver-plated, laminated-glass mirrors, arranged 5 × 5. However, the Sandia unit is for short-time tests, whereas the Barstow unit is rated for round-the-clock operation. It must collect enough energy during the day to operate during the night also.

Referring back to Fig. 7.2, the direct sunlight reaching the earth is at most 83% of the solar constant. The diffused sunlight cannot be focused. Also, from Table 7.16 we can conclude that almost this amount *does* reach the earth in the sunny desert area of the Southwest. In Phoenix in June, the mean daily solar radiation is 2728 Btu/ft^2, or 2728/12 = 227 Btu/ft^2 hr, on a horizontal plane—12 hr average. On June 21, the sun's rays at noon are 10° from vertical in Phoenix. The solar insolation then is nearly a cosine function from sunrise to sunset, and its average value is 0.637 of maximum. Hence, the insolation at noon is about 227/0.637 = 356 Btu/ft^2 hr. But 88% of the solar constant is 429 × 0.88 = 377 Btu/ft^2 hr. Thus the actual received radiation on a surface facing the sun in the sunny southwest, and particularly in the Mojave Desert, is close to 88% of the solar constant; 83% can be focused.

To focus the sun on the boiler, the heliostat must point halfway between and hence does not present its full face to the sun. The actual reduction due to this factor depends on the location of the heliostats and is therefore empirical.

Each heliostat has a reflecting surface of 400 ft². Thus, 1818 heliostats represent 727,200 ft². If directed at the sun, they would receive direct sunlight of approximately 83% of the solar constant, or 0.83 × 125.7 W/ft = 104 W/ft², or 75.6 MW altogether for approximately 12 hr, or 37.8 MW average for the day. The steam plant efficiency may be as high as 40%, giving a potential continuous output of 15.12 MW. As mentioned, this must be reduced by the empirical factor of effective heliostat area. It is also reduced by the mirror and storage efficiencies. Thus, 10 MWe is a nominal rating.

Economics. It is early to discuss the economics of solar power plants. Coal and nuclear energy should be relatively inexpensive for hundreds of years. EPRI (Table 3.7) projects a coal cost of $1.70/million Btu in the year 2000, and considerably lower equivalent cost by nuclear generation. Fuel at $1.70/million Btu costs (10,000/0.34 × 8760 × 0.85 × 3413 × 10⁻⁶ × $1.70 = $1.27 million per year of operation of a 10 MWe, 34% efficient plant at 85% plant factor. At a financing rate of 18% for maintenance, depreciation, taxes, insurance, return, a normal utility financing rate, an investment of 1.27/0.18 = $7.11 millions could be afforded for a facility that would supply the heat "free." DOE funding for this experimental plant has been estimated at $100 million,[11] with $20 million additional allocated by the Southern California Edison Co. for site and conventional equipment required for the plant. Naturally, the economics of such plants remains in doubt at this point.

Nevertheless, the repowering of *existing* power stations in the Southwest with solar energy as oil and natural gas runs out may extend the life of a very large investment in plant facilities. Studies were being made in 1979–1980 of some 14 specific applications, eight for repowering existing power plants and six for supply of energy to various industries. These studies are expected to provide the basis for several demonstration projects to be operational by 1985.[21]

Solar Chemical System

An alternative solar collector system is under development in which parabolic reflectors replace the heliostats and each one contains at its focus a small catalytic reactor that splits SO_3 into SO_2 and O_2. The process is endothermic and the energy is thus stored in chemical form. The gases are collected at the central power plant, where they are recombined, releasing the stored energy to a heat exchange fluid and heat storage system. This fluid produces steam in a steam generator much like the hot-water loop of a pressurized-water reactor (see Chapter 6).

Economic Perspective

To put these DOE costs in perspective, $100 million for the Barstow plant, and so on, one must be reminded that there are over 200,000,000 of us and that $200 million represents $1 apiece. Considering the public clamor for solar energy, the uncertainty of the future energy picture, the rapidly rising energy costs, and the likelihood that some economic applications will develop in the next few decades, as well as much useful scientific knowledge, the government would be remiss indeed if it did not thoroughly run down the possibilities at these costs.

This can only be done by actual large-scale tests, using nearly life-size equipment. This brings in industry and really challenges the ingenuity and creativeness of the engineers and scientists of our country.

Solar Sea Power Plants (SSPP) or Ocean Thermal Energy Conversion (OTEC)

The idea of a heat engine that would use the warm surface waters of the tropical ocean as a heat source and the colder deeper waters as a sink was first proposed by d'Arsonval in 1881.[12] In 1930, a 40 kW plant was attempted off Cuba by the French engineer Georges Claude[13] but failed for a number of technical reasons. Today (1979), the idea is receiving intensive study,[14,15] and new trials with ample engineering and government support will soon be underway. A tanker converted to an OTEC test ship for testing various system components was to be deployed off Hawaii in mid-1980,[16] and a pilot OTEC plant is under design for deployment about 1984 pending successful tests of components in the interim. This program will be drastically curtailed in 1982.

The ocean is a huge solar collector, and the surface waters, about the top 656 ft (200 m), are warmed to about 77°F (25°C) over the entire tropical and subtropical ocean. The cold water at lower depths comes from the Artic regions and can be as low as 41°F (5°C), at a depth of about 2624 ft (800 m). Suitable waters for the operation of SSPPs are

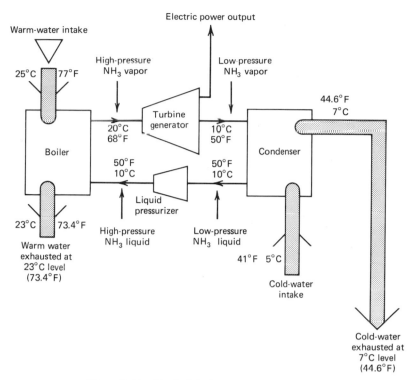

Figure 7.14. Typical design concept of a SSPP.

found[14] offshore at distances varying from 0.6 mile (1 km) in Florida to 188 miles (300 km) for most of Australia. In most of the Caribbean, cold deep waters are found within 1.2 mile (2 km) offshore. For these shorter distances, floating power plants linked to shore by power cables are feasible.

The temperature range between source and sink is at most 36°F (20°C). This drop must be allocated among the heat exchanges and the temperature drop of the working fluid. A typical design concept, allocating the temperature drops, is given in Fig. 7.14, 18°F (10°C) drop in the turbine and 18°F (10°C) drop in the heat exchanges.

As shown in the diagram, the warm water gives up about 3.6°F of its energy in passing through the boiler. This is the same amount that would be given up in falling 3.6 × 778 = 2800 ft, quite a waterfall! Unfortunately, it cannot be used as efficiently as falling water. It is limited by the Carnot efficiency of the heat engine to 18/528 = 3.4% of this amount. The energy, converted to work, cannot exceed 0.034 × 3.6 = 0.1224 Btu/lb of warm water. This is equivalent to a fall of 95 ft, which would certainly be developed if it were firm, year-round power, as is ocean solar.

The 3.4% Carnot efficiency has the further connotation that 1/0.034 = 29 times as much heat must

be transferred into and out of the working fluid as is converted to work. This requires a tremendous volume of water (as does a 95 ft head hydro plant). But in this case, the water must be pumped and uses up about one third of the power generated.

This enormous heat transfer, about 16 times that of a steam plant with the same net output, does present a very great heat transfer problem. However, there are some compensations. First, the working pressure is much lower. With either propane or ammonia as the working fluid, the pressure, at the boiler temperature of 68°F, is not over 125 psia, compared with 2400 or 3200 psia in a modern steam plant. This means thinner walls, better heat transfer, and drastically cheaper materials and construction. It is necessary however to deal with sea water corrosion, fouling, and cleaning problems, and much of the current DOE testing concerns this problem (1981).

In one proposed design,[15] the boiler and condenser are located 250 and 170 ft, respectively, below the surface, where the water pressure equals the pressure of the working fluid (ammonia) and the differential pressure on the boiler or condenser walls is very low. Other designs have the boiler and condenser on the barge.[17] As a result of the much lighter construction, some cost estimates[14] for the

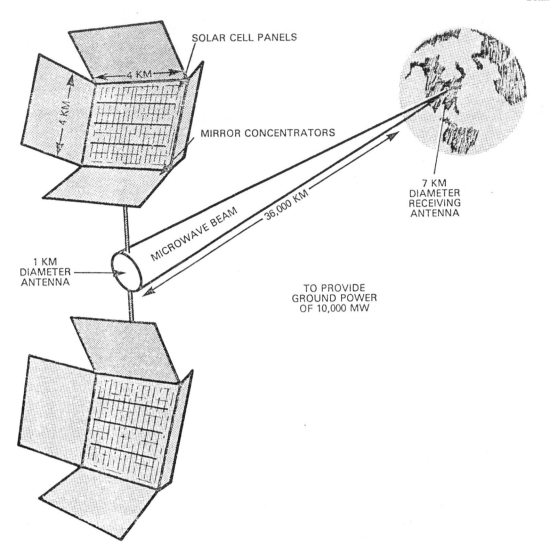

SOLAR CELL PANELS

MIRROR CONCENTRATORS

7 KM DIAMETER RECEIVING ANTENNA

36,000 KM

MICROWAVE BEAM

1 KM DIAMETER ANTENNA

TO PROVIDE GROUND POWER OF 10,000 MW

Figure 7.15. Space satellite power system with concentrators. *Source.* Martin Wolf, hearings before the Subcommittee on Energy of the Committee on Science and Astronautics, U.S. House of Representatives, June 6 and 11, 1974. Courtesy DOE.

SSPP are below the prevailing costs of fossil-fuel and nuclear land-based plants.

The DOE program in progress (1981 through 1984) involves concept and cost studies in the 100 to 1000 MWe range and sea trials of an offshore system. This program will provide a more realistic cost basis and a clearer knowledge of the problems and their solutions.

Solar Pond Power Plant

Similar in principle to OTEC, but more efficient, is the solar pond power plant.[18] In a pond of 1 to 2 m depth with increasing salinity (and density) from top to bottom, no convection currents form, even with 80°C higher temperature at the bottom than at the surface. The pond is an efficient (16 to 18%) collector of solar energy, but in this case the hot water is at the bottom.

Compared with the OTEC proportions of 20°C range overall, 10°C turbine drop, and 3.4% Carnot efficiency, a solar pond in the tropics may develop a 60°C range, a 42°C turbine drop, and 12% Carnot efficiency. The realized gross efficiency may be 8.3%, which with 25% auxiliary power leaves 6.2% net.

Some 20×10^6 kWh/km² per year of net output can be obtained in favorable locations, an average

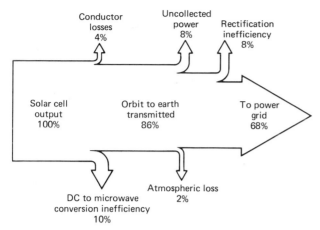

Figure 7.16. Space Satellite Power System Efficiency Estimate. *Source.* Peter E. Glaser, "Space Satellite Power System," Solar Energy Lecture III, IEEE, Washington, D.C. March 28, 1974. Courtesy DOE.

load of 2280 kW, or 20 MW for 1000 hr of peaking from a 1 km² pond.

Space Satellite Solar Power System[19,20]

Thus far, we have been speaking of solar energy systems in use or under test currently. What of the future? One system under consideration is the solar satellite (Fig. 7.15). Some 15 times as much solar energy is received by a cell facing the sun in earth orbit as a fixed-position cell on earth: about 1.25 higher density, three times the hours of use per day, a factor of 2 for clear skies, and another factor of 2 for angle of incidence, altogether a factor of 15. With reflectors, as shown in Fig. 7.15, the factor is still higher.

At microwave frequencies, with a 1 km diameter transmitting antenna and a 7 km diameter receiving antenna, a transmission efficiency of 92% is calculated. Other losses, shown in Fig. 7.16, result in an overall calculated efficiency, space–solar cell output to earth–power grid, of 68%. The cost of orbiting such a satellite and the environmental consequences of such a beam directed to earth remain unanswered questions.

From Fig. 7.2, the maximum sunlight reaching a 7 km diameter circle on the earth's surface is 45,800 MW. Thus, the 11,200 MW beam has an average intensity of 24% of maximum sunlight, a peak of possibly 40%. It is however of microwave frequency and more penetrating.

Reflector satellites are also under preliminary consideration for reflecting the solar energy of remote desert areas, via microwave, to load centers tens of thousands of miles distant.[22]

REFERENCES

1. "Solar Energy," in D. N. Lapedes, Ed. in Chief, *McGraw-Hill Encyclopedia of Science and Technology*, New York: McGraw-Hill, p. 522.

2. J. Keyes, *Harnessing the Sun*, Dobbs Ferry, N.Y.: Morgan and Morgan, 1975.

3. W. A. Beckman, S. A. Klein, and J. A. Duffin, *Solar Heating Design by the f-Chart Method*, New York: Wiley, 1977.

4. *ASHRAE Handbook-Systems*, New York: American Society of Heating, Ventilating and Air Conditioning Engineers, 1976, p. 43.2.

5. G. Daniels, *Solar Homes and Sun Heating*, New York: Harper and Row, 1976, p. 120.

6. J. F. Elliot, "Photovoltaic Energy Conversion," in *Direct Energy Conversion*, G. W. Sutton, Ed., New York: McGraw-Hill, 1966, p. 21.

7. "Silicon 'Ribbon' One Step Toward Solar Energy Goal," *Westinghouse News*, October 1979.

8. *Solar Augmented Heat Pump System*, SP-SAHP-1, Falls Church, Va.: Westinghouse Electric Corp., 1978.

9. E. L. Harder, "Specific Collector Areas for Solar Heating and Cooling," Unpublished, 1979.

10. *ERDA's Pacific Region Solar Heating Handbook*, Los Alamos Lab., Univ. of Calif., Washington, D.C.: Supt. of Documents, 1976.

11. "Costs of Barstow Solar Plant," *Sunworld*, November 1977.

12. J. d'Arsonval, *J. Revue Scientifique*, September 17, 1881.

13. G. Claude, "Power from the Tropical Seas," *Mechanical Eng.*, 1930, p. 1039.

14. A. Lavi and C. Zener, "Plumbing the Ocean Depths," *Spectrum*, October 1973, p. 22.

15. J. H. Anderson and J. H. Anderson, Jr., "Thermal Power from Sea Water," *Mechanical Eng.*, April 1966, p. 41.

16. *Solar Energy Program Summary Document*, FY 1981, DOE, January 1980, p. III96.

17. *Westinghouse Solar Energy Systems—Ocean Thermal Energy Conversion*, Pittsburgh, Pa.: Westinghouse Electric Corp., June 1978.

18. Y. L. Bronicki, "A Solar-Pond Power Plant," *Spectrum*, February 1981, p. 56.

19. *Dept. of the Navy Energy Fact Book*, 0584-LP-200-1420, Philadelphia, Pa.: Navy Publication Form Center, 1979.

20. R. W. Johnson, "Solar Power Satellite: Putting it Together," *Spectrum*, September 1979, p. 37.

21. *Central Receiver Systems–Solar Thermal Repowering*, San Francisco: DOE, 1980.

22. T. F. Rogers, "Reflector Satellites for Solar Power," *Spectrum*, July 1981, p. 38.

8

Wind Energy

INTRODUCTION

Wind power was used over 8000 years ago to drive sail boats and for centuries provided most of the power for mills in level country. With the advent of more convenient forms of energy, its use has diminished to an insignificant fraction in the United States. However, currently (1981), with impending shortage of oil and natural gas and steeply rising energy prices, there is renewed interest in all renewable energy sources, wind power among them.

From 1970 to 1980, while the GNP price index doubled, the producer energy price quadrupled (see Chapter 1, Price Indices). In 10 years, this halved the *relative* cost of wind power and greatly increased its chances of being a viable alternative.

Extensive federal, state, and private development programs are now underway (1981), involving industry and the scientific community, that will delineate the areas of economic application more fully within a few years. These developments are outlined in this chapter.

POWER IN THE WIND

The power in the wind coming to a windmill is

$$P_w = \frac{(1/8)\,\pi\,\rho\,D^2V^3}{550}\ \text{Hp} \qquad (8.1)$$

where D is the blade-tip circle diameter in ft, V is the wind velocity in ft/sec, and ρ is the density of the air,[1] that is, 0.002378 lb/ft^3 at sea level, 59°F (15°C),* 0.00224 lb/ft^3 at 2000 ft elevation, 51.9°F (11.1°C),* and 0.00211 lb/ft^3 at 4000 ft elevation, 44.7°F (7.1°C).*

*U.S. average, latitude 40°.

160

Thus,

$$P_w = 1.70\ D^2\ V^3 \times 10^{-6}\ \text{Hp at sea level} \quad (8.2)$$

$$\times\ 0.94\ \text{at 2000 ft elevation}$$

or

$$\times\ 0.887\ \text{at 4000 ft elevation}$$

Or, expressing D and V in ft and mph, the power in kW is

$$P_w = 4.00 D^2V^3 \times 10^{-6}\ \text{kW at sea level} \quad (8.3)$$

with the same elevation multipliers.

For example, the power in the wind approaching a 175 ft diameter wind turbine at 24 mph is 1693 kW at sea level, or 1502 kW for the *same* wind speed at 4000 ft elevation. Actually, at a given location, the wind speed normally increases with distance above ground level, as will be explained later.

Of this power, Eq. 8.3, only a part can be transmitted to the turbine blades. The theoretical maximum is 59.3%.[1] However, practical blades fall short of that amount, even when operating at the optimum wind speed. The power output of the turbine may be expressed as

$$P_{\text{out}} = 4.00 P_c D^2V^3 \times 10^{-6}\ \text{kW at sea level} \quad (8.4)$$

for D and V in ft and mph; P_c is an experimentally determined coefficient, such as given in Fig. 8.1 for typical blade shapes and configurations. For the high-speed two-blade propeller, at its best speed, $P_c = 0.46$, and the corresponding turbine output is 46% of the power in the wind. By proper blade design, the optimum speed can be placed at any desired wind speed.

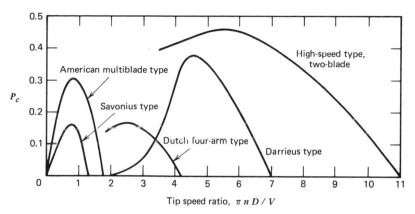

Figure 8.1. Power coefficients of windmill blades (power output/power in the wind).

WIND TURBINE GENERATOR (WTG)

Most large windmills (100 kW or over) built or considered in the last 50 years (1977) have been horizontal axis machines, equipped with electric generators, and arranged for parallel operation with a power system. They "cut in" at some wind speed, and when they come up to speed they are connected to the system (synchronized). Wind power increases as the cube of the wind speed, Eq. 8.4; and when the generator rating is reached, the power is regulated to this value for all higher speeds, for example, by blade pitch control. Connection to a large battery or to any system that could absorb *all* of the generated energy would be equivalent. This results in the power output-versus-wind speed characteristic shown in Fig. 8.2, curve (a). The wind speed at which full rating is reached is termed the rated wind speed V_r. The cut-in speed is usually about 50% of the rated speed.

A gear train is normally used between the turbine and the generator in order to use a high-speed generator, although a multipole, permanent-magnet generator could be used without gears.

Vertical axis or Darrieus-type machines, to be described later, are self-regulating, and have a power output-versus-wind speed characteristic as shown in Fig. 8.2, curve (b).

Specific Output—Plant Factor

The kilowatt hours generated in a year by any given installation depends on the wind regime—how many hours the wind blows at various speeds throughout the year—and of course on the power output-versus-wind speed characteristic of the WTG and the availability. The kWh divided by the kW of generator rating is called the "specific output" and is expressed in kilowatt hours per kilowatt year of rating, kWh/kWyr.

Alternatively, the "plant factor" is used, the ratio of energy actually generated to that which would be generated at full load all year. Obviously,

$$\text{Plant factor} = \frac{\text{specific output}}{8760} \qquad (8.5)$$

This plant factor is a theoretical value based on 100% availability. The actual plant factor is *approximately* the theoretical plant factor multiplied by the availability factor.

Wind Regime

It is well known that the mean annual velocity of the wind varies widely from place to place on the earth and that its velocity distribution varies to some extent. Because of this, it has always been assumed[1-4]

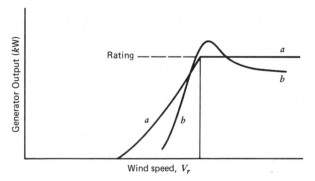

Figure 8.2. Power output versus wind speed of constant speed wind turbine generators. (*a*) horizontal axis. (*b*) Vertical axis.

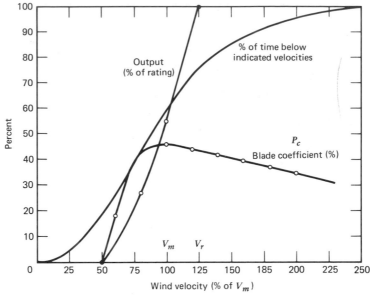

Figure 8.3. Wind velocity distribution, Dayton, Ohio,[1] blade coefficient, generator output.

that the specific output could only be determined from the particular wind regime involved at the site and the power output-versus-wind speed characteristic of the particular WTG.

WTG Example. The author shared this view, and a specific example was worked out for this chapter, using the wind velocity distribution for Dayton, Ohio, shown in Fig. 8.3, which was available in a handbook.[1]

The mean annual velocity is designated as V_m. Note that velocities over $2V_m$ occur less than 5% of the time, whereas velocities above V_m occur about 40% of the time. If the rated wind speed V_r is equal to $2V_m$, the generator rating would be eight times as large as for $V_r = V_m$ (wind power is proportional to V^3), but it would be up to rating only 5% of the time. Based on these considerations, it was decided for the example to have the generator come up to rating at $1.25V_m$, that is, $V_r/V_m = 1.25$.

The blade power coefficient-versus-speed characteristic of Fig. 8.1 (high speed, two blades) was used with the maximum set at V_m, as shown in Fig. 8.3. The power would then be nearly maximum up to V_r, above which full power is generated regardless of the blade coefficient. A constant efficiency of gear train and generator was assumed, so that the output characteristic, Fig. 8.3, was a cubic, modified only by the blade coefficient.

A specific output of 4000 kWh/kWyr resulted

from the detailed numerical integration for this example (100% availability). To see if it was "in the ballpark," it was compared with specific outputs determined by ERA in England[2] in an extensive study of 23 sites over several years. Surprisingly, the Dayton, Ohio, example, intended only as a crude example, agreed almost exactly with the ERA data at the same V_r/V_m ratio of 1.25, both for a site having a mean annual velocity of 24 mph and another with V_m of 15 mph. At Dayton, Ohio, $V_m \cong 11$ mph.

Specific Output Discovery

This unexpected coincidence led to the discovery that the specific output of a horizontal-axis WTG is practically independent of the wind velocity distribution over the earth. It depends only on the ratio of the rated speed V_r to the mean annual velocity V_m.

This is shown on the curve (Fig. 8.4) for constant-speed, horizontal-axis WTGs, on which the following have been plotted:

1. 75 outputs determined by ERA[2] for 23 locations in England, with mean annual velocities from 25.3 to 6.2 mph (11.3 to 2.8 m/sec).

2. Two outputs from Putnam's studies[3] in "interior New England," at mean annual

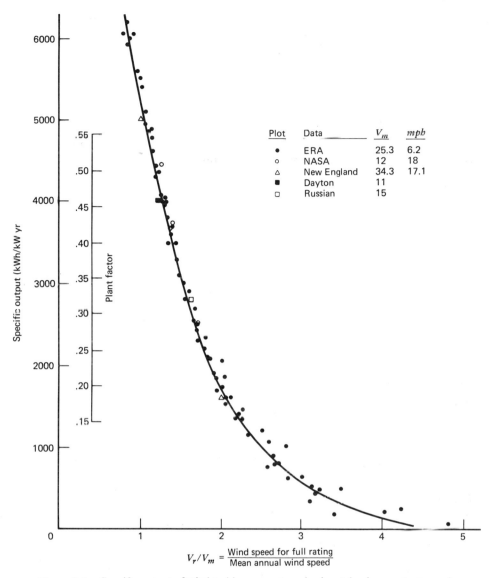

Plot	Data	V_m	mph
●	ERA	25.3	6.2
○	NASA	12	18
△	New England	34.3	17.1
■	Dayton	11	
□	Russian	15	

$$V_r/V_m = \frac{\text{Wind speed for full rating}}{\text{Mean annual wind speed}}$$

Figure 8.4. Specific output of wind turbine generators, horizontal axis, constant speed.

velocities of 34.3 and 17.1 mph (15.3 and 7.6 m/sec).

3. Experimental results from the Russian windmill[3] at Yalta on the Black Sea, where $V_m = 15$ mph (6.7 m/sec).

4. Four recent NASA-sponsored industry designs,[4] two by the General Electric Co. and two by the Kaman Co., using wind regimes provided by NASA and designed for $V_m = 12$ and 18 mph (5.4 and 8 m/sec).

5. The author's empirical design, based on the wind regime at Dayton, Ohio, $V_m \cong 11$ mph (4.9 m/sec), described earlier.

The close agreement of all points with a single curve is apparent. For example, in the important range, $V_r/V_m = 1$ to 2, none of the four NASA plant factors shown deviates from the curve more than the equivalent of 3% in mean annual velocity (which varies from year to year at any given location by 10 to 12%).

At $V_r/V_m = 2$, the WTG would be at full rating only 5 or 6% of the time, and designs much above this are not generally considered. Also the cut-in speeds are normally 50% of V_r or higher, so that at $V_r/V_m > 2$, the WTG cuts in above the mean annual velocity. Small variations in wind velocity distribution then become important. For these reasons, the curve is neither valid nor useful above approximately $V_r/V_m = 2$. Note the increasing spread of the plotted points.

Typical Uses of the Specific Output Curve

This result, Fig. 8.4, is useful, for example, in optimizing a design. Consider V_r/V_m ratios of 1.25 and 2.0, for example. The power output of a horizontal-axis WTG, up to rating, is given by

$$P_{\text{gen}} = 4.00\, P_c\, \text{n}\, D^2\, V^3 \times 10^{-6}\ \text{kW at sea level}$$
(8.6)

$$\times\ 0.94\ \text{at 2000 ft elevation}$$

or

$$\times\ 0.887\ \text{at 4000 ft elevation}$$

where P_c is the blade coefficient (0.46 max. for high-speed, two-blade turbine); n is the efficiency of generator, gears, and drive (0.75 is typical at 100 kW); D is the blade tip-circle diameter in ft; and V is the wind velocity in mph.

For D and V in m and m/sec, the constant is 481, in place of 4.00, and the multipliers of 0.94 and 0.887 apply to 610 and 1220 m elevation.

For the optimizing example, if P_c and n are taken to be the same for the two designs, a V_r/V_m ratio of 2 corresponds to a generator $(2/1.25)^3 = 4.1$ times as large (rating) as for $V_r/V_m = 1.25$. The specific outputs, from Fig. 8.4, are 4000 and 1700 kWh/kWyr, and the comparative energies are 4000 and $4.1 \times 1700 = 7000$.

If the cost of the smaller generator, including gears, is approximately 25% of the WTG cost and increases as the 0.6 power of the rating, then the larger generator costs $25 \times 4.1^{0.6} = 58\%$, and the larger WTG costs $75 + 58 = 133\%$.

That is, 175% energy is obtained for 133% cost by going to the larger generator. (This assumes production cost proportional to capital cost.) This result is obtained without the labor of calculating the specific outputs for the two cases from the wind regime and the WTG output characteristics, as has been done in the past.

The generator size might be optimized to obtain (1) the lowest cost per kWh, as above, (2) the most energy below a fixed competitive cost, or (3) the lowest cost per kWh for the total energy requirements from the WTG and a supplementary source.

In any case, the specific output curve, Fig. 8.4, saves most of the work. In fact, Eq. 8.6 and the curve of Fig. 8.4 constitute together an empirical design or estimating method about as accurate as is warranted, since the possible annual variation of 10 to 12% in mean annual velocity causes much greater uncertainties.

If an installation is predicated on $V_m = 18$ mph (8 m/sec) and $V_r/V_m = 1.5$, and the actual mean velocity turns out to be 90% of 18 or 16.2 mph (7.2 m/sec), then actually $V_r/V_m = 1.67$. From Fig. 8.4, the specific output is 2500 kWh/kWyr instead of 3100 kWh/kWyr, about 80% of the expected energy. This result also would be difficult to obtain without the newly discovered relation.

This discovery was published in 1977[5] and is now in use by DOE in their "design and field testing activities" and by various other investigators.[6,7] It will be used in the following empirical designs rather than the lengthy numerical integrations which it replaces. It is also easily interpreted[7] into the specific outputs obtainable with different generator ratings for a given diameter WTG at a location of known mean annual velocity, V_m.

Empirical Design

Based on Eq. 8.6 and the specific output curve of Fig. 8.4, the performance of WTGs can now be given empirically as a function of the blade tip diameter D and the mean annual wind speed V_m. For the empirical design, assume that:

1. At the rated speed V_r, the efficiency of the generator, gears, and drive is 75%. This is typical of 100 kW and higher. The efficiency of the NASA Plum Brook 100 kW unit is 75%. Smaller units would be generally less efficient.

2. The blade coefficient P_c at the rated speed V_r is 0.43.

3. The ratio V_r/V_m is 1.50.

The generator rating, or energy generated at V_r, with these assumptions is, from Eq. 8.6,

$$P_{\text{gen}} = 4.00 \times 0.43 \times 0.75 \times 10^{-6}\, D^2 V_r^3$$

$$= 1.29 \times 10^{-6} D^2 \left(\frac{V_r}{V_m}\right)^3 V_m^3$$
(8.7)

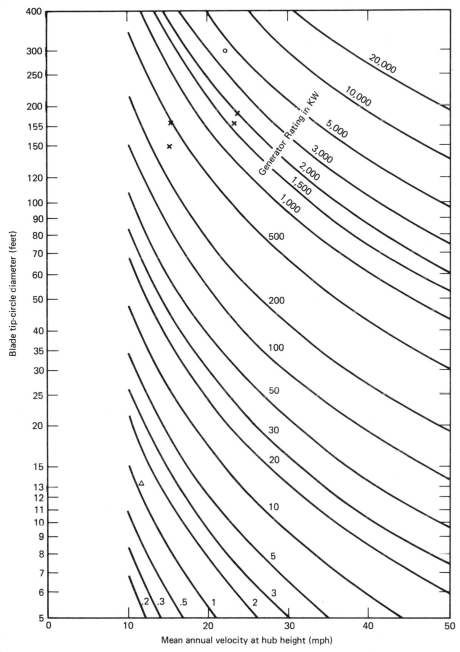

Figure 8.5. Empirical generator rating for 3100 kWh/kWyr at 100% availability, $V_r/V_m = 1.5$.

For the empirical machine with $V_r/V_m = 1.5$, the rating is:

$$10^6 P_{\text{gen}} = 1.29 \times 1.5^3 D^2 V_m^3 = 4.354 D^2 V_m^3$$

or

$$D = 479 \left(\frac{P_{\text{gen}}}{V_m^3}\right)^{1/2} \qquad (8.8)$$

This is plotted in Fig. 8.5. The specific output is 3100 kWh/kWyr* in all cases. For $V_r/V_m = 2$, the ratings are 2.37 times those shown in Fig. 8.5, and the specific output is 1700 kWh/kWyr.* For any V_r/V_m between 1 and 2, the generator rating is $[(V_r/V_m)/1.5]^3$ times those shown in Fig. 8.5, and the specific output* can be read directly from Fig. 8.4.

*To be multiplied by availability, typically 80–90%.

Typical Design Examples

The NASA-sponsored designs of Table 8.4, to be described later, two at 500 kW and two at 1500 kW, are indicated by crosses on Fig. 8.5. These units have V_r/V_m of 1.25 to 1.71, the curves being for 1.5.

The circle represents the 300 ft diameter Mod-2 WTGs considered for sites[7] of 22 mph at hub height (17 mph at 45 ft) in the Goodnoe Hills, near Goldendale, Washington. At a V_r/V_m ratio of 1.5, these units could have a rating of about 4000 kW and a specific output of about 3100 kWh/kWyr. Other V_r/V_m ratios are also considered with correspondingly different ratings and specific outputs. (The actual Mod-2 has a rating of 2500 kW.) The optimum ratio would be selected for any particular installation.

The triangle is representative of the 12 to 14 ft American multiblade, or turbine-type, windmill, with 1 or 1.5 kW generator, widely used in the past in areas having a 10 to 12 mph mean annual wind velocity.

Hub Height Correction

The use of Fig. 8.5 requires a knowledge of the wind speed at hub height. The wind speed varies approximately as the 0.167 power of the height above ground. Thus, if the mean wind speed is known to be 17 mph at 45 ft and the hub height is 200 ft, the mean wind speed at hub height is

$$V_H = V_h \left(\frac{H}{h}\right)^{0.167} = 17 \left(\frac{200}{45}\right)^{0.167} = 21.8 \text{ mph}$$

$$(8.9)$$

Generally, a clearance of about 50 ft is allowed below the blades for large WTGs. The hub height is usually $D/2 + 50$ ft. For 190 ft blades, the hub height would normally be $95 + 50 = 145$ ft. Wind measurements are frequently made at 30 to 45 ft above ground level.

Small Wind Turbine Generators

Small WTGs, of the order of 5 to 10 kW, cannot afford the precise governors used on the larger units 100 kW and up. Instead of a sharp break in the power output curve at rated load, as shown in Figs. 8.2 and 8.3, the breakpoint is likely to be rounded, as in Fig. 8.6. However, if the rising part of the curve is approximated by the nearest fit to a cubic

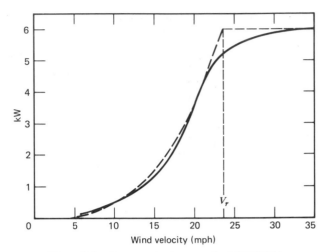

Figure 8.6. Power output curve of a 6 kW WTG.

(for example, through the 4 kW point of Fig. 8.6, as shown dotted), an equivalent V_r is obtained as shown and can be used with Fig. 8.4 for approximate results.

The machine shown in Fig. 8.6 has a virtual V_r = 23.7 mph. In a location having a mean annual velocity V_m of 15 mph, it has a V_r/V_m ratio of 1.58. From Fig. 8.4, it has a specific output of 2800 kWh/kWyr. It will generate $6 \times 2800 = 16,800$ kWh in an average year (100% availability).

However, being a small machine, in the speed range of 22 to 30 mph (1.47 to 2.0)V_m, it is short an average of 0.4 kW due to the rounding. From Fig. 8.3, a machine spends about 11% of its time between $1.47V_m$ and $2.0V_m$. The shortage is thus 0.11 $\times$ 8760 $\times$ 0.4 = 385 kWh, or 2.3% of the kWh generated. It is too small to be considered.

Dutch and American Windmills

To most people, the word "windmill" conjures up a mental image of the picturesque Dutch windmill or, if from America's farming country, the multiblade or "turbine-type" windmill widely used for pumping water or the more recent and simpler propeller-type windmill for limited electric service.

The Dutch windmill dates from the twelfth century. The American multiblade or turbine type was developed about 1850, primarily for pumping water. Windmill electric plants became commercial about 1920 using this type of wheel, although individual units were built much earlier. The propeller type, for electric generation, started about 1925. Both were in widespread use throughout the American farm country until the advent of small farm electric sets and rural electrification in the late 1930s. All of these are still in limited use (1981), as are other types in various lands.

The Dutch Windmill

The Dutch windmill is usually ascribed "about 2 Hp."[8] But this was hardly on the same basis as the ratings of Fig. 8.5, with $V_r/V_m = 1.5$. These ratings are reached with winds that occur only 15% of the time (Fig. 8.3). The Dutch miller would certainly want to serve his customers more regularly than that. And indeed he could.

Dutch windmills were characteristically 60 to 80 ft in diameter. From Eq. 8.4 and Fig. 8.1, a Dutch windmill of 70 ft diameter, turning at 13 rpm in a 15 mph wind, would develop about 10 kW (13.4 Hp), to be multiplied by some unknown efficiency of transmission parts. Tests were made on just such a mill, under these conditions, by C. A. Coulomb[9] in 1821, resulting in an output of 8 brake Hp. In view of the probable low efficiency and the limitations of measurements in those days, these figures are in fair agreement. Such a mill would develop 2 Hp at about half of this wind speed or 7.5 mph, which might be available about 70% of the time.

The Early American Windmills

The American multiblade or turbine-type windmill, for electric generation, was characteristically 12 to 14 ft in diameter with a 1 to 1.5 kW generator.

A typical propeller-type unit in Iowa[9] in 1929 was a two-blade (or three-blade), 10 ft diameter (6 to 12 ft) unit on a 60 ft tower (40 to 80 ft), with a 250 amp hr battery (150 to 400), 32 V, 16 cell (few at 112 V, 56 cells). It cost $735 average [$250 (6 V) to $1600, (12 V)], exclusive of labor for installation. The minimum cost in 1930, including wheel and 1 kW generator, 60 ft tower, 250 amp hr battery, and control panel, exclusive of labor, was $760. GNP inflation 1940 to 1980 was about 5.7:1; a corresponding 1980 price would be $4000 to $5000.

The mean annual wind velocity for the United States is about 10 mph, with variations from place to place as shown in Table 8.1. With $V_r/V_m = 1.5$, the empirical generator rating for a 10 ft diameter, two-blade unit, with $V_m = 10$ mph, is given in Fig. 8.5. It is 0.5 kW. With a specific output of 3100 kWh/kWyr, it would produce 1550 kWh/yr. With a battery efficiency[9] of 60%, between 930 and 1550

kWh would be useful, depending on the amount passed through the battery. The 250 amp hr, 32 V battery would store about 8 kWh, or about two days' supply. Thus, it could easily smooth out daily fluctuations of load and wind and, if fully charged, would help tide over a bad week.

Grandpa's Knob

In the fall of 1941, a wind turbine generator rated 1250 kW was operated atop the 2000 ft elevation Grandpa's Knob, 12 miles west of Rutland, Vermont. It consisted of a high-speed, two-blade turbine, 175 ft in diameter, with its hub 125 ft above the ground. It turned at 28.7 rpm in winds above 17 mph. It was connected through hydraulic coupling and gears to a 600 rpm generator, which was connected synchronously to the lines of the Central Vermont Public Service Co.

This was the largest wind-powered generator built up to 1977. It incorporated the most advanced knowledge of aerodynamics, structures, and winds of its day. Although the wind estimate proved high and the energy generated was only 27% of that expected, the unit operated for several years and provided firm design data for a bank of six 1500 kW units to be located on Lincoln Ridge in Vermont. However, at that time the cost would have been $205/kW installed, whereas the value to the Vermont Public Service Co. was determined to be $125/kW, and the project was dropped.

Inflation over the next 25 years would not have changed this picture materially. However, during the 1970s, producer energy prices quadrupled while general prices doubled. The same project *might* now be economic (1981).

Developments Since 1945

Since the Grandpa's Knob effort of 1941–1945 vintage, there have been a number of wind power developments, large and small. In the brief space available here, attention will be focused primarily on current (1981) U.S. and Canadian developments. For earlier developments, see Ref. 10.

England—ERA.[2] In the early 1970s, wind regimes were measured at 23 locations in England, including four prospective wind power sites. Specific outputs were determined for three generator ratings at each location, corresponding to $V_r = 20, 25$, and

Table 8.1. Wind Velocities in U.S. Cities[a]

Station	Avg. Velocity, (mph)	Prevailing Direction	Fastest Mile	Station	Avg. Velocity, (mph)	Prevailing Direction	Fastest Mile
Albany, N.Y.	9.0	S	71	Louisville, Ky.	8.7	S	68
Albuquerque, N.M.	8.8	SE	90	Memphis, Tenn.	9.9	S	57
Atlanta, Ga.	9.8	NW	70	Miami, Fla.	12.6	—	132
Boise, Idaho	9.6	SE	61	Minneapolis, Minn.	11.2	SE	92
Boston, Mass.	11.8	SW	87	Mt. Washington, N.H.	36.9	W	150
Bismarck, N. Dak.	10.8	NW	72	New Orleans, La.	7.7	—	98
Buffalo, N.Y.	14.6	SW	91	New York, N.Y.	14.6	NW	113
Burlington, Vt.	10.1	S	72	Oklahoma City, Okla.	14.6	SSE	87
Chattanooga, Tenn.	6.7	—	82	Omaha, Neb.	9.5	SSE	109
Cheyenne, Wyo.	11.5	W	75	Pensacola, Fla.	10.1	NE	114
Chicago, Ill.	10.7	SSW	87	Philadelphia, Pa.	10.1	NW	88
Cincinnati, Ohio	7.5	SW	49	Pittsburgh, Pa.	10.4	WSW	73
Cleveland, Ohio	12.7	S	78	Portland, Maine	8.4	N	76
Denver, Colo.	7.5	S	65	Portland, Ore.	6.6	NW	57
Des Moines, Iowa	10.1	NW	76	Rochester, N.Y.	9.1	SW	73
Detroit, Mich.	10.6	NW	95	St. Louis, Mo.	11.0	S	91
Duluth, Minn.	12.4	NW	75	Salt Lake City, Utah	8.8	SE	71
El Paso, Tex.	9.3	N	70	San Diego, Calif.	6.4	WNW	53
Galveston, Tex.	10.8	—	91	San Francisco, Calif.	10.5	WNW	62
Helena, Mont.	7.9	W	73	Savannah, Ga.	9.0	NNE	90
Kansas City, Mo.	10.0	SSW	72	Spokane, Wash.	6.7	SSW	56
Knoxville, Tenn.	6.7	NE	71	Washington, D.C.	7.1	NW	62

Source.   Reprinted from *Mark's Standard Handbook for Mechanical Engineers*, Copyright 1966, McGraw Hill Publishing Co.

[a]U.S. Weather Bureau records of the average wind velocity, and fastest mile, at selected stations. The period of record ranges from 6 to 84 years, ending 1954. No correction for height of station above ground.

30 mph wind speeds. The published results of this survey[2] provided most of the data for the specific output curve of Fig. 8.4, as has been mentioned.

"Study Group" Windmill, Stuttgart, Germany.[11-13]   A German "Study Group" was set up in the mid-1950s to study the use of wind power in Germany, particularly in connection with electric utility systems.

This group, under the direction of Dr. Hutter, built a large windmill, 112 ft in diameter, three-blade, on a 72 ft mast, with a 100 kW generator, which they operated from September 1957 to October 1968. This unit was connected synchronously to the public power supply system from 1959 to 1968. Together with Dr. Hutter's reports[11-13] it served as a starting point for the NASA designs in the 1970s.

NSF–ERDA–DOE–NASA Program—Large Windmills[4,10,14-18]

The funding of $30,000,000 for five-year development of the technology needed to build reliable and cost-effective wind generator systems was originally placed under the National Science Foundation (NSF) in 1973. In 1975, it was transferred to the newly formed Energy Research and Development Administration (ERDA).

Since the formation of the Department of Energy (DOE) in 1977, the program has been greatly expanded to $80 million in fiscal year 1981. The DOE Wind Systems Program is managed at the program level from headquarters, while projects are managed from DOE field offices, national, NASA, and U.S. Department of Agriculture laboratories. The development of large and intermediate horizontal-axis machines is centered at the NASA Lewis Research Center, that of vertical-axis Darrieus machines at the Sandia Laboratory, and that of small systems at the Rocky Flats Plant, near Denver.

This program was designed to explore quite thoroughly the possible uses of wind power in the United States to augment other energy sources. The outstanding characteristics of wind energy—inexhaustible, clean, and "free"—were as always the enticing lure to any one "who dared to enter here." Only one real contender[11] for these prizes had graced the scene since Putnam's[3] valiant attempt in 1941–1945.

Now there was new technology, large helicopter blades, Fiberglas construction, and better understanding of the wind. More important, there was a great need and a public clamor to use this great natural reservoir of clean energy. There was an oil embargo that underscored the imminent exhaustion of oil and gas reserves in the United States, and soon in the world. There were impending fuel shortages and soaring fuel prices, with the end not yet in sight. In fact, the stakes were so high that the government could hardly do other than exhaust all possibilities.

And there was a new generation of Don Quixotes eager to give battle to the windmills. Unfortunately, it was the same old wind, with power varying as the cube of velocity, with which they must contend. The high rate of inflation of energy prices relative to the general economy was however in their favor, and would become more so.

The planned program was in three parts, as follows:

1. Design and operation of a 100 kW experimental wind turbine generator for use in test, development and demonstration.
2. Sponsor a series of industry-designed and user-operated wind turbine generators in the range of 50 to 3000 kW.
3. Support R&D to advance the state of the art in all phases that would make wind power more economic.

It is now (1981) too early to assess the outcome. However, a few facts about the program can be stated.

Figure 8.7. Plum Brook 100 kW experimental wind turbine generator. Courtesy DOE/NASA.

NASA–Plum Brook 100 kW Experimental Wind Turbine Generator

A 100 kW wind turbine generator was placed in operation at the Plum Brook site of NASA–Lewis, near Sandusky, Ohio, on Lake Erie, in July 1975. While this was not an economic site for a production unit, since the mean annual velocity V_m was under 10 mph, it was quite adequate as a test site. Winds of all desired values were known to exist for adequate durations for the tests. There were 10 years of on-site wind records. The site was well equipped to handle the numerous mechanical modifications involved in the development.

Figure 8.7 shows the test unit at Plum Brook. Note size of the man at the base. The entire upper housing with blades can be lifted into place or lowered to the ground for major modifications by a crane available at the Lewis Plum Brook site. The transmission components inside the housing are shown in Fig. 8.8. The 100 kW generator is physically a small part of the installation. The yaw drive keeps the blades facing the wind. The blade pitch

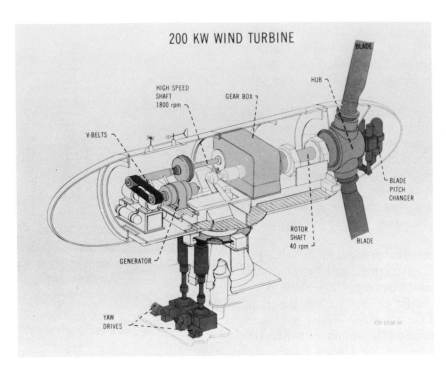

Figure 8.8. Transmission components of a wind turbine generator. Courtesy DOE/NASA.

changer responds to a power governor, not shown, to maintain full load power above rated speed or to feather the blades in a hurricane.

The principal characteristics of the installation are given in Table 8.2. The rated speed, $V_r = 18$ mph, makes it typical of a unit that might be applied in locations having a mean annual velocity V_m of 12 mph ±, that is, some of the windier parts of the United States (Table 8.1). The turbine power coefficient, 0.375, is the fraction of the wind power extracted by the turbine at rated wind speed (P_c in Eqs. 8.4 and 8.6). The whirling blades describe a cone facing the wind. The cone angle is measured between this cone and a plane normal to the axis of rotation.

The costs of the unit (1975), shown in Table 8.3, were given by NASA with the caution that some of the figures may include some development. However, were this windmill applied in a location of 12 mph mean annual velocity, $V_r/V_m = 1.5$, it would generate 3100 kWh/kWyr (Fig. 8.4), or 310,000 kWh in an average year (279,000 kWh at 90% availability). Using the cost of $650,000 (high, as mentioned) with a normal utility financing rate of 18% for maintenance, depreciation, taxes, insurance, and return results in an annual cost of $64,800, or 23.2¢/kWh. The cost of a similar production unit would

Table 8.2. Specification of NASA Plum Brook 100 kW Wind Turbine Generator

Blade circle diameter	125 ft
Number of blades	2
Tower height	100 ft
Turbine output	133 kW
Generator rating (synchronous, 60 Hz, 3 phase, 1800 rpm, 125 kVa, 0.8 power factor, 480 V)	100 kW
Wind turbine speed	40 rpm
Rated wind speed V_r	18 mph
Turbine power coefficient P_c	0.375
Cut-in speed	8 mph
Feather at	60 mph
Withstand hurricane of	150 mph
Cone angle	7°
Gear box, conventional	40/1800 rpm
Generator belt driven	1/1
Blade rigidity	3%
Turbine downwind from tower	

Source. DOE/NASA.[14]

Table 8.3. Projected Costs of NASA Plum Brook 100 kW Wind Turbine Generator

	Dollars
Blades (includes engineering) (for three blades)	300,000
Hub, pitch change	95,000
Machinery, shafts, couplings, etc.	43,000
Gear Box (includes engineering)	16,000
Generator, electric controls	68,000
Tower, foundation, and housing	128,000
Total	650,000

Source. DOE/NASA.[14]

presumably be less, and application in a higher mean wind velocity improves the economy rapidly.

Since beginning operation in July 1975, the generator output in winds of various velocities has been as predicted. For these tests, the output is simply absorbed in a resistor bank.* The stresses in the blade roots due to vibrations—shock excited as the blades pass through the "dead air space" back of the tower—have been much greater than anticipated and have necessitated some modifications. Thus, the unit has performed its planned function in providing experience and knowledge necessary for the "industry designs" to follow.

*Later supplied to Ohio Edison System.[23]

Industry-Designed Wind Turbine Generators

For the second phase of the program, several manufacturers are building WTGs over a range of sizes for application in a large number of typical situations. They are being turned over to the user for operation in each case.

The General Electric Co. and Kaman Co. produced designs optimized for 500 kW with 12 mph mean wind speed and for 1500 kW with 18 mph mean wind speed. With but slight loss in economy (10% above minimum power cost), the 500 kW unit could be used in winds with mean velocities from 9 to 15 mph, and the 1500 kW units from 15 to 21 mph. The principal data for these four designs are given in Table 8.4.

The energy costs given are the same as would be obtained using 100% availability and approximately 15% fixed charge on the capital cost to cover maintenance, depreciation, taxes, insurance, and return, and also any operating costs. At 90% availability and 18% fixed charge the costs would be 1.33 times those shown. The NASA report states that the energy costs are competitive with some utility costs. However, this will be in rather special situations. The average energy component of the busbar cost at U.S. fossil fuel-fired plants in 1979 was 1.8¢/kWh (Table 1.3). This is the only part of the

Table 8.4. Preliminary Industry Designs for Large Wind Turbine Generators[17]

	500 kW Rating		1500 kW Rating	
	G.E. Co.	Kaman	G.E. Co.	Kaman
Mean annual wind velocity V_m (mph)	12	12	18	18
Rated Wind Velocity V_r (mph)	16.3	20.5	22.5	25
Rotor diameter D (ft)	183	150	190	180
Rotor speed (rpm)	23	32.3	40	34.4
Capital cost[a]				
$	486,000	450,670	674,000	720,800
$/kW	974	901	499	481
Energy cost[a] (¢/kWh)	4.18	5.55	1.65	2.02
Plant factor (at 100% availability)	0.42	0.29	0.51	0.43
Specific output (kWh/kWyr)	3679	2540	4468	3767
V_r/V_m	1.36	1.71	1.25	1.39
Hub height $H = D/2 + 50$ (ft)	141	125	145	140
Mean wind velocity at hub height (mph)	15.5	15.2	23.4	23.3
Wind regime				

(As supplied by NASA, Ref. 16, Fig. 13)

Source. DOE/NASA.[17]
[a]Based on production runs of 100 to 1000 units.

cost that would normally be replaced by wind power. The costs of Table 8.5 would be subject to at least the GNP inflation of 29% in 1975 to 1979.

The Plant Factors or specific outputs arrived at by the manufacturers, using the wind regime provided by NASA, agree quite well with the curve of Fig. 8.4 and have been plotted thereon as circles. With the wind power increasing as the cube of the wind speed, any constant speed, horizontal axis, machine will realize full power above rated speed, and regardless of the blade efficiency it will drop off *about* as the cube of the speed below rating. There is very little room for any improvement in specific output for a given V_r/V_m ratio. This is borne out by the fact that the "new" points fall close to the "old" curve. Hence the *cost* of obtaining a given amount of energy in a year from a given wind regime is a better measure of the overall design job.

Vertical-Axis (Darrieus-Type) Wind Turbine (VAWT)[18,22]

The Darrieus-type vertical-axis wind turbine (Fig. 8.9) is being developed concurrently with the horizontal-axis wind turbine (1977) both in the United States and Canada. A 200 kW machine of this type is being installed in the Magdalen Islands in the Gulf of St. Lawrence by Dominion Aluminum Fabricating, Ltd., with R&D support by the National Research Council of Canada. The two-blade unit is 80 ft (24.4 m) in diameter and 120 ft (36.6 m) high. The mean annual wind velocity at this location is 19 mph.

The wind turbine will operate in parallel with the 26 MW diesel-electric power plant now supplying the islands. The contract price for the first unit was $230,000, with estimates in production quantities of $120,000 including installation. This competes favorably with diesel oil at 3¢/kWh.

The DOE-sponsored VAWT (vertical-axis wind turbine) program is being carried out at the Sandia Laboratory in Albuquerque, N. Mex. A 17 m (55.8 ft) unit was being built (about 1976) as a test bed for studying system performance and evaluating components. This corresponds to the NASA 100 kW (125 ft diameter) HAWT (horizontal-axis wind turbine) at Plum Brook. The overall program contemplates the development of units from 5 to 60 m (16.4 to 196.8 ft) which would develop power as shown in Table 8.5.

The curious shape of the Darrieus blade is the troposkien shape taken by a flexible rope whirled

Figure 8.9. Vertical-axis wind turbine (VAWT). Courtesy DOE/Sandia.

about an axis. Thus, bending moments are a minimum. The theory of operation is shown in Fig. 8.10, in which a rotor blade is shown in four successive positions as it rotates with a velocity ω at a radius R. Because of its rotation, there is a counterwind velocity $R\omega$ which, added to the actual wind velocity V, produces a resultant wind velocity W acting on the blade. There are drag and lift forces D and L acting along and normal to the resultant wind direction W. It can be seen that in each of the four positions shown, the lift produces a torque in the positive direction of ω, although the magnitude varies. Actually it varies from 0, when the blade is moving directly up or down wind (not shown), to a maximum about a quarter revolution later. The

Table 8.5. Sandia Development Program—VAWT

Diameter				
m	5	17	40 and/or	60
ft	16.4	55.8	131.2	196.8
kW at 18 mph	1.6	20	150	350
kW at 25 mph	4.4	50	350	1000

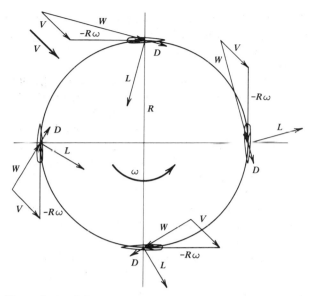

Figure 8.10. Principle of operation of a Darrieus wind turbine, V = wind velocity, R = radius, ω = rotational speed, W = resultant velocity, L = lift, "lift driven"; D = drag.

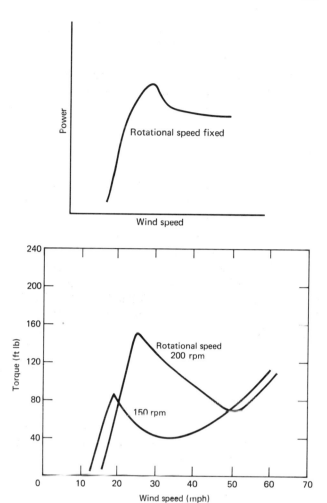

Figure 8.11. Typical Darrieus wind turbine characteristics. (*a*) Theoretical characteristic of design A (Ref. 18, p. I-43). (*b*) Test characteristic of design B (Ref. 18, p. II-15).

torque thus makes two complete excursions from 0 to maximum and back each revolution, both in the positive direction.

The pulsations in shaft torque can be minimized by the use of three blades, at the expense of greater blade root forces. However, at the Magdalen Islands installation (two-blade), starting with a 100% torque pulsation at the blades, the flexibility in the blades reduces this to 12% in the shaft. Acting through an induction generator into the particular power system involved at Magdalen Islands, the resulting voltage ripple is below the permissible limit of 0.05%.[19] Note: the limit is not usually under 0.5% at this frequency.[20]

The blade coefficient of the Darrieus-type wind turbine is shown in Fig. 8.1. At wind speeds above an optimum, that is, at lower tip speed ratios, the power coefficient falls off rapidly, resulting in the power-versus-wind speed curves of Fig. 8.11, for a fixed rotational speed. Because of this inherent limitation of power output, a power regulator is not required.

Thus, the advantages of the Darrieus-type wind turbine are considered to be:

1. Vertical symmetry eliminates the need for yaw control.
2. Delivers mechanical power at ground level.
3. Simple support tower construction.

4. Reduced blade fabrication cost.
5. Requires no pitch control for synchronous operation.

The two-blade design has the advantage over the three-blade design of lower manufacturing and erection costs, the latter because it can be assembled flat on the ground and erected as a unit. However, the torque pulsations must be considered as mentioned.

The specific output, 3000 kWh/kWyr, given[18] for the Magdalen Islands vertical-axis machine for $V_r/V_m = 30/19 = 1.58$, falls practically on the curve (Fig. 8.4) for horizontal-axis machines. This coincidence is no doubt due to above-rating operation at low speeds compensating for below-rating operation at high speeds, as suggested in Fig. 8.2. It cannot be expected in general.

Note: In later ALCOA VAWT designs the power drops off very little from maximum (rating) at higher wind speeds and the specific output is close to that given in Fig. 8.4 for horizontal axis machines.

1981 Update.

This chapter was written in 1977. Between then and 1981 great progress has been made in the development of cost effective WTGs.[21-26]

The DOE sponsored horizontal axis WTG development is described by Robbins and Thomas.[21] For a site having 14 mph mean annual wind velocity at 30 ft elevation, the "second unit" costs per kWh (18% capital charge, 1978 dollars) are 41¢ for the Mod-OA, 200 kW, 125 ft diameter WTG; 30¢ for the Mod-1, 2000 kW, 200 ft diameter WTG (economy of scale); and about 8¢ for the Mod-2, 2500 kW, 300 ft diameter WTG (improved technology), with about 3¢ forecast for the latter with further improvements and mass production (100th unit). 90% availability is assumed.

The Mod-OA was installed at Clayton, N. Mex., Block Island, R. I., Culebra, Puerto Rico, and Oahu, Hawaii, the Mod-1 in Boone N. Carolina, and three Mod-2's in the Goodnoe Hills near Goldendale, Wash.

A 3 MW Bendix/Wind Power (Schachle) horizontal axis machine is under test (1981) at the Southern California Edison Palm Springs site, and a number of 2–4 MW machines are on order (20 for Hawaiian Electric), and some larger machines.[23]

Although the development is not as far advanced (1981), a number of vertical axis units are operating successfully. In addition to the 5 m (16.4 ft) and 17 m (55.8 ft) diameter test units at Sandia, and the 5.5 m (18 ft) and 25 m (82 ft) diameter test units at ALCOA (Pittsburgh), 12.8 m (42 ft) diameter units (100 kW), are operating at Bushland, Tex., near Amarillo, on Martha's Vineyard, and at Rocky Flats. A 25 m (82 ft) diameter unit (500 kW) is operating at Agate Beach, Oregon, as well as several other units of various sizes in the United States and Canada.

Unfortunate accidents have wrecked three large units due to out-of-control operation above design speed. While expensive "lessons," these occurrences should be eliminated by improved controls and operating procedures. In particular, while

theoretically not self-starting, the Darrieus does self start on occasion, with unequal wind gusts on the widely separated blades. It must be secured at all times.

The blade height of the Darrieus is typically 50% greater than the diameter; 82 ft × 123 ft for the 500 kW unit, 42 ft × 63 ft for the 100 kW unit.

The 1981 price for a single 100 kW ALCOA unit installed is about $130,000. At 90% availability it will produce 200,000 kWh/yr at a site having 14 mph mean annual wind velocity. At 18% capital charge the cost of electricity is 65¢/kWh.

The 1981 price of a single 500 kW ALCOA unit, optimized for a higher wind velocity site, is about $500,000 installed. At a site having 18 mph mean annual velocity, it will produce 1,127,000 kWh/yr at 90% availability. At 18% capital charge this represents 44.4¢/kWh.

In quantities of 10 or more, as might be used for a "wind farm," the costs would be substantially less, generally equivalent to comparable horizontal axis machines.

A large number of small WTGs, up to 100 kW, are in operation. The DOE Rocky Flats Test Center coordinates this development and carries on related research. Some 12 test towers were in operation in 1980, and 18 more are being installed. Most small machines are uniformly tested at these facilities. Ref. 25 is a guide to commercially available small machines, and it lists over 100 WTGs under 100 kW. SERI has prepared a Wind Energy Directory[26] that lists the principle sources of wind energy information.

REFERENCES

1. T. Baumeister, *Marks' Standard Handbook for Mechanical Engineers*, 6th and 7th eds., New York: McGraw-Hill, 1966.

2. E. W. Golding, *Generation of Electricity by Wind Power*, London: E. and F. N. Spon, 1976.

3. P. C. Putnam, *Power From the Wind*, New York: Van Nostrand Reinhold, 1967.

4. R. L. Thomas and J. E. Sholes, *Preliminary Results of the Large Experimental Wind Turbine Phase of the National Wind Energy Program*, NASA TMX-71796, Cleveland, Ohio: Lewis Research Center, 1975.

5. E. L. Harder, "Specific Output of Windmills—A Discovery," *Proc. IEEE, November 1977*, p. 1623, Corr. Aug. 1978, p. 987.

6. E. McGlinn, *Wind Energy Workshop—MERRA*, University of Michigan, February 22, 1978.

7. N. G. Butler, "Wind Energy from a Utility Planning Per-

spective," *Conference on Wind Characteristics—BPA*, Portland, Ore., June 19–21, 1979.

8. H. Z. Tabor, "Power for Remote Areas," *Sci. Technol.*, May 1967, p. 52.

9. "Windmill Electric Plants," in *Standard Handbook for Electrical Engineers*, 6th ed., New York: McGraw Hill, 1933, p. 2695.

10. *Wind Energy Developments in the Twentieth Century*, NASA Lewis Research Center, Cleveland 1975.

11. U. Hutter, *Operating Experience with a 100-kW Wind Power Plant*, NASA, TT-F-15608, Cleveland, Ohio: Lewis Research Center, 1973.

12. U. Hutter, *Influence of Wind Frequency on Rotational Speed of Windmill Generators*, NASA, TT-F-15184, Cleveland, Ohio: Lewis Research Center, 1973.

13. U. Hutter, *The Development of Wind Power Installations for Electrical Power Generation in Germany*, NASA TT-F-15050, Cleveland, Ohio: Lewis Research Center, 1973.

14. R. L. Puthoff and P. J. Sirocky *Preliminary Design of 100-kW Wind Turbine Generator*, NASA, TMX-71585, 1976.

15. J. M. Savino *A Brief Summary of the Attempts to Develop Large Wind-Electric Generating Systems in the United States*, NASA, TMX-71605, Cleveland, Ohio: Lewis Research Center, 1974.

16. R. W. Vernon, *Summary of NASA–Lewis Research Center Solar Heating and Cooling and Wind Energy Programs*, NASA TMX-71745, Cleveland, Ohio: Lewis Research Center 1975.

17. R. L. Thomas, *Large Experimental Wind Turbines–Where We Are Now*, NASA TMX-71890, Cleveland, Ohio: Lewis Research Center 1976.

18. *Proc. of Vertical-Axis Wind Turbine Technology Workshop*, Sandia Labs., Albuquerque, N. Mex., 1976.

19. R. J. Templin and P. South. "Canadian Wind Energy Program," Ref. 18, p. I 57.

20. *Electrical Transmission and Distribution Reference Book*, 4th ed., Pittsburgh, Pa.: Westinghouse Elec. Corp., 1950, p. 720, Fig. 4.

21. W. H. Robbins and R. L. Thomas, *Large Horizontal-Axis Wind Turbine Development*, NASA TM-79174, Cleveland, Ohio: Lewis Research Center, 1979.

22. S. E. Johnston, Jr., Ed., *Proceedings of the Vertical-Axis Wind Turbine (VAWT) Design Technology Seminar for Industry*, Albuquerque, N.M.: Sandia Labs., 1980.

23. M. G. McGraw, "Wind Turbine Generator Systems," *Electrical World*, May 1981, p. 97.

24. *Large Wind Turbine-Generator Performance*, NASA N80-31960, EPRI AP-1317, Palo Alto, Ca.: EPRI, 1980.

25. *Commercially Available Small Wind Systems and Equipment*, Rockwell International, Rocky Flats, Washington, D.C.: U.S. Govt. Printing Office, 1981.

26. *Wind Energy Information Directory*, SERI/SP-69-290R, Washington, D. C.: Supt. of Documents, 1980.

9

Geothermal Energy

INTRODUCTION

In this chapter, we endeavor to answer the same questions as for other energy sources: What is geothermal energy? What are the principal facts about it relevant to energy? What are the processes of converting it to electricity? Its uses solely for low-temperature heat are not considered here.

To answer these questions, we review the current installations and discuss the resources and likely rates of development. Then, the earth crust properties, of vital importance in this technology, are considered. The Geysers field, a "dry steam field," is discussed in some detail as an example. Finally "hot-water dominated" systems and the future possibilities of geothermal energy from hot dry rock are discussed.

The earth is a far more complex boiler than any of the man-made systems described throughout this book. There are no neat evaporators, superheaters, and heat exchangers, and no distinct boundaries. In spite of this, a number of crude calculations are presented throughout this chapter, bringing together the known properties of water, steam, and heat, with the variable earth crust properties and the known final results in developed fields. Even these crude calculations are quite revealing and are presented with the apology that some quantitative understanding, however crude, is better than none.

Geothermal power refers primarily to the generation of electricity from the stored heat in the earth's crust. Other uses for process or space heating are termed "direct heat." Geothermal power may be "natural," that is, from existing emissions of steam or hot water, including shallow wells up to 1.8 mile (3 km) in depth, to near-surface reservoirs of heat and fluid. Or it may be "drilled" or "artificial" deep wells up to 50,000 ft, 10 miles (16 km). In the latter case, cold water is injected into deep wells, heated, and brought back up.

Geothermal resources may be classified as:

1. Hydrothermal resources—underground reservoirs of hot water and steam.
2. Geopressured resources—hot saline fluids found at very high pressures in porous formations, as beneath the coastal areas of Louisiana and Texas. These are believed to contain large amounts of dissolved natural gas, and the production of natural gas and heat may be equally important.
3. Hot dry rock resources—typically at about 3-mi depth, with little or no fluid present.

EXISTING FIELDS, RESOURCES, AND GROWTH RATES

All current geothermal power in the world (1979), and most active exploration, is of the natural type. Development started with a 1904 installation in Lardarello, Italy. Operation was started at Wairekai, New Zealand, in 1959, at Geysers, California, in 1960, and at Matzukawa, Japan, in 1966. Work on artificial geothermal includes work on hot, dry rock,[8] using forced circulation between two wells, theoretical studies, and tests of stimulation methods.

Nearly half of the world geothermal generation is currently (1980) at a single location, the Geysers field in northern California, where 800 MW is operating and additional 1200 MW is planned. There is extensive exploration and relatively small-scale use in all the western states.[1] Elsewhere in the world, the principal users are: Italy, 400 MW plus 200 MW planned; New Zealand, 200 MW plus 500 MW planned; Mexico, 75 MW plus 200 MW planned; Japan,* 50 MW plus 150 MW planned; and the Soviet Union, 25 MW plus 400 MW planned.

Total geothermal generation in the world in 1980

---

*165 MW by later reference,[10] with goal of 48,000 MW by year 2000.

was about 1.6 GW (including 0.8 GW in the United States), with about 2.9 GW planned (including 1.5 GW in United States).

For reference the total installed generating capacity in the United States in 1980 was about 625 GW, the 0.8 GW at Geysers thus representing 0.13%. The rather extravagant "conservative estimates" of some of the proponents,[2] 2 to 3% by 1989 (made in 1974), must be reduced by about 10:1 to be realistic. The National Petroleum Council in 1971 also estimated 51 billion kWh geothermal electricity in the United States by 1985,[1] which would be about 2% (of 2600 billion kWh). Hickel (1972)[1] forecast a *possible* 132 GW of U.S. geothermal by 1985. This would be about 19% (of 700 GW). Needless to say, the higher estimates assumed a climate highly favorable to geothermal development, which has been almost totally lacking (1981).[3*]

The Salton Sea area of Southern California has recently been measured[4] to contain 6×10^9 acre ft of water (7.4×10^{12} m^3) at temperatures above 500°F at high pressure. At an enthalpy of 487 Btu/lb (500°F saturated water), this corresponds to 2.31×10^{15} kWh. At 15% efficiency, it would generate 40,000 GWyr electricity. This is viewed[4] as a sufficient resource to produce 20 GW for Southern California. The amount is certainly adequate. Its economic recovery is the problem, as with most geothermal energy.

Similar estimates for all of western United States indicate the possibility of eventually supplying 25% of the western demand by geothermal energy.[4]

Developed Geothermal Fields

At the *Geysers field*, dry (superheated) steam is found at depths of 800 to 7000 ft (240 to 2100 m). Typical well "shut-in" pressures are 400 to 500 psia (27 to 34 atm). At full flow, this drops to 115 psia (7.8 atm) at the wellhead. The turbine inlet pressure is 95-115 psia.

At *Lardarello*, dry (superheated) steam is found at depths of 1000 to 3000 ft (300 to 900 m), at shut-in pressures up to about 400 psia (27 atm), which drops to 80 psia (5.4 atm) at the wellhead, the turbine design pressure.

At *Wairekai, N.Z.*, wet steam is found at depths of 574 to 3200 ft (175 to 976 m). The wellhead and

---

*The reference refers to oil shale; substitute "geothermal." The factors are almost identical.

turbine design pressure used is about 196 psia (13 atm). Moisture separation is required.

These and other installations[9,10] have been developed with private capital and produce electricity at costs fully competitive with other sources available to those regions.

At *Matzukawa, Japan*, dry (superheated) steam is obtained from wells of 2625 to 3940 ft (800-1200m) depth, at a working pressure of 441 kPa (64 psia) at the turbine. At other fields in Japan developed 1967–1977, wet steam is used with single or double flash.[10] R&D is shared between government and industry on a 90/10 ratio.

Geothermal Resources[1,9,10]

Estimates of U.S. geothermal resources vary from 40,000 GW centuries[1] to relatively insignificant amounts. Current use of electricity is 2300 billion kWh/yr (1980), or 262.6 GW average. The high estimates assume volcanic and hot dry rock and rock–water systems to a depth of 35,000 ft (10.7 km) to be "technically exploitable." They also calculate the total heat in place above 212°F (100°C) rather than that extractible. However, if reserves are defined as with coal, oil, and gas, "recoverable economically with present technology," the geothermal reserves shrink to a relatively insignificant amount. The potential of the Geysers field will be discussed later.

A good geothermal area for electric power generation may be defined as one having reservoirs of at least 356°F (180°C), and preferably above 392°F (200°C), at depths of less than 1.8 mile (3 km). It should have natural fluids, (steam or water) for transferring the heat from the earth to the power plant. It should have adequate reservoir volume,[1] 10 to 100 mi^3 (42 to 420 km^3), to supply 1 to 10 GW electrical generation for a lifetime of 40 yr. It should have sufficient reservoir permeability to ensure delivery to wells at adequate rates. And it should have no major unsolved problems.

The geothermal fields at Lardarello, Wairekai, and Geysers, and several others, all currently meet these requirements, although all *did* have major unsolved problems which had to be solved in bringing them to this status. However, such a favorable combination of circumstances is extremely rare in the earth, and if geothermal power generation is ever to be more than a curiosity, methods must be developed for economically using less favorable areas. These include hot water or brine systems at hydrostatic pressure, now under intensive development, and wells into dry hot rock, which are receiving study.[8]

Location of Geothermal Fields

Good geothermal locations are found principally along fault lines, frequently evidenced by volcanoes, hot springs, and fumaroles. Here, the hot magma from the interior has pushed up through the crust and formed pockets or intrusions of high heat content and high temperature. Such high-temperature reservoirs of heat, relatively close to the surface (essentially all fields explored to date, 1980), are near the margins of crustal plates, on which the continents rest. The fields in California, Italy, New Zealand, and Japan all fall in this category.

Elevated temperatures of less intensity are also found in some interior areas, where the heat flow is above normal and the upper layers of rock are of low conductivity. These conditions, alone or in combination, result in above-normal gradients in certain areas such as the Hungarian Basin.

At Geysers, the heat source is apparently an igneous mass at a depth of perhaps 3 to 5 miles (5 to 8 km). Heat from this mass has supplied the reservoir of highly fractured and permeable rock at depths of 800 to 7000 ft (244 to 2130 m), which is now being tapped. Since dry steam enters the wells, the theory has been advanced[1] that much of the water in this reservoir is vaporized (dry steam) due to a reservoir pressure less than one fourth of the hydrostatic pressure, coupled with temperatures about four times normal at 1 mile depth. That is, it is "vapor dominated." (It will be shown later that it is more likely hot water flashing to steam before it enters the wells.) Most of the geothermal fields in the world are "water dominated." At the reservoir pressure, the temperature is below saturation and the fluid is water. As it is brought to the surface and the pressure is lowered, some of it flashes to steam, the proportion depending on the wellhead pressure.

There is also the possibility of producing this hot water at the reservoir pressure, extracting its heat in a heat exchanger, and reinserting it. This technique is discussed later.

EARTH CRUST CHARACTERISTICS

The average earth crust characteristics and heat flows are given in Table 9.1. Rock has a specific gravity of 2.73 times water, or a density ρ of 170 lb/ft³. It has a specific heat C_p of 0.17 Btu/lb °F and a specific heat capacity ρC_p of 28.9 Btu/ft³ °F.

Its thermal conductivity k is 1.45 Btu/ft hr °F, which results in a diffusivity α ($= k/\rho C_p$) of 0.05 ft²/hr. Diffusivity may be thought of as the heat flow out of unit volume for unit decrease in gradient in passing through it; or, more precisely, it is the coefficient in the La Place transient heat flow, Eqs. 9.1 through 9.3.

The mean heat flow through the earth's crust, 0.02 Btu/ft² hr, may be compared with the mean solar energy reaching the earth's surface, about one eighth of the solar constant, or 429/8 = 53.6 Btu/ft² hr. The energy permeating out from the interior is tiny in proportion, 1 part of 2680, but nonetheless important. It is the source of all geothermal energy. Note: Since the earth surface area is four times its projected area, the average insolation of the earth's atmosphere is one fourth of the solar constant. About 50% of this reaches the surface, or one eighth of the solar constant.

All of these constants have a substantial range of variation from the average.[1] In good geothermal areas, the temperature gradient and heat flow may be up to 10 times the average value. In the Jemez, N. Mex., volcanic region, for example, over large areas the measured gradient to 2300 ft (700 m) is about 9.9°F/100 ft (18°C/100 m), or 6.6 times the average. Similarly high gradients are found at the currently developed fields—Geysers, Lardarello, and so on.

Porosity and Permeability

A necessary condition for a good geothermal field is sufficient porosity to hold a fluid (water, steam, or brine) and sufficient permeability for it to flow into the wells at an adequate rate.

Porosity is the storage capacity for fluids, the hollow volume per unit volume of the material. Porosities of 0 to 30% are found in various rocks and rock formations of the earth.

Permeability is the flow through unit area per unit of pressure drop in the direction of flow. In large earth sections, this includes any natural channels, supported crevices, interstices in fractured rock, and the like. It is the total path for the flow of fluids due to pressure differentials. There may be alternate permeable and impervious layers, faults, in fact, every conceivable combination. Permeability is the general term used to designate the ability of fluids to get from here to there in the earth. Highly fractured rock generally has good permeability.

Table 9.1. Average Earth Crust Characteristics

Quantity	Symbol	Value	
		English	Metric
Density (average)	ρ	170 lb/ft^3	2.73 g/cm^3
Specific gravity (water = 1)		2.73	2.73
Specific heat	C_p	0.17 Btu/lb °F	0.17 cal/g °C
Specific heat capacity	ρC_p	28.9 Btu/ft^3 °F	0.464 cal/cm^3 °C
Thermal conductivity	k	1.45 Btu/ft hr °F	0.0060 cal/cm sec °C
Thermal diffusivity[a]	α	0.05 ft^2/hr	0.013 cm^2/sec
Mean heat flow through crust		0.02 Btu/ft^2 hr	1.5 μ cal/cm^2 sec[b]
Mean thermal gradient		1.5 °F/100 ft	2.75 °C/100 m

[a]The thermal diffusivity α $(= k/\rho C_p)$ is the coefficient in the La Place transient heat flow equation for a homogeneous medium:

$$\frac{\partial T}{\partial t} = - \alpha \frac{\partial^2 T}{\partial x^2} \text{ in one dimension} \tag{9.1}$$

or

$$\frac{\partial T}{\partial t} = - \alpha \nabla^2 T \text{ in three dimensions} \tag{9.2}$$

where

$$\nabla^2 T = \frac{\partial^2 T}{\partial x^2} + \frac{\partial^2 T}{\partial y^2} + \frac{\partial^2 T}{\partial z^2} \tag{9.3}$$

in which T is temperature; t is time; and x, y, and z are the space coordinates.
[b]Also called 1.5 hfu (geothermal heat flow units).

A porous material is not necessarily permeable. (The term porous is sometimes used to mean permeable, e.g., porous, membrane.) Often, in geothermal writing the terms porosity and permeability are used together, indicating the ability to hold a substantial amount of fluid as well as a structure that permits it to circulate.

Hydrostatic and Rock Pressures

Water in the land surface of the earth has come mostly from the atmosphere and is termed "meteoric water." Since it has permeated down from the surface, there must be some permeability (either now or in earlier geologic eras), although the water may have traveled considerable distances horizontally in the more permeable layers.

An imperfect but very useful concept is that there is an impervious layer somewhere at the bottom, several miles down, and a continuous water path, however torturous, from there to the surface. There

is then a continual gradation of "hydrostatic pressure," the pressure of the water, from 0 psig at the surface to about 2300 psi (156 atm) at 1 mile depth, and so on, to whatever depth water is found. The rock pressure at 1 mile depth is 2.73 times as great, or about 6280 psi (427 atm).

These two pressures may exist simultaneously, just as they would if a 1 mile deep well were filled with water. The water pressure is termed "hydrostatic pressure" to distinguish it from the rock pressure, or from the actual fluid pressure when no continuous water path to the surface exists. Water or steam may then be found at more or less than the hydrostatic pressure. Thus, the reservoir pressure of 400 to 500 psi 1 mile down at the Geysers field signifies that there is no continuous path from the surface to the reservoir and that the pressure there is less than a quarter of what it would be if there were such a path. We say it is less than a quarter of the hydrostatic pressure.

Water in the lower strata may have been trapped

Table 9.2. Normal Boiling Temperature at Depths

	Pressure		Boiling Temperature	
	psia	atm	°F	°C
At the surface	14.7	1	212	100
At 1000 ft depth (305 m)	448.2	30.5	445.9	235.5
At 1 mile depth (1.61 km)	2,303	156.6	655.9	346.6
At 1.4 mile depth (2.25 km)	3,206	218	705	374
At 10 mile depth (16.1 km)	23,030[a]	1,566		

[a]Actual pressure is less due to lower density of water at high temperature.

Table 9.3. Normal Earth Temperatures at Depths

	Temperature	
Depth	°F	°C
Surface	50	10
1000 ft (305 m)	65	18
1 mile (1.61 km)	129	54
1.4 mile (2.25 km)	161	72
2.05 miles (3.29 km)	212	100
10 miles (16.1 km)	842	450

there since earlier geologic times, with no present path to the surface until wells are drilled through the impervious layers. A typical well in the Geysers field may extend 2000 ft through water-bearing rock (permeable rock), then through 2000 ft of dry, impervious, hard rock, and then into the steam-bearing strata.

Heat, Water, and Steam

It can readily be shown that at the normal earth temperature gradient of 1.5°F/100 ft, water at hydrostatic pressure is far from boiling. It is hot water, not steam. This is significant up to a hydrostatic pressure of 3206 psi, the critical pressure of water, which occurs about 1.4 mile down. At higher pressures, water and steam are indistinguishable.

When there is a continuous water path from the surface, with the "water table" near the surface, the fluid pressure at every level is the hydrostatic pressure. There is a corresponding boiling temperature at each level, as shown in Table 9.2. As mentioned, boiling has no significance above the critical pressure and temperature of water.

The pressure column of Table 9.2 is calculated with normal-density cold water. Pressure at a given level would of course be less with lower-density hot water. However, this picture is adequate for present purposes.

Now, turning to the mean earth temperature gradient of 1.5°F/100 ft (2.75°C/100 m) and starting with an average temperature of 50°F (10°C) at the surface, the normal earth temperatures at various depths can be calculated and are given in Table 9.3. Note that these temperatures are all well below the

boiling points at hydrostatic pressure at the corresponding depths. Thus, the fluid is hot water with normal earth temperatures and hydrostatic pressure, at all depths of interest.

This is shown in Fig. 9.1, in which both normal earth temperature and the boiling temperature at hydrostatic pressure are plotted. Thus, only if the temperature gradient were several times normal, or the pressure and the associated boiling temperatures were far below hydrostatic, or both, would steam be found in the earth. As stated earlier, most geothermal systems in the world are "hot water dominated." For reasons to be presented later, it is doubtful if any appreciable part of geothermal energy exists as steam before wells are drilled.

Energy from Hot Water

On the average, water in the earth more than 2 miles down is above 212°F (100°C) (Table 9.3). At 10 miles deep, it is at 842°F (450°C). This may be water or brine already present in the earth, or water injected into a deep well from the surface. The energy in this water may be brought to the surface and used in one of two ways: (1) Under a pressure sufficient to prevent flashing to steam, the fluid remains water. Its heat can be transferred to a working fluid in a heat exchanger. (2) At reduced pressure, some of the water flashes to steam for direct use in turbines.

In good geothermic areas, hot water occurs much closer to the surface. These ideas are amplified and a few rough calculations presented in the next few paragraphs.

The pressurized hot water system is very similar to the pressurized-water nuclear reactor shown in Fig. 6.3a. Hot water is supplied by the earth in place of the nuclear reactor. For example if hot water were brought to the surface at 600°F (316°C) at a pressure of 2250 psi (153 atm), it could be used to generate steam at 520°F, 800 psi (271°C, 54.4

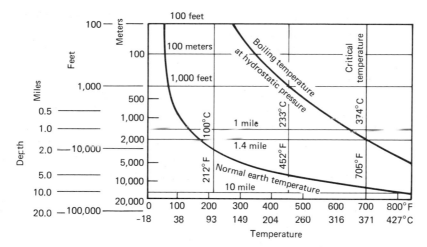

Figure 9.1. Normal and boiling temperatures at various depths.

atm), as shown in Fig. 6.3a. The turbine would then be similar to that in the nuclear system, except much smaller, possibly 100 MW instead of 1000 MW. This is the basic principle of deep-well geothermal or hot, dry rock systems (injected water) and is one of the methods used in shallower, hot water- or brine-dominated geothermal fields at lower temperature.

Water Flashing to Steam. The second scheme, in which *some* of the hot water flashes to steam, is the basis of the Wairekai, New Zealand, operation and other hot water-dominated systems.

As an example, suppose saturated water at 500 psi, 467°F is to be flashed to 200 psi, 382°F wet steam, and after moisture separation, saturated steam at 200 psi, 382°F is supplied to the turbine. The enthalpy of the wet steam will be approximately the same as the saturated water at 500 psi, neglecting losses. From the tables of properties of saturated water, the enthalpies are:

Sat. liq.   at 500 psi 467°F =   449Btu/lb
Sat. vapor at 200 psi 382°F = 1198 Btu/lb
Sat. liq.   at 200 psi 382°F =   355 Btu/lb

Let x be the fraction that is vapor. Then,

$$1198x + 355(1 - x) = 449, \quad x = 0.11 \ (11\% \text{ steam})$$

Of the initial enthalpy in the 500 psi water, 29% is now in the 200 psi steam and the remainder in hot water. The 89 weight % of water separated out,

containing 71% of the energy, could be used for other heating tasks; or a lower pressure could be selected to get more of the enthalpy in steam (at the expense of a larger turbine). That is, an optimization study would be needed.

Both the hot-water and wet-steam systems are discussed later under Water-Dominated Systems and Deep-Drilled Wells. However, in order to better picture the phenomenon taking place in the earth, we next discuss heat flow and water supply in geothermal fields.

Heat Flow in the Earth

Heat flow may be by conduction or convection. The average gradient cited earlier, 1.5°F/100 ft, is characteristic of conduction through rock, the usual situation. However, in certain circumstances convective systems may be in operation, with hot water rising in a central area over a heat source and cooler water falling in peripheral areas to complete the flow path. In this case, there may be relatively little temperature difference between the top and bottom of the convective system, far less than if the same amount of heat were flowing through rock by conduction. A convective system may thus bring heat from considerable depths and result in higher surface gradients.

Water Supply to Geothermal Fields

When heat and water meet, the phenomenon that takes place depends on the water supply and on the pressure and temperature. Normally, in the earth the water is just heated and is far from boiling, as shown in Fig. 9.1. If the temperature is high enough for boiling at the pressure encountered and the water supply is ample to absorb all the heat at saturation temperature, the conditions are the same as in a boiler (evaporator). Part of the water changes to steam, absorbing all the heat not needed to bring the water up to boiling. Part remains water. The result is steam coexisting with water.

However, it should be carefully noted that while saturated vapor at 500 psi has 2.68 times the enthalpy per pound saturated water, it occupies 47.1 times the volume per pound saturated water. Therefore *if the space is confined*, it acts like a pressure cooker. A small amount of steam fills the voids. Then the pressure must rise, raising the boiling point, until the saturated water and surrounding rock can absorb practically all the heat. Any fluid escaping (as through wells) will be dry superheated steam, as in a pressure cooker.

As shown in Fig. 9.1, boiling will exist only in reservoirs in which the pressure is far below the hydrostatic pressure and the temperature considerably above normal. It is found at only a few commercially producing fields in the world. Dry steam is obtained from the wells only at the Lardarello field in Italy, the Geysers field in northern California, and a few others. Whether it is dry steam in the reservoir or water flashing to dry steam in the ground en route to the wells is a matter of conjecture. Some calculations given later under Geysers bear on this point.

It is of course possible that *all* water available in a reservoir is evaporated and superheated to the temperature of the region.

When the dominant phase in the reservoir is hot water well below the boiling temperature, water or wet steam enters the wells, with part of the water flashing to steam as it rises in the wells, and the pressure is lowered. This is the condition encountered at Wairekai, New Zealand, and other fields where wet steam is encountered.

Evidence indicates that water in the Geysers geothermal reservoir is only that which was trapped there in earlier geologic times. It is not being renewed in this geologic period (contrary to earlier expectations); and when it is exhausted to the economic limit, that field will be exhausted as a natural reservoir. The extent to which it might be extended by reinsertion of water is not known, but this possibility is discussed later.

GEYSERS GEOTHERMAL FIELD

While conditions in various geothermal fields vary widely, the actual experience in the Geysers field in northern California will serve to introduce many of the problems and their solutions.

Prospecting and "firming up" steam supply is done by an oil company experienced in exploration and drilling. This has been the Union Oil Co. of California since entering into a joint venture with the original developers, the Magma Power Corp. and Thermal Power Co., in 1967.

The power plants are built and operated by the Pacific Gas and Electric Co., the utility serving that area. The initial installation was a 12,000 kW unit in 1960. Currently (1980), 55,000 kW (55MW) units are being used. About 800 MW of generating capacity is currently installed, with about 1200 MW additional planned as steam supplies are firmed up and the necessary permits obtained. This is expected to be at about 100 MW/yr.

As judged by the "gravity anomaly," the Geysers field is about 100 square miles (64,000 acres) in extent, although its actual limits will not be known until wells have been drilled over the full extent.

An average well in the better parts of this field will supply about 150,000 lb/hr of steam when new. It will decline to about half flow in 5 yr. Seven such wells when new will supply about 1,000,000 lb/hr steam, enough for a 55 MW unit. The steam has an enthalpy of about 1200 Btu/lb. Thus, the overall thermal efficiency is about

$$\frac{55 \times 10^3 \times 3413}{1200 \times 10^6} = 15.6\%$$

Elsewhere, 9 kg steam/kWh is given, which results in a thermal efficiency of 14.2%.

On the average, seven wells supply a 55 MW unit for 5 yr, or 56 wells for a 40 yr plant life. A plant usually contains two 55-MW units (1980) and will need about 112 wells in its 40 yr lifetime. About 5 acres are needed for each well; and when a plant is sited, provisions must be made for enough wells in its vicinity to keep it going for its lifetime of about 40 yr, with a minimum length of steam piping. Thus, 112 × 5 acres minimum are needed per plant, or roughly 5 acres per MW.

In the 64,000 acre field, some 12,800 MW might be supplied for 40 yr if all were usable, which it is not. In addition to geologic features which preclude drilling, experience to 1973 had been 85 producing wells out of 110 drilled. Estimates of this resource in 1973 ranged from "1200 MW to 4800 MW and even more."[1] Using the larger figure, this would be 2 GW centuries.

Steam Supply

Steam or hot water direct from the earth contains a variety of gaseous, liquid, and solid impurities not usually encountered when using clean treated water in steam plants. In the first installation at Geysers, steam from the earth was led directly to the turbine. Dirt, including rocks the size of one's fist, soon destroyed the blades.

Skipping over several intervening stages, the current treatment (1973)[1] is as follows. A new well is first blown to atmosphere through a "blow-down T," clearing out rocks and dirt until it has settled down. Next, it is cut over to a cleaning system (without interruption) that removes remaining dirt, liquid water, and gases. Since the steam is dry and superheated, moisture separation is necessary only in unusual situations, such as the mechanical failure of a well casing.

The steam is then carried around a 90° turn, with a straight projection pipe at the corner into which dirt and small rocks are carried by inertia. The extension pipe, normally closed, is blown out at intervals to clear it of accumulated dirt. This is repeated at strategic points in the collecting system. The steam is then passed through a centrifuge to separate out water, gases, and remaining dirt. These centrifuges are 99% efficient in removing particles over 10 μ in diameter. The steam then enters the collecting lines. Every effort is made to keep the flow steady.

Each generating unit also has a final separator after the steam has passed through all the collecting lines. It also has a relief valve with muffler to handle full flow at 170 psig when the unit is shut down for any reason, and also rupture discs to protect the lines. Only for long shut-down periods are the wells "shut in."

Starting is a gradual process of clearing the wells, then the lines, warming up to 2 to 3 hr, and finally, when clean steam at full flow is obtained, cutting it over to the turbine.

From this, it can be seen that natural geothermal energy is best suited to base load operation rather than to a load-follow or peaking pattern. Steady flow from the wells is essential.

Even with the treatment described, austenitic steels must be used in all turbine and condenser parts that come in contact with the steam. There is obviously no way to remove or neutralize all harmful impurities as is done in the treated and purified makeup water of a closed-cycle steam plant.

Condensate. The steam is used in condensing turbines; and since no adequate source of water for "once-through" cooling is available, wet cooling towers are used. Most of the condensate is used for makeup water for the cooling system. Some 20 to 25% remains, and this was initially discharged to neighboring streams. However, as the size of the installation grew, the high boron and ammonia content of this water has precluded discharging it into the streams, which are of highly variable flow. Instead, it is now being reinjected into old wells at approximately 6500 ft (2000 m) depth. This may serve to recharge the reservoir, allowing some secondary recovery, although the extent is not known at present (1980). The low reservoir pressure and high permeability allow high injection rates to be sustained, over 1000 gpm in some wells. That is, one reinjection well may care for the leftover condensate from all the wells feeding a plant.

Developing a Plant[1]

The general procedure for a new 110 MW plant requires about 5 yr after preliminary exploration, plant siting, and granting of permits.

The first phase involves the drilling and measuring of test wells to determine the production characteristics likely to be encountered in the general area for the new plant. This determines the actual acreage necessary for new and replacement wells over the plant life and defines the boundaries of the area that must be committed. It also provides data for optimizing the well designs and for establishing the production plan.

The wells conform to the same flow equations as for gas wells. That is,

$$W = C(P_s^2 - P_f^2)^n \tag{9.4}$$

where W is the steam flow rate in lb/hr; C is a constant depending on the reservoir, well bore, and fluid characteristics which change slowly with time; P_s is the static pressure in the reservoir; P_f is the well bore pressure at the steam entry; and n is a constant between 0.5 and 1.0 for usual production rates. The constants C and n are obtained by measurements on the test wells and from previous experience.

Based on the performance, which can now be estimated, with different well bore diameters and the costs of drilling, casing, and surface equipment, the system is then optimized (minimum cost to obtain the steam energy needed for the plant, weighing present and future costs). Typically, this may result in a turbine inlet pressure of 80 to 100 psig and an equivalent well bore of 9 to 11 in. The actual well would be made up of several bore sizes. For example:

Top 300 ft	20 in. casing	Water-bearing strata
Next 1700 ft (to 2000 ft depth)	13¾ in. casing	Drill with mud
Next 2000 ft (to 4000 ft depth)	9⅝ in. casing	Drill with air or mud as needed
		Top of probable steam zone
Next 2000 ft (to 6000 ft depth)	8¾ in. open hole.	Drill with air

The top 2000 ft approximately through water-bearing strata is drilled with mud to keep water out of the bore hole. Below significant water levels, drilling through the exceptionally hard rock (graywicke sandstone) is done better with air. This also eliminates the lost circulation in the steam zone that occurs if mud is used, plugging some of the steam passages.

The cost of drilling increases rapidly with depth. Also, other factors being fixed, a well completed at 10,000 ft delivers almost 20% less steam than the same well completed at 5000 ft.

From this optimization study, both the plant design and operating plan are roughly determined, and a license can be applied for. This must be at least 1 yr before the beginning of construction (in California) or about 4 yr prior to commercial operation. Including the exploration work and optimization study, prior to the license, approximately 5 yr are required between confirmation drilling and commercial operation (1973).

Some Conclusions from the Geysers Experience

A few observations may be made based on the experience to date at the Geysers geothermal field.

1. The "shut-in" reservoir pressures reported,[1] 450 to 500 psig, and the reservoir temperatures reported, 473°F (245°C), are near saturation. Saturation conditions are 450 psia, 456°F or 500 psia, 467°F. Thus, the fluid could be either steam or water or a mixture. As it flows into the wells, it is dry superheated steam.

2. Based on experience, 85 of 110 wells drilled are productive. The others evidently do not have sufficient local permeability to produce or are lacking in the necessary stored fluids. All wells encounter high temperatures, around 473°F (245°C).

3. The average 110 MW station draws 2.0 million lb of steam per hour, from an area of $560 \times 110/85 = 728$ acres (1.13 square miles). In 40 yr it will have drawn $2.0 \times 10^6 \times 8760 \times 40 \times 0.85 = 6.0 \times 10^{11}$ lb water or steam, at 85% plant factor.

4. In order to test whether the reservoir fluid is water or steam (vapor), we may ask how deep it must extend in either case to contain the 6×10^{11} lb fluid that was taken out. An average porosity of 15% will be assumed. The specific volume of 450 psia saturated steam is 1.032 ft³/lb, saturated water 0.0195 ft³/lb.

If the fluid is water, it occupies $6 \times 10^{11} \times 0.0195 = 1.17 \times 10^{10}$ ft³. If the fluid is steam, it occupies $6 \times 10^{11} \times 1.032 = 6.19 \times 10^{11}$ ft³. Since the area is 728 acres, or 1.13 square miles or 31.5×10^6 ft², and the porosity is 15%, the depth of fluid must be

$$\text{If water: } \frac{1.17 \times 10^{10}}{31.5 \times 10^6 \times 0.15} = 2476 \text{ ft}$$

$$\text{If steam: } \frac{6.19 \times 10^{11}}{31.5 \times 10^6 \times 0.15} = 131,000 \text{ ft or } 24.8 \text{ mi}$$

Since the wells extend only 2000 to 3000 ft into the "steam bearing" strata, the volume of fluid drawn out in 40 yr can only be accounted for if it is water. This strongly suggests that the reservoir fluid is saturated hot water, flashing to steam, and becoming slightly superheated as the pressure is lowered in its passage to the wells.

The same conclusion is reached if one well on 5 acres is fully depleted in 5 yr (equivalent).

5. The wells average about 500 ft apart and have relatively little effect on each other. Each producing well communicates effectively in its lifetime only with the rock within 200–300 ft of it.

6. The enthalpy of the steam used at the plant is about 1200 Btu/lb. The enthalpy of saturated water at 450 psia is 437 Btu/lb. The heat in this water accounts for only 437/1200 = 36.4% of the heat that got to the plant. Where did the rest of the heat come from? Altogether we have to account for

$$6 \times 10^{11} \times 1200 = 7.2 \times 10^{14} \text{ Btu} \quad \text{(total)}$$

The hot water taken out accounts for

$$6 \times 10^{11} \times 437 = 2.62 \times 10^{14} \text{ Btu}$$
$$\text{(water taken out)}$$

There are two other sources of heat—the rock and the remaining water that was not drawn out. As the reservoir pressure drops from 450 psia to 100 psia in the 40 yr life of the plant, the saturation temperature drops from 456°F to 328°F. The mass of rock and the remaining water drop 128°F.

Since all the water was not drawn out, the depth involved is greater than 2476 ft. Consider a nominal depth of 3000 ft of solid rock of specific heat capacity 28.9 Btu/ft³ °F. Its heat capacity is

$$31.5 \times 10^6 \text{ ft}^2 \times 3000 \text{ ft} \times 28.9$$
$$= 2.73 \times 10^{12} \text{ Btu/°F}$$

In dropping 128°F, it will give up

$$2.73 \times 10^{12} \times 128 = 3.49 \times 10^{14} \text{ Btu} \quad \text{(rock)}$$

The enthalpy of 100 psia saturated water is 298 Btu/lb. Each pound of water remaining gives up 437 − 298 = 139 Btu. Assume an amount remaining equal to that removed (in increased porosity and extent). In dropping 128°F, this water will give up

$$6 \times 10^{11} \text{lb} \times 139 \text{ Btu/lb} = 0.83$$
$$\times 10^{14} \text{ Btu} \quad \text{(water remaining)}$$

Altogether, water drawn out, water remaining, and rock account for

$$2.62 + 0.83 + 3.49 = 6.94 \times 10^{14} \text{ Btu}$$

This accounts for the heat received by the plant as closely as could be expected with such crude calculations. It indicates that the saturated water in flashing to steam receives its additional heat from the rock and the remaining water whose temperature is falling.

7. Could any appreciable amount of heat have been fed up from below during the 40 yr? With a surface temperature of 50°F and a reservoir temperature of 450°F at a depth of about 4000 ft, the gradient is about 10°F/100 ft. or about 6.7 times normal (initially). If we assume this high gradient below the reservoir as well, the rate of heat flow is

$$0.1°\text{F/ft} \times k = 0.1 \times 1.45 = 0.145 \text{ Btu/ft}^2 \text{ hr}$$

In 40 yr, this will supply

$$0.145 \text{ Btu/ft}^2 \text{ hr} \times 31.5$$
$$\times 10^6 \text{ ft}^2 \times (40 \times 8760) \text{ hr} = 0.016$$
$$\times 10^{14} \text{ Btu} \quad \text{(from below)}$$

or 0.2% of the heat extracted. It is negligible.

Summary. Only in the water phase could the amount of fluid drawn out in 40 yr be within range of the wells. The heat to flash this saturated water to steam could have come from the surrounding rock and the remaining water as their temperature dropped. Negligible additional heat reaches the field from below during the lifetime of the plant.

These observations are from very meagre and approximate average data but are presented to show at least one mechanism to account for the phenomenon being observed. Future experience may confirm or alter these speculative results.

HOT WATER-DOMINATED SYSTEMS

While a few geothermal fields, such as Geysers, Lardarello, and Matzukawa in Japan, supply dry superheated steam to the wells, the fluid in the rock strata into which the wells project must be mainly saturated water. In no other way can the amount of fluid actually extracted be accounted for. However, since the fluid actually entering the wells is dry, superheated steam, the field is referred to as "vapor dominated." Such situations are extremely rare on the earth, the three mentioned being the principal examples. Saturation (boiling) temperature or higher can only exist when an impervious layer between the reservoir and the surface makes possible pressures far below hydrostatic, where a high thermal gradient exists, and where, at the same time, adequate porosity, permeability, and fluids exist in the reservoir.

In most cases, if a fluid exists at all, it is water or

brine at near-hydrostatic pressure and well below saturation temperature. It enters the wells as liquid; and only with substantial reduction in pressure, as it nears the surface, does some of it flash to steam. This can be separated out and used in a turbine. Usually, the percent of steam is small, 10 to 20%, and the heat in the water is lost (see Water Flashing to Steam earlier) or used for nearby space heating.[10] Furthermore, a considerable part of the energy in the steam is used up in raising the useless hot water to the surface. However, there are many fields where the wet steam, produced in this fashion, produces lower-cost electricity (the final measure) than other available sources. The extensive field at Wairekai, New Zealand, is of this nature.

For many fields now being explored, there is hot water or brine fairly near the surface. For example, in the Imperial Valley, California, 390°F has been recorded at a depth of 2600 ft. Assuming a surface temperature of 70°F, this represents a temperature gradient of 320/26 = 12.3 °F/100 ft, or over eight times normal.

Bringing up large volumes of hot water along with the steam presents several problems: loss of energy, wet steam, release of dissolved gases into the atmosphere, turbine corrosion, disposal of the polluted water. bigger well bores for the same steam flow, and low efficiency. To solve these problems, two principal developments are effective: (1) a pressurized recirculation of the well water and transfer of its heat through heat exchangers, and (2) the vapor turbine.

If the hot water is kept under pressure, it remains in the water phase and its heat can be transferred to a Rankine cycle, either steam or other vapor, through heat exchangers. If transferred to a steam cycle, the system is very similar to the pressurized-water reactor nuclear system (PWR) of Fig. 6.3a, as mentioned earlier (Energy from Hot Water).

However, the temperature of ground water from shallow (up to 3 km) wells is very low compared with that from a nuclear reactor, typically 300°F instead of 600°F, and the steam turbine becomes relatively large and expensive. If, instead of steam, a vapor such as isobutane or freon is used, the turbine is much smaller for the same mass flow (same power). For example, at 80°F condensing temperature, the specific volume of steam is 633 ft³/lb, whereas that of isobutane is only 1.68 ft³/lb. The last-row blades of the turbine, which primarily determine its size and cost, are correspondingly smaller. The vapor turbine can efficiently utilize much lower temperature well water without the turbine size becoming excessive.

Vapor Turbine. One arrangement of a vapor turbine system is shown in Fig. 9.2. Hot water is pumped from the geothermal well. This energy loss is much less than if steam lifted the water in an open system. The entire ground-water system is pressurized above the saturation pressure. After being used to heat, evaporate, and superheat the vapor (in counterflow), the waste water is injected into another well. However, now a considerable part of the heat brought up from the earth is used effectively. Typically, the ground water temperature may drop from 325°F to 170°F in transferring heat to the vapor, and its cooling curve has a good match with the heating curve of the liquid–vapor. This is essential for efficiency (see Chapter 13, Counterflow Heat Transfer Principle).

The diagram in Fig. 9.2 shows an optional vapor turbine boiler–feed pump drive, as this function requires relatively more energy than in the steam cycle.

The cost of wells, to be discussed later under Deep-Drilled Wells, is a major part of the total cost, and a considerable amount of research is being devoted to increasing the heat delivery per well, by fracturing the surrounding rock and increasing the fluid and heat flow, or by circulating water between two wells in hot dry rock.[8]

DEEP-DRILLED WELLS—ARTIFICIAL GEOTHERMAL

Hot dry rock at about 800°F occurs at a 10 mile depth nearly everywhere in the world, and at much less depth in many areas. With present technology, the cost of drilling to a 10 mile depth is prohibitive for the amount of power that can be produced. However, drilling techniques are being continually developed for other purposes, high temperatures are found closer to the surface in geothermal areas, and research is underway on methods of increasing the energy output per well.[8] These include hydraulic or explosive fracturing and the use of percolation between down- and up-wells.

The simplest concept of a deep well is a single hole of, say, 24 in. equivalent bore, with an inner insulated pipe for the return water as shown in Fig. 9.3. Water is fed down the outer cylinder, heated

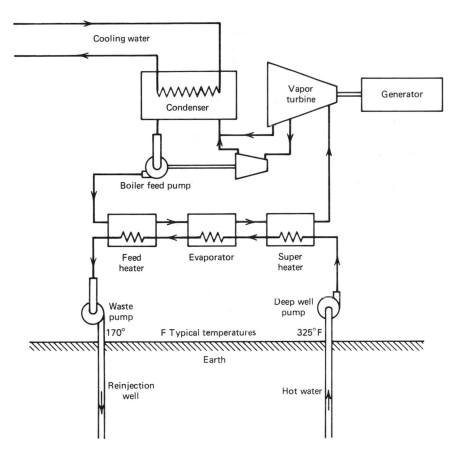

Figure 9.2. Geothermal generation using vapor turbine.

on the way, and is brought back up through the insulated center pipe.

Economics of Deep-Well Geothermal Systems

The only thing the geothermal heat source replaces is some alternate heat source for the plant. Thus, to get an indication of its viability, we need to know the annual cost of the well and the amount of heat it can supply in a year.

Cost of Wells. The drilling costs for wells of different depths are shown in Fig. 9.4.[1] These are 1971–1972 costs. A 28,000 ft well completed in 1971 cost $6 million. Oil companies are citing $1 million per mile in 1979. Costs increase rapidly with depth. As a rough estimate, consider $15 million for a 50,000 ft well in 1979.

At a normal utility financing rate of 18% to cover maintenance, depreciation, taxes, insurance, and return, the annual cost is $2.7 million. It will be shown later that the well of Fig. 9.3 would supply about 5000 kW average load, or 150,000 million Btu/yr. The cost of heat energy from the well is thus 2.7/0.15 = $18/million Btu.

How Much Can Be Afforded? The geothermal plant for a deep well supplying 600°F hot water resembles closely the steam generator and turbine generator sections of a pressurized water-reactor nuclear plant Fig. 6.3a (except for size), which has a thermal efficiency of about 33%. Both use 600°F hot water. Assume that the geothermal plant will have an efficiency of 33%.

The equivalent of residual oil from coal was about $4.40/million Btu in 1978, Table 3.9. A plant using it would have about 40% efficiency. Thus, geothermal heat must be available at 33/40 × $4.40 = $3.63/million Btu (1978 basis) to be competitive. The geothermal cost is about five times what could be afforded. For geothermal heat available at lesser depths, or if developments in progress lead to a

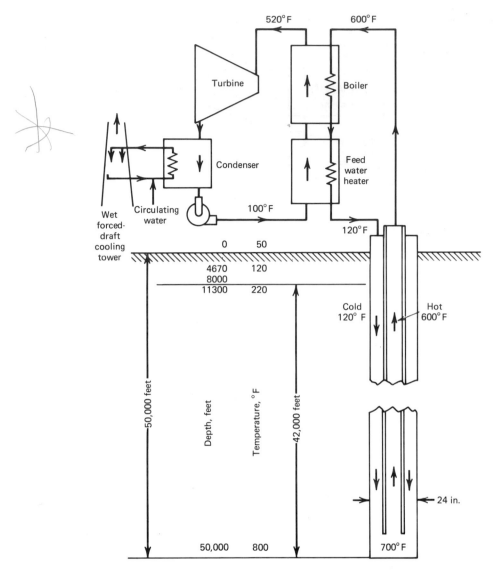

Figure 9.3. Drilled deep-well concept.

much greater output per well, or both, the economic picture is much brighter.

Hot Dry Rock. At the Fenton Hill site of Los Alamos Scientific Laboratory (LASL) wells drilled into the hot dry rock of the Jemez Mountains encounter temperatures of 200°C (392°F) at a depth of 3 km (1.86 miles), and 250–275°C (483–527°F) at a depth of 4 km (2.49 miles). Hydraulic fracturing between two wells at these depths exposes relatively large areas of contact with the rock. Water circulated in a closed loop through the earth would transfer its heat to a fluid suitable for a vapor turbine. An extensive test program is under way[8] to develop the technology and establish the economics. Costs competitive with alternate generation are anticipated.

Geothermal Deep-Well Example. In order to obtain a quantitative understanding of the phenomenon taking place in the earth surrounding a geothermal deep well, also of the best proportions of outer and inner pipe and insulation, and of the effect of base-load or peak-load operation, the heat flow was calculated for the configuration of Fig. 9.3, using average earth crust properties from Table 9.1.

Assuming the water to be heated uniformly as it descends, with a gradient at every level of 100°F

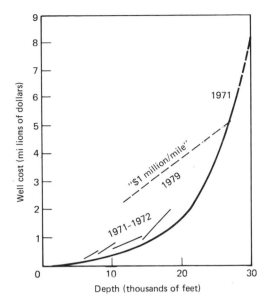

Figure 9.4. Cost of wells.

between "remote earth" and the pipe, the problem becomes two-dimensional. The surface effect is accounted for by considering 42,000 ft of the uniform pattern.

The calculation was carried out for a century, with results as shown in Fig. 9.5. The initial transient is unimportant. At the end of a week, the flow

had decreased to 5800 kW, with a penetration of 13 ft. Less than 5% of the heat had come from beyond the penetration. At the end of a month, the flow was 4500 kW, penetration 25 ft; at the end of 1 yr, 3000 kW, 80 ft; at the end of 10 yr, 2300 kW, 200 ft; at the end of 40 yr (plant life), 2000 kW; and at the end of a century, 1800 kW.

The system is linear, and a 200°F gradient would double the heat flow but cut the water temperature by 100°F. However, a mere 12.5% increase in earth gradient would result in an earth temperature at 50,000 ft depth of 900°F and almost double the heat flow (1.83 times) with no decrease in delivered water temperature. Based on these possibilities, the 5000 kW average flow was used in the earlier economic comparison.

Geothermal Peaking Power. While a natural geothermal-generating station is unsuited to peaking or intermittent loads, the deep-well system is particularly well suited, since the water flows in a closed loop and also provides the needed storage. The energy supplied per day is practically the same whether used in 3 or 24 hr. The plant capacity must however be eight times as large for the 3 hr operation.

Penetration in a day is about 5 ft. Beyond 5 ft, the heat flow in the earth is unaffected by daily varia-

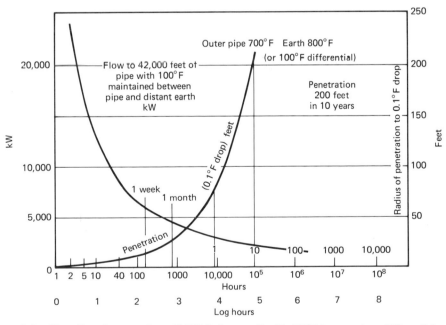

Figure 9.5. Heat flow from earth to 50,000 ft deep well with 100°F temperature differential maintained between remote earth and lower 42,000 ft of pipe, 2 ft OD pipe. Initial condition: normal earth temperature throughout.

tion in water flow. After the first few weeks, practically all the energy comes from beyond 5 ft.

Note that even if the cost of geothermal heat were equivalent to clean distillate from coal (upgraded liquids, Table 3.9), the geothermal plant would be noncompetitive for peaking because of the low capital cost of the combustion turbine (compare with "oil–steam," Table 3.8). Note also that by 1993, advanced batteries are expected to displace the combustion turbine for peaking (Fig. 3.2).

1980 DEVELOPMENTS[5-7]

A cooperative government (DOE)–industry geothermal development program is continuing in 1980 involving both the direct use of geothermal energy (not covered in this book) and new developments in geothermal-electric generation. As mentioned earlier the DOE Los Alamos Scientific Laboratories is involved in the research aspects of the development. LASL manages the national program. Some of the projects underway are the following:

1. A 50 MW geothermal power generation demonstration plant is being constructed in Sandoval Co., New Mexico, by the Public Service Co. of New Mexico. Power will be transmitted to a substation at the LASL 20 miles to the east.

2. The DOE Hot Dry Rock Geothermal Energy Program research site is at Fenton Hill in New Mexico. Part of the energy for the site (at times) comes from turbines run by heat from hot granite located 2 miles below the site.

3. A binary cycle of geothermal-electric generation is being tested in California, Arkansas, and Idaho. Two different working fluids circulating in different loops are used.

4. Two hundred megawatts of geothermal generation from hot water are expected to go on line in the Imperial Valley of Southern California in 1983.[7]

5. There are also numerous experiments in progress involving direct use of geothermal heat.

REFERENCES

1. P. Kruger and C. Otte, Eds., *Geothermal Energy*. Stanford, Calif.: Stanford University Press, 1973.

2. L. Burton, "Geothermal Energy." *Orange Disc*, January–February 1974, p. IFC.

3. F. A. L. Halloway, "Tar Sands and Oil Shale," in *The Bridge*, National Academy of Engineers, Summer 1979, p. 7.

4. "Geothermal Energy," in D. N. Lapedes, Ed. in Chief *McGraw-Hill Encyclopedia of Science and Technology*, New York: McGraw-Hill.

5. *Geothermal Energy*, DOE/OPA-0051, Oak Ridge, Tenn.: Energy, P.O. Box 62, 1979.

6. *U.S. Department of Energy, Secretary's Annual Report to Congress, January 1980*, DOE/S-0010 (80), Washington, D.C.: Supt. of Documents, 1980.

7. "Energy," *National Geographic*, February 1981, special issue.

8. J. Ahearne, "Expanding the Hot, Dry Rock Program," *The Atom*, June 1979.

9. *Geothermal Energy*, LASL-77-23, Washington, D.C.: U.S. Govt. Printing Office, 1978.

10. R. DiPippo, *Geothermal Energy as a Source of Electricity*, DOE/RA/28320-1, Washington, D.C.: Supt. of Documents, 1980.

10

Energy From Current Growth— (Biomass)

INTRODUCTION

The current growth or biomass of the earth must provide all of the food for man and animals. It also provides a substantial part of our clothing and material for building and for manufactured articles. It provides a substantial amount of energy. To some extent, the use for energy is in competition with the use for food, clothing, and building material. Both of these uses are in competition with the use of these resources for recreation, for natural beauty, for protection of the land, and for wildlife habitat.

Some uses are compatible. Watershed protection also provides wildlife habitat, recreation, and natural beauty. The production of electricity from municipal waste solves a waste disposal problem and produces useful energy. The production of useful methane gas from manure in a digester also improves the fertilizer.

However, with limited land area suitable for useful crops and limited food supply from the sea, we are faced with improving land and sea use, and allocating them wisely to provide for an increasing world population.

This chapter provides some of the basic information about current growth or biomass. First, the forests of the world and the production and uses of wood are discussed. This is followed by the potential energy production from waste in the United States and the technology of converting municipal waste to energy. The food crop of the world is treated briefly from an energy point of view. Finally, the use of biomass for energy, including alcohol from plant life and methane from manure, is covered, particularly the current (1981) U.S. developments.

ENERGY FROM WOOD AND PLANTS

The historical sources of energy in the United States from 1850 to 1975 are shown in Fig. 10.1. At the end of the Civil War, 80% of U.S. energy was supplied by wood. In many parts of the world, wood is still highly important as fuel. In Africa (1955), 90% of the wood harvested was for fuel.

However, the increasing demand for wood for lumber, pulpwood, and other industrial uses has made it generally too valuable to use for fuel in the more highly industrialized areas, in competition with other available fuels. Wood waste is, however, increasingly used for fuel.

The primary wood products of the world, valued at $30 billion per year (1955), include 51% lumber, 31% pulpwood, 7% fuel, and 12% of other products.[1] In actual bulk, however, fuel constitutes 46% of the wood harvested from the forests of the world.

The forests also supply such secondary products as tar, pitch, rubber, cork, tanbark, dyestuffs, maple syrup, lac, and oils, having a combined value about 10% of the primary wood products.

Distribution of Wood and Products

Wood is widely distributed throughout the world. Table 10.1, adapted mainly from FAO[1] information, shows the distribution of the world's forests, land, and population. Only 47% of the world's forests were accessible at that time (1955). The distribution of these is shown in column 5. Only 30% of the world's forests were used. The distribution of these is shown in column 6. The wood actually harvested is shown in columns 7, 8, and 9.

The USSR, with 7% of the world population, har-

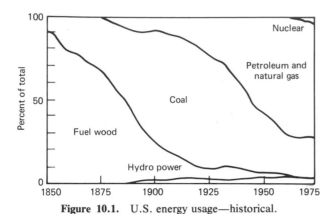

Figure 10.1. U.S. energy usage—historical.

Growth Rate

In the 30% of the world's forests that are used, the standing growth is estimated to be 121 billion m³. The growth rate is estimated at 2.4 billion m³/yr. Both figures include wood and bark, but not roots, leaves, twigs, and other droppings and unusable brush from trees. Taking into account the loss from wind, fire, ice, disease, and insects and the bark removal, some 1.4 billion m³ wood is harvested annually, distributed as shown in Table 10.1. About 54% is for industrial uses and 46% for fuel (other sources give 58% industrial, 42% fuel.) Two thirds of the wood is harvested in N. America (United States and Canada), USSR, and Europe and is used 84% for industrial and 16% for fuel purposes. One third is harvested in Asia, Africa, Latin America, and the Pacific Area and is used 78% for fuel and 22% for industrial purposes.

vests 21% of the wood; North America, with 6.3% of the world population, harvests 26%; and Europe, with 12.5% of the world population, harvests 19% of the wood, although it has only 4% of the forests. They are essentially all accessible and used.

On the other hand, Latin America (Mexico and Central and South America), and Africa, with 16% of the world population, harvests 18% of the wood, although they have 34% of the accessible forest. Asia, with 58% of the world population, has only 17% of the accessible forest and harvests only 14% of the wood.

If the annual growth of 2.4 billion m³ in the 30% of the forests that are used is extrapolated to all forests, there results some 8 billion m³ annual growth in the forests of the world. Using an approximate specific gravity[2] of 0.7, this represents 5.6 billion metric tons growth, or 25% of the 22 billion metric tons green growth in the forests of the earth annually. The actual tree growth is considerably greater than this, since the annual growth fig-

Table 10.1.   Distribution of the World's Forests and Products[a]

Region 1	% of World Pop. (1975) 2	% of World Land 3	% of World Forest 4	% of Accessible World Forest 5	% of Utilized World Forest 6	Distribution of Fuel Wood Used (%) 7	Distribution of Industrial Wood Used (%) 8	Distribution of Total Wood Harvested (%) 9
Europe	12.5	4.0	4	8	13	16	21	19
USSR	7.1	16.0	19	23	30	16	24	21
North America	6.3	15.5	17	17	19	9	39	26
Latin America	7.9	16.5	23	18	7	23	3	11
Africa	7.9	22.0	21	16	9	14	1	7
Asia	57.8	20.0	14	17	21	21	10	14
Pacific Area	0.5	6.0	2	1	1	2	2	2
World	100.0	100.0	100.0	100.0	100.0	100.0	100.0	100.0

Source.   Ref. 1.

[a]Value of primary wood products (fuel and industrial wood) $30 billion per year.

Table 10.2. Approximate Chemical Composition and Heating Value of Moisture- and Resin-Free Wood

	Percent by Weight
Carbon	49
Hydrogen	6
Oxygen	44
Ash	1
Total	100%
Wood higher heating value HHV	8,300 Btu/lb
Resin higher heating value	16,900 Btu/lb

Table 10.3. Approximate Moisture Content of Wood

	Moisture Content (%)
Green	50
Air dried or seasoned	12–25
Kiln dried	8

ure does not include roots, leaves, brush, and other wood waste. Kuiper[20] gives 8.8 billion metric tons of carbon fixed in the forests of the world annually, corresponding to 22 billion metric tons of glucose.

Characteristics of Wood

All wood starts out as glucose molecules, which build into the large cellulose molecules that form the structure of trees, and the lignin that glues them together. It might be expected therefore that, except for the resin, the composition would not vary greatly from wood to wood. It should be close to that of glucose.

Chemical Composition of Wood. The approximate composition of moisture- and resin-free wood is given in Table 10.2. This composition corresponds closely to the chemical formula $C_6H_9O_4$. The formula for glucose is $C_6H_{12}O_6$. They differ by *about* two water molecules, $2 H_2O$.

Higher Heating Value. The sun's energy added to glucose in the process of its formation is 696 kcal per gram mole of glucose formed, that is, per 180 g. It is thus $696 \times 454/180 \times 1/0.252 = 6950$ Btu/lb glucose.

The energy density thus increases in passing to dry wood, 8300 Btu/lb, and of course still more in the metamorphosis changing it to coal, about 13,000 Btu/lb, or to oil, about 19,000 Btu/lb, or to natural gas. Methane, the principal constituent of natural gas, has 23,890 Btu/lb.

Note that 1 lb glucose, having 6950 Btu, contains 72/180 lb carbon, which has $72/180 \times 14,096 = 5650$ Btu. Stripping H and O from a carbohydrate in the proportion of water, H_2O, leaves most of the heat,

as would be expected, since water has no heat value.

Moisture in Wood. The variation in the heating value of wood as fired can be accounted for almost entirely by the moisture and resin content. The approximate moisture in wood is given in Table 10.3.

Resin in Wood. The woods most influenced in heating value by resin content are pine, cedar, juniper, fir, hemlock, redwood, and cypress. For example, white and yellow pine have about a 4% larger heating value at 12% moisture than all the other woods in Table 10.5, column 8.

Heating Values and Analyses of Selected Woods. The approximate compositions and heating values of selected woods as fired are given in Table 10.4. Note that the heating value is less than for dry wood primarily due to the moisture (6300/0.76 = 8300, first column). Also most of the wood wastes used in the western part of the United States are resinous woods having a HHV above 8300 Btu/lb dry (bottom row).

Charcoal. Charcoal is made by heating wood to its charring temperature in the absence of air. It loses up to 75% in weight and 50% in volume in charring, depending on the moisture content of the wood and the temperature used. It thus has a much higher heating value than wood.

Fuel Wood. Fuel wood is generally sold by the cord. A standard cord is 128 ft³, that is, 4 ft × 4 ft × 8 ft. It contains about 70% solid wood, or 90 ft³. A "cord run," sometimes called a cord, is 8 ft long and 4 ft high, but the width corresponds to the length of pieces being supplied, for example 16 in. for stove wood or block wood, 4 ft for cordwood. Table 10.5a gives the approximate weights and heating values for various woods. Note that all have the same heating value per pound at 12% moisture

Table 10.4. Analyses of Wood Fuels, as Fired (Percent)

| Constituent | Wood, typical nonresinous, seasoned | Charcoal willow | Wood Waste | | | Hogged fuel, Douglas fir | Sawdust, green, Douglas fir | Sawdust briquets, Douglas fir | Tanbark |
			California redwood	Western hemlock	Douglas fir				
Proximate Analysis									
Moisture	24.0	3.2	50.4	57.9	35.9	47.2	44.9	10.3	71.8
Volatile matter	65.5	14.7	40.9	31.3	52.5	42.9	44.9	78.3	22.4
Fixed carbon	9.5	80.2	8.6	9.9	11.1	8.9	9.5	11.2	4.5
Ash	1.0	1.9	0.1	0.9	0.5	1.0	0.7	0.2	1.3
Ultimate Analysis									
Hydrogen	7.2	2.7	8.5	8.9	8.0	—	—	—	9.6
Carbon	37.9	85.0	26.5	21.2	33.5	—	—	—	14.2
Nitrogen	0.1	0.2	0.1	0	0.1	—	—	—	0
Oxygen	53.8	10.1	64.8	69.0	57.9	—	—	—	74.9
Sulphur	0	0.1	0	0	0	—	—	—	0
Ash	1.0	1.9	0.1	0.9	0.5	—	—	—	1.3
High Heat Value (Btu/lb)	6300	13,530	4570	3630	5800	4670	4910	8130	2600
High heat value, (Btu/lb) —Dry Basis	8300	13,980	9210	8630	9040	8860	8910	9050	9210

(7670 to 7700 Btu/lb, column 8), except for pines, which are resinous and have about 4% higher heating values.

The Fireplace. An open wood fire in a fireplace, while inefficient, has always been, and remains today, one of man's most cherished pleasures. (In the average fireplace, about 90% of the heat goes up the chimney. This includes warm room air drawn up the chimney and replaced by colder air from outside.) A number of considerations enter, in addition to pure heating value. For example the wood should burn easily and be easy to split. It should not produce heavy smoke, and preferably should not pop or throw sparks. The fragrance of certain woods is particularly pleasant and enhances their effect. Some woods are long lasting and form a fine bed of coals. Others are suitable for a quick fire or kindling but soon burn out.

The U.S. Department of Agriculture Ratings for Firewood[3] are given in Table 10.5b. In general, the "excellent" hardwoods in the first line are the woods having 26 million Btu/cord, green, or higher (column 4, Table 10.5a). These heavy woods last longer, give off more heat, and usually make a fine bed of hot coals.

Woods with pleasant fragrance are apple, wild cherry, sassafras, beech, and hickory. Woods that pop and throw sparks are hemlock, cedar, larch, spruce, and pine, and to a degree birch and hickory. Apple wood burns with a colored flame (including crab apple), and birch with a bright intense flame. Ash and yellow birch burn almost as well when first cut as after seasoning.

If the goal is heat rather than the pleasure of an open fire, then a stove is to be preferred, since it can generate 5 to 10 times as much useful heat with the same amount of fuel.

Other Fuels of Plant Origin

A large number of other fuels of plant origin are used where readily available, or in emergencies. For example, bagasse, the sugar cane after the juice has been extracted, supplies practically all the needed energy in the cane sugar industry. Even corn is burned in emergencies, though not generally considered a by-product fuel. Refuse is certainly a "by-product" of civilization but is treated separately. The heating values of a number of by-product fuels are given in Table 10.6 (see also Alcohols later).

ENERGY FROM ORGANIC WASTE

Organic waste generated in the United States is a potential source of up to 13% of its energy require-

Table 10.5. Firewood

a. Approximate Weights and Heat Values of Fuel Woods[a]

Variety of Wood 1	Weight per cord containing 90 cu ft of solid wood (lb)		High heat value per cord (million Btu)		Equivalent in Heat Value to tons of Coal[b]		Btu/lb[c]
	Green Wood 2	Wood with 12% moisture 3	Green wood 4	Wood with 12% moisture 5	Green wood 6	Wood with 12% moisture 7	Wood with 12% moisture 8
Ash, white	4,320	3,690	26.0	28.3	1.00	1.09	7,670
Beech	4,860	4,050	27.1	31.1	1.04	1.20	7,700
Birch, yellow	5,130	3,960	27.2	30.4	1.05	1.17	7,680
Chestnut	4,950	2,700	19.2	20.7	0.75	0.80	7,670
Cottonwood	4,410	2,520	18.0	19.4	0.69	0.75	7,700
Elm, white	4,860	3,150	22.2	24.2	0.85	0.93	7,680
Hickory	5,670	4,590	29.0	35.3	1.12	1.36	7,690
Maple, sugar	5,040	3,960	27.4	30.4	1.05	1.17	7,680
Maple, red	4,500	3,420	23.7	26.3	0.91	1.01	7,690
Oak, red	5,760	3,960	27.5	30.4	1.06	1.17	7,680
Oak, white	5,670	4,230	28.7	32.5	1.10	1.25	7,680
Pine, yellow[d]	4,770	3,240	23.7	26.0	0.91	1.00	8,020
Pine, white[d]	3,240	2,250	17.3	18.1	0.67	0.70	8,040
Walnut, black	5,220	3,420	24.8	26.3	0.95	1.01	7,690

b. Ratings for Firewood

	Relative Amount of Heat	Easy to Burn	Easy to Split	Does it Have Heavy Smoke?	Does it Pop or Throw Sparks?	General Rating and Remarks
Hardwood Trees						
Ash, red oak, white oak, beech, birch, hickory, hard maple, pecan, dogwood	High	Yes	Yes	No	No	Excellent
Soft maple, cherry, walnut	Medium	Yes	Yes	No	No	Good
Elm, sycamore, gum	Medium	Medium	No	Medium	No	Fair—contains too much water when green
Aspen, basswood, cottonwood, yellow-poplar	Low	Yes	Yes	Medium	No	Fair—but good for kindling
Softwood Trees						
Southern yellow pine, Douglas-fir	High	Yes	Yes	Yes	No	Good but smoky
Cypress, redwood	Medium	Medium	Yes	Medium	No	Fair
White cedar, western redcedar, eastern redcedar	Medium	Yes	Yes	Medium	Yes	Good—excellent for kindling
Eastern white pine, western white pine, sugar pine, ponderosa pine, true firs.	Low	Medium	Yes	Medium	No	Fair—good kindling
Tamarack, larch	Medium	Yes	Yes	Medium	Yes	Fair
Spruce	Low	Yes	Yes	Medium	Yes	Poor—but good for kindling

Source.   U.S. Department of Agriculture. (*a*) Bull. 753. (*b*) Leaflet No. 559.

[a] Compiled from Use of Wood for Fuel, *U.S. Dept. Agr. Bull.* 753.

[b] Based on high heat values; 2,000 lb coal with a heat value of 13,000 Btu per lb.

[c] Column 8 = column 5 divided by column 3.

[d] Resinous.

Table 10.6. Heating Values of Miscellaneous By-Product Fuels

	Moisture (%)	Ash[a] (%)	Heat value[a] (Btu/lb)
Black liquor			
Soda	20–25	45	6200
Sulfate	35	40–45	6500
Sulfite (Ca)	45	10	8000
Guayule fiber	—	12.0	9000
Bark (spruce)	60	5	9000
Bark (pine)	40–50	5–10	9500
Rice hulls	3–5	25	6000
Tung-oil hulls	6.0	3–4	8000
Cottonseed cake	8–10	8	9500
Linseed cake	10	12–14	8750
Flax straw	10	2.0	8250
Furfural waste	20–25	3–4	8200
Lampblack	3.0	0.5	14800
Wheat straw	10	4	8500
Spent coffee	65	1.5	10000
Corn on cob	15–18	1–1.5	8200
Shelled corn	10–12	1–1.5	9300
Bagasse	42–53	1.3–3	8000–8700

Source. Ref. 5.
[a]Dry basis.

ments in 1980. Waste is increasing at about 2% per year. The historical rate of increase in the use of energy is 4%, but a rate of 1.9% is projected for 1977 to 2000 (Chapter 1).

The total amount of waste generated in the United States in 1971 was in excess of 2 billion tons, of which the moisture- and ash-free organic material was about 880 million tons.[4] The sources of these wastes in 1971 and 1980 (estimated) are shown in Table 10.7. If these wastes were collected and converted to oil, the potential shown amounts to 6.4 quads (6.4 × 10^15 Btu) in 1971 and 7.7 quads in 1980, compared with the national energy requirements of about 70 quads and 79 quads for those two years, that is, about 9.7% in 1980. If they were converted to fuel gas, the potentials shown are 8.8 quads in 1971 and 10.6 quads in 1980, that is, 13.4% of the U.S. energy requirements in 1980.

None of the conversion methods has been fully developed (1976), although they are known techniques. Also, most of these wastes are not collected.

The most likely use of this energy source is of course the use of those wastes that are collected and which represent serious disposal problems. The

Table 10.7. Estimates of Organic Wastes Generated, 1971 and 1980

Source	1971	1980
Manure (million tons/yr)	200	266
Urban refuse (million tons/yr)	129	222
Logging and wood manufacturing residues (million tons/yr)	55	59
Agricultural crops and food wastes[a] (million tons/yr)	390	390
Industrial wastes[b] (million tons/yr)	44	50
Municipal sewage solids (million tons/yr)	12	14
Miscellaneous organic wastes (million tons/yr)	50	60
Total (million tons/yr)	880	1061
Net oil potential[c]		
million barrels	1098	1330
quads	6.4	7.7
Net gas for fuel potential[d]		
trillion cubic feet	8.8	10.6
quads	8.8	10.6
U.S. energy usage (quads)	70	79 (est.)
U.S. natural gas usage (quads)	22	21 (est.)

Source. Ref. 4.
[a]Assuming 70% dry organic solids in major agricultural crop waste solids.
[b]Based on 110 million tons of industrial wastes per year in 1971.
[c]Quantities of oil are based on conversion of wastes to oil by reacting carbon monoxide and water. Net oil produced based on 1.25 barrel per ton dry organic waste.
[d]Gas estimate is based on 5.0 ft^3 methane produced from each pound of organic material.

Table 10.8. Estimates of Available Organic Wastes, 1971

Source	Total Organic Wastes Generated	Organic Solids Available
Manure (million tons/yr)	200	26.0
Urban refuse (million tons/yr)	129	71.0
Logging and wood manufacturing residues (million tons/yr)	55	5.0
Agriculture crops and food wastes (million tons/yr)	390	22.6
Industrial wastes (million tons/yr)	44	5.2
Municipal sewage solids (million tons/yr)	12	1.5
Miscellaneous organic wastes (million tons/yr)	50	5.0
Total (million tons/yr)	880	136.3
Net oil potential		
Million barrels	1098	170
Quads	6.4	1.0
Net gas for fuel potential		
Trillion cubic feet	8.8	1.36
Quads	8.8	1.36

Source. Ref. 4.

combined incentive of waste disposal and energy production is most likely to be attractive economically. Table 10.8 shows which wastes are collected and available. Based on 1971 estimates, these represent a potential of 1.0 quad if converted to oil and 1.36 quad if converted to gas, that is, 1.4 and 1.9% respectively of the national energy requirements.

Energy Recovery from Municipal Waste[6,18,19]

One of the more promising processes that has been tried at essentially full scale[6] is the burning of residential solid waste, along with pulverized coal in the suspension-fired boilers of an electric utility. Some 10% of the heat input is provided by shredded waste blown into the flame pattern through separate ports, in parallel with the pulverized coal burners. The boilers must have the bottom ash- as well as the fly ash-handling facilities customary with pulverized coal.

In this cooperative venture of the City of St. Louis and the Union Electric Co., two 125 MW boilers were used and disposed of a total of 600 tons of waste per day (waste from 170,000 people). Scaled up, all of the suspension-fired boilers of the utility could handle about twice the residential solid waste of the metropolitan area which it serves.

The refuse used had a heating value of 4171 to 5501 Btu/lb,[6] compared with 11,600 Btu/lb average for the coal being used. It was run through a conventional hammermill, which reduced it to particles under 1½ in. (50% under ⅜ in.), then through a magnetic separator that removed the magnetic materials. Glass and ceramics were not removed (they were removed in a later modification).[6]

This project is still experimental (1973), with a few problems under study. For example, the dwell time in the flame, 1 to 2 sec, is not sufficient for complete combustion of the larger particles, and hence the ash from the waste is about double that from an equivalent amount of coal.

The cost of modifying the two 125 MW boilers and providing the surge bin, conveyor to feeder, and pneumatic feeder was $550,000, supplied by the utility, or about $2.20 per kW of plant capacity. This is of the order of 0.5% of the 1976 cost of such a plant (Fig. 1.4). Hammermill, magnetic separator, and receiving facilities were provided by the city.

The economic factors are evident:

1. The total cost to both city and utility of transporting, processing, and firing the refuse as supplementary fuel.

2. The total cost of alternate means of residential solid waste disposal.

3. The value of the magnetic materials recovered.

4. The value of milled residential solid waste as fuel, that is, the value of the fuel displaced.

The alternatives being considered would generally be incinerator, open dump, or sanitary land fill. Pollution controls may increase the costs of existing methods, or even rule out open dumps. Generally, the new system would be an alternative to an addition to the existing system, and this means operating two systems for some time. The *total* costs would include all of these factors. Energy and materials conservation and elimination of unsightly facilities may introduce other than purely economic incentives.

The full scale commercial embodiment of this system is described in Ref. 19.

ENERGY FROM FOOD

There are as yet no synthetic energy-giving foods. All food energy starts out as plant life, the first trophic (nutritional) level. It may be eaten directly or pass through several additional levels of the food chain before being eaten by man or large animals.

Only a small part of the solar energy reaching the surface of the earth (about 0.15%) is fixed into plant carbohydrates. With each trophic step in the food chain, there is a 90% energy loss (to man as food). Each trophic level has about $\frac{1}{10}$ the energy of the next lower level. This greatly reduces the available food supply for man, depending on the trophic level eaten.

Food produced on land is generally eaten in the first or second trophic level—first for vegetables and grains (green plant growth) and second for meat from animals, which in turn have lived and grown on green plant food, including hay, grain, and so on.

Seafood, on the other hand, is generally eaten in the third or fourth trophic levels, involving an enormous energy loss from the energy initially fixed in plant life by photosynthesis.

Fish eaten by man have generally started with plankton (minute drifting organisms, rarely more than a few millimeters in length). The photosynthesizing plant plankton, called phytoplankton (first trophic level), are eaten by the grazing animal plankton, called zooplankton (second trophic level). These in turn are eaten by tiny fish such as sardines, herring, and menhadden (third trophic level), which constitute the food for larger, fast-swimming precacious fish such as salmon and mackerel (fourth trophic level). Thus, fish normally consumed by man are in the third trophic level, about 40%, or fourth trophic level, about 60%.

A second aquatic chain starts with attached plants, either submerged or emergent (first trophic level). Larvae, crustations, and other herbivores (plant eaters) feed on these (second trophic level). From here on, the chains merge, small fish of the third trophic level eating these larvae and crustations as well as the zooplankton and phytoplankton.

Thus, the primary energy fixed in phytoplankton or attached plants in water (seas, lakes, rivers, etc.) suffers about a 1000 to 1 reduction to the fish generally eaten by man. In the chain phytoplankton–zooplankton–herring–mackerel–tuna, the 100 lb tuna (fifth trophic level) requires 500,000 lb of phytoplankton for its production (5000 to 1 reduction).[7]

If we start with roughly 126×10^{12} kg of carbon fixed annually in marine plant life,[20] corresponding to 1.22×10^{18} kcal, we arrive at 1.22×10^{15} kcal in fish of the varieties normally eaten by man. Since the total food consumed by the world in 1955 contained about 3.0×10^{15} kcal (calculation later), including about 1% seafood, or 0.03×10^{15} kcal, it can readily be seen that food from the sea is not infinite. It can, however, be dramatically increased by using lower trophic levels.[8]

Food Crop of the World

The recommended daily intake of food energy in the United States[9] varies from 490 kcal for a 9 lb baby to 3000 kcal for a 14 to 18 yr old boy (130 lb, 67 in. tall). For a 23 to 50 yr old man, it is 2700 kcal.

The food energy per capita in 1954, in kcal, was about:

New Zealand	3289
United States	3092
Chili	2488
India	1837

Other countries were intermediate.

The food crop of the world in the 1950s, less USSR, is given in Table 10.9. In Table 10.10, the food crop is reduced to equivalent cereal, using energy values from Refs. 10 and 16. The energy content is

$$862 \times 10^{12} \text{ g/yr} \times 3.3 \text{ kcal/g} = 2.84 \times 10^{15} \text{ kcal}$$

Table 10.9. Annual World Food Crop, Less USSR, 1950–1955, Approximate[a]

Cereals	636
Rice	162
Wheat	151
Corn	137
Other[b]	186
Tubers and roots	296
Potatoes	167
Other[c]	129
Pulses (peas, beans, etc.)	25
Oil-bearing crops	60
Sugar	41
Meat	43
Beef and veal about $\frac{1}{2}$	
Pork less than $\frac{1}{2}$	
Mutton and lamb about $\frac{1}{10}$	
Other meat (poultry, offals, horse)	9
Edible pig fat	3
Milk	254
Cow	225
Goat	8
Sheep	5
Buffalo	16
Eggs	10
Seafood	27[d]
Fruit	75[d]

Source. Encyclopaedia Britannica.
[a]In metric megatons (1 metric ton = 2200 lb).
[b]Includes rye, barley, oats, millet, sorghum, etc.
[c]Includes sweet potatoes, yams, cassava, etc.
[d]Estimates from other sources.

Since the world population in the mid 1950s, less USSR, was 2.6 billion, the food energy per capita was

$$\frac{2.84 \times 10^{15}}{2.6 \times 10^9 \times 365} = 3000 \text{ kcal/day}$$

In view of the vagaries of reporting, the loss in storage and use, and the fact that some of the food listed is eaten by animals, and other inaccuracies, this agrees as closely as could be expected with the food consumption in various countries given earlier.

Note that most of the food of the world (about 74%) is grain (cereal). In 1979, the world grain crop was about 1200 megatons (30% feed).

Allowing for the population of the USSR, the world food crop in 1955 contained about:

$$\frac{2.8}{2.6} \times 2.84 \times 10^{15} = 3.0 \times 10^{15} \text{ kcal,}$$

as used earlier. Note that the world food energy was about 12 quads in 1955, 18 quads in 1978; United States food energy was about 1.0 quad in 1980.

In addition to energy, human nutrition requires proteins, minerals, and vitamins in certain proportions (see U.S. Recommended Daily Dietary Allowances[9]). These requirements are beyond the scope of this chapter.

THE BIOMASS PROGRAM

Biomass, or current growth, is all recent vegetable and animal growth and its residues. It is distinct

Table 10.10.   Calculation of Equivalent Cereal

Food	Energy Value (kcal/100 g)	Actual (megatonnes)	Equivalent Cereal (megatonnes)
Cereals	330	636	636
Tubers and roots	60	296	54
Pulses	60	25	5
Sugar	400	41	50
Meat	250	52	40
Milk	70	253	50
Eggs	150	10	5
Fruit	62	75	14
Seafood	100	27	8
Total equivalent cereal			862

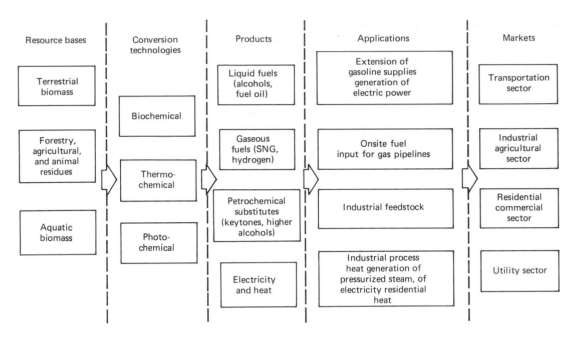

Figure 10.2. Scope of markets and applications for products from biomass resources. *Source:* Doe.[15]

from fossil fuels, which are the remains of vegetable and animal growth hundreds of millions of years ago. Biomass has been the principal source of food, energy, and materials for man in earlier times. In 1850, 90% of U.S. energy usage was fuel wood (Fig. 10.1). Since then, it has shifted to over 90% coal, oil, and gas and, more recently, some nuclear. Biomass still provides practically all food, except salt and a few minerals, and is used in limited applications, as convenient, for energy in fireplaces and wood stoves, wood and waste boilers, bagasse on sugar plantations, and the like. This has served the joint purpose of waste disposal and economic energy. However, with ample low-cost fossil fuels, there has been little incentive to extend beyond the most obvious and economic or otherwise desirable uses. Now (1980), with oil and gas supplies limited and rapidly rising energy prices, the maximum energy and feedstock contribution from biomass has become a national goal.

The overall program is shown in Fig. 10.2. All vegetable and animal growth on land and sea and the residues from primary forestry, agriculture, and animal products can be dried and burned directly for cooking, heating, or power generation, or can be converted by a variety of processes to more refined forms—gas, oil, alcohols, and chemical feedstocks. Some of these processes have been in use for many years, such as the direct burning of wood and cer-

tain wastes; and the conversion of the starches in corn, grain, and potatoes or the sugar in beets or cane to ethyl alcohol by fermentation and distillation. Anaerobic decay in digesters converts plant or animal residues into methane gas, as is occurring continually in nature. Some municipal wastes are being burned along with pulverized coal to produce electricity, or separately to produce process steam.

Typical applications are shown in which these biomass-produced fuels can supplement or replace conventional fuels in every market sector.

With the greatly increased economic incentives and the national security need to reduce dependence on foreign oil, the DOE–Industry program is aimed at improving every aspect of biomass utilization for energy, increasing the base by developing more productive plants, and determining the best biomass utilization formula for each region of the country. The fermentation, distillation, and digestion processes are all under intensive development (1980), consistent with the rapidly increasing economic value of the products. The use of alcohols in automobiles, tractors, and combines is being intensively studied and the devices improved.

Emphasis is on near-term developments that can reduce oil imports, but long-term possibilities such as using plant growth from the sea are also receiving consideration.

The current (1980) contribution of biomass to

U.S. energy consumption is estimated to be 1.5 quad/yr.[15] This includes process heat in the wood products industry, the use of bagasse in the sugar industry, domestic use of wood, and so on. The 0.07 quad of "other" listed in Fig. 1.2 and Ref. 4 of Chapter 1 includes only geothermal electric power and electricity produced from wood and wastes. The principal biomass fuels have not been included in commercial energy records in the past. The contribution of ethanol, at 80 million gal/yr (1978), is still negligible, under 0.01 quad. The potential from biomass in the year 2000 has been estimated at 3.1 to 5.4 quads.[11]

In this chapter, information is given on some of the currently most important uses of biomass. The use of wood for energy and materials, the food of the world, and the energy potential from industrial, domestic, and agricultural waste in the United States has already been covered. The production of ethanol from plants and the production of methane by digesters follows.

The production of oil from shale is described in Chapter 2. In this process, the kerogen in the shale is converted by pyrolysis to an oil which is hydrorefined to a synthetic crude, which is suitable as a refinery input. The kerogen is organic material, as in wood and plants. Thus, the process of extracting (retorting and hydrorefining) oil from other organic material is not too different from that described for oil shale and will not be repeated in this chapter.

ALCOHOLS

The two principal alcohols are methyl alcohol, or methanol, CH_3OH, known as wood alcohol, and ethyl alcohol, or ethanol, C_2H_5OH, known as grain alcohol.

Methanol, or Wood Alcohol. This was originally extracted from wood but since the mid-1920s has been mostly synthesized from natural gas in the United States and from naptha in Europe. It is the familiar alcohol fuel for camp stoves and lanterns and solvent for paints and varnishes, frequently retailed as "shellac thinner." It is widely used as an antifreeze. It is poisonous if taken internally and can lead to blindness. Since the 1920s, its use has expanded greatly. Over 3 million tons of it were used in the United States in 1974, amounting to 4.5

gal per capita. Over 9 million tons were used worldwide.

Since it is now produced mainly from fossil fuels, it is not of primary interest in biomass utilization. However, it is readily formed from syngas from coal according to Eqs. 10.1 and 10.2, and thus represents a possible automotive fuel from coal, as petroleum supplies are exhausted:

$$2 H_2 + CO \rightarrow CH_3OH \qquad (10.1)$$

$$3 H_2 + CO_2 \rightarrow CH_3OH + H_2O \qquad (10.2)$$

Ethanol, or Grain Alcohol. This is the same alcohol as in alcoholic beverages. It is produced by the fermentation of grain and other farm crops, which thus constitute a renewable energy source. It is suitable for use in internal combustion engines and heaters with suitable modifications.

Gasohol. Ten percent by volume of pure, 200-proof ethanol can be blended with gasoline and used in spark-ignition engines without modification. This is known as Gasohol.

The Alcohol Family. The alcohols are alkyl groups with the highly polar OH group attached (see Chapter 11). Thus, methanol is the methyl group, CH_3, with OH attached, becoming CH_3OH. Ethanol is the ethyl group, C_2H_5, with an OH group attached, becoming C_2H_5OH. Similarly, all the higher alkyl groups and their isomers with more than 2 carbon atoms per molecule form alcohols by the addition of the OH group. Only the two simple alcohols, with one or two carbon atoms per molecule, are of interest as fuels. Because of the highly polar OH group, the simple alcohols are infinitely miscible with the polar fluid water. In this respect, they differ from the hydrocarbon liquids.

Proof. The proof of an alcohol–water mixture is twice the percent alcohol; 200-proof is 100% or "neat" alcohol.

Characteristics of Alcohol Fuels

The comparative properties of several fuels are given in Table 10.11. The properties of methanol and ethanol are compared with gasoline and diesel oil. The latter are commercial mixtures, whereas the pure alcohols are single species, having definite

Table 10.11. Typical Alcohol and Hydrocarbon Fuels

	Unleaded Gasoline (Approx.)	Octane	Propane	Methanol	Ethanol	No. 1 Diesel Oil (Approx.)
Chemical formula	C_4-C_{12}	C_8H_{18}	C_3H_8	CH_3OH	C_2H_5OH	
Molecular weight	126	114	44	32	46	170
HHV (Btu/lb)	20,260	20,590	21,646	9,758	12,800	19,240
LHV (Btu/lb)	18,900	19,100	19,916	8,577	11,500	18,250
Specific gravity	0.739	0.702	0.493	0.796	0.794	0.876
Density (lb/gal)	6.17	5.86	4.12	6.65	6.63	7.31
LHV (Btu/gal)	116,500	111,824	81,855	56,866	76,152	133,300
LHV/gal relative to gasoline	1.00	0.960	0.703	0.488	0.654	1.14
Research octane	85–94	100	112		106	10–30
Motor octane	77–86	100	97		89	10–30
Cetane number	10–20	—	—		−20 to 8	45
Latent heat of vaporization (Btu/lb)	142	141	147		361	115
Stochiometric mass, A/F ratio	14.7	15.1	15.6	6.4	9.0	
Boiling temp. (°F)	90–410	258.2	−44.5	149	173	340–560
Flammability limit volume (%)	1.4–7.6				4.3–19	

Source. Ref. 12.

chemical formulas. Octane, a single species approximating gasoline, and propane, the principal constituent of U.S. LPG, are included for comparison.

The useful energy in a fuel for engines and most heaters is the lower heating value (LHV), since water is exhausted as a vapor and its latent heat of vaporization is lost. However, the higher heating value (HHV) is most frequently given for U.S. fuels and has been included in Table 10.11. The only significant comparison is between LHVs.

The LHV of ethanol is 11,500 Btu/lb. It is 61% of the LHV of unleaded gasoline, 18,900 Btu/lb. Ethanol is heavier, sp. gr. 0.794 versus 0.739 for unleaded gasoline. (It sinks to the bottom of a mixture with gasoline if it separates.) Hence, on a volume basis ethanol has 65.4% of the LHV of unleaded gasoline (76,152 versus 116,500 Btu/gal).

The octane number of ethanol is higher than that of unleaded gasoline, which means that it can be used in an engine of higher compression ratio and consequent higher thermal efficiency.

The cetane number is correspondingly very low, which makes it unusable in a diesel engine directly, until cetane improvers are developed.

The latent heat of vaporization is 2.5 times that of unleaded gasoline, per pound, or 4.2 times per unit of energy. This, together with the lack of volatile components, requires a heater for starting at low temperatures, when used unblended.

Production of Ethanol from Organic Materials[12]

The following feedstocks can be used in making ethanol: sugars—sugar beets, sugar cane, sweet sorghum, and ripe fruits; starches—grains, potatoes, and Jerusalem artichokes; and cellulose—stover, grasses, wood, straw, and paper pulp. Grains include corn, wheat, barley, rye, and the like. Stover is the residue of farm crops.

Regardless of the feedstock used, ethanol production consists of three main steps:

1. The formation of a solution of fermentable sugars.
2. The fermentation of the sugar to ethanol.
3. The separation of the ethanol from the mixture, usually by distillation.

The process is simplest for *sugars.* In this case, fermentation with a yeast is needed, as in making wine, followed by fractional distillation to separate out the ethanol. Theoretically, from 100 lb sugar,

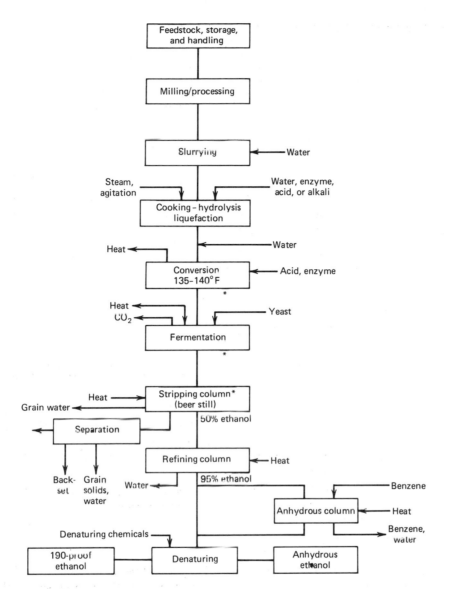

Figure 10.3. Schematic ethanol production diagram. The asterisk indicates where spent grain may be removed. Courtesy D. Shelton, University of Nebraska.[12]

7.7 gallons (57.1 lb) ethanol and 48.9 lb CO_2 are produced. The CO_2 is vented in small stills.

In the case of *starches*, several additional steps are required, as shown in Fig. 10.3. The grain must be milled to a fine meal, then slurried with 10 to 33 gal water, for each bushel of grain, to form a mash.

In the next step, with the pH adjusted to typically 5.2 to 6 and with the application of heat and agitation and a suitable enzyme, the mash gelatinizes. The starch is then transformed into soluble, high-molecular-weight sugars called dextrins. They dissolve into the liquid in the process of liquefaction.

Next, with proper temperature, pH, and enzyme, the dextrins are "converted" into fermentable sugars. The process is exothermic, and heat must be removed.

Fermentation and distillation follow as with sugars. The fermentation requires 48 to 72 hr depending on temperature, the optimum being about 86°F (30°C), with a pH of 4 to 5. Since the process is exothermic, heat must generally be removed to maintain the optimum temperature for the yeast. The resulting ethanol–water mixture is known as "beer." The spent solids can be removed at several stages as indicated.

With *cellulosic feedstocks*, an acid pretreatment

Table 10.12.   U.S. Department of Agriculture Prototype Small-Scale Stills

	Pot Still	Small On-Farm	Large On-Farm	Small Community	Small Community	Large Community
Proof	160–190	190	190	200	200	200
Annual production (gal)	16,000	60,000	360,000	1,000,000	1,000,000	2,000,000
Type of stillage	Wet	Wet	Wet	Wet	DDGS	DDGS
Type of plant	Package	Package	Package	Custom	Custom	Custom
1979 Production costs ($/gal)						
Feedstock	1.04	1.00	1.00	1.10	1.10	1.10
Labor and supplies	0.37	0.42	0.33	0.32	0.39	0.35
Maint., tax, insur., admin.	0.11	0.16	0.11	0.10	0.13	0.11
By-product credit	(0.25)	(0.54)	(0.48)	(0.48)	(0.46)	(0.46)
Capital cost charge	0.36	0.30	0.17	0.17	0.22	0.19
Total	1.63	1.34	1.13	1.21	1.38	1.29

Source.   Ref. 12.

is necessary to break the cellulose material down to starch before the slurrying and further steps of Fig. 10.3 can be carried out. This process is currently being studied at the DOE Livermore Berkeley Laboratory (1980).

Distillation

Distillation requires a stripping column, which removes most of the water, and a refining column, whose output is about 95% ethanol. The azeotrope of ethanol and water is 95.6% ethanol. Above this, no further separation by distillation is possible. 190-Proof ethanol is normally obtained.

200-Proof or Anhydrous Ethanol.   In order to remove the rest of the water, a third ingredient, benzene, is introduced. It forms a ternary azeotrope with water and ethanol, boiling at a low temperature. Thus, the rest of the water, the benzene, and some ethanol can be removed, leaving pure 200-proof ethanol. This requires both stripping and rectifying columns. It is a fairly expensive process but is necessary for 200-proof ethanol.

Denaturing

The ethanol, whether to be used as 190-proof from the refining column or 200-proof from the anhydrous column, must first be denatured by the addition of prescribed amounts of gasoline or ketones according to the requirements of the Bureau of Alcohol, Tobacco, and Firearms (BATF).

Distillers Dried Grain Solids (DDGS)

The spent grain, rich in proteins and other nutrients, is dried and sold as DDGS, a feed supplement similar to soybean meal. Or it may be sold and used as a liquid for livestock feed within 24 hr. The latter procedure is in an early investigative stage, not fully proved and standardized (1980).

DDGS has 27% protein; soybean meal has 44%. DDGS and lysine replace soybean meal for swine and poultry. DDGS is sold at about 78% of the price of soybean meal.

Costs of Production

The Department of Agriculture has considered six prototype stills[12] for small-scale ethanol production and has developed the cost of ethanol from each of these facilities. The principal characteristics and costs of these stills are shown in Table 10.12. Output and costs are based on practically 100% plant factor, as no substantial experience is available with small installations of this type. Costs and volumes will have to be adjusted after actual operating experience is available.

The prototype stills range from a small "moonshiner type" pot still, producing about 160 gal per batch, two batches a week, year round, or 16,000 gal/yr, to a large community still of 2,000,000 gal/yr 125 times as large.

Feedstock is a principal cost. As shown in Table 10.13, corn, a principal feedstock, runs 91 bushels/acre and produces 2.36 gal/bushel. The average

Table 10.13. Average Feedstock Cost per Gallon of Ethanol[12]

Feedstock	Typical Range of Yield per Acre 1977	Gal/Unit	Avg. Price to Farmers 1963–1977 (1979 Dollars)	$/Gallon
Corn	90.8 bu (29–116)	2.36/bu	2.69/bu	1.14
Grain sorghum	56.2 bu (16–80)	2.24/bu	2.40/bu	1.08
Wheat	31.0 bu (22–72)	2.55/bu	3.46/bu	1.36
Rye	24.5 bu (16–31)	2.20/bu	2.36/bu	1.07
Oats	55.6 bu (36–70)	1.03/bu	1.46/bu	1.43
Barley	43.8 bu (27–76)	1.89/bu	2.35/bu	1.24
Rice, average	44.12 cwt	3.47/cwt	11.44/cwt	2.88
Potatoes, white	261 cwt	1.15/cwt	4.98/cwt	4.35
Potatoes, sweet	111 cwt	1.71/cwt	9.78/cwt	5.72
Sugar beets	20.6 T	20/T	31.58/T	1.43
Sugar cane	36.5 T	15.2/T	23.68/T	1.56

Source. Reference [12].
[a]bu = Bushel; cwt = hundred weight; T = tons.

1963–1970 price to farmers, in 1979 dollars, was $2.69/bushel. Thus, corn feedstock is $1.14/gal.

Similarly, wheat, using three times as much land per bushel, results in a feedstock cost of $1.36/gal, only 19% more than corn. Thus, wheat farmers would probably use wheat, and corn farmers would use corn. An average feedstock cost of $1.10 was used for the larger community stills and $1.00 for the on-farm stills, except $1.04 for the less efficient pot still.

A by-product credit of 25¢ to 54¢/gal is given for the stillage, depending on the expected value of wet and dry stillage, on farm and off. The pot still does not produce daily stillage, and its value is less.

The resulting cost of ethanol (Table 10.12) varies from $1.13 to $1.63/gal of 160- to 190-proof ethanol from the on-farm stills. For the large community still, 2,000,000 gal/yr of 200-proof ethanol, it is $1.29/gal.

Note that a fractional distillation column in a petroleum refinery may process 3 billion gal/yr oil (Chapt 2, Refining Methods). While the large community still is very small compared with an oil refinery, the cost as shown in Table 10.12 is dominated by feedstocks and credit for by-products and would be about the same in larger quantities. Feedstocks could not be less than the average price to farmers in any event.

Thus, the cost, $1.29/gal (1979 basis), must be compared with the cost of bulk gasoline. Using a nominal oil cost of $18 per 42 gal barrel, or 43¢/gal, and adding 7¢/gal for refining it to gasoline, results in a bulk cost of 50¢/gal (Table 1.3). Supplier and dealer markups would be the same with either fuel. Thus, a large subsidy is needed to use ethanol. Remission of the 4¢/gal federal excise tax provides 40¢ per gallon ethanol in the 10% mixture, gasohol. Further reduction of some state taxes (total tax is about 12¢/gal, 1980), or a premium price for gasohol, is necessary to balance on economics alone. As oil prices increase, the price of ethanol will increase more slowly, and less subsidy will be needed.

The use of straight ethanol on farms and in rural transportation, without subsidy, would require a very low valuation of labor and on-farm crops and a very high valuation on having an independent supply (1980).

Energy Required

The LHV of ethanol is 78,152 Btu/gal, or 65.4% of that of unleaded gasoline. The energy required to produce a gallon of ethanol in the 360,000 and 1,000,000 gal/yr stills of Table 10.14 is*:

190-Proof, wet stillage,	43,000 Btu/gal
200-Proof, wet stillage,	61,000 Btu/gal
200-Proof, DDGS,	82,000 Btu/gal

Only the DDGS is proved technology at present (1980). Thus, the energy to produce 1 gal ethanol at the prices shown exceeds its LHV. The use of ethanol cannot be considered as an energy conservation method. It must be viewed as a way of converting the renewable biomass energy into a liquid fuel for vehicles and other uses, at the expense of an equal or greater amount of energy from coal or other fuel. The energies given are the process requirements. The boiler inputs would need to be considerably more than these values. Small, efficient boilers for biomass fuels are currently undeveloped (1980).

As shown in Table 10.13, about 214 gal ethanol per year can be made from 1 acre of corn (90.8 × 2.36), or 79 gal per acre of wheat. The 16,000 gal/yr pot still would require 75 acres of corn or 203 acres of wheat to keep it going continuously. Its production, 16,000 gal/yr, corresponds roughly to the energy usage of a 1400 acre farm.

ENERGY FROM ANIMAL WASTES (METHANE FROM DIGESTERS)[13,14,17]

All forms of biologically degradable vegetable and animal matter and their residues can be converted to biogas by digestion in the absence of air, that is, by anaerobic decay. This includes crop wastes, human and animal wastes, wastes from agriculture-based industries, forest litter, and aquatic growth. The biogas is about 60% methane, CH_4 (50 to 70%), and 40% carbon dioxide, CO_2 (30 to 50%), with a little hydrogen, H_2 (0 to 4%) and a small amount of other gases such as H_2S. Its HHV is over 500 Btu/ft³ (HHV of 0.6 methane + 0.4 inert = 978 × 0.6 = 587 Btu/ft³).

This is the simplest and most practical method

known for treating human and animal wastes to minimize the public health hazard associated with their handling and disposal. The residue left after removal of the gas is a valuable fertilizer and soil conditioner that contains all the essential nutrients present in the raw materials.

The digestion process is the natural source of "marsh gas." It has been used commercially at least since 1895 when a "carefully designed" septic tank was used for street lighting in Exeter, England.

The practice of burning animal dung directly for domestic heating is widespread where other fuels are scarce. To avoid this loss of nutrient to the soil, a number of governments have promoted the use of small methane digesters in order to both use the fuel value and conserve the nutrients. As a result, many thousands of such plants have been built in rural India, Taiwan, and elsewhere, typically serving one to several families. Tens of thousands of such generators are operating in the Peoples Republic of China (PRC) and in Korea.

Interest in the United States is recent (1980). The Department of Energy (DOE) is installing a demonstration methane generator at a cattle feedlot in Florida to supply methane fuel to a nearby packing house. Methane is being produced from cattle manure at a project in the Imperial Valley[17]. BuMines had earlier made estimates of the available waste in the United States and its potential if converted to gas (see Tables 10.7 and 10.8). The BuMines estimates were based on 5 ft³ methane per pound organic material.

Typical Small Digester

As an introduction, assume nominal parameters as follows:

5 ft³ Methane per pound organic material.

Biogas = 60% methane, 40% inert, by volume; HHV = 587 Btu/lb.

Average cow, 1000 lb.

Dung per 1000 lb cow per day, 77 lb wet, 6.1 lb volatile solids (United States, less in India).

Biogas needed per person per day for cooking in developing countries, 15 ft³, 75 ft³ for five persons.

Digester size, 50 ft³ per 1000 lb cow.

Temperature of digester, 52°F winter, 86°F summer, pH = 7.

---

*Divide by boiler efficiency for energy input.

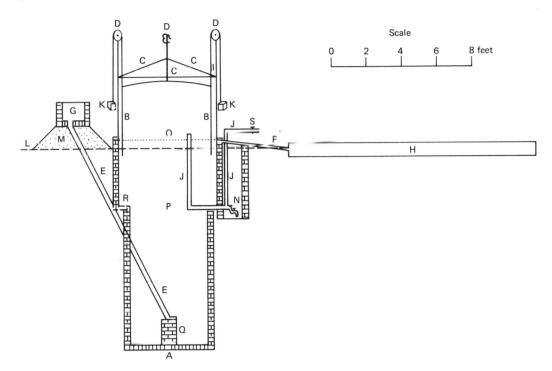

Figure 10.4. Model of a cowdung gas plant. A, Brick wall; B, Gas holder; C, Iron rod; D, Pulley; E, Cowdung inlet pipe; F, Slurry exit channel; G, Cowdung mixing tank; H, Drying bed; I, Angle iron posts; J, Gas outlet pipes; K, Counterpoise heights; L, Ground level; M, Earth platform; N, Gas moisture exit tap; O, Slurry level; P, Fermentation tank; Q, Platform; R, Ledge; S, Gas cock. *Source*: National Academy of Science.[13]

Digestion time, 10 to 20 days.

Dung from one cow in India produces 35 ft³ of biogas/day.

Figure 10.4 shows an Indian design for a three cow digester, 150 ft³. The dung is mixed to a slurry of the proper consistency in mixing tank G. Night soil (human feces and urine) would no doubt be added. No extra water is usually needed with dairy cattle dung and liquid. It flows down the inlet pipe E, about 3 × 77 = 231 lb or 3.7 ft³/day, $\frac{1}{40}$ of the digester capacity. A similar amount of spent slurry is removed each day through exit channel F. Thus, the average dwell time is 40 days, well in excess of the digestion time of 10 to 20 days.

The gas collects under the gas holder B, a simple bell cover, counterweighted by weights K, to result in slightly above atmospheric pressure. About 100 ft³ biogas is produced daily and can be drawn off through gas cock S.

Since 75 ft³ is needed for cooking for a family of five, and a little for lighting, the output supplies only these bare necessities, with none left for space heating.

A 10 cow digester of 500 ft³ and 350 ft³/day output would supply an additional 250 ft³/day biogas at 587 Btu/ft³, or 147,000 Btu/day, equivalent to about 12 lb coal per day, enough to run a small heater: 147,000 Btu = 43 kWh and would run an 1800 W (gas) heater continuously.

Deceptively simple though it seems, the bacteria that do the work are rather demanding in their requirements. Separate organisms decompose cellulose, proteins, and fats into soluble compounds. This requires most of the digesting time. A second stage, acid bacteria, converts these compounds to organic acids, which are quickly converted by the third stage, methane bacteria, into methane gas. The second stage duration is short so that the overall pH is close to 7, or neutral. In some cases, if it becomes too acid, buffering with lime is necessary.

All of these bacteria function without oxygen but require nitrogen. In fact, the presence of any oxygen would degrade the process. The digester construction must be leakproof.

Table 10.14. Manure Production per 1000 lb Animal per Day[a]

Animal	Volume (ft³)	Wet Weight (lb)	Volatile Solids (% Wet Weight)
Dairy cattle	1.33	76.9	7.98
Beef cattle	1.33	83.3	9.33
Swine	1.00	56.7	7.02
Sheep	0.70	40.0	21.5
Poultry	1.00	62.5	16.8
Horses	0.90	56.0	14.3

[a]Abbreviated from Ref. 13.

With too much nitrogen, that is, low C/N ratio, ammonia may be toxic to the bacteria, and the production falls off. With too little nitrogen (high C/N ratio), the process is limited by nitrogen availability to support the bacteria.

The C/N ratio of cow dung is about 18. A C/N ratio of about 30 by weight is needed for optimum production. Straw or sawdust has a higher C/N ratio and must be mixed to provide the optimum C/N. Central bodies, like the Indian Agricultural Research Institute (IARI), would no doubt advise on the best mix of available raw materials in each region and the correction of problems, as well as the general construction and operation of the digesters.

Production of methane also depends on the temperature, being greater at higher temperatures. However, the temperatures realizible without heat in mild climates, with the digester buried below ground level, provide close to optimum production with larger dwell time than would be needed at a higher temperature. Other designs use thick insulation (3 ft) both below ground and clear about the gas holder, to conserve the exothermic heat of digestion.

Relative Energy

How much energy is extracted as biogas compared with burning the organic material directly (as dung)? The heating value of dry resin-free wood is 8300 Btu/lb, corn with 12% moisture, 9300 Btu/lb, domestic waste, 4171 to 5501 Btu/lb (includes glass and noncombustibles), wheat straw, 8500 Btu/lb. If dung be taken as 8000 Btu/lb and the 5 ft³ methane obtained from it as $978 \times 5 \cong 5000$ Btu/lb, the heat extracted is roughly 62% of the heat that could be obtained by burning directly. No doubt, the greater efficiency of a small gas flame, directly on a utensil,

Table 10.15. Yield of Biogas from Various Waste Materials[a]

Raw Material	Dry Solids (ft³/lb)
Cow dung	5.3
Cattle manure	5.0
Cattle manure (India)	3.6–8.0
Cattle manure (Germany)	3.1–4.7
Beef manure	13.7
Beef manure	17.7
Chicken manure	5.0
Poultry manure	4.0–4.7
Poultry manure	8.9 (based on volatile solids)
Swine manure	11.1–12.2
Swine manure	7.9
Swine manure	16.3
Sheep manure	5.9–9.7
Forage leaves	8.0
Sugar beet leaves	8.0
Algae	5.1
Night soil	6.0

[a]Abbreviated from Ref. 13. Differing values for the same raw material are from different references.

more than compensates for this reduction for cooking. A gas space heater is probably also more efficient.

Economics

Studies by government institutes in developing countries have indicated that these small digesters are economic in terms of human and material resources for the existing conditions. Government subsidies and help are given for their construction. The economics of larger digesters for use with collected waste in the United States is yet to be worked out and is a part of the DOE program.

Large-scale municipal digesters are used in the treatment of municipal sewage sludge, with the evolved gases satisfying part of the energy requirements of the municipal treatment plant.

Production

Tables 10.14 and 10.15 give typical data on manure and biogas production. Data from various sources vary widely. For more complete information, see Ref. 13.

REFERENCES

1. *Report of the Food and Agriculture Organization (FAO) of the United Nations*, New York: Publishing Services, United Nations.

2. T. Baumeister, *Marks' Standard Handbook for Mechanical Engineers*, 7th ed., New York: McGraw-Hill, 1966.

3. U.S. Department of Agriculture, Forest Service, Leaflet No. 559, Washington, D.C.: Supt. of Documents, 1978.

4. L. L. Anderson, *Energy Potential From Organic Wastes: A Review of the Quantities and Sources*. I.C. 8549, Pittsburgh, Pa.: BuMines, 1972.

5. C. G. R. Humphreys, "Heating Values of Miscellaneous By-Product Fuels," *Combustion*, January 1957, p. 35.

6. Horner and Shifrin, Inc., *Energy Recovery from Waste—A Municipal-Utility Joint Venture*, EPA Report. EP 1.17 d.i, 1972, and R. A. Lowe, *Energy Recovery From Waste*, EP 1.17 d.ii, 1973.

7. G. A. Borgstrom, *Too Many: The Biological Limitations of Our Earth*, New York: Macmillan, 1969, p. 379.

8. C. P. Idyll, *The Sea Against Hunger*, New York: Crowell, 1970, p 24.

9. "U.S. Recommended Daily Dietary Allowances." *World Almanac*. Pittsburgh, Pa.: Pittsburgh Press, 1981, p. 229.

10. "Food Energy Values," U.S. Department of Agriculture, HDBK. No. 8, Washington, D.C.: Supt. of Documents, 1950.

11. U.S. Department of Energy, *Secretary's Annual Report to Congress, January 1980*. Washington, D.C.: Supt. of Documents, 1980, p. 3.2, Table 11.

12. *Small Scale Fuel Alcohol Production*, U.S. Department of Agriculture, March 1980, Washington, D.C.: Supt. of Documents, 1980.

13. *Methane Generation from Human, Animal, and Agricultural Wastes*, Nat. Acad. of Science, Golden Colo.; Solar Energy Research Inst., 1977.

14. E. Ashare and E. H. Wilson, *Analysis of Digester Design Concepts*, DOE-COO-2991-42, Springfield, Va.: National Technical Information Service, 1979.

15. *Solar Energy Program Summary Document, FY-1981*, Washington, D.C.: U.S. Department of Energy, 1980, p. III-6.

16. *Food Industries Manual*, London: Leonard Hill Books, Ltd., 1957.

17. S. Schellenbach, "Imperial Valley Bio-Gas Project, Operations and Methane Production from Cattle Manure, Oct. 1978–Nov 1979," *Energy from Biomass and Wastes, IV, Symposium Papers*, Chicago, Ill.: Inst. of Gas Technology, 1980.

18. *The Conversion of Refuse to Energy*, Proc. First Int. Conf., Montreux Sw., New York: IEEE Order Dept., 1975.

19. D. L. Klumb, "Union Electric Company's Solid Waste Utilization System—SWUS," paper 0-3 in Ref. 18.

20. G. P. Kuiper, Ed., *The Solar System, II The Earth as a Planet*, Chicago: Univ. of Chicago Press, 1954, p. 379.

11

Chemistry of Energy—
Selected Topics

INTRODUCTION

This chapter provides a brief introduction to the families of hydrocarbons for the understanding of petroleum, gas, coal conversion, and other hydrocarbon fuels. Covered also are the subjects of combustion and heating values. The chemistry of batteries and fuel cells is found in Chapter 13, under Energy Conversion. This division is purely arbitrary. The energy conversion in Chapter 13 is mainly from fuels to mechanical or electrical energy. Combustion, covered in this chapter, is also energy conversion, but from fuels to heat.

The subject of equilibrium in a chemical reactor or process is also treated in Chapter 13, under Thermodynamics. Adiabatic flame temperature is treated here along with combustion.

THE HYDROCARBONS

Unlike the other 100-odd known chemical elements, the element carbon forms a very large number of compounds—well over a million known at present. It forms many times as many compounds as all the other elements put together. The chemistry of these compounds is called organic chemistry.[1] These compounds are "organic" in distinction to all other compounds, which are "inorganic."

The compounds containing only carbon and hydrogen, the hydrocarbons, alone number well over 5000. The fossil fuels are mixtures of these, together with some impurities.

The hydrocarbons occur in families or "homologous series," and in order to understand their structures and the nomenclature in common everyday use, the basic concepts involved in these families of hydrocarbons will have to be explained.

Atomic Structures. The structures of the atoms hydrogen, carbon, and oxygen are shown in Fig. 11.1. The electronic orbitals occur in concentric shells. Recall that the innermost shell can hold only two electrons, the next shell eight electrons, and so on, as indicated by the numbers at the right of the periodic table, Table 11.1. Each inner shell must be filled before any electrons can be added to the next outer shell.

Thus, hydrogen, the lightest element, has one electron in its outer shell. Two would fill it. It has a valence of 1. Carbon, the sixth element, has four electrons in its outer shell. Eight would fill it. It has a valence of 4. Oxygen, the eighth element, has a valence of 2.

Formation of Compounds. Stemming from minimum energy considerations, compounds are formed according to three rules:

1. Atoms try to fill their outer shells by sharing electrons with other atoms.
2. Each bond orbital shared by two electrons tries to get as far as possible from each other bond.
3. Each bond forms between just two atoms.

The Methane Molecule. Putting these rules together for methane, CH_4, we obtain the structural formula seen in Fig. 11.2. Each of the four valence electrons of the carbon atom joins with a valence electron of a hydrogen atom to form a shared pair of electrons, that is, a bond. Each common orbital belongs to both the C atom and one of the H atoms, and the two electrons of a bond orbital spend most of their time in the general area between the C and the H atom. That is, the negative charge density of this bond is greatest in, in fact is largely confined to, a sausage-shaped volume symmetrical with the axis

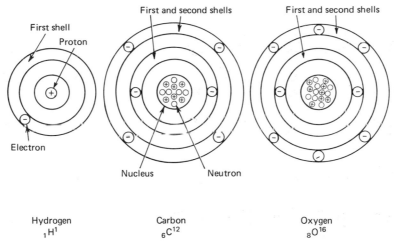

First shell

First and second shells

First and second shells

First shell
Proton

Electron

Nucleus   Neutron

Hydrogen
$_1H^1$

Carbon
$_6C^{12}$

Oxygen
$_8O^{16}$

Figure 11.1. Atoms—common isotopes.

of these two atoms and lying largely between them. This bond is thus directional. Its length is 1.10Å between the two nucleii involved. This compares with the typical radius of a molecule of 1 to 2Å.

The bond is often indicated by a line joining the two atoms, as shown in Fig. 11.2a, or the two shared electrons may be indicated as in Fig. 11.2b. Note that in the methane molecule the carbon atom has its outer or valence shell filled with eight electrons. Also, each of the hydrogen atoms has its outer or valence shell filled with two electrons. These bonds are called covalent because of the sharing of the valence electrons.

Polar Covalent Bonds. When covalent bonds join elements of somewhat different electronegativity, the electronic charge in the bond tends more toward the electronegative element. The bond is polar. Such bonds are more susceptible to being broken in chemical reactions. Molecules having polar bonds, called polar molecules, tend to align with other polar molecules and be attracted to them, and so are soluble in polar liquids such as water. For example, alcohols are soluble in water.

Carbon and hydrogen are so near together in electronegativity that their bonds are nonpolar. This is a factor in their great stability, which has caused these hydrocarbons to remain unchanged for millions of years. It is the reason for their insolubility in water. It is a factor in the large number and size of hydrocarbon molecules.

Molecular Shape. The rule that bond orbitals try to get as far apart from each other as possible does not result in a "flat" molecule, as might be inferred from Fig. 11.2; but rather the four hydrogen atoms are at the four corners of a regular tetrahedron. The bond angles are all 109.5° instead of 90°, as appears in Fig. 11.2. It will be convenient to indicate structural formulas as in Fig. 11.2, but it must always be remembered that these are only shorthand representations of actual three-dimensional models, with the bonds separated as far as possible.

Homologous Series

The hydrocarbons form five principle families, or homologous series, and several other less common ones:

The alkanes, such as methane, CH_4 (Fig. 11.3a), and ethane, C_2H_6 (Fig. 11.3b) (formerly called paraffins), have the general formula C_nH_{2n+2}. They are termed "saturated" hydrocarbons because they contain all the hydrogen atoms that could possibly be accommodated. The length of the chains is practically unlimited. The compound $C_{110}H_{222}$ has actually been isolated in pure form.

The alkenes, such as ethene, C_2H_4 (Fig. 11.4a), and propene, C_3H_6 (Fig. 11.4b) (formerly called olefins), have the general formula C_nH_{2n}. They are unsaturated hydrocarbons and are characterized by having a double bond between two carbon atoms. They also form an extensive series of compounds.

Table 11.1. Periodic Chart of the Elements

Used with the permission of the Fischer Scientific Co.

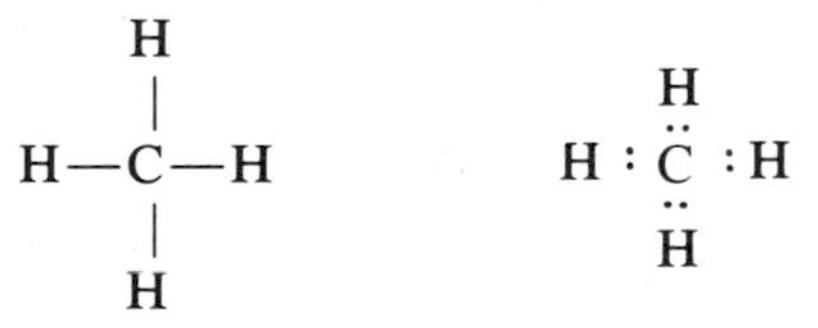

H
|
H—C—H   H : C̈ : H
| H
H   H
(a)   (b)

Figure 11.2. Methane molecule: (a) lines join atoms; (b) shared electrons indicated.

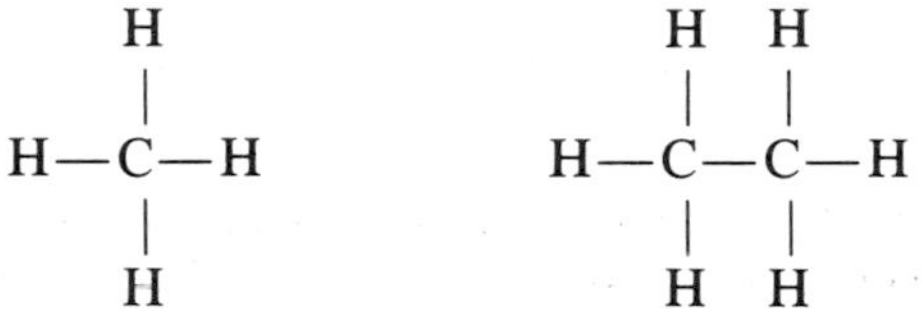

H   H H
| | |
H—C—H   H—C—C—H
| | |
H   H H
(a)   (b)

Figure 11.3. Alkanes: (a) methane; (b) ethane.

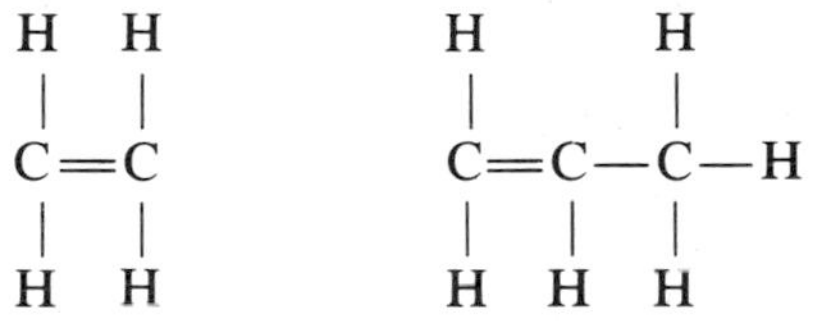

H H   H   H
| | | |
C=C   C=C—C—H
| | | | |
H H   H H H

(a)   (b)

Figure 11.4. Alkenes: (a) ethene; (b) propene.

H
|
H—C≡C—H   H—C—C≡C—H
|
H
(a)   (b)

Figure 11.5. Alkynes: (a) acetylene; (b) propyne.

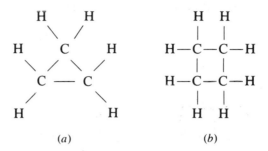

Figure 11.6. Cycloalkanes: (a) cyclopropane; (b) cyclobutane.

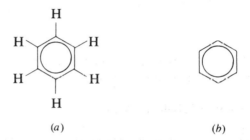

Figure 11.7. Benzene molecule representation: (a) hydrogen atoms indicated; (b) hydrogen atoms not shown.

They are much more reactive than the alkanes, that is, less stable.

The alkynes, such as acetylene, C_2H_2 (Fig. 11.5a), and propyne, C_3H_4 (Fig. 11.5b), have the general formula C_nH_{2n-2}. They are unsaturated hydrocarbons and are characterized by having a triple bond between two of the carbon atoms. Like the alkenes, they are relatively reactive.

The cycloalkanes, such as cyclopropane (Fig. 11.6a) and cyclobutane (Fig. 11.6b), have the general formula C_nH_{2n}. They are saturated hydrocarbons formed by joining the two ends of an alkane chain on itself, which eliminates two of the hydrogen atoms. Cycloalkanes as large as $C_{54}H_{108}$ have been prepared. Cycloalkanes are known as naphthenes in the petroleum industry.

The preceding four families of compounds, together with some other less common families having more than one double or triple bond are known as the "aliphatic" compounds (formerly fatty compounds).

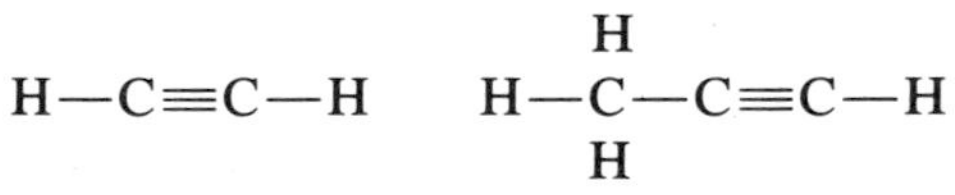

In addition, there is a very stable group of compounds, the *aromatic hydrocarbons*, such as benzene, C_6H_6, in which, with six carbon atoms in a ring and one hydrogen attached to each, the three remaining bonds join each carbon atom to its two neighbors equally. It is represented as in Fig. 11.7a or Fig. 11.7b. It is understood that one hydrogen atom is attached at each corner unless another is indicated. Additional members of this family are as shown in Fig. 11.8. These compounds were originally called "aromatics" because some of their derivatives were quite fragrant, for example, oil of wintergreen.

Isomers. The first three alkanes each have only one possible structure, but the fourth, butane, could be as shown in Fig. 11.9a or Fig. 11.9b. Such compounds of the same chemical formula but having different structures and different chemical and

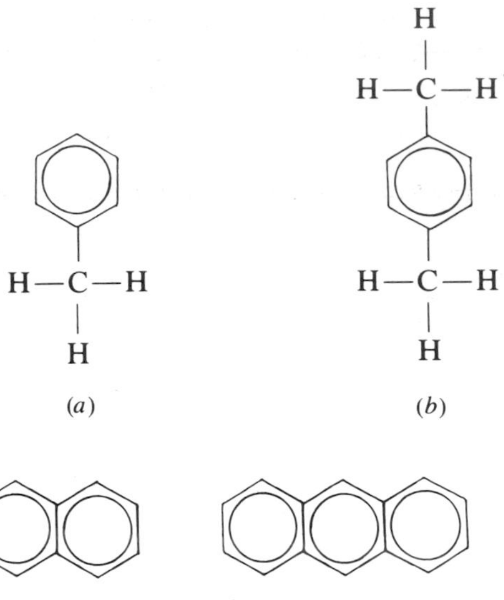

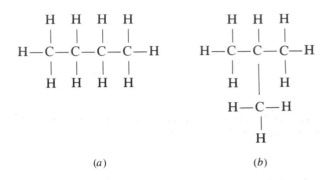

Figure 11.9. Isomers of butane: (a) n-butane; (b) isobutane.

Systematic names have been adopted for all the larger molecules, as will be evident from Table 11.2 giving the names of the normal molecules with up to 10 carbon atoms.

Isomer Names. As shown in Fig. 11.9, butane has two isomers, n-butane and isobutane. Pentane has three isomers designated *n*-, iso-, and neo-. There are five hexanes, nine heptanes, 75 decanes, and so on. Obviously a systematic method of designation is required.

A straight-chain alkane is always designated as *n*-, or normal. Alkanes with six or fewer carbon atoms, with all but one in a straight chain and that one attached to next to the last carbon atom, are designated as iso-, as in Figs. 11.9 and 11.10. Note the abbreviated, yet unambiguous, representations of the structure in Figs. 11.10a and 11.10b.

Alkyl Groups. For convenience, commonly recurring groups of atoms in the large molecules are

Figure 11.8. Aromatics: (a) toluene, or methylbenzene, C_7H_8; (b) xylenes, C_8H_{10}, three isomers; (c) polycyclic aromatics.

physical properties are called "isomers" (from Greek *isos,* equal; and *meros,* part).

Names of Hydrocarbons

It can readily be seen that as the chains become longer, the number of possible isomers increases rapidly. Only for the simplest compounds have the isomers been given discrete names. In fact, only for the simplest structures have the compounds themselves been given common nonsystematic names.

Table 11.2. Names of Normal Aliphatic Hydrocarbons

Number of Carbon Atoms	Alkanes (Paraffins)	Alkenes (Olefins)	Alkynes	Cycloalkanes (Naphthenes)
1	Methane			
2	Ethane	Ethene (or ethylene)	Acetylene (or ethyne)	
3	Propane	Propene (or propylene)	Propyne	Cyclopropane
4	Butane	Butene	Butyne	Cyclobutane
5	Pentane	Pentene	Pentyne	Cyclopentane
6	Hexane	Hexene	Hexyne	Cyclohexane
7	Heptane	Heptene	Heptyne	Cycloheptane
8	Octane	Octene	Octyne	Cyclooctane
9	Nonane	Nonene	Nonyne	Cyclononane
10	Decane	Decene	Decyne	Cyclodecane

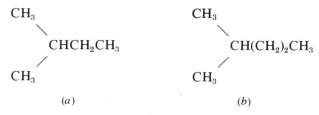

Figure 11.10. (a) Isopentane; (b) isohexane.

$$CH_3-OH \qquad\qquad C_2H_5-OH$$
$$(a) \qquad\qquad\qquad (b)$$

Figure 11.11. (a) Methanol (methyl alcohol); (b) ethanol (ethyl alcohol).

given names. For example, the alkyl groups are the same as the corresponding alkanes, except with the final H dropped. The suffix -yl replaces the -ane:

methyl	CH_3	butyl	C_4H_9
ethyl	C_2H_5	pentyl	C_5H_{11}
propyl	C_3H_7	hexyl	C_6H_{13}

Hydroxyl Group. If one or more hydroxyl groups (OH) replace hydrogen atoms in the chain, the compound becomes an alcohol, as in Fig. 11.11.

The Ethers. The ethers may be considered as formed by replacing the hydrogen atoms of water by alkyl groups, as shown in Fig. 11.12.

There are several other functional groups that modify the hydrocarbons into the 1 million odd organic compounds of biology, life, food, and synthetic materials of all kinds; tens of thousands of new compounds are produced every year. For example, the amino group, NH_2, leads to the numerous amino acids that form the proteins. Out of this huge field of organic chemistry, we restrict ourselves here to the tiny subfield of hydrocarbons—of importance as *primary* fuels—with a few characteristics of alcohols and ethers added because of their importance as fuels.

IUPAC Names. With so many compounds, any arbitrary assignment of common names would lead to endless confusion. Consequently, the International Union of Pure and Applied Chemists (IUPAC) has adopted the following systematic method of naming the compounds of carbon:

$$CH_3CH_2-O-CH_2CH_3 \qquad CH_3-O-CH_3$$
$$(a) \qquad\qquad\qquad\qquad (b)$$

Figure 11.12. Ethers: (a) ethyl ether (the familiar anesthetic); (b) methyl ether.

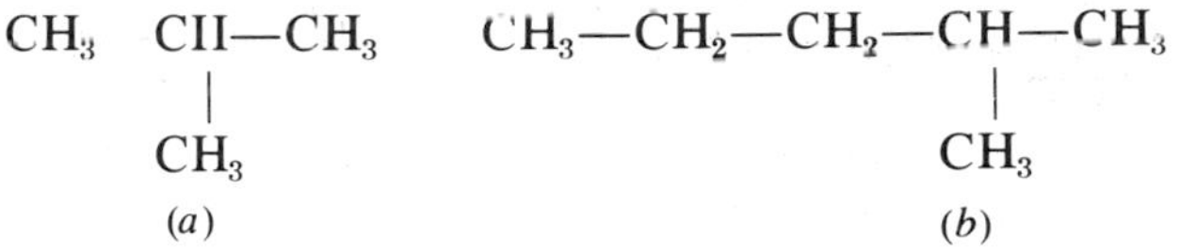

$$CH_3 \quad CH-CH_3 \qquad CH_3-CH_2-CH_2-CH-CH_3$$
$$\qquad\quad | \qquad\qquad\qquad\qquad\qquad\qquad\qquad | $$
$$\qquad\quad CH_3 \qquad\qquad\qquad\qquad\qquad\qquad\quad CH_3$$
$$(a) \qquad\qquad\qquad\qquad\qquad\qquad (b)$$

Figure 11.13. Branched-chain hydrocarbons: (a) methylpropane (the isobutane of Fig. 11.9b); (b) 2-methylpentane.

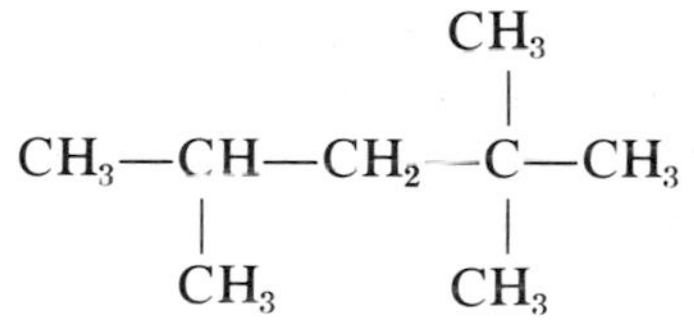

$$\qquad\qquad\qquad\qquad CH_3$$
$$\qquad\qquad\qquad\qquad\ | $$
$$CH_3-CH-CH_2-C-CH_3$$
$$\qquad\quad | \qquad\qquad\ | $$
$$\qquad\quad CH_3 \qquad\quad CH_3$$

Figure 11.14. 2,2,4-Trimethylpentane, commonly, though wrongly, called "isooctane."

1. Select as the parent structure the longest continuous chain of carbon atoms, and then consider the compound to have been derived from this structure by the replacement of hydrogen by various alkyl groups, as shown in Fig. 11. 13.

2. Where necessary, indicate by a number the carbon atom to which the alkyl group is attached, as in Fig. 11.13b.

3. In numbering the parent chain, start at whichever end results in the lowest numbers, as in Fig. 11.13b.

4. If the same alkyl group occurs more than once as a side chain, indicate this by the prefix di-, tri-, tetra-, and so on, to show how many of these alkyl groups there are, and indicate by various numbers the positions of each group, as in Fig. 11.14.

Selected Properties of Hydrocarbons

Octane Rating. In spite of the foregoing rules, the 100-octane hydrocarbon used, along with *n*-heptane, to determine the octane rating (the no-knock quality) of a gasoline is commonly called "isooctane" instead of its correct IUPAC name,

Table 11.3.   Methods of Designating Octane Rating

	Typical Values	
Method	Regular	Premium
Research (in laboratory at low speeds)	94	100
Motor (in engines at higher speeds)	86	92
(Research + motor)/2, or (R + M)/2 (most common in 1980)	90	96

given in Fig. 11.14. The percentage of "isooctane" in a mixture of *n*-heptane and isooctane to give the same antiknock performance as any particular gasoline is the octane rating of that particular gasoline. Thus, *n*-heptane has an octane rating of 0, and isooctane has an octane rating of 100.

There are three ways of designating the octane rating of gasoline, as shown in Table 11.3 (DOT—1977).

Cetane Rating. Detonation, which leads to knocking and loss of power in a gasoline engine, is a desirable property in a diesel engine which depends on the heat of compression for ignition. It is desirable for the fuel to start burning spontaneously at many places as soon as injected. Thus, straight-chain hydrocarbons, of low octane rating, make the best diesel fuels. Cetane (normal hexadecane, $C_{16}H_{34}$) gives very good performance; α-methylnaphthalene, $C_{10}H_7CH_3$ (the alpha isomer), very poor performance. The "cetane rating" of a diesel fuel is the percentage of cetane in a mixture of cetane and α-methylnaphthalene that gives equivalent performance to the fuel being rated. High-quality diesel fuels have cetane ratings around 50.

Boiling Points. The boiling points of the straight-chain alkanes and some of their isomers are given in Fig. 11.15. The intermolecular attraction increases with the size of the molecule. The boiling and melting points thus increase with increasing number of carbon atoms. Except for the very small alkanes, the boiling point rises 20 to 30°C for each carbon atom that is added to the chain. This increment of 20 to 30°C per carbon atom holds also for the other homologous series of hydrocarbons.

It can readily be seen from Fig. 11.15 that the first four *n*-alkanes—methane, ethane, propane, and butane—are gases at normal atmospheric pressure.

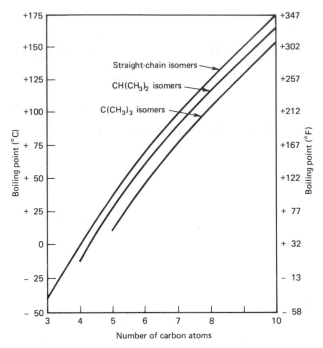

Figure 11.15.   Boiling point of some alkanes at 1 atm.

Methane constitutes a large part of natural gas. For example, a sample taken from a pipeline supplied by a large number of Pennsylvania wells contained methane, ethane, and propane in the ratio of 12:2:1, with higher alkanes making up only 3% of the total. Propane and butane, easily liquefied by pressure, are the bottled gases. The next 13 alkanes from C_5 to C_{17} are liquids, and those containing 18 carbons or more are solids.

The melting point increases similarly to the boiling point with increasing size of molecule, but not quite so regularly. The branched-chain isomers have lower boiling and melting points (see Fig. 11.15). The greater the amount of branching, the more spherical becomes the molecule, with consequent less surface area to hold the molecules together; hence the lower boiling and melting points.

COMBUSTION

The principal combustile elements are carbon and hydrogen. Sulfur, present in small quantities in most fuels, is also a combustible. These are burned with air or oxygen. A wide variety of other combustibles and oxidizers are used in special applications such as rocket fuels and explosives. This discussion is limited to the more common fuels.

Combustion Equations

For the common combustibles, the equations of complete combustion are:

$$C + O_2 \rightarrow CO_2 \qquad (11.1)$$

$$12 + 32 = 44$$

$$2 H_2 + O_2 \rightarrow 2 H_2O \qquad (11.2)$$

$$2 \times 2 + 32 = 2 \times 18$$

$$S + O_2 \rightarrow SO_2 \qquad (11.3)$$

$$32 + 32 = 64$$

The molecular weights in these equations are as shown below the equations. These, multiplied by the coefficients, are the combining weights or masses of the elements and compounds involved. For example, 12 lb carbon combines with 32 lb oxygen to form 44 lb carbon dioxide. The exact amount of air (or oxygen) for complete combustion is called the stoichiometric amount (or the theoretical air), and the coefficients in Eqs. 11.1 to 11.3 are the "stoichiometric coefficients."

The combining volumes are given by the relative number of gas molecules on the two sides of the equation. Thus, in Eq. 11.1, the volume of oxygen and the volume of the product of combustion, CO_2, are equal at constant temperature and pressure.

If air is used, nitrogen also appears on both sides of Eq. 11.1. The volume proportion of air (78.03% N_2, 20.99% O_2, 0.94% Ar, 0.04% CO_2) is frequently approximated in combustion calculations by lumping the noncombustibles in nitrogen, as 79% N_2, 21% O_2, with a ratio of N_2 to O_2 of 3.76 (sometimes 4 is used). The equivalent molecular weight of air is then $(32 + 3.76 \times 28)/4.76 = 28.84$, or about 29. Thus, if air is used in stoichiometric proportion, each side of Eq. 11.1 would include the term $+3.76$ N_2:

$$C + O_2 + 3.76 N_2 \rightarrow CO_2 + 3.76 N_2 \qquad (11.4)$$

In general, more or less than the stoichiometric amount of air or oxygen may be used in a combustion process. For example, in the high-temperature combustion for MHD, at about 4400°F, 95% of stoichiometric air may be used to prevent the formation of nitrous oxides (see Chapter 13, MHD).

Normally, in a boiler some "excess air" is used to ensure nearly complete combustion, in spite of imperfect mixing. In internal combustion engines, much higher air–fuel ratios are used (200 to 400%) to limit the temperatures to those permissible in the engines. The three terms "excess air," "percent of theoretical air," and "air–fuel ratio" are all commonly used.

Determination of Combustion Ingredients From Flue Gas Analysis

Much can be learned about the fuel, air (or oxygen), and products of combustion from an analysis of the flue gas and the mass balances of the elements involved.[2] No elements are changed in a chemical process.

The Orsat apparatus is generally used to analyze a sample of the combustion gas. Starting with a 100% volume at known temperature and atmospheric pressure, the gas is first contacted with a liquid which absorbs CO_2, and the decrease in volume is noted. Then, other liquids are introduced in succession which absorb O_2 and CO selectively. The remaining gas volume is assumed to be nitrogen. Thus, the volume percentages of CO_2, O_2, CO, and N_2 in the flue gas are determined. If other products are present in the flue gas, the process is extended to include them.

The analysis of the combustion from the flue gas analysis may be illustrated by an example. Suppose a hydrocarbon fuel of unknown composition is being burned in air and that the volume composition of the flue gas is: CO_2, 11.9%; O_2, 4.1%; CO, 1.0%; and N_2 the remainder, 83%. Find the composition of the fuel being burned, the air–fuel ratio, and the amount of excess air.

Since the flue gas is composed of essentially perfect gases, the mole percentages are in the same proportion as the volumes. Thus the equation of combustion per 100 moles of dry flue gas becomes:

$$C_aH_b + c\,O_2 + 3.76c\,N_2 \rightarrow 11.9\,CO_2 \qquad (11.5)$$

$$+ 4.1\,O_2 + 1.0\,CO + 83.0\,N_2 + d\,H_2O$$

where a, b, c, and d are to be determined. The average composition of the fuel, the air–fuel ratio, the percent excess air, and the number of moles of water can now be determined.

Since N_2 is inert, the same amount appears before and after combustion. Thus, $c = 83.0/3.76 = 22$

moles oxygen, and the amount of entering oxygen has been determined.

An oxygen mass balance reveals that

$$22\,O_2 = 11.9\,O_2 + 4.1\,O_2 + 0.5\,O_2 + \left(\frac{d}{2}\right)O_2 \qquad (11.6)$$

$d = 11.0$ moles water in the flue gas. From a hydrogen mass balance, $b = 2d = 22.0$.

A carbon balance shows that $a = 11.9 + 1.0 = 12.9$. Thus, Eq. 11.5 becomes

$$C_{12.9}H_{22} + 22\,O_2 + 83\,N_2 \rightarrow 11.9\,CO_2 + 4.1\,O_2$$
$$+\,1.0\,CO + 83\,N_2 + 11\,H_2O \qquad (11.7)$$

The formula $C_{12.9}H_{22}$ is the average composition of the fuel, which may be a mixture of gases or may include some solid or liquid hydrocarbons.

Air–Fuel Ratio.   Adding the 22 moles O_2 and 83 moles N_2, there were 105 moles air used per mole fuel. The air–fuel ratio is therefore

$$AF = \frac{105\text{ moles air}}{\text{mole fuel}} \times \frac{29\text{ lb air}}{\text{mole air}} \times \frac{\text{mole fuel}}{177\text{ lb fuel}}$$

$$= 17.2\text{ lb air/lb fuel}$$

Note: Average molecular weight of fuel is $12.9 \times 12 + 22 \times 1 = 177$

Excess Air.   To just burn the fuel to CO_2 and H_2O requires:

For the $C_{12.9}$	to 12.9 CO_2	—	12.9 O_2
For the H_{22}	to 11 H_2O	—	5.5 O_2
	Total		18.4 O_2

Since 22 O_2 was used, the excess air contains 3.6 O_2. Thus, there is $3.6/18.4 = 19.6\%$ excess air.

Higher and Lower Heating Values

The heat released in the basic reactions Eqs. 11.1 to 11.3 depends on the conditions. Heating values of most fuels are determined experimentally. For example, a sample of combustible may be burned in a bomb calorimeter[3] with sufficient oxygen for complete combustion. The amount of heat removed to bring the calorimeter to its initial temperature is measured and is the "higher heating value."

If one of the combustibles is hydrogen, then one of the products of combustion is water, Eq. 11.2. In the bomb calorimeter, when the temperature is returned to its initial value, the water gives up its latent heat of vaporization as it condenses, and a high heat value results. However, in most applications, the exhaust gases are at a high enough temperature that the water exits as a vapor. The effective heating value is lower. The Higher Heating Value (HHV) has been defined as including all the heat of vaporization of the water formed, and the Lower Heating Value (LHV) none of it. (Heat value is synonymous with heating value or calorific value.) Enthalpy of combustion, a more precise value for pure chemical species, is defined later.

Thus, in burning carbon to CO_2, Eq. 11.1, no water is formed, and there is but one heating value, 14,096 Btu/lb, as given in Table 11.4, whereas in burning hydrogen to water, Eq. 11.2, there is a higher heating value, 61,031 Btu/lb, and a lower heating value, 51,648 Btu/lb.

In general, the lower heating value is less than the higher heating value by 1050 Btu for each pound water formed (or initially present) per pound fuel. The enthalpy of vaporization of water at 77°F (25°C) and 1 atm, h_{fg}, is 1050 Btu/lb.

For a wet fuel, such as wood with 25% moisture, the higher heating value per pound fuel is the same as for 0.75 lb dry wood. However, the lower heating value is less than this by 1050 Btu/lb of all the water, both that formed and that initially present.

Dry Basis.   For fuels with highly varying moisture content the heating value is frequently given on a "dry basis." Then for any particular moisture content, the HHV is the HHV (dry basis) $\times$ (1 − the fraction of moisture). For example, the HHV of typical, nonresinous, wood with 24% moisture is

$$8300(\text{dry basis}) \times (1 - 0.24) = 6300\text{ Btu/lb}$$

(See first column of Table 10.4.)

Enthalpies of Formation and Combustion for Pure Chemical Species

The heats of combustion given in Table 11.4 to five significant digits are obviously more precise than the heating values measured by calorimeter, with external work and initial temperature neglected. These are all pure chemical species, for which very

Table 11.4. Heats of Combustion in Air or Oxygen[a]

Fuel	Chemical Symbol	Molecular Weight	HHV (Btu)		LHV[b] (Btu)	
			Per lb	Per ft^{3d}	Per lb	Per ft^{3d}
Carbon to CO_2	C(s)	12.011	14,096			
Carbon to CO	C(s)	12.011	3,960			
CO to CO_2	CO(g)	28.011	4,346	310.6		
Sulfur to SO_2	S(s)	32.064	3,984			
Hydrogen	H_2(g)	2.016	61,031	314.0	51,648	265.7
Methane	CH_4(g)	16.043	23,890	978.0	21,532	881.5
Ethane	C_2H_6(g)	30.070	22,329	1713.3	20,442	1568.5
Propane	C_3H_8(g)	44.097	21,670	2438.4	19,954	2245.3
n-Butane[c]	C_4H_{10}(g)	58.124	21,316	3161.5	19,689	2920.3
n-Pentane	C_5H_{12}(g)	72.151	21,095	3883.8	19,522	3594.2
n-Hexane	C_6H_{14}(l)	86.178	20,783		19,246	
n-Heptane	C_7H_{16}(l)	100.20	20,682		19,172	
n-Octane	C_8H_{18}(l)	114.23	20,604		19,114	
n-Decane	$C_{10}H_{22}$(l)	142.29	20,371		18,909	
Ethylene (ethene)	C_2H_4(g)	28.054	21,646	1549.6	20,297	1453.0
Propylene (propene)	C_3H_6(g)	42.081	21,053	2260.7	19,704	2115.8
Acetylene (ethyne)	C_2H_2(g)	26.038	21,477	1427.4	20,750	1378.7
Benzene	C_6H_6(g)	78.114	18,188	3625.3	17,461	3480.4
Toluene (methylbenzene)	C_7H_8(g)	92.141	18,441	4335.8	17,611	4140.7
Methanol (methyl alcohol)	CH_3OH(l)	32.043	9,758		8,577	
Ethanol (ethyl alcohol)	C_2H_5OH(l)	46.070	12,770		11,538	
Naphthalene	$C_{10}H_8$(s)	128.7	17,310		16,720	

[a]Derived from enthalpies of combustion and vaporization, Table 11.6, where possible; Carbon to CH_4 (burned in hydrogen), HHV = 2680 Btu/lb[4]; HV/ft$^3_{(77°F, 1\,atm)}$ = HV/lb × (mol. wt.)/391.89.

[b]LHV = HHV − 1050 × (lb water formed per lb fuel burned). For $C_nH_bO_c$, LHV = HHV − 1050 (9.008b)/(12.011a + 1.008b + 16.000c).

[c]HVs of isomers of C_4 and higher differ by a fraction of a percent. Enthalpies of vaporization, h_{fv}, given in Table 11.6 to convert fuels from liquid to gas, and vica versa. For example, for butane, C_4H_{10}(l), HHV = 21316 − 9060/58.124 = 21,160 Btu/lb.

[d]At 77°F (25°C), 1 atm.

precise determinations have been made by theoretical thermodynamics. They apply under very precisely defined conditions which are given in the following paragraphs.

Standard State. The standard state, usually denoted by the superscript zero, is taken as 1 atm pressure. The temperature is always indicated. Most authoritative references[5] use 298.15°K (77°F, 25°C) as the standard temperature reference. This has been used in Table 11.4, while recognizing that the U.S. natural gas industry bases all volume measurements on 60°F and the STP (standard temperature and pressure) means 0°C (32°F) and 1 atm for many physical constants, and that enthalpies in the steam tables[6] are based on the triple point of water,

and so on. As will be seen, it is best to adhere to the 77°F standard for enthalpies of formation, of combustion, and of reaction, to be consistent with the principle sources of data for pure chemical species.

The standard state of an element or compound is its normal state at 77°F (25°C) and 1 atm. Under these conditions, H_2O is a liquid; CH_4, methane, is a gas; carbon, C, is a solid; H_2 and O_2 are gases, and so on.

Enthalpy of Formation. The enthalpy of formation of all elements in their standard states is taken as zero. For example, for O_2(g), $\Delta H^\circ_{f\ 298.15} = 0$. The need for precise definition of the standard state is evident. The enthalpy of formation of each compound from its elements can then be determined by

Table 11.5.   Values of the Enthalpy of Formation ΔH_f^0 at 77°F (25°C) and 1 Atm

Substance	Formula	Mol. Wt.	ΔH_f^0 kcal/g mole	ΔH_f^0 Btu/lb mole
Carbon	C(s)	12.01	0	0
Hydrogen	$H_2(g)$	2.018	0	0
Oxygen	$O_2(g)$	32.00	0	0
Carbon monoxide	CO(g)	28.01	− 26.417	− 47,540
Carbon dioxide	$CO_2(g)$	44.01	− 94.154	− 169,290
Water	$H_2O(g)$	18.02	− 57.798	− 104,040
Water	$H_2O(l)$	18.02	− 68.317	− 122,970
Hydrogen peroxide	$H_2O_2(g)$	34.02	− 32.53	− 58,550
Ammonia	$NH_3(g)$	17.04	− 11.04	− 19,870
Methane	$CH_4(g)$	16.04	− 17.895	− 32,210
Acetylene	$C_2H_2(g)$	26.04	+ 54.19	+ 97,540
Ethylene	$C_2H_4(g)$	28.05	+ 12.496	+ 22,490
Ethane	$C_2H_6(g)$	30.07	− 20.236	− 36,420
Propylene	$C_3H_6(g)$	42.08	+ 4.879	+ 8,790
Propane	$C_3H_8(g)$	44.09	− 24.82	− 44,680
n-Butane	$C_4H_{10}(g)$	58.12	− 30.15	− 54,270
n-Pentane	$C_5H_{12}(g)$	72.15	− 35.00	− 63,000
n-Hexane	$C_6H_{14}(g)$	86.17	− 39.96	− 71,930
n-Heptane	$C_7H_{16}(g)$	100.20	− 44.89	− 80,800
n-Octane	$C_8H_{18}(g)$	114.22	− 49.82	− 89,680
Benzene	$C_6H_6(g)$	78.11	+ 19.82	+ 35,680
Toluene	$C_7H_8(g)$	92.13	+ 11.95	+ 21,510
Methyl alcohol	$CH_3OH(g)$	32.05	− 48.08	− 86,540
Ethyl alcohol	$C_2H_5OH(g)$	46.07	− 56.24	− 101,230

Source: Primarily from Refs. 5, 7, and 8.

thermodynamic methods. These have been carefully scrutinized by competent scientists and are set forth in tables such as the JANAF Thermochemical Tables[5] and other selected[7] or specialized[8] tables. For example,[5] $\Delta H_{f\,298.15}^0$ of $CO_2(g)$ is −94.054 kcal/g mole. This is the enthalpy of formation of carbon dioxide gas from solid carbon and gaseous oxygen at 77°F and 1 atm.

The enthalpy of formation may be tabulated for two phases of a compound. For example, for $H_2O(g)$, it is −104,040 Btu/lb mole; and for $H_2O(l)$, it is −122,970 Btu/lb mole. These obviously must differ by the enthalpy of vaporization of water at 77°F and 1 atm, which is 1050 Btu/lb, or 1050 × 18.02 = 18,920 Btu/lb mole, as indeed they do.

The *total enthalpy* of an element or compound is its enthalpy of formation plus its sensible enthalpy above the standard state.

Enthalpy of Reaction. Once the enthalpies of formation of all the compounds of interest have

been determined, the enthalpy change in any reaction involving these compounds can be determined. From the First Law of Thermodynamics, it is simply the enthalpy of formation of the products minus the enthalpy of formation of the reactants, that is, the gain in enthalpy in the reaction. Thus, in Eq. 11.1 the enthalpy increase in the reaction is −94.054 − 0 − 0 = − 94.054 kcal/g mole. The energy going into the system for this reaction is −94.054 kcal/g mole. That is, 94.054 kcal/g mole is given off by the reaction.

Enthalpy of Combustion. Combustion is simply a particular type of reaction. The enthalpy of combustion of carbon in oxygen or air is −94.054 kcal/g mole. The positive direction is heat flow into the "system," the usual thermodynamic convention.

Tables 11.5 and 11.6 give the enthalpies of formation and combustion for a number of pure chemical species. The enthalpies of vaporization are also in-

Table 11.6.   Values of the Enthalpy of Combustion, ΔH_c^0, and the Enthalpy of Vaporization, h_{fg}, 77°F (25°C) and 1 Atm[a]

Substance	Formula	ΔH_c^0 kcal/g mole	ΔH_c^0 Btu/lb mole	h_{fg} (Btu/lb mole)
Hydrogen	$H_2(g)$	$-$ 68.317	$-$ 122,970	
Carbon	$C(s)$	$-$ 94.054	$-$ 169,290	
Carbon monoxide	$CO(g)$	$-$ 67.636	$-$ 121,750	
Methane	$CH_4(g)$	$-$ 212.80	$-$ 383,040	
Acetylene	$C_2H_2(g)$	$-$ 310.62	$-$ 559,120	
Ethylene	$C_2H_4(g)$	$-$ 337.23	$-$ 607,010	
Ethane	$C_2H_6(g)$	$-$ 372.82	$-$ 671,080	
Propylene	$C_3H_6(g)$	$-$ 491.99	$-$ 885,580	
Propane	$C_3H_8(g)$	$-$ 530.60	$-$ 955,070	6,480
n-Butane	$C_4H_{10}(g)$	$-$ 687.65	$-$ 1,237,800	9,060
n-Pentane	$C_5H_{12}(g)$	$-$ 845.16	$-$ 1,521,300	11,360
n-Hexane	$C_6H_{14}(g)$	$-$ 1002.57	$-$ 1,804,600	13,563
n-Heptane	$C_7H_{16}(g)$	$-$ 1160.01	$-$ 2,088,000	15,713
n-Octane	$C_8H_{18}(g)$	$-$ 1317.45	$-$ 2,371,400	17,835
Benzene	$C_6H_6(g)$	$-$ 789.08	$-$ 1,420,300	14,552
Toluene	$C_7H_8(g)$	$-$ 943.58	$-$ 1,698,400	17,176
Methyl alcohol	$CH_3OH(g)$	$-$ 182.61	$-$ 328,700	16,092
Ethyl alcohol	$C_2H_5OH(g)$	$-$ 336.82	$-$ 606,280	18,216

Source: Primarily from Refs. 7 and 8.

[a]Water appears as a liquid in the products of combustion.

cluded for species that are liquid in the standard state or, in the case of propane and butane, liquid at 77°F (25°C) and a few atmospheres pressure.

If liquid propane is burned, the heat of vaporization must be supplied and reduces the heat given off. However, if it is supplied from an LPG tank, heat from the atmosphere normally keeps the space above the liquid filled with vapor, and it is this *gas* that is burned.

Higher Heating Value.   The atomic weight of carbon is 12.011. Thus, noting that Btu/lb = kcal/g $\times$ 1800, and changing to the sign convention for heating values (positive out), the HHV of carbon is 94.054 $\times$ 1800/12.011 = 14,096 Btu/lb, as given in Table 11.4. Other HHVs in the table were determined similarly from the enthalpies of combustion and thus are precise values.

Obviously, the heating values of mixed fuels, such as coal, oil, and natural gas, with varying composition, must be determined by calorimeter, and only average values, or values for specific samples, can be given (see the respective chapters or appendix 1 for these fuels).

Lower Heating Value.   The lower heating values of Table 11.4 are less than the higher heating values by the enthalpy of vaporization at 77°F (25°C) and 1 atm for each pound of water formed. Thus, in Eq. 11.2, 9 lb water is formed for each pound hydrogen burned (more precisely, 18.016/2.016 = 8.9365 lb). Since $h_{fg77°F}$ (water) is 1050 Btu/lb, the LHV of hydrogen is 61,031 $-$ 8.9365 $\times$ 1050 = 51,648 Btu/lb.

Heating Values per Unit Volume.   At STP of 0°C (32°F) and 1 atm, 1 lb mole perfect gas occupies 359.03 ft³; at 77°F and 1 atm, it occupies 391.89 ft³. Thus, for hydrogen, molecular weight 2.016, 1 lb occupies 391.89/2.016 = 194.39 ft³. The HHV of hydrogen is therefore 61,031/194.39 = 314.0 Btu/ft³ at 77°F and 1 atm. The other heating values per unit volume in Table 11.4 were determined similarly.

Adiabatic Combustion (Flame) Temperature

The temperature reached in a combustion is of interest for a number of reasons, such as the thermal drop available for heat transfer, the permissible inlet temperature of gas turbines, and the formation of nitrous oxides. (Little is formed below 1800°F.)

The theoretical temperature that would be reached if all of the heat of combustion went into heating the combustion products is called the "adiabatic combustion temperature." This theoretical temperature is never reached[9] because of:

1. Heat losses in the system.
2. Incomplete combustion of the fuel.
3. Dissociation of the products, which absorbs some of the energy.

Nevertheless, the adiabatic flame temperature is a good guide, an upper limit, and the reduction can be estimated or calculated if needed.

The adiabatic flame temperature can be calculated from the temperatures of the reactants and the enthalpies of formation of the species involved, as shown in the following example. The enthalpies of formation are generally tabulated at 77°F (25°C) and 1 atm as shown in Table 11.5. Thus, the sensible enthalpy of the reactants above this condition must be added to determine the total enthalpy available to raise the products of combustion from 77°F (25°C) to the flame temperature. At the flame temperature, the water would be vapor. Thus, the enthalpy of formation of water vapor is used.

Example 11.1. Liquid butane, C_4H_{10}, at 77°F (25°C) is burned with 400% of theoretical air at 600°F in a steady flow process, at atmospheric pressure. Determine the adiabatic flame temperature.

Solution. For 100% air the reaction equation is

$$C_4H_{10}(l) + 6.5\ O_2(g) + 24.4\ N_2(g) \rightarrow 4\ CO_2(g) + 5\ H_2O(g) + 24.4\ N_2(g) \qquad (11.8)$$

Air is considered as 21% O_2, 79% N_2, ratio 3.76, 29 lb/lb mole. Thus, for 400% air, the equation becomes

$$C_4H_{10}(l) + 26\ O_2(g) + 97.6\ N_2(g) \rightarrow 4\ CO_2(g) + 5\ H_2O(g) + 19.50\ O_2(g) + 97.6\ N_2(g) \qquad (11.9)$$

The energy *input* for this reaction is the enthalpy of formation of the products minus the enthalpy of formation of the reactants. Thus, the energy input per lb mole fuel is

$$4(-169,290) + 5(-104,040) + 19.5(0) + 97.6(0) - 1(-63,330) - 26(0) - 97.6(0) = -1,134,000\ Btu$$

Note: $-63,330 = -54,270 - 9060$. The air was supplied at 600°F, a total of $26 + 97.6 = 123.6$ lb moles, or $29 \times 123.6 = 3584.4$ lb air per lb mole fuel. For air,[10] $h_{600°F} - h_{77°F}$ (at 1 atm) $= 255.96 - 128.10 = 127.86$ Btu/lb. Since the fuel is supplied at 77°F, the total heat to bring the reactants to 77°F is that for the air, namely, $-127.86 \times 3584.4 = -458,000$ Btu/lb mole fuel.

Thus, the total enthalpy input is

$$-1,134,000 - 458,000 = -1,592,000\ Btu/lb\ mole\ fuel$$

That is, 1,592,000 Btu/lb mole fuel is available to raise the products of combustion from 77°F to the flame temperature.

At 77°F, the enthalpies of the products in Btu per lb mole fuel are[11,12]:

4 CO_2(g)	4×4027.5	= 16,100
5 H_2O(g)	5×4258	= 21,300
19.5 O_2(g)	19.5×3725.1	= 72,600
97.6 N_2(g)	97.6×3729.5	= 364,000
	Total	474,000

At the flame temperature, they will have a combined enthalpy of

$$474,000 + 1,592,000 = 2,066,000\ Btu/lb\ mole\ of\ fuel$$

Altogether, there are $4 + 5 + 19.5 + 97.6 = 126.1$ lb moles of products. If they were all nitrogen,[2] the enthalpy per mole would be

$$2,066,000/126.1 = 16,384\ Btu/lb\ mole$$

and the flame temperature would be about 2230°R.[11] Since the polyatomic gases, CO_2 and H_2O, have higher specific heats than N_2 and O_2, the temperature will be a little less. Thus, by trial, the following enthalpies per lb mole fuel are obtained[11]:

	2100°R	2200°R	Adiabatic Temp.
4 lb moles CO_2(g)	89,500	94,700	
5 lb moles H_2O(g)	92,400	97,500	
19.5 lb moles O_2(g)	312,000	328,500	
97.6 lb moles N_2(g)	1,497,000	1,576,000	
Total	1,990,900	2,096,700	2,066,000

The temperature 2200°R (1740°F, 950°C) is the adiabatic flame temperature as close as warranted in view of the three reductions mentioned earlier. At this temperature, there would be very little dissociation, and the approximation of no dissociation is warranted.

Typical Combustion Temperatures. In a similar way, the adiabatic flame temperature can be determined for various fuels, with oxygen or air as the oxidant and with various amounts of excess air. Typical values may be found in handbooks.[13]

Methane burned with 100% air produces a flame temperature, corrected for dissociation, of 4010°R (3550°F, 1954°C); and with 120% air, of 3200°F (1760°C). Methane is the principal constituent of natural gas and has a HHV of 1010 Btu/ft³ at 60°F, close to that of natural gas (U.S. average 1020 Btu/ft³). Similar flame temperatures are achieved with pulverized coal or oil burned with air.

Gas turbine inlet temperatures are limited to 2000 to 2200°F (1976), and inlet temperatures for steam turbines are seldom over 1000°F, the approximate metallurgical limit. Thus, the use of air is not limiting in these processes. However, for MHD, to get combustion temperatures of 4400°F (see Chap. 13), the air must be preheated. Where still higher temperatures are needed, as in oxy-acetylene cutting and welding, oxygen must be used to avoid dilution with nitrogen.

In internal combustion engines, gas temperature above the metallurgical limit can be used since the cylinder walls are water (or air) cooled and the piston is cooled by oil from below.

REFERENCES

1. R. T. Morrison and R. N. Boyd, *Organic Chemistry*, 3rd ed., Boston: Allyn and Bacon, 1975.

2. K. Wark, *Thermodynamics*, New York: McGraw-Hill, 1966.

3. "Bomb Calorimeter," *ANSI/ASTM Standard*, D2015-'66/72, Philadelphia, Pa: American Socy. for Testing and Materials, 1980.

4. F. C. Schora, "IGT Gasification Process," *Proc. 1st AGA Symp. on Pipeline Gas,* Arlington, Va.: American Gas Assn., 1966.

5. *JANAF Thermochemical Tables*, 2nd ed., U.S. Bureau of Standards, Washington, D.C.: Supt. of Documents, 1970.

6. J. H. Keenan and F. G. Keyes, *The Thermodynamic Properties of Steam*, New York: Wiley, 1936.

7. *Selected Values of Thermodynamic Properties*, U.S. National Bureau of Standards Circ. 500, Washington, D.C.: Supt. of Documents, 1956.

8. *Selected Values of Physical and Thermodynamic Properties of Hydrocarbons and Related Compounds*, API Project 44, Pittsburgh, Pa.: Carnegie Press, 1933.

9. Reference 2, p. 434.

10. Reference 2, Table A15.

11. Reference 2, Tables A-16 to A-21.

12. J. H. Keenan, J. Chau, and J. Kaye, *Gas Tables*, 2nd ed., New York: Wiley, 1979.

13. T. Baumeister, Ed., *Marks Standard Handbook for Mechanical Engineers*, 7th ed., New York: McGraw Hill, 1967, p. 4–76, Table 48.

12

Nuclear Physics of Energy

INTRODUCTION

The purpose of this chapter is to provide an introduction to the physics concepts involved in nuclear energy, or a brief review if one has already been introduced. It will cover for example the designation of nucleii, the equations of typical nuclear reactions, and the nature of the elementary particles entering into them. The well-known mass-energy relation will be illustrated by a few examples. The fission chain reaction, fast and slow neutrons, the moderator, the conversion from fertile to fissile material, fission products, and delayed neutrons, will be explained and defined.

The functioning of these elements in typical reactors, or the neutron balance, provides a rough quantitative concept of their relative importance in breeder and nonbreeder reactors. The basic parameters of a nuclear reactor are defined and the concepts of breeding ratio, doubling time, and fuel burnup are outlined.

Typical fusion reactions are also given. However, the fusion development, in spite of its enormous import if successful, was judged too long range and speculative to include in this book, a decision which if made today (1982) might be reversed.

NUCLEAR STRUCTURE AND DEFINITIONS

Element Number—Protons and Electrons. Each element of the periodic table has one more proton than the preceding one. The 92 naturally occurring elements thus vary from hydrogen, the first and lightest, with one proton, to uranium, the heaviest, with 92 protons. Still heavier elements can be artificially produced, starting with neptunium with 93 protons, plutonium with 94, and several others. All elements with more than 92 protons are radioactive and decay, even though some have quite long half-lives and hence are not

found in nature. The number of protons is the element number.

Some elements with relatively short half-lives are found in nature. For example, radium 226 is found even though its half-life is only 1600 yr. It is continuously being produced by the decay of uranium 238 (through intermediates) whose half-life of 4.56×10^9 yr is about the age of the earth.

In the neutral state, the number of negatively charged planetary electrons in orbitals about the nucleus is exactly equal to the number of positively charged protons. Thus, the element number also indicates the number of electrons. Details of the electron arrangement are given in Chapter 11.

The unit charge of the proton is 1.6×10^{-19} coulombs, and that of the electron is -1.6×10^{19} coulombs.

The rest mass of the proton is 1.00727 amu (atomic mass units), based on the standard, since 1961, in which carbon 12 is taken as 12.00000 amu. Each amu is thus 1.66057×10^{-27} kg, and when converted to energy according to the Einstein relation, it is 931.49 MeV (millions of electron volts), or approximately 931 MeV. (1 eV is 1.6×10^{-19} joules, or watt-sec).

The rest mass of the electron is 0.0005486 amu. The mass of the proton is about 1836 times that of the electron.

Neutrons. In addition to protons, the nucleus contains neutrons, which are uncharged particles of about the same mass as protons. The neutron rest mass is 1.008665 amu. The number of neutrons contained in an element is not fixed. The hydrogen nucleus may contain zero, one, or two neutrons, in which case it is called hydrogen, deuterium, or tritium, respectively. Hydrogen is also the generic name for all three. These are called isotopes of hydrogen—the same chemical element with different numbers of neutrons in the nucleus. They cannot be distinguished chemically but have different masses in the ratio of about 1:2:3. Deuterium, with one neutron, is also called "heavy hydrogen," and water, H_2O, in which the hydrogen atoms are deuterium, is called "heavy water."

As found in nature, an element contains a rather fixed proportion of its different isotopes. For example, about one hydrogen atom in 6600 is deuterium (0.015%, $_1H^2$, Table 12.1) or heavy hydrogen. The rest are ordinary hydrogen, $_1H^1$. Some 0.72% of all uranium atoms found in nature are U-235; 99.27% are U-238. Charts of the nuclides are available[1,2] showing all isotopes together with their proportions and properties. The properties of a few selected isotopes are given in Table 12.1.

As the element number increases, the average number of neutrons increases, the most common isotope of carbon, the sixth element, having six neutrons, silver, the 47th element, having 60 neutrons, and uranium, the 92nd element, having 146 neutrons. Above helium, the element tends to have 1 to $1\frac{1}{2}$ times as many neutrons as protons, the relative number increasing with the element number.

Mass Number. The total number of protons and neutrons is called the mass number of the isotope. It is about equal to the atomic mass, since protons and neutrons both have a mass of about 1 amu, and the electrons are negligible.

Designation of Isotope. In designating an isotope, the element number is usually given as a subscript preceding the symbol, and the mass number as a superscript following it. Thus, $_{92}U^{235}$ is the isotope of uranium, the ninety-second element, having a total number of protons and neutrons of 235. It has 92 protons and 92 planetary electrons. It has $235 - 92 = 143$ neutrons.

Since uranium *is* the ninety-second element, the subscript 92 conveys no additional information, and is usually dropped, leading to the common designations U^{235} or U-235. The designation ^{235}U is used in some later references,[11] but the designation given here is used in the 1981/1982 table of nuclides[2] and is used throughout this book.

Note: In extending this designation system to particles, the subscript indicates the charge. For an atom, the number of protons, the element number, and the nuclear charge are all the same. Thus, the negative electron, $_{-1}e^0$, has a charge of -1, that is, -1.6×10^{-19} coulombs.

Fission and Fusion. When heavy elements such as U-235 or Pu-239 are broken apart (fissioned) into two nearly equal-sized pieces, the combined mass of these fission fragments is slightly less (about 0.1%) than the mass of the heavy element. The difference in mass appears as equivalent energy according to the Einstein relation $e = mc^2$ and is called fission energy. It is 9×10^{10} joules or watt-sec per kilogram, or 931 MeV per amu.

Similarly, when several nucleii of a light element such as hydrogen are combined (fused) into a heavier nucleus such as helium, the mass of the combined or fused nucleus is slightly less than the sum of the masses of the lighter nucleii from which it was formed (about 1%). This loss of mass appears as fusion energy. Fission energy and fusion energy together are called nuclear energy.

Fissile and Fertile Material, Chain Reaction. The isotope U-235 can be made to fission by the entry of a neutron into its nucleus. In fissioning it gives off two to three neutrons. If proper arrangements are made so that at least one of these neutrons enters another U 235 nucleus and causes it to fission, a sustaining chain reaction is set up.

An isotope that can fission and that gives off neutrons to make a chain reaction possible is called a fissionable or "fissile" material. The isotope U-235 is the only fissile material found in nature. It is the key that unlocks the door to nuclear energy.

Two other materials, U-238 and Th-232, can be converted or "bred" into the fissile materials Pu-239 and U-233, respectively, by bombarding them with neutrons. Thus, U-238 and Th-232 are called "fertile" materials. They are not fissile materials in themselves but can be used to produce fissile materials.

To summarize, there are five isotopes of principal importance in producing nuclear energy by fission. U-235 is the only naturally occurring fissile material. U-238 and Th-232 are fertile materials from which fissile material can be bred. Pu-239 and U-233 are artificially produced fissile materials. As shown in Table 12.4 later, Pu-241 is also fissile, but the five isotopes mentioned above are the most important.

While U-238 and Th-232 are not fissile materials by the foregoing definition, since their probability of fission is too low, they do fission, as do other heavy isotopes, into smaller fragments. A substantial part of the power in reactors using uranium is produced by fissioning of U-238, as is explained later.

Table 12.1. **Properties of Selected Isotopes**[a]

Isotope	% Natural Abundance	Atomic Mass	Half-Life	Modes of Decay[b]	Decay Energy (MeV)	Thermal Neutron Capture Cross Section and Fission Cross Section[c]
$_0n^1$		1.008665	12 min	β^-	0.7825	
$_1H^1$	99.985	1.007825				
$_1H^2$	0.015	2.0140				0.51 mb
$_1H^3$		3.01685	12.26 yr	β^-	0.01861	6 b
$_2He^4$	100	4.00260				0
$_4Be^9$	100	9.01218				9.2 mb
$_5B^{10}$	19.78	10.0129				0.5 b, $\sigma(n,\alpha)$ 3836 b
$_5B^{11}$	80	11.00931				5 mb
$_6C^{12}$	98.89	12.00000				3.4 mb
$_6C^{13}$	1.11	13.00335				0.9 mb
$_7N^{14}$	99.64	14.00307				75 mb
$_7N^{15}$	0.36	15.00011				24 μb
$_{11}Na^{23}$	100	22.9898				400 mb
$_{11}Na^{24}$			15 hr	β^-	5.51	
$_{12}Mg^{24}$	78.99	23.98504				52 mb
$_{26}Fe^{56}$	91.8	55.9349				2.5 b
$_{26}Fe^{57}$	2.1	56.9354				2.5 b
$_{36}Kr^{97}$		96.9197				
$_{47}Ag^{107}$	51.83	106.90509				35 b
$_{53}I^{137}$			23 sec	β^-, n		
$_{54}Xe^{136}$	8.9	135.9072				280 mb
$_{54}Xe^{137}$			3.9 min	β^-	4.0	
$_{56}Ba^{136}$	7.9	135.9044				10 mb, 0.4 b
$_{56}Ba^{137}$	11.2	136.9056				5.1 b
$_{86}Rn^{222}$		222.0175	3.823 days	α	5.587	0.72 b
$_{88}Ra^{226}$		226.0254	1600 yr	α		20 b
$_{90}Th^{230}$		230.0331	8×10^4 yr	α	4.767	33 b
$_{90}Th^{232}$		232.0382	1.41×10^{10} yr	α	4.08	7.4 b
$_{90}Th^{233}$			22.2 min	β^-	1.246	1500 b, f15 b
$_{90}Th^{234}$		234.0436	24.1 days	β^-	0.263	1.8 b
$_{91}Pa^{233}$		233.040	27.0 days	β^-	0.571	22 b, 21 b
$_{92}U^{232}$		232.0372	73.6 yr	α, SF	5.414	75 b, f 75 b, σ(abs.) 150 b
$_{92}U^{233}$		233.0395	1.62×10^5 yr	α	4.909	
$_{92}U^{234}$	0.005	234.0409	2.47×10^5 yr	α	4.856	95 b, f 0.65 b, σ (abs.) 95 b
$_{92}U^{235}$	0.720	235.0439	7.1×10^8 yr	α, SF	4.681	100.5 b, f579.5 b, σ (abs.) 679.9 b
$_{92}U^{238}$	99.275	238.0508	4.51×10^9 yr	α, SF	4.268	2.72 b
$_{92}U^{239}$			23.5 min	β^-	1.28	22 b f14 b
$_{92}U^{240}$			14.1 hr	β^-	0.5	
$_{93}Np^{239}$			2.35 days	β^-	0.723	35 b, 25 b, f < 1 b
$_{94}Pu^{239}$		239.0522	24,400 yr	α, SF	5.243	266 b, f 742 b, σ (abs.) 1008 b
$_{94}Pu^{240}$		240.0540	6580 yr	α, SF	5.255	290 b

[a]Complete Table in Ref. 2, p. B-236.
[b]SF denotes spontaneous fission.
[c]In barns (b), millibarns (mb), or microbarns (μb) (1 barn = 10^{-24} cm^2); f = fission.

The capture of a neutron by a fertile material such as U-238 results in breeding a fissile nucleus. Thus, the cross section for capture by a fertile material is the cross section for breeding.

Fusion. To bring about the fusion of hydrogen into helium requires extremely high temperatures—hundreds of millions of degrees. To date, such temperatures have only been produced artificially by an atomic bomb (except for minute quantities in a test device). The hydrogen bomb, or thermonuclear bomb, requires the fission bomb (atomic bomb) to set it off.

Since material at such temperatures cannot be long contained by any man-made device to date (1977), there has been no controlled fusion with net power output, only the uncontrolled bomb. Power from controlled fusion awaits a solution to this problem.

Basic Elements of Nuclear Physics. Small particles enter into all nuclear reactions. They are the projectiles that enter the nucleus, causing fission, and that are given off in fission, fusion, and radioactive decay processes. In all known nuclear reactions—fission, fusion, and radioactivity—the total number of neutrons and protons is the same before and after the reaction. The energy absorbed or given off is associated with the arrangement of these particles. The mass change is of the order of 0.1% in fission and 1.0% in fusion. The particles generally entering into the reactions are as follows:

Table 12.2. Basic Elements of Nuclear Physics

Name	Symbols	Mass (Approx.)	Charge
Electron	$_{-1}e^0$, $-e$, β^-	0^a	-1
Neutron	$_0n^1$, n	1.0	0
Proton (also hydrogen nucleus)	$_1H^1$, p	1.0	$+1$
Deuteron (heavy hydrogen nucleus) = n + p	$_1H^2$, d	2.0	$+1$
Positron	$_{+1}e^0$, $+e$, β^+	0	$+1$
Alpha particle (helium nucleus)	$_2He^4$, α	4	$+2$
Gamma ray	γ	0	0
Meson		$0–0.3^b$	$+1$ or 0

[a]Small compared to neutron; actually, 0.0005486, 1/1839 that of the neutron.

[b]In different forms appears to have different mass; pi meson approximately 300 times that of the electron; mu meson about 200 times that of the electron.

specifically in the conventional equations of nuclear reactions. Other stable particles are the neutrino and antineutrino, having zero mass and zero charge, which annihilate each other upon combining, and antiprotons of the same mass as protons but opposite charge. The remaining unstable particles decay

	Symbol
Proton, or hydrogen nucleus, defined above	$_1H^1$ or p
Neutron, defined above	$_0n^1$ or n
Negative electron, or β particle, defined above	$_{-1}e^0$ or $-e$ or β^-
Positive electron, or positron, having the same mass as the electron but positive charge. This is a transitory particle of very short life that quickly combines with a negative electron, annihilating mass and charge, and the corresponding energy appearing as γ radiation	$_{+1}e^0$ or $+e$ or β^+
Gamma ray, or photon, having zero mass or charge, and having energy e = $h\nu$, where h is Planck's constant and ν is the frequency of the radiation.	γ
Alpha particle, or helium nucleus	$_2He^4$ or α
Deuteron, or heavy hydrogen nucleus	$_1H^2$ or d

The properties of these nuclear particles are summarized in Table 12.2. These are but a few of the known nuclear particles, namely those that enter

into the seven stable ones in times ranging from a few millionths to less than a quadrillionth of a second. Hence they do not appear in the usual nuclear equations. Even the neutron, when not bound in a atom, has a half-life of 12 minutes, decaying into a proton, an electron, and a neutrino.

TYPICAL NUCLEAR REACTIONS

Typical nuclear reactions are given below. The explanations follow.

Bombardment of the Element Beryllium with Deuterons

$$_4Be^9 + {}_1H^2 \rightarrow {}_5B^{10} + {}_0n^1 + 4.35 \text{ MeV} \quad (12.1)$$
$$\text{(boron)}$$

Typical Fission Reactions: U-235, U-233, Pu-239

$$_{92}U^{235} + {}_0n^1 \rightarrow {}_{56}Ba^{137} + {}_{36}Kr^{97}$$
$$+ 2\ {}_0n^1 + 200 \text{ MeV} \quad (12.2)$$

$$_{92}U^{233} + {}_0n^1 \rightarrow {}_{56}Ba^{136} + {}_{36}Kr^{96}$$
$$\text{(barium)} \quad \text{(krypton)}$$
$$+ 2\ {}_0n^1 + 200 \text{ MeV} \quad (12.3)$$

$$_{94}Pu^{239} + {}_0n^1 \rightarrow {}_{56}Ba^{137} + {}_{38}Sr^{100}$$
$$\text{(strontium)}$$
$$+ 3\ {}_0n^1 + 210 \text{ MeV} \quad (12.4)$$

Typical Breeding Reactions: U-238 → Pu-239-, Th-232 → U-233

$$_{92}U^{238} + {}_0n^1 \rightarrow {}_{92}U^{239} + \gamma \xrightarrow{24 \text{ min}} - e$$
$$+ {}_{93}Np^{239} \xrightarrow{2.3 \text{ days}} - e + {}_{94}Pu^{239} \quad (12.5)$$
$$\text{(neptunium)}$$

$$_{90}Th^{232} + {}_0n^1 \rightarrow {}_{90}Th^{233} + \gamma \xrightarrow{22 \text{ min}} - e$$
$$+ {}_{91}Pa^{233} \xrightarrow{27 \text{ days}} - e + {}_{92}U^{233} \quad (12.6)$$
$$\text{(protactinium)}$$

Typical Delayed Neutrons (The Fission Fragment, Iodine, Decays to Xenon)

$$_{53}I^{137} \xrightarrow{23 \text{ sec}} -e + {}_{54}Xe^{137} \xrightarrow{0 \text{ sec}} {}_{54}Xe^{136} + {}_0n^1$$
$$(12.7)$$

Parasitic Reactions (Which Capture Neutrons Unproductively)

$$_{94}Pu^{239} + {}_0n^1 \rightarrow {}_{94}Pu^{240} \quad (12.8)$$

$$_{11}Na^{23} + {}_0n^1 \rightarrow {}_{11}Na^{24} \xrightarrow{14.8 \text{ hr}} - e + {}_{12}Mg^{24} \quad (12.9)$$
$$\text{(sodium)} \qquad\qquad\qquad \text{(magnesium)}$$

$$_{26}Fe^{56} + {}_0n^1 \rightarrow {}_{26}Fe^{57} \quad (12.10)$$
$$\text{(iron)}$$

Bombarding with Neutrons. Equation 12.1 describes the reaction that takes place when beryllium is bombarded with high-speed deuterons from a particle accelerator, producing boron, neutrons, and energy.

Fission. Equations 12.2, 12.3, and 12.4 represent the most typical fission reactions, using U-235, U-233, and Pu-239, respectively, in a flux of neutrons. Many other fission reactions are also possible, giving fission products ranging from zinc to gadolinium, covering a span of 35 elements and 200 isotopes.[3] These elements, numbers 30 to 64, are all near the middle of the periodic table (see Table 11.1).

Breeding. The breeding reactions for producing the fissile material Pu-239 from the fertile U-238 or the fissile U-233 from the fertile Th-232 are shown in Eqs. 12.5 and 12.6. The electrons, -e, produced in each step are assumed to fly off.

Delayed Neutrons. The ability to control a fission nuclear reactor depends to a considerable extent on delayed neutrons. If all neutrons were prompt, the slightest withdrawal of the control rods beyond criticality would result in an extremely rapid buildup of power. However, with a small part of the neutron production delayed, there is a reasonable time to move the control rods. There are other stabilizing factors, such as negative temperature coefficient, to be described later.

A typical source of delayed neutrons is radioactive iodine, produced as a fission product, which decays to xenon with a half-life of 23 sec and then emits a neutron, as shown in Eq. 12.7.

Parasitic Reactions. Typical parasitic reactions which capture neutrons unproductively are shown in Eqs. 12.8, 12.9, and 12.10. Pu-239 may capture a neutron without fissioning and become

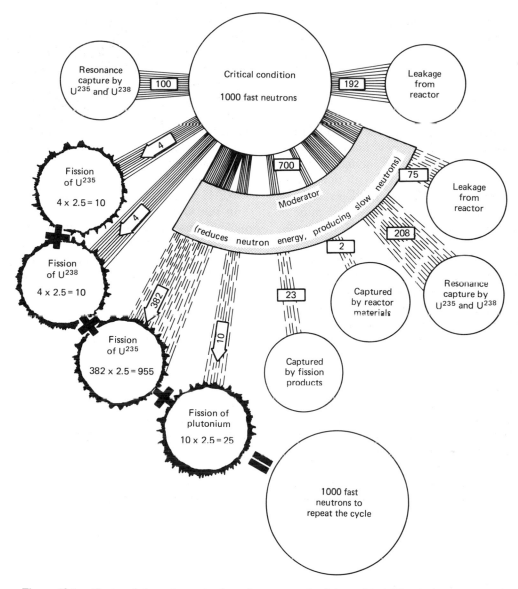

Figure 12.1. Neutron balance in a natural uranium reactor that is just critical. The sketch shows what might happen to 1000 representative neutrons out of the many million present within a reactor. Courtesy W. E. Shoupp.[3]

Pu-240. While Pu-240 is fissionable, this reaction is undesirable because of the loss of neutrons. This is shown in Eq. 12.8. Or the coolant, such as sodium, may absorb a neutron as in Eq. 12.9 and then decay to magnesium. Or light water, with $_1H^1$ atoms, may capture neutrons, becoming heavy water with $_1H^2$ atoms.

The structural materials of the reactor may also capture neutrons. Equation 12.10 illustrates the capture of a neutron by Fe-56 to become Fe-57.

Both fast and slow neutrons may be captured

parasitically, as shown in Fig. 12.1. This includes resonance capture.

In addition to the reactions of Eqs. 12.8 to 12.10, the fission products absorb neutrons, and this steadily increasing loss as the reactor operates is a principal factor in determining the refueling cycle of a nuclear reactor.

Controlled Fusion.[4,12] With the high temperatures and enormous gravitational forces of the sun, hydrogen is fused into helium, probably by the following four-step process[3]:

$$_6C^{12} \longrightarrow {}_6C^{13} \longrightarrow {}_7N^{14} \longrightarrow {}_7N^{15} \longrightarrow {}_6C^{12} + {}_2He^4$$
$$\uparrow \qquad \uparrow \qquad \uparrow \qquad \uparrow$$
$$_1H^1 \qquad {}_1H^1 \qquad {}_1H^1 \qquad {}_1H^1$$

A hydrogen nucleus is fed in at each step. Carbon acts somewhat as a catalyst but is not consumed.

On earth, the problem is to maintain a high enough temperature for a long enough time with a sufficient concentration of fusible particles, to break even or better on energy. Of course, this must be done competitively with the cost of alternative energy sources.

No man-made materials can contain the high temperatures needed. Magnetic and inertial confinements are the options. In the thermonuclear bomb, inertia holds the fusible material together until it can fuse and produce its prodigious energy output. Inertial confinement is also possible with laser, electron, or ion beams, impinging on tiny pellets of fuel, allowing them to explode in chambers and produce fusion energy.

However, the best results so far (1981) have been with magnetic confinement, using the tokamak reactor and numerous other approaches to magnetic confinement.[4] The encouraging success of these experiments in the last few years leads most scientists to believe that controlled-fusion electric power generation can become economic in the next two or three decades.

When controlled fusion becomes a reality, deuterium from the sea is a fusion fuel resource of unbelievable magnitude, "64 billion years at 1979 world energy-consumption rates" (30 billion years by Ref. 4).

However, other fuels are used in current development reactors and may be best ultimately, because they are easier to use—lower temperature requirements and reduced confinement requirements for breakeven on energy (output equal to input). Altogether, there are several factors, shown in Table 12.3, in addition to the fuel resource.

In the first reaction, a deuterium and a tritium nucleus fuse to produce a helium nucleus, a neutron, and 17.6 MeV energy. This reaction is the easiest. The reactor must achieve a temperature of 10 keV, which is equivalent to 110 million °K, or Kelvins. A density–confinement time product ($n\tau$) of about 10^{14} particle sec/cm^3 would be needed for useful output.

In the other reactions shown, deuterium is fused with helium-3 or deuterium, or hydrogen (p for proton) is fused with boron.

A high fraction of output energy in the form of charged particles makes direct conversion to electricity at high efficiency a possibility and reduces radiation damage to the reactor structure.

The confinement requirement is expressed as fusible particles per cm^3 times seconds, and is least for the deuterium–tritium reaction (D–T) and greatest for the deuterium–deuterium (D–D) reaction.

Table 12.3. Principal Fusion Reactions[a]

Reaction	Energy Released (MeV)	Fraction of Energy in Charged Particles	Typical Reaction Temperature (keV)	Density–Confinement Time Product for Breakeven, (sec/cm³)
D + T → He⁴ + n	17.6	0.2	10	1.7×10^{13}
D + He³ → He⁴ + p	18.4	1.0	80	4.5×10^{13}
D + D → He³ + n	3.3	0.66	100	9.0×10^{14}
T + p	4.0			
p + B¹¹ → 3 He⁴	8.7	1.0	250	2.5×10^{14}

Source. Reprinted with permission from "Magnetic Fusion Power," by E. J. Lerner, *IEEE Spectrum*, December 1980.
[a]Deuterium–tritium is the easiest fusion fuel to burn, but it produces mainly neutrons. Hydrogen–boron requires 20 times greater temperatures and densities, but it produces no neutrons. He³ must be bred from the deuterium–deuterium reaction. The symbol keV is the equivalent of 11 million Kelvins.

The required temperature is highest for the p–B reaction, but there is no neutron output. Obviously, a compromise will be needed.

As to fuel resources, deuterium and boron are constituents of seawater and are adequate for 30 and 2 billion yr, respectively, at the current world rate of energy consumption. Tritium would have to be bred from lithium (a constituent of seawater) in a fusion reactor. The cost of extracting any of the fuels from seawater is trivial compared to the value of the energy resulting.

With success practically assured, full support is now being given to the fusion development. The Fusion Act of 1980, voted unanimously by the U.S. Congress, sets as a national goal a demonstration commercial-scale fusion reactor by the year 2000 and the construction of a prototype test facility that would produce some electricity by 1990.

Fission Energy.[3] The energy from nuclear fission may be illustrated by Eq. 12.2 for the fission of U-235:

$$_{92}U^{235} + _{0}n^{1} \rightarrow _{56}Ba^{137} + _{36}Kr^{97} + 2 \, _{0}n^{1} + Q \quad (12.2)$$

The masses in amu from Table 12.1 are

$$235.0439 + 1.008665 \rightarrow 136.9056 + 96.9197$$
$$+ 2(1.008665) + \Delta m,$$

from which $\Delta m = 0.2100$ amu. The loss of mass in the reaction is 0.2100 amu per atom of U-235 fissioned. The energy Q released is, therefore, $Q = 931 \times 0.2100 = 196$ MeV. This energy is imparted to the barium and krypton nucleii, which repel each other. The initial stored energy is

$$E = \frac{q_1 q_2}{r} \text{ erg} \quad (12.11)$$

where the charges are $q_1 = 56$ (barium), $q_2 = 36$ (krypton), and r is the separation in cm. (Multiply charges by 4.8×10^{-10} for cgs ES units.)

An empirical expression for the radius of a nucleus is

$$R = 1.5 \times 10^{-13} A^{1/3} \text{ cm} \quad (12.12)$$

where A is the mass number. Thus,

For Barium

$$R_1 = 1.5 \times 10^{-13} \times 137^{1/3} = 7.7 \times 10^{-13} \text{ cm}$$

For Krypton

$$R_2 = 1.5 \times 10^{-13} \times 97^{1/3} = 6.9 \times 10^{-13} \text{ cm}$$

The initial separation of the two nucleii is the sum of their radii:

$$r - R_1 + R_2 = 14.6 \times 10^{-13} \text{ cm}$$

from which

$$E = \frac{56 \times 36 \times (4.8 \times 10^{-10})^2}{14.6 \times 10^{-13}}$$
$$= 320 \times 10^{-6} \text{ erg} = 200 \text{ MeV}$$
$$(1 \text{ MeV} = 1.6 \times 10^{-6} \text{ erg})$$

This is in good agreement with the energy that must appear due to loss in mass. It demonstrates that this energy first appears as stored electrostatic energy and is quickly converted to kinetic energy of the nucleii as they fly apart. As they strike other nucleii, their velocity is reduced and the energy appears as random motion of the molecules, that is, as heat.

Fission products penetrate about 0.0005 in. into the surrounding reactor materials.[3] This means that the heat energy is generated practically at the point of fission. Also, some 99.99% of the fission products are confined within the cladding of the fuel rods,[5] a most important consideration in disposal of the radioactive waste.

THERMAL AND FAST REACTORS

From the three fission equations, Eqs. 12.2, 12.3, and 12.4, one might suppose that the three fissile materials, U-235, U-233, and Pu-239, all work about alike and are more or less interchangeable, especially since all produce about 200 MeV energy per nucleus fissioned. Nothing could be further from the truth.

Similarly, the breeding reactions, Eqs. 12.5 and 12.6, for producing Pu-239 and U-233 look deceptively similar. Here again, there are vast differences in mechanism. A look at the nuclear parameters for fast and thermal reactors will begin to unfold these differences (see Table 12.4).

Fast and Thermal Neutrons. All neutrons produced in a fission process are "fast" neutrons, having energies of the order of 1,000,000 electron volts (1 MeV) and speeds of some 5000 miles per

Table 12.4. Nuclear Parameters[a] for Fast and Thermal Reactors[b]

| | Fissile Fuel | | | | | | | | Fertile Material | | | |
| | U-235 | | U-233 | | Pu-239 | | Pu-241 | | U-238 | | Th-232 | |
	Fast	Thermal	Fast	Thermal	Fast	Thermal	Fast	Thermal	Fast	Thermal	Fast	Thermal
Fission cross sect. σ_f (barns, 10^{-24} cm²)	1.4	577	2.20	527	1.78	790	2.54	1000	0.112	0	0.025	0
Neutrons per fission (ν)	2.5	2.4	2.59	2.51	3.0	2.90	3.04	2.98	2.60	—	2.4	0
Capture-to-fission ratio $\alpha = \sigma_c/\sigma_f$	0.15	0.17	0.068	0.10	0.15	0.5	0.114	0.4	—	—	—	—
Neutrons per absorption η	2.2	2.06	2.42	2.28	2.6	1.93	2.73	2.13	2.27	—	2.0	—
Delayed neutron fraction β	0.0065	—	0.0027	—	0.002	—	0.0053	—	0.0147	—	0.0204	—
Fission threshold (MeV)	—	—	—	—	—	—	—	—	1.4	—	1.4	—

Other Representative Materials

| | Fe | | Na | | Zr | | Effective Fission Product Pair | |
	Fast	Thermal	Fast	Thermal	Fast	Thermal	Fast	Thermal
σ_c/σ_f (U^{235}) 10^{-3}	6.1	4.4	1.8	0.87	—	0.31	0.08	0.1
σ_c/σ_f (U^{233}) 10^{-3}	4.0	4.8	1.18	0.96	—	0.34	—	0.1
σ_c/σ_f (Pu^{239}) 10^{-3}	4.9	3.4	1.46	0.68	3.3	0.24	—	0.1
σ_c/Σ_f Yankee 10^{-24} PWR	—	15	—	3	—	1	—	236
σ_c/Σ_f fast breeder (Pu) 10^{-24} FBR	4	—	1	—	3	—	120	—

Source. Courtesy R. E. Creagan.[6]

[a]These constants are approximate and change with neutron spectrum.

[b]There are about four times as many fissile atoms per cm³ in the fast breeder as in Yankee thermal reactor. $\Sigma_f = N\sigma_f$, where N is number of fissile atoms per cm³.

second. In order to cause U-235 to fission effectively, these must be slowed down to "thermal" energies of less than 1 eV and speeds of about 1 mile/sec. This is usually accomplished in a water moderator, where the neutrons give up their energy in billiard ball-like collisions with the hydrogen nuclei in the water molecules. The neutrons are then in essentially thermal equilibrium with the surrounding medium and are thus called "thermal" neutrons. The predominant neutron energies used in fission and in breeding characterize the reactor and lead directly to the names "thermal reactor" and "fast reactor."

Fission Cross Section σ_f. The probability that a neutron moving in the vicinity of a nucleus will cause it to fission is expressed as a cross section in "barns," units of 10^{-24} cm². The fission cross section is the probability that a neutron passing through a cubic centimeter, containing one nucleus, will collide with it and cause it to fission. Similarly, there are probability cross sections for each kind of nuclear event of importance, for nonfission capture, for scattering, and so on.

The radius of a nucleus is given in Eq. 12.12. If A is 235, for U-235, then the radius is $R = 9.27 \times 10^{-13}$ cm, and the physical cross section is $\pi R^2 = 2.7 \times 10^{-24}$ cm², or 2.7 barns. A neutron is relatively

small, 1.5×10^{-13} cm radius. Thus, 2.7 barns is about the probability of a physical collision of a neutron with a U-235 nucleus in 1 cm³. The fission cross section of U-235 with fast neutrons is $\sigma_f = 1.4$ barns (Table 12.4). This is less than the physical cross section of 2.7 barns. Simply hitting it is not enough to cause fission. It must hit pretty square.

The corresponding cross section for fission with thermal neutrons is 577 barns. Evidently, if the neutron passes through an area of 577/2.7 = 214 times that of the U-235 nucleus, and centered on it, it will be captured and cause fission. It appears to have a proximity fuse.

The same order of magnitude of fission cross sections for fast and slow neutrons applies to the three fissile materials, U-235, U-233, and Pu-239. For Pu-241, the cross sections are somewhat higher. However, Pu-241 contains two more neutrons than Pu-239. In the process of breeding it, two more neutrons have been absorbed than for Pu-239, so that the overall fissions per neutron are poorer.

The fertile materials have very low or zero cross sections for fission. Nevertheless, even though the fission cross section of U-238 with fast neutrons is only 0.112 barn, compared with 1.78 barns for Pu-239, the fission of U-238 produces about 20% of the total power (20% of total fissions) in a fast breeder reactor. The lower cross section is partially compensated by a much higher exposure to the fast neutrons.

Number of Neutrons Emitted per Fission, ν. For a self-sustaining chain reaction, the neutrons per fission, ν, must be greater than one, so that at least one is available to continue the chain in spite of losses, nonproductive captures, and breeding. For the several fuels shown in Table 12.4, ν is between 2 and 3. The neutrons per fission are very slightly higher for fast neutrons.

Equations 12.2 and 12.3 show two neutrons per fission, Eq. 12.4 shows three neutrons per fission. However, as pointed out, these reactions are simply typical of a large number that are possible, and Table 12.4 gives the experimental value, which is the statistical average of all fission processes occurring. These constants are approximate and change with the neutron spectrum.

Capture-to-Fission Ratio $\alpha = \sigma_c / \sigma_f$. Another important parameter is the capture-to-fission ratio, and here important differences appear among the different fuels that profoundly influence their possible use. For plutonium, with thermal neutrons, half as many will be captured unproductively as will cause fission. When using U-235, with a capture-to-fission ratio of only 0.17, careful design is necessary to obtain a chain reaction and still greater economy of neutrons to increase the breeding ratio. With plutonium, a thermal breeder reactor would be virtually impossible. Nevertheless, plutonium is the most logical breeder fuel, since vast amounts of U-238 are available for breeding. Consequently, the breeder reactor is a "fast breeder reactor." With fast neutrons, the capture-to-fission ratio of Pu-239 is only 0.15.

Since the cross section for fission with fast neutrons is relatively low, a much higher neutron flux density is required than in a thermal reactor. Also, some four or five times more fissile nuclei are required per unit volume in a practical fast reactor than in a thermal reactor.

For the same power density (power per unit volume) in a fast and thermal reactor, the product $\phi \sigma_f N$ must be the same, where ϕ is the neutron flux density, σ_f is the cross section for fission, and N is the number of fissile nuclei per unit volume. Thus, with the fission cross section only 1.78 compared with 577, even with five times as many fissile nuclei per cm³, N, the neutron flux, must be 65 times as great in a fast reactor using plutonium as in a thermal reactor using U-235, in order to have the same power density. Compared with Pu-239, the fuel U-233 bred from Th-232 has a low capture-to-fission ratio with thermal neutrons. Thus, thorium provides the basis for a thermal breeder reactor.[6]

Neutrons per Absorption, η. An atom may capture a neutron productively, causing fission, or unproductively, simply capture it. The neutrons produced per absorption are the neutrons per fission multiplied by the fissions per absorption. Thus, in the first column of Table 12.4, with the capture-to-fission ratio $\alpha = 0.15$, the fissions per absorption are $1/1.15 = 0.87$, and the neutrons per absorption are $0.87 \times 2.5 = 2.2$, as shown in the table.

Delayed Neutron Fraction β. As shown in Eq. 12.7, some neutrons are produced by the decay of fission products many seconds after the initial fission and the "prompt" neutrons shown in Eqs. 12.2, 12.3, or 12.4. While small in number as shown by the fraction in Table 12.4, these are extremely important in making possible controlled fission. For example, in the fission of U-235, 0.65% of the neutrons are delayed.

Fission Threshold. A final parameter, fission threshold, is important in the fast breeder reactor. The number of fissions is very low for both fertile materials with thermal neutrons, the fission threshold being much above these energies. With fast neutrons, this threshold is 1.4 MeV for both fertile materials. Typical neutron flux spectra in fast and thermal reactors are shown in Fig. 12.2. Most of the neutrons are below this threshold. However, the greater proportion of the power from fission of U-238 in the fast reactor is due to the larger proportion of neutrons above 1.4 MeV. (Remember that the total neutron flux may be $\frac{1}{65}$ as great in the thermal reactor.)

Parasitic Capture. Similar parameters are defined to express the ratio of neutrons captured in structural, coolant, and control materials compared with those effective in fissioning the fuel. For example, in a reactor using U-235 fuel, 6.1×10^{-3} of the fast neutrons are absorbed in the iron, and 4.4×10^{-3} of the thermal neutrons are similarly absorbed. The first three lines at the bottom of Table 12.4 show these values for Fe, Na, and Zr in reactors fueled with U-235, U-233, and Pu-239.

The last two lines compare the relative loss by

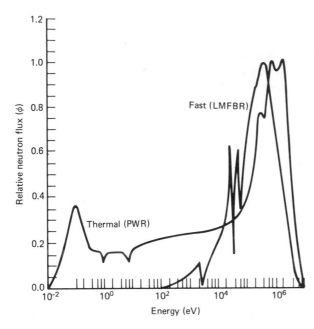

Figure 12.2. Typical neutron flux spectra in fast and thermal reactors. Courtesy R. E. Creagan.[6]

parasitic capture in the original Yankee PWR and the FBR. With four times the fissile fuel density in the latter, the parasitic captures are relatively less, even though the ratio of cross sections is higher in the fast reactor. Note: The Yankee Nuclear Power Station at Rowe, Massachusetts, started in 1961, was the third large nuclear power plant in the United States and the first full-scale, economically viable plant using a pressurized-water reactor (PWR).

In general, all parasitic reactions (Eqs. 12.8, 12.9, and 12.10) are undesirable. The Pu-240 production has the net effect of reducing the available neutrons, as does capture in the Na or the Fe. The Na reaction not only uses up neutrons but creates radioactive Na, which is an impediment to reactor maintenance and requires careful shielding until the Na-24 decays, with a 14.8 hr half-life.

Breeding. The most significant indicator of the breeding possibility with a nuclear fuel is the number of neutrons per absorption, η. This ratio must be at least 2, 1 to continue the chain reaction, and 1 to produce new fuel equal to that used. Thus, two neutrons per absorption are needed with zero neutron loss by leakage or parasitic capture, and with a breeding ratio of just barely 1. More than two neutrons per absorption provides for some loss of neutrons and for a breeding ratio above unity.

In order to have at least two neutrons per absorption, $\eta = 2$, plutonium would have to be used in a fast reactor, whereas U-233 could be used in either a fast or thermal reactor (see Table 12.4).

The *breeding ratio* (BR) is defined as the ratio of new fissile material produced to fissile material destroyed (by fission or capture). In terms of nuclear parameters, it may be expressed as:

$$BR = \eta - 1 + \frac{F_e(\nu_c - 1) - C}{F_i(1 + \alpha)} \quad (12.13)$$

with the symbols defined as follows:

		Typical Values
η	is the number of neutrons per absorption in fissile material	2.34
ν_c	is the neutrons per fast fission in the fertile material	2.14
α	is the ratio of capture to fission cross section for the fissile material	0.219
F_e	is the fraction of fissions (or power) from fertile atoms	0.224
F_i	is the fraction of fissions (or power) from fissile atoms	0.776
C	is the number of neutrons absorbed in other than fertile or fissile material, per fission	0.197

Substituting the typical values shown, derived from Table 12.5 for a fast breeder reactor using Pu-239 and U-238,

$$BR = 1.34 + \frac{0.224(1.14) - 0.197}{0.776(1.219)} = 1.40$$

Or, using the definition directly,

$$BR = \frac{\text{captures in fertile material}}{\text{absorptions in fissile material}}$$

$$= \frac{494}{287 + 63} = 1.40$$

Equation 12.13 is useful in showing how BR varies with the basic nuclear parameters.

Typical Neutron Balance. As a result of numerous calculations and experiments, the phe-

Table 12.5. Typical Neutron Balance

	Thermal Pressurized-Water Reactor (PWR)	Fast Breeder Reactor (FBR)
Neutrons Produced		
Fissions in plutonium-239 $\times v_{239}$	—	822
Fissions in uranium-238 $\times v_{238}$	80	178
Fissions in uranium-235 $\times v_{235}$	920	—
Total	1000	1000
Neutrons Absorbed		
Fissions		
Fissle material (U^{235}, Pu^{239}, Pu^{241})	383	287
Fertile material (U^{238}, Pu^{240})	31	83
Captures		
Fissile material	65	63
Fertile material	270	494
Structure and coolant	89	23
Fission products	100	20
Leakage and control	62	30
Total	1000	1000
Breeding Ratio	0.6	1.4

Source. Courtesy R. J. Creagan.

nomenon taking place within a nuclear reactor has been determined. It can be expressed as a neutron balance. Figure 12.1 shows the neutron balance in a natural uranium reactor that is just critical. Starting with 1000 fast neutrons from fission, the picture shows where they go and the manner in which enough of them cause further fissions to just produce 1000 more neutrons, that is, to be "critical."

Note that 955 of the new neutrons come from the fissioning of U-235 by thermal neutrons, the principal phenomenon. A small number, 10 each, come from fissioning of U-235 and U-238 by fast neutrons. Twenty-five come from fissioning of plutonium fuel that has been bred from U-238 in past operation of the reactor.

Leakage of neutrons from the reactor, particularly fast neutrons, is considerable. This leakage has been greatly reduced in later (and larger) reactors, as is evident in Table 12.5. However, the neutrons producing fission are remarkably the same (400 versus 414 in a PWR), as they must be, since this is the principal phenomenon in both cases.

The capture by fission products, while small, is extremely important, since it is over 10 times the capture by reactor materials, including the control

rods, and dictates the refueling cycle. Full power cannot be attained with the control rods fully withdrawn if this loss becomes too great.

While there are vast changes in reactor technology from the early natural uranium reactor proportions depicted in Fig. 12.1 to the PWR and FBR of 1981, the basic principles of what goes on in any reactor are well depicted in Fig. 12.1.

PWR and FBR Neutron Balances. Typical neutron balances of a PWR and an FBR (design) of about 1967 are shown in Table 12.5. The pressurized-water reactor (PWR) is operating with enriched uranium fuel, about 3 to 4% of U-235. The fast breeder reactor (FBR) is operating with plutonium fuel with a blanket of the fertile U-238 material.

Assuming nearly equal energy released by each fission (200 MeV), it can be seen that over 22% of the power in the FBR ($\frac{83}{370}$) is supplied by fission of the fertile material, compared with 7.5% ($\frac{31}{414}$) in the PWR. More U-238 is fissioned in the greater neutron flux above the fission threshold. Note also that the fissile material burned in the FBR to produce 1000 neutrons is only 75% of that in the PWR ($\frac{287}{383}$), whereas 1.8 times as much fertile material is converted to fissile in the FBR ($\frac{494}{270}$).

As pointed out earlier, the breeding ratio of the FBR from Table 12.5 is 1.4. Similarly in the PWR, there are 270 captures in fertile material, each of which produces a fissile nucleus. There are 383 fissions of, and 65 captures by, fissile material, or altogether 448 fissile nuclei destroyed. Thus, the breeding ratio is $\frac{270}{448} = 0.60$. These breeding ratios of 1.4 and 0.6 are typical of the two types of reactor. Note: ERDA[7] gives 1.3 as typical of LMFBR.

MODERATORS

The three moderators normally used in power reactors are light water, heavy water, and graphite. Fast reactors, of course, do not require a moderator. They use a fuel such as plutonium which has a low capture-to-fission ratio with fast neutrons.

Light-Water Moderator. Since the hydrogen nuclei of light water have approximately the same mass as the neutrons being slowed down, namely 1 amu, the collisions are most effective. One billiard ball striking another head on imparts all of its kinetic energy to the second ball and stops dead in its tracks. So also would a neutron striking a hydrogen nucleus head on. While not many collisions are "head on," the average energy loss per collision is high. Thus, from this standpoint, ordinary water, or light water, is an ideal moderator, and results in by far the most compact reactor.

However, there is a finite probability that $_1H^1$ hydrogen nuclei will capture neutrons, becoming $_1H^2$ nuclei. This loss of neutrons is so great that a natural-uranium (0.72% U-235) reactor using light-water moderator is impractical. Enrichment to some 3 to 4% U-235 is needed for practical light-water reactors.

Heavy-Water Moderator. By using heavy water, in which the hydrogen nuclei are already the $_1H^2$ or deuterium isotope, this possibility of capture is eliminated, and a natural-uranium reactor is practical. However, the moderator nuclei are now twice the mass of the neutrons being slowed down, and a much greater number of collisions is required. Roughly three times the volume of moderator is needed. The reactor is considerably larger and about 25% more expensive. However, no uranium enrichment plant is needed. This alterna-tive has therefore been adopted by some countries, for example, the CANDU reactors in Canada. The usual arrangement is a tube-type reactor (Fig. 6.4). A coolant separate from the heavy-water moderator is passed through fuel tubes in the core and conducts the heat to an external steam generator.

Graphite Moderator. Graphite (carbon) has certain advantages that tend to compensate for its much higher nuclear mass (12 amu) and consequent greater number of collisions required. Its neutron capture probability is very low. It is an excellent structural material, easily formed into blocks with ducts or spaces for fuel elements, circulating coolant and control rods. It is suitable for very high temperatures. Its strength increases above the highest temperatures considered for nuclear reactors (about 2000°F for process heat reactors). Thus, higher steam temperatures and thermal efficiencies comparable with the best fossil fuel-fired plants (about 40%) are obtained. This compares with about 33% for light-water or heavy-water reactor plants. The high-temperature gas-cooled reactor, using a graphite moderator, is thus being pursued (1977),[7] both as a breeder and a nonbreeder, although most nuclear power in the United States is from light-water reactors.

The graphite moderator and gas cooling combined result in a reactor of quite large size compared with light-water reactors. The high-temperature gas-cooled reactor is described in Chapter 6 (see Fig. 6.5).

REACTOR CONTROL

Control Rods and Shims. The reactivity of nuclear reactors is controlled by inserting rods containing "poisons"—substances such as boron and cadmium that have a very high coefficient for neutron absorption. Hafnium has also been used.[8] The usual choices are silver–indium–cadmium, hafnium, or boron stainless steel. Boron–carbide–graphite control rods are used in the gas-cooled reactors.

In the pressurized-water reactor, the control material, such as boron, can also be dissolved in the moderator–coolant water.[9] However, it cannot be added or withdrawn quickly. In general, the fast reactivity changes involved in start-up or shut-down, or large load changes of short duration, must be handled by the control rods. The very slow control requirements to compensate for reactivity changes that occur gradually during the lifetime of a fuel loading can well be handled by the chemical shim.

An intermediate category consists of the reactivity effects of xenon and samarium and coolant temperature effects. The shim can assist the control rods in handling these changes, in which times of the order of hours are involved.

In a chemical shim-controlled reactor, the control rod requirements are reduced by a factor of more than 2.[9] Also practically unlimited cold shut-down control is provided.[10]

Delayed Neutrons and Control. The neutron balances given earlier have been for critical conditions. Exactly the same number of neutrons are produced and used in a given time. This assumes steady conditions in which both the prompt and delayed neutrons are counted. The delayed neutrons come from processes with half-lives of 0.4 to 55 sec (10 sec average) following fission, which is long enough for neutron-absorbing control rods to move and control the neutron level. Prompt neutrons are emitted within 0.1 μsec after fission.

The fraction of delayed neutrons is about 0.25% in the fission of plutonium, about 0.7% with U-235 (Table 12.4 gives 0.2% and 0.65%). Withdrawal of the control rods to provide this much additional reactivity would result in the reactor being "prompt critical," in contrast with "delayed critical," which requires the delayed neutrons to balance.

This band of reactivity between "delayed critical" and "prompt critical," 0.25% to 0.7%, is called the control band. It is about one third as large with plutonium fuel as with U-235. However, in the fast breeder, some 20% of the fissions and power come from fast fission of the U-238, which has 1.47% delayed neutron fraction, some six times that of Pu-239. The weighted average delayed neutron fraction is thus 0.5%, or double that of the plutonium alone.

Slightly above prompt critical, the reactor would "take off" and increase its power very rapidly were it not for the temperature coefficient and other stabilizing characteristics to be described later. The reactivity band between prompt and delayed critical is referred to as a "dollar of control," and the magnitudes of various control effects are referred to in "dollars and cents."

Temperature Coefficient (Doppler Effect). Neutrons in the 1 to 100,000 eV energy range have an increasing capture cross section with temperature. This is a stabilizing effect of considerable importance, since an increase in power raises the temperature and reduces the available neutrons (more are captured). Vice versa, a decrease in power lowers the temperature and increases the available neutrons. A 1000°F temperature change is half as important as delayed neutrons, that is, it has 50¢ of reactivity in the PWR, $1.00 in the FBR.

The capture in this region has sharp resonances which are broadened by the increased atomic motion in the crystal lattice as the temperature is increased. Since this effect has to do with relative velocities of neutrons and nuclei, it is termed "Doppler effect."

Sodium Void Coefficient. In a sodium-cooled reactor, if the sodium starts to boil and create a void, the reactivity of the reactor is altered by several factors. The net effect may be an increase in reactivity (positive void coefficient) or a reduction in reactivity (negative void coefficient). The effect is usually small for a local void but can be quite large for a major loss of sodium. For example, in one design of fast breeder reactor, the loss of sodium from the central 10% of the core volume would increase the reactivity by 17¢, whereas for complete loss of sodium from the core there would be a reactivity decrease of $4.00, and for complete loss from the reactor vessel an additional decrease corresponding to $6.00.[6]

Depending on geometry, a sodium void may have two opposite effects: (1) increased neutron leakage and reduced reactivity, and (2) increased average neutron energy because of reduced scattering and consequent increase in reactivity. Thus, the designer can control the design to obtain either positive or negative void coefficient but generally limits any positive void coefficient to a small amount.

Expansion Coefficients. Mechanical expansion of fuel elements, support structures, and coolant with increased power (and temperature) has two effects: (1) the surface-to-volume ratio of the core is increased causing greater neutron leakage, and (2) sodium is squeezed out of the core with effects similar to a sodium void, but smaller. Typically, the combined effect in a fast reactor is of the order of 0.4¢/°F. It can be controlled by design.

REACTOR OPERATING RATIOS

The concepts of fuel burnup, breeding ratio, and, for breeder reactors, the fuel doubling time are useful in characterizing reactor performance. The exact definitions and expressions relating them to fundamental reactor parameters are given, in Eq. 12.13 for the breeding ratio, in Eq. 12.16 for fuel burnup, and in Eqs. 12.14 and 12.15 for doubling times. Many current developments (1981) are aimed at increasing fuel burnup to extend the use of uranium fuel in nonbreeder reactors (see Chapter 6).

Fuel Doubling Times, DT_s and DT_c. For the breeder reactor, which produces more fissile material than it uses, it is important to know how many years will be required to double the fuel supply. Will the fuel supply grow as fast as the requirement for it? The inventory of fuel associated with a reactor includes that in the core and that being processed or in storage. This gives rise to two doubling times, based on core inventory or total inventory (in core and out).

Also one can consider the bred fuel as simply stockpiled until it equals the original inventory. This results in a simple doubling time, DT_s. Or one may think of a large system in which fuel, as soon as produced in many reactors, goes into new reactors. This results in a compounding similar to compound interest and is designated DT_c.

Thus, altogether there are four possibilities, the expression for simple doubling time being:

$$DT_s = \frac{1000}{0.365 \, P_{st} \, g \, f \, F_p \, (1 + \alpha)(BR - 1)} \quad (12.14)$$

with the symbols defined as follows:

		Typical Values
DT_s	is the simple doubling time in yr	—
P_{st}	is the specific power in MW thermal per tonne fissile material in and outside of the reactor (or P_{st} can be based on in-core fuel if that case is of interest)	750
g	is the grams fissioned per MW-day	1.0

		Typical Values
f	is the plant factor	0.85
F_p	is the fraction of the power from fissile atoms	0.8
α	is the ratio of capture-to-fission cross sections	0.15
BR	is the breeding ratio	1.4

Using the typical values shown, $DT_s = 11.7$ yr.

The compound doubling time is based on the bred fuel being used immediately in another reactor, similar to compound interest:

$$DT_c = \frac{0.693}{\ln (1 + 1/DT_s)} \quad (12.15)$$

Using a simple doubling time DT_s of 11.7 yr, the compound doubling time is $DT_c = 8.5$ yr. Both values are theoretical limits that would not be fully realized due to the economics of reprocessing blanket fuel elements having little plutonium, and so on.

Fuel Burnup. The fuel burnup B in MW-days, thermal, per tonne total fuel (fissile and fertile) loaded into the reactor, is given by the expression

$$B = \frac{P_s R_c F_p P_f}{F_c} \times 1000 \text{ MW-days/tonne} \quad (12.16)$$

With the symbols defined as follows:

		Typical Values
P_s	is the specific power in MW$_t$/kg fissile material loaded	1.0
F_p	is the ratio of fissile to total atoms loaded	0.15
P_f	is the plant factor	0.85
R_c	is the reloading period in days	365
F_c	is the fraction of fuel reloaded	0.5
	(for a fast breeder reactor)	

Using the typical values given,

$$B = \frac{1 \times 365 \times 0.15 \times 0.85 \times 1000}{0.5}$$

$$= 93,000 \text{ MW-days/tonne}$$

At 1.0 gm per MW-day this represents 0.093 tonnes

fissioned per tonne fuel used. Nearly 10% of the atoms are fissioned. Assuming that 78% of these, or 0.072 tonnes, are fissile atoms (see Table 12.5), this represents 48% of the 0.15 tonnes fissile atoms loaded per tonne fuel.

Since 10% of the fuel is fissioned, and the fission-product atoms, such as barium and strontium, each occupy nearly as much space as the original plutonium or uranium atoms, the fuel elements must be designed to provide for this much expansion, 10% for metal fuel, 5% for carbide, and 3% for oxide.

REFERENCES

1. R. Weinstein, *Nuclear Physics*, New York: McGraw Hill, 1964.

2. R. C. Weast and M. J. Astle, Eds., *Handbook of Chemistry and Physics*, 61st ed., Boca Raton, Fla.: CRC Press, 1981.

3. W. E. Shoupp, "The Physics of Nuclear Power," *Westinghouse Engineer*, September 1954, p. 162.

4. F. J. Lerner, "Magnetic Fusion Power," *Spectrum*, December 1980, p. 44.

5. J. Hogerton, *Atomic Fuel*, Oak Ridge, Tenn.: U.S. Atomic Energy Commission, Division of Technical Information, 1964, p. 23.

6. R. J. Creagan, "Fast-Breeder Power Reactors—Some Basic Concepts," *Westinghouse Engineer*, January 1968, p. 8.

7. *Advanced Nuclear Reactors, An Introduction*, ERDA 76-107, Oak Ridge, Tenn.: Tech. Inf. Center, May 1976.

8. "Shippingport," *Westinghouse Engineer*, March 1958, p. 39, special issue.

9. P. Cohen and H. W. Graves, "Chemical Shim Control for Nuclear Reactors," *Westinghouse Engineer*, March 1964, p. 90.

10. R. J. Creagan, "Experience with Water Reactors," *Westinghouse Engineer*, January 1968, p. 35.

11. *Nuclear Standards, ASTM Part 45*, Philadelphia, Pa.: Amer. Soc. for Testing and Materials, 1981.

12. S. Glasstone, *Fusion Energy*, Washington, D.C.: U.S. Govt. Printing Office, 1980.

13

Physics of Energy, Energy Conversion—Selected Topics

INTRODUCTION

This chapter is intended as an introduction to some of the important physical concepts that are involved in energy conversion from one form to another and in heat transfer. For those readers who have already been "introduced," it will constitute a brief review.

What are the cycles used in steam turbines, gas turbines, and in gasoline or diesel engines? How does a fuel cell work? An electrolytic cell? A light cell? A battery? A thermoelectric junction? Or a thermionic converter? What is the principle of magnetohydrodynamics (MHD) or electrogasdynamics (EGD)? What about refrigeration and air conditioning cycles and the heat pump?

In a gasifier, or other chemical reactor, what is the equilibrium composition at a given temperature and pressure? What proportion of the various possible ingredients will be present at equilibrium. What, in fact, *is* equilibrium?

What are the characteristics of the various "working fluids"—steam, air, various gases, metal vapor, liquid sodium, ammonia, or organic fluids such as methane and ethylene—that lead to their selection in various cycles or heat transfer systems?

How do combustion-air and feed-water heaters, economizers, recuperators, and steam reheaters and separators increase the efficiency? What is accomplished by "dual cycles" such as gas turbine/steam turbine or metal-vapor Rankine topping/steam bottoming?

What are some of the basic concepts and laws of thermodynamics that govern these processes—without going too deep? Naturally, with this range of topics, each one will have to be condensed to its bare essentials. Readers desiring only an understanding of the basic principles of the various conversion cycles can read this chapter omitting the equations and calculations. These are needed for a more quantitative understanding.

ELEMENTARY THERMODYNAMIC PRINCIPLES

An elementary knowledge of thermodynamics is required for a full understanding of this chapter, although anyone with a technical education will understand the system descriptions, the general principles stressed, and the conclusions drawn regarding the various systems. A brief statement of the thermodynamic principles used throughout the chapter is first given.

Thermodynamic Definitions

The "System." Just as in mechanics the "free body" is used to show all the forces and torques necessary to balance, so in thermodynamics the "system" is the isolated entity, with the net heat flow "in" and the net work "out" balancing in a complete cycle (return to exactly initial conditions). (Quotes are used to differentiate this special thermodynamic meaning of "system" from the general usage of the term system.)

The usual example is the gas in a closed cylinder, with a piston, as shown in Fig. 13.1. Note that it is the *gas* which is the "system," not the cylinder or the piston. It is the *gas* into which heat is supplied, or from which work is withdrawn. It is the *gas* that does work or has work done on it. The *gas* is the "working fluid" in this case.

The Process. If heat is added to the gas, or the piston is allowed to move, or both, the gas undergoes a "process" in which its temperature,

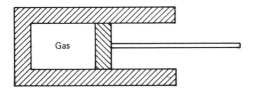

Figure 13.1. Example of a "system"—the gas in a closed cylinder. Due to the heat insulation, the "process" undergone by the gas as the piston moves is "adiabatic" (no heat flow in or out).

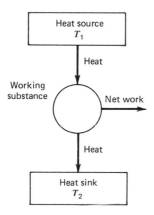

Figure 13.2. Idealized heat engine.

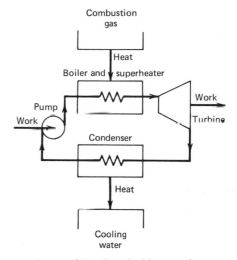

Figure 13.3. Practical heat engine.

The Heat Engine. The heat engine converts heat to work. It is an idealized engine to which all real engines that convert heat to work may be compared. It is shown in Fig. 13.2. In a heat engine, the "system" (the working substance) is subjected to a cyclic process. The ideal heat engine is shown in Fig. 13.2, and a practical example is given in Fig. 13.3.

The Carnot Cycle

In 1824, the French military engineer Sadi Carnot discovered the maximum possible efficiency of a heat engine operating between infinite heat reservoirs at temperatures T_1 and T_2. The Carnot cycle is shown in Fig. 13.4. All heat, Q_1, enters at T_1. All waste heat, Q_2, leaves at T_2. Expansion and compression are adiabatic, and there are no internal friction or turbulence losses. The work done, W_{12}, is

pressure, volume, energy, and other variables change. There are various kinds of processes. For example, if the cylinder is heat insulated, as in Fig. 13.1, so that no heat can enter or leave, the process is said to be "adiabatic." The process may also be isothermal, constant volume, constant pressure, isenthalpic (constant enthalpy), isentropic (constant entropy), on so forth. These are, of course, ideals, only approached by real processes. The process is "cyclic" if the "system" is returned exactly to its initial condition.

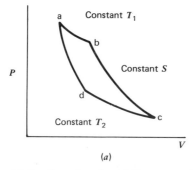

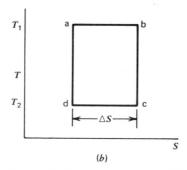

Figure 13.4. Carnot cycle. (*a*) Pressure-volume diagram. (*b*) Temperature-entropy diagram.

therefore $Q_1 - Q_2$. Carnot discovered that under these conditions,

$$\frac{Q_2}{Q_1} = \frac{T_2}{T_1} \qquad (13.1)$$

The resulting Carnot efficiency is

$$\eta_c = \frac{W_{12}}{Q_1} = \frac{Q_1 - Q_2}{Q_1} = \frac{T_1 - T_2}{T_1} \qquad (13.2)$$

Any heat loss through the cylinder walls would lower the efficiency. Any internal friction would do likewise. Thermal drops would lower the effective T_1 and raise the effective T_2 and thus lower the efficiency.

Carnot could see clearly that only if all heat was admitted to the "system" at the highest possible temperature and all waste heat exhausted at the lowest possible temperature, with no heat loss to the surroundings and no internal friction losses during the expansion and compression, could this maximum theoretical efficiency be obtained.

Furthermore, the efficiency had a definite upper limit depending only on the source and sink temperatures. With T_2 equal to 60% of T_1, the thermal efficiency could not exceed 40%.

Since there is no loss in this theoretical Carnot cycle, it is reversible. Starting from any point on the cycle, it can be followed just as well in either direction as long as

$$\frac{Q_1}{T_1} = \frac{Q_2}{T_2} \qquad (13.3)$$

Heat Pump. For example, suppose it is desired to remove heat from a low-temperature reservoir and deliver it to a higher-temperature reservoir. This can be done by the expenditure of the same work W_{12}. The gas starts at the low temperature T_2. Heat is added at this temperature, an amount Q_2. The gas is compressed to T_1, and the heat Q_1 is discharged. Expansion to T_2 completes the cycle. The coefficient of most interest in this case is the heat delivered at the higher temperature divided by the work required to do it. This theoretical maximum heat pump efficiency is

$$\eta_{HP} = \frac{Q_1}{W_{12}} = \frac{Q_1}{Q_1 - Q_2} = \frac{T_1}{T_1 - T_2} \qquad (13.4)$$

To pump heat from the cold outdoors at 32°F to a warm indoors at 100°F (radiator or hot air) (absolute temperatures of 492 and 560°R), the theoretical maximum efficiency is 560/68 = 8.2. The work required is one eighth of the heat delivered.

Practical efficiencies over 3.0 (300%) are achieved under favorable conditions. The annualized efficiency of 170% given in Table 4.11 is typical of domestic applications in temperate parts of the United States (Pittsburgh Natural Gas Co. uses 160% in comparisons). At 170% efficiency, the heat pump uses 59% of the electric energy that would be required with electric space heating.

Refrigeration. In refrigeration, we are interested in the work required to remove heat from a low-temperature reservoir and discharge it at a higher temperature. The maximum theoretical refrigeration efficiency is

$$\eta_{REF} = \frac{Q_2}{W_{12}} = \frac{Q_2}{Q_1 - Q_2} = \frac{T_2}{T_1 - T_2} \qquad (13.5)$$

In pumping heat from a 20°F freezer unit (480°R) into an 80°F room, the efficiency could not exceed 480/60 = 8.0 (800%). The practical efficiency or coefficient of performance is of course much less, but still may be of the order of 3.0 (300%) (see Example 13.1 later).

Practical Importance. Carnot's discovery and the principle it embraces have tremendous importance and guide all practical heat engine design today. The source and sink temperatures are separated as widely as possible economically, within the constraints of available technology and materials. High-temperature heat is never used to do a job that could be done by lower-temperature heat somewhere in the system.

Carnot's ideas were formulated into the Second Law of Thermodynamics by Clausius in 1850, and the concept of entropy was developed.

Entropy. The equivalence of Q_2/T_2 and Q_1/T_1 in a reversible process indicated that Q/T was a fundamental property; it is called "entropy" S. Entropy is defined by

$$dS = \frac{dQ}{T} \qquad (13.6)$$

The concept of entropy is far broader than indicated here. However, this will serve as a definition at this point.

In the Carnot cycle, the entropy gained by the

"system" at T_1 is the same as that lost at T_2. There is no gain or loss of entropy, either by the "system" or by its surroundings.

Plots of the Carnot Cycle. On a plot of temperature T versus entropy S, the Carnot cycle appears as a rectangle, as shown in Fig. 13.4b. It is bounded by two isothermal and two isentropic processes. The S reference is arbitrary, but the entropy increase S_{ab} is equal to the entropy decrease S_{cd}. In the steam tables, ASME 1967, S is assigned the value zero at the triple point of water, a precisely reproducible physical reference.

The $P-V$ diagram of the Carnot Cycle is shown in Fig. 13.4a for comparison. In either case, the area included in the diagram is the work W_{12}. The integral of $P\ dV$ around the cycle of Fig. 13.4a obviously gives the area enclosed as well as the work per cycle. For the $T-S$ diagram,

$$W_{12} = Q_1 - Q_2 = T_1\ \Delta S - T_2\ \Delta S$$
$$= \Delta S(T_1 - T_2) = \text{area } abcd \quad (13.7)$$

The Laws of Thermodynamics

The Laws of Thermodynamics are formulations of the physical observations in this field, both the "laws" themselves and any corollaries of them, that are believed never to have been violated.

Zeroth Law. When two "systems" are each in thermal equilibrium with a third, they are in thermal equilibrium with each other. This is the principle of the thermometer. If two bodies ("systems") are each measured by the same thermometer and are found to have the same temperature, then if they are placed in contact with each other no heat will flow from one to the other. They are already in thermal equilibrium with each other.

First Law (Conservation of Energy). Energy cannot be created or destroyed. It can only be changed from one form to another. For example, in a heat engine, over a complete cycle, the net heat "in" equals the net work "out."

In view of the Einstein relation $e = mc^2$, it is clear that actually $e + mc^2$ is conserved. However, in all systems where conversion of mass to energy is not involved, the First Law is taken to have the meaning given above.

Second Law. No real heat engine can be 100% efficient. In any real process, the entropy increases. Heat cannot be transferred from a lower temperature to a higher temperature without the expenditure of work. An example will be given shortly.

Third Law. There is a Third Law, which states that "it is impossible to lower the temperature of any 'system' to absolute zero in a finite number of steps."

We will use only the Zeroth, First, and Second Laws, which have become "second nature" to most engineers and scientists today.

Enthalpy

When heat is added to a working substance at constant pressure, there is usually expansion. Part of the heat goes into internal energy, the kinetic energy of the molecules or particles, and part into expansion against the pressure p. The differential increase per unit mass of a homogeneous working substance is $du + p\ dv$, where u is the internal energy and the quantity of $u + pv = h$ is the enthalpy per unit mass, above some arbitrary reference. It is a state property, independent of the path by which it got there.

Steady Flow Process

Nearly all practical processes with which we shall be concerned in this chapter are steady-state flow processes as distinct from the closed system discussed earlier. Figure 13.5 illustrates an arbitrary control volume in steady-state flow. The mass flow rate is necessarily the same at both ports, although three or more ports are possible as long as the total flow "in" is zero. This control volume could be a steam boiler, a turbine, a refrigerant condenser, and so on—any element of a power cycle. Note that the positive direction is taken "in" for heat, both q_1 and q_2, rather than the convention of Fig. 13.2.

With these conventions the First Law as applied to the control volume is

$$q_1 + q_2 = P_x + m\ \Delta\left(h + \frac{V^2}{2} + gz\right) \quad (13.8)$$

where q_1 and q_2 are the heat flow rates *from* the indicated (infinite) reservoirs, P_x is the rate of *delivery* of shaft work, m is the mass-flow rate, h is the

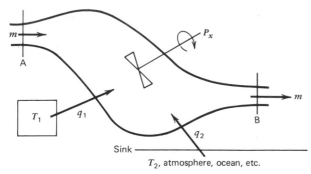

Figure 13.5. Control volume in steady-state flow. Courtesy Prof. Fletcher Osterle.

enthalpy, just defined, and $V^2/2$ is the kinetic energy of unit mass at the velocity V; gz is the potential energy of unit mass at a height z, with acceleration of gravity g.

In all the applications in this chapter, $\Delta m\, V^2/2$ and $\Delta m\, gz$ are taken as zero and the First Law expression becomes simply

$$q_1 + q_2 = P_x + m\, \Delta h \qquad (13.9)$$

The Second Law as applied to the control volume may be expressed as

$$\frac{q_1}{T_1} + \frac{q_2}{T_2} + \theta = m\, \Delta s \qquad (13.10)$$

where θ is the entropy production in the process and is never negative; s is the entropy of the flowing stream per unit mass.

Substituting the value of q_2 from the Second Law equation, 13.10, into the First Law equation, 13.9, yields

$$P_x + T_2\theta = \frac{q_1(T_1 - T_2)}{T_1} - m\,\Delta(h - T_2 s) \qquad (13.11)$$

In the ideal case of a reversible process, $\theta = 0$ and

$$(P_x)_m = \frac{q_1(T_1 - T_2)}{T_1} - m\,\Delta(h - T_2 s) \qquad (13.12)$$

and q_2 changes to satisfy Eq. 13.9. Comparing Eqs. 13.11 and 13.12,

$$(P_x)_m = P_x + T_2\theta \qquad (13.13)$$

where $T_2\theta$ is defined as the dissipation, the power that could have been produced but wasn't because of irreversibilities in the process; and $(P_x)_m$ is the *maximum* power that could have been produced.

In the special case where q_1 is zero, Eq. 13.11 becomes

$$T_2\theta = -m\,\Delta(h - T_2 s) - P_x \qquad (13.14)$$

The quantity $h - T_2 s$ has been defined as "availability" b:

$$b \equiv h - T_2 s \qquad (13.15)$$

since

$$(P_x)_m = P_x + T_2\theta = -m\,\Delta b \qquad (13.16)$$

Expressed in words, the maximum power obtainable from the process is the decrease in "availability."

Second Law efficiency is defined as the ratio of power delivered to the availability drop, $P_x/(P_x)m$, or $P_x/-m\,\Delta b$.

The Gibbs Free Energy

The Gibbs free energy, also known as the "free enthalpy," or the Gibbs function, is

$$G = H - TS = U + PV - TS \qquad (13.17)$$

If the temperature and pressure are constrained, it can be shown[1] that the system is in thermodynamic equilibrium when the Gibbs free energy is minimized. This result is particularly useful in determining the equilibrium concentration of various possible chemical species in a gasifier or chemical reactor as a function of the temperature and pressure used. In Eqs. 13.18 to 13.23, it is shown that when only expansion and chemical work are allowed, G is simply the sum of the chemical work terms. Later in the chapter it will be shown how this sum can be minimized (the equilibrium condition) in the special case of perfect gases.

Start with the known relation of heat input to internal energy and work:

$$dQ = dU + dW \qquad (13.18)$$

or

$$T\, dS = dU + dW \qquad (13.19)$$

There are many kinds of work, expansion, strain, magnetization, chemical, and so on, and the differential form of the internal energy equation, 13.19, can therefore be expressed as

$$dU = T\,dS - \sum_i F_i\,dX_i \qquad (13.20)$$

where F_i are generalized forces, and X_i are generalized displacements.

It can be shown[1] that the integrated form of the internal energy equation, 13.20, is

$$U = TS - \sum_i F_i X_i \qquad (13.21)$$

If the work terms are expansion and chemical only, Eq. 13.21 takes the form

$$U = TS - PV + \sum_j \mu_j n_j \qquad (13.22)$$

where the μ_j are the chemical species present, and n_j are the mole proportions. Substituting this value of U into Eq. 13.17 leads to

$$G = \sum_j \mu_j n_i \qquad (13.23)$$

This is the function that must be minimized, then, to determine the equilibrium conditions when temperature and pressure are fixed (or known). This result will be used towards the end of the chapter (see Equilibrium in a Gasifier or Chemical Reactor).

The Kinetic Theory of Gases

In taking up the various cycles of energy conversion, the Brayton cycle, used in the gas turbine, is one of the simplest and will be presented first. The working fluid involves no phase changes and is sufficiently close to a "perfect gas" with constant specific heats that the cycle can be analyzed theoretically to a fair approximation. A good physical picture of the specific heats is given by the kinetic theory of gases.

The theory is based on a few simple assumptions, and the results are in remarkable agreement with macroscopic measurements. The assumptions are:

1. The gas consists of point particles called molecules.
2. The molecules are in random motion and obey Newton's laws of motion.
3. The total number of molecules is very large.
4. The combined volume of all the molecules is a negligibly small part of the volume occupied by the gas.
5. No forces act on the molecules except during collisions with other molecules or with the walls.

6. All collisions are perfectly elastic and of negligible duration.

If these assumptions are made, it develops that

$$P = \rho \bar{v}_x^2 = \rho \frac{\bar{v}^2}{3} \qquad (13.24)$$

That is, the pressure P, of the gas is equal to the density ρ times the mean square velocity in any one direction, or times one third of the mean square total velocity. Thus,

$$PV = V\rho\frac{\bar{v}^2}{3} = M\frac{\bar{v}^2}{3} \qquad (13.25)$$

where $M = \rho V$ is the mass.

For a perfect gas,

$$PV = nRT = M\frac{\bar{v}^2}{3} \qquad (13.26)$$

which shows that T is proportional to $\bar{v}^2$.

The kinetic energy is

$$\text{K.E.} = \tfrac{1}{2}M\bar{v}^2 = \tfrac{3}{2}nRT \qquad (13.27)$$

or

$$\text{K.E.} = \tfrac{3}{2}RT \text{ per mole} \qquad (13.28)$$

The kinetic energy is the internal energy U of the gas. That is,

$$U = \tfrac{3}{2}nRT \qquad (13.29)$$

Since U and PV are each functions of T only, for a perfect gas, the enthalpy $H = U + PV$ is a function of T only for a perfect gas. Note: At temperatures and pressures where the gases are imperfect, enthalpy is a function of both T and P and isenthalpic cooling is possible (see LNG, Chapter 4, Thompson–Joule effect).

Specific Heats of an Ideal Gas. Since the internal energy of a perfect monatomic gas is $\tfrac{3}{2}RT$ per mole, and since at constant volume all heat goes into internal energy, the constant volume specific heat C_v is $\tfrac{3}{2}R$.

Monatomic molecules (point particles) have three energy modes, x, y, and z. The equipartition-of-energy theorem requires that the energy be divided

Table 13.1.   Typical Specific Heats[b] **(joules/g-mole °K) ($R = 8.32$)**

Type of Molecule	Molecule	C_v	C_p	$C_p - C_v$	$k = C_p/C_v$
Monatomic	Theoretical[a]	12.48	20.80	8.32	1.67
	He	12.6	21.0	8.4	1.67
	A	12.6	20.8	8.2	1.65
Diatomic	Theoretical[a]	20.8	29.12	8.32	1.4
	O_2	20.7	29	8.3	1.4
	N_2	20.7	29	8.3	1.4
Polyatomic	Theoretical[a]	24.86	33.16	8.32	1.33
	NH_3	27.8	36.8	9.0	1.31
	CO_2	28.4	37	8.6	1.30

[a]Considering translational and rotational modes of K.E. fully excited, and no other modes.
[b]Multiply by $1.98/8.32 = 0.238$ for Btu/lb mole °R.

equally among the available modes. C_v is thus $R/2$ per mode.

Diatomic molecules have five modes, two rotational in addition to the three translational modes, and hence $C_v = \frac{5}{2}R$. Polyatomic molecules have six modes, three translational and three rotational; $C_v = 3R$.

From Eq. 13.26, for one mole of a perfect gas,

$$P\, dV = R\, dT \qquad (13.30)$$

At constant pressure, if T increases, V increases proportionally, and R must be added to the specific heat:

$$C_p = R + C_v \quad \text{or} \quad R = C_p - C_v \quad (13.31)$$

The theoretical and some actual specific heats at near-normal temperatures and pressures are given in Table 13.1. Note the close agreement of measured specific heats with theoretical values from the simple kinetic theory.

At very high temperatures, vibrational and other modes may be excited, and at low temperatures also the theory breaks down. There is gradual transition so that specific heats vary over wide temperature ranges. However, the good agreement of Table 13.1 lends substantial credence to the theory of the phenomenon taking place.

Isentropic Process in an Ideal Gas. In an isentropic process, external work is at the expense of internal energy:

$$nC_v\, dT + P\, dV = 0 \qquad (13.32)$$

From the perfect gas law,

$$P\, dV + V\, dP = nR\, dT \qquad (13.33)$$

Eliminating dT between these two equations and integrating yields

$$PV^k = \text{a constant} \qquad (13.34)$$

and, as corollaries,

$$T^k P^{1-k} = \text{a constant} \qquad (13.35)$$

$$TV^{k-1} = \text{a constant} \qquad (13.36)$$

where $k = C_p/C_v$. These equations can be used to analyze approximately the "air standard" Brayton, or gas turbine cycle, in which k ranges from 1.4 to 1.3 over the temperature range involved. A fixed average value can be used. Note that air, being primarily nitrogen and oxygen, would be expected to have k about like these two gases in Table 13.1, as indeed it does.

Counterflow Heat Transfer Principle

The Second Law shows clearly that high-temperature heat can do things that the same amount of low-temperature heat cannot do. In the limit, heat at the same temperature as the surroundings is worthless. It can do no work. It can heat nothing that the surroundings could not heat. Thus, heat at high temperatures must be conserved in heat engines.

In our daily experience, we are very wasteful of

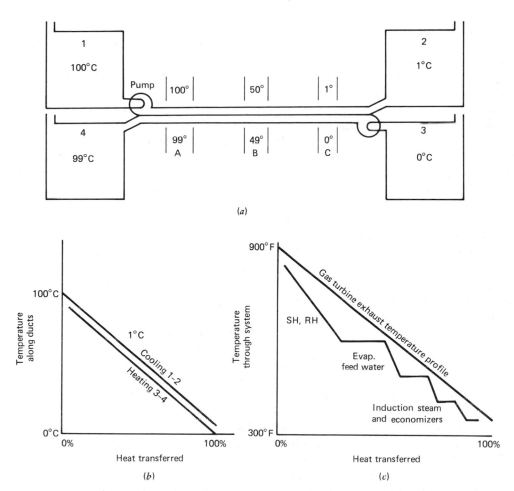

Figure 13.6. Counterflow heat transfer principle. (*a*) Ideal system. (*b*) Heating and cooling curves of ideal system. (*c*) Heating and cooling curves of a practical system.

this precious commodity. We burn gas with a flame temperature of several thousand °F, to heat water at 200°F, to warm our house at 70°F (hot-water system). There is no practical way to use the low entropy of the high-temperature gas flame. We can only use the heat contained in it. However, in a heat engine we cannot afford this loss of capability to do work.

For example, in Fig. 13.6*a*, suppose tank 1 is full of 100°C water and tank 3, of equal volume, is full of 0°C water. Tanks 2 and 4 are empty. If we were to simply bring tanks 1 and 3 into contact, insulated from all else, the water in both would eventually reach 50°C.

However, if the water in tank 1 is slowly pumped into tank 2, and that in tank 3 is pumped into tank 4 at the same rate, a steady condition would soon be set up with, say, 1°C drop across the wall between

the two ducts. At A, 100°C water is heating 99°C water; and at C, 1°C water is heating 0°C water. The water from tank 3 arrives at tank 4 heated to 99°C. Water from tank 1 arrives at 2 cooled to 1°C.

Perhaps the water in tanks 1 and 2 is dirty, or radioactive, and that in tanks 3 and 4 is clean. Or suppose they are gases, with tanks 1–2 containing combustion products and tanks 3–4 clean air. We have managed by this "counterflow" heat transfer principle to transfer nearly all the heat from one working fluid to another and, more importantly, at nearly the same temperature. That is, there is very little increase in entropy.

This principle is used all through heat engine design to conserve the precious ability of high-temperature heat that lower temperature heat does not have. It is expressed in the rule, "Never use high-temperature heat to do a job that can be done by lower-temperature heat somewhere in the system.'' A corollary is, ''Never heat a working substance by burning fuel (which has the capability of producing high-temperature heat) if it can be heated by low-temperature heat available in the system.'' These rules invariably lead to higher efficiency.

In the simple theoretical case of Figs. 13.6a and 13.6b, the heating and cooling curves have an excellent ''fit.'' Similarly, in a complex plant, such as a gas turbine-topping, steam-bottoming plant,[4] the profile of the water–steam heating and the gas cooling are compared, as in Fig. 13.6c. ''The close correspondance of the two profiles is responsible for the good efficiency of this arrangement.''[4] Here, the high-temperature exhaust gas is used for superheating and reheating, the next highest for feed-water heating and evaporating, and the lowest for generating low-pressure steam for ''induction'' into the last stage of the turbine, and in economizers.

THE BRAYTON OR GAS TURBINE CYCLE

In the simple gas turbine, air is first compressed adiabatically (nearly isentropically) to a high pressure. Fuel is then injected and burned at nearly constant pressure, raising the volume and temperature proportionally to the ''inlet temperature'' of the turbine. The gas is then expanded adiabatically in the turbine and discharged at atmospheric pressure.

This is represented by the ''air standard'' Brayton cycle, shown in the $P–V$ diagram of Fig. 13.7a, and the $T–S$ diagram of Fig. 13.7b. The ''air standard'' cycle is similar to the actual cycle, except that pure air is carried through the four processes. It is compressed along 1–2 and heated along 2–3, instead of burning a fuel. It thus remains pure air. This is a good approximation since actual air-fuel ratios are of the order of 50:1.

After expansion through the turbine along 3–4, heat is removed along 4–1 to complete the cycle. The heat removal at atmospheric pressure is a good approximation to exhaust at 4 and picking up fresh air at 1 in the actual system. No work is performed in this cooling, since dP is zero.

The turbine output is sufficient to drive the compressor and also provide useful work to a generator. In an aviation gas turbine, a jet engine, the expansion through the turbine is just sufficient to drive the compressor, with the balance of the output work going into acceleration of the exhaust gas through a nozzle or jet, to produce thrust and driving power. The ''fan jet'' is intermediate, with the turbine output driving a propeller or ''fan'' in addition to the compressor, but still with a substantial part of the expansion work going into acceleration of the exhaust gas to produce thrust and power. Each of these arrangements is best suited to a particular duty: the straight jet to the highest-altitude, highest-speed planes; the fanjet to slower, shorter-range, lower-flying planes; and the gas turbine engine to driving stationary loads such as electric generators or surface vehicles such as ships or automobiles.

Recuperators. If the gas temperature at 2, Fig. 13.7, is less than the exhaust temperature at 4, the exhaust gases may be used to partially heat the compressed gas at 2 in a recuperator, before fuel injection and ignition. This obviously reduces the fuel requirements and improves the efficiency. In the automotive gas turbine engine, it also serves the important function of cooling the exhaust gases, which would otherwise be dangerously hot to discharge from the vehicle.

Note: With the 20:1 compression ratio illustrated in Fig. 13.7, T_2 is higher than T_4, and a recuperator could not be used. It is, however, applicable at lower ratios, as will be shown.

The Gas Turbine Engine. The ''recuperated'' gas turbine engine is shown in Fig. 13.8, the ''simple'' gas turbine being the same except with the omission of the recuperator. The turbine is shown driving a generator, this being a common form of electric power generation for peaking and intermediate-load duty. Being lower in capital cost than the systems used for base load generation but usually higher in operating cost, it frequently provides the lowest overall cost in peaking duty and sometimes in intermediate-load duty as well.

A typical central station gas turbine is shown in Fig. 13.9. Such units are supplied by manufacturers as fully assembled, rail-shippable modules having outputs to about 100 MW (1978).

Theoretical and Practical Brayton Cycles

The Brayton cycle is completely defined by the values of P, V, and T at point 1, Fig. 13.7, the compression ratio r_p, and the turbine inlet temperature T_3. Using the air standard cycle and the approximation

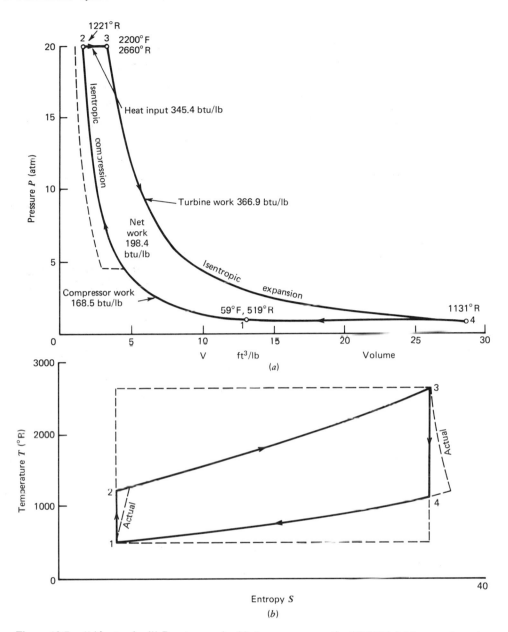

Figure 13.7. "Air standard" Brayton cycle, 20:1 compressor ratio, 2200°F inlet temperature, $k = C_p/C_v = 1.4$, efficiency 57.5%.

of constant specific heats, the turbine and compressor work can be calculated from the P–V relation of Eq. 13.34, noting that for steady-flow processes work is given by $\int V \, dP$. The results shown on Fig. 13.7 are on this basis, using $k = C_p/C_v = 1.4$ for air. Note: For air, $k = 1.4$ at 100°F, 1.3 at 3000°F.

Also with the approximation of constant specific heats, it can be shown[2] that the thermal efficiency is:

$$\eta = 1 - \frac{1}{r_p^{(k-1)/k}} \qquad (13.37)$$

For a compression ratio of 20 and $k = 1.4$, this yields an efficiency of 57.5%. This result is the same as that obtained by calculating the input and output from the P–V diagram of Fig. 13.7, since the assumptions are the same. However, the expression 13.37 shows that the theoretical efficiency is inde-

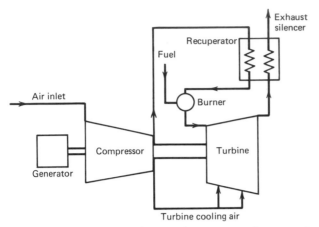

Figure 13.8. Schematic diagram of a recuperated open-cycle gas turbine. The recuperator is omitted in a simple gas turbine. Courtesy NSF–ERDA–NASA.

h_2 = 295.2 from table

T_2 = 1203.9°R from table

T_3 = 2660°R (2200°F inlet)

h_3 = 691.8 from table

p_{r3} = 565.42 from table

p_{r4} = 28.271 (20:1 compression ratio)

h_4 = 305.0 from table

T_4 = 1253.7°R from table

Compressor work = $h_2 - h_1$ = 171.0 Btu/lb air

Turbine work   = $h_3 - h_4$ = $\underline{386.77}$ Btu/lb air

Net work   =   215.77 Btu/lb air

Heat added   = $h_3 - h_2$ = 396.59 Btu/lb air

Efficiency = net work/heat added
            = 215.77/396.59 = 54.4%

pendent of the volume ratio and inlet temperature, except as they restrict the compression ratio.

Somewhat greater theoretical accuracy can be obtained by the use of the gas table for air,[2,3] as shown by the following calculation:

Calculation From Gas Table[3]

T_1 = 519°R, 59°F

h_1 = 124.2 Btu/lb of air, from table

p_{r1} = 1.2069 from table

p_{r2} = 24.1383 (20:1 compression ratio)

Comparison of Efficiencies. Note that the efficiency is slightly less than was obtained using k = 1.4, but the specific output (net work per pound air) is somewhat more. At the highest temperature, 2200°F, k = 1.313; at the lowest temperature, 59°F, it is 1.4. The use of an average value between 1.4 and 1.3 would give results quite close to the gas table. As will be shown later, the overall efficiency of the system depends on many factors in addition to the theoretical efficiency of the cycle.

These theoretical results are summarized in Table 13.2. As mentioned, with T_2 higher than T_1 (Fig.

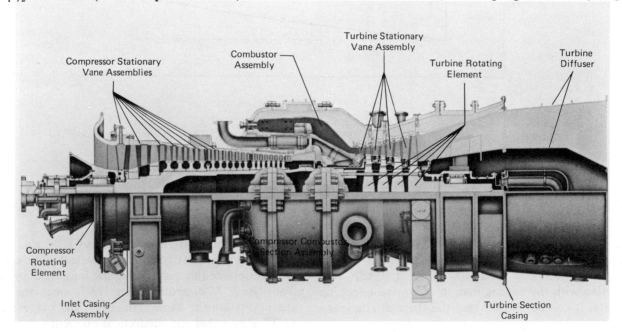

Figure 13.9. Gas turbine major functional groups. Courtesy Westinghouse Electric Corp.

Table 13.2. Brayton Cycle. Theoretical Work, Heat, and Efficiency by Different Methods[a]

	Heat Input (Btu/lb)	Comp. Work (Btu/lb)	Turb. Work (Btu/lb)	Net Work (Btu/lb)	Efficiency (%)
By Eq. 17.51	—	—	—	—	57.5
By Eq. 17.51 $r_p = 10$	—	—	—	—	48.2
By integration, $k = 1.4$ ($T_2 = 1221°R$)	345.4	168.5	366.9	198.4	57.5
By gas table ($T_2 = 1204°R$)	399.6	171.0	386.8	215.8	54.4

[a]$P_1 = 14.7$ psia, $T_1 = 59°F$, $T_3 = 2200°F$, compressor ratio 20:1, except as noted.

13.7), no heat could be transferred from the exhaust gas to the compressed gas in a recuperator. In an actual cycle, there would be losses in all processes as indicated by the entropy increases shown dotted in Fig. 13.7. Even so, as shown in Fig. 13.10, for the actual gas turbines considered in the ECAS study,[4] no increase in efficiency can be obtained with recuperators for pressure ratios as high as 20 (except at an inlet temperature of 2500°F, well beyond the state-of-the-art in 1978).

Whether the heat along 2–3 (Fig. 13.7) comes from recuperation or from the burning of fuel makes no difference in the appearance of the theoretical cycle. However, it reduces the heat input from fuel.

For comparison with the theoretical efficiency of 54.4%, the open cycle efficiency of an actual simple gas turbine, with 20:1 compression ratio and inlet temperature of 2200°F, from Fig. 13.10, is 33.5%. The Carnot efficiency with upper and lower temperatures of 2200 and 59°F (2660 and 519°R) is 80.5%.

The reason for the reduction from the Carnot efficiency of 80.5% to the theoretical efficiency of 54.4%, for the case cited, is best shown on the T–S diagram of Fig. 13.7b. The input heat is not all supplied at the maximum temperature, nor is the waste heat discharged at the lowest temperature. This is a characteristic of the Brayton cycle.

The reduction from the theoretical efficiency of 54.4% to the actual efficiency of 33.5% is due to a number of losses:

1. First, there are losses in each process of the cycle. These can only be determined experimentally from tests on similar structures. The ECAS[4] calculations were based on a turbine polytropic stage efficiency of 90%. The compressor efficiency is an empirical function of the compressor ratio, derived from tests.

2. The cooling air for the blades introduces substantial loss. Uncooled blades and vanes (not ceramic) could be operated to about 1650°F. The use of about 14% cooling air permits operation to 2200°F, which increases the efficiency and the specific output more than the decrease due to cooling, as shown in Fig. 13.10 (compare 1800°F and 2200°F, simple cycle). However, the use of 14% of the compressed air for this purpose does create a large loss compared with the theoretical cycle at the same inlet temperature. As this air passes into the turbine gas stream, there are both pressure and temperature losses.

3. The actual efficiency is based on the HHV of the fuel, the standard U.S. procedure, whereas the difference HHV − LHV is lost in the water vapor in the exhaust. For the fuel distillate from coal used in the ECAS study, HHV = 18,700 Btu/lb, LHV = 17,700 Btu/lb. Thus, 5.4% of the input energy is lost in this way.

4. Generator and mechanical losses are included in the ECAS efficiency, as well as all auxiliaries except unrelated station power.

5. There are pressure drops in the inlet duct and in the recuperator, if used, which deduct from the output energy. Note: The recuperator also adds heat not counted in the theoretical cycle.

Recuperated Gas Turbines[4]

As shown in Fig. 13.10, the efficiency of a gas turbine can be increased by the use of a recuperator. In the "base case" of the ECAS study, representative

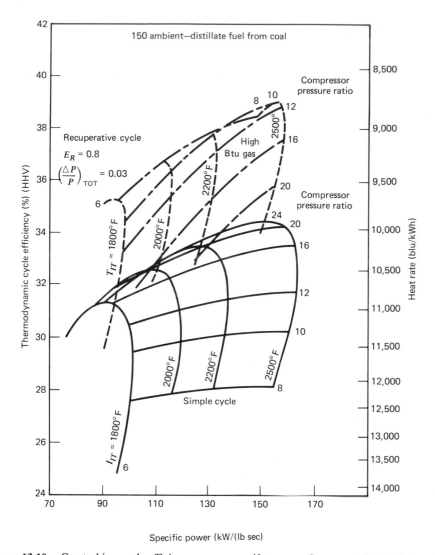

Figure 13.10. Gas turbine cycle efficiency versus specific power. Courtesy NSF–ERDA–NASA.

of the advanced state of the art in 1976, the following proportions were specified:

Inlet temperature	2200°F
Compressor pressure ratio	10 : 1
Recuperator effectiveness E_R	80%
Recuperator pressure drops $(\Delta P/P)_{TOT}$ (total hot and cold sides)	3%

Operating in the ISO ambient (59°F, 14.7 psia, 60% humidity), with the distillate fuel mentioned earlier, this unit, typically 100 MW, would have a cycle efficiency of 37.8% and a specific output of 131.6 kW/lb sec of air. As shown in Fig. 13.10, this is about top efficiency with a 2200°F inlet temperature.

Recuperator Designs. The conventional shell and tube recuperator for stationary gas turbines (includes locomotives and ships) is shown in Fig. 13.11. The compressed inlet gas flows down through tubes, where it is heated by counterflow exhaust gases rising through the shell. Alternatively, "plate fin" recuperators are used with the compressed air and exhaust gases in counterflow on opposite sides of a plate, with corrugated material attached to the plate to improve the heat transfer. In a "tension braze" design, the corrugations also provide the strength that holds the outer walls to the plate.[4]

For automotive applications, a slowly moving disc of heat-absorbing material is used. The disc material is alternately heated as it passes through

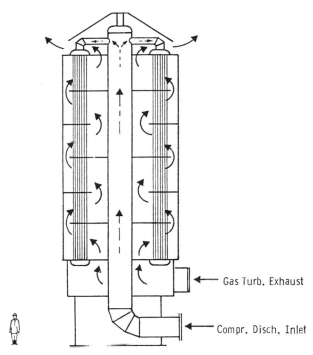

Figure 13.11. Shell and tube recuperator. Courtesy NSF–ERDA–NASA.

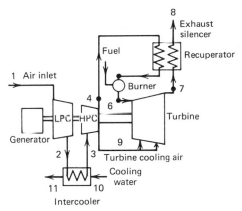

Figure 13.12. Sample recuperated intercooled open-cycle results.

CYCLE DATA SUMMARY (Point 69)

Station	Pressure, psia	Temperature, °F	Flow, lb/s
1	14.7	59	750
2	147.0	367	750
3		96	750
4	294.0	425	642
5		777	642
6	283.0	2200	657
7	15.0	865	765
8	14.7	581	765
9		425	108
10		66	
11		89	

Fuel: distillate from coal (18,700 Btu/lb HHV); Flow = 15.4 lb/sec; Cycle efficiency = 41.0%; Specific power = 165.7 kW/(lb/sec). Sample recuperated intercooled open-cycle results. Courtesy NSF–ERDA–NASA.

the hot exhaust gases and cooled as it gives up this heat to the compressed air. Some 80 to 90% of the exhaust heat can be transferred to the compressed air by this means.

Intercooler. By cooling the air when partly compressed, as shown in Fig. 13.12, both efficiency and specific power can be increased at the expense of greater complication and cost. For example, if the 20:1 compressor ratio of Fig. 13.7 were split into two √20:1 sections, that is 4.5:1 each, and the air between were cooled to within 30°F of the available cooling water temperature, for example, 80°F, the volume and temperature at the 4.5 atm pressure level would be reduced as shown dotted in Fig. 13.7*a*, and the 20:1 compression would be at a considerably lower temperature. A recuperator would then be effective even with compressor ratios up to 20:1. (10:1 and 2:1 ratios-Fig. 13.12.)

The ECAS study showed that the optimum cycle efficiency would be increased by about 3.2 points to 41% by this measure, with the optimum pressure ratio 16:1, as shown in Fig. 13.13.

Cost of Electricity

The cost of electricity is generally the final criterion in selecting a plant. Because of its complication and

cost, the most efficient plant may not produce electricity at the lowest cost, particularly for peaking or intermediate load duty. The cost of electricity depends on three main factors: the total capital cost of the plant, the fuel and other operating costs, and the plant factor (percent of full capacity generated in a year).

For the ECAS studies, fixed charges were taken as 18%, including depreciation, taxes, insurance, return (cost of money), and working capital, but not maintenance. These costs are typical of the power industry. The median fuel cost for clean distillate from coal, expected (in 1966) to be available in quantities about 1980, was $2.60/million Btu (1966 basis). Using these values for the simple and recuperated gas turbine systems considered thus far, the cost of electricity as a function of capacity (plant) factor was calculated and is given in Table 13.3 for a gas turbine inlet temperature of 2200°F.

At the low capacity factor of 0.12, the simple gas turbine, with no frills, produces the lowest-cost

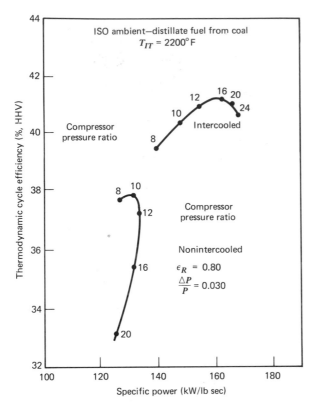

Figure 13.13. Recuperative gas turbine cycle efficiency versus specific power for cycle with and without intercooling. Courtesy NSF–ERDA–NASA.

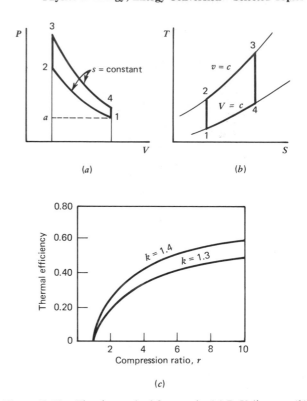

Figure 13.14. The air-standard Otto cycle: (a) P–V diagram; (b) T–S diagram; (c) thermal efficiency as a function of the compression ratio and specific heat ratio. From Wark, *Thermodynamics*.[2] Used with the permission of the McGraw Hill Book Co.

electricity at a compressor ratio of 12. As shown in Fig. 13.10, its thermal efficiency at this ratio is about 31%, whereas its highest efficiency of 32.5% occurs at a ratio of about 20.

At 0.45 capacity factor, the added complication of a recuperator is justified.

The capacity factors of 0.12 and 0.45 are representative of peaking and intermediate duty and are

the field of application of simple and recuperated gas turbines. Capacity factors of 0.50, 0.65, and 0.80 are characteristic of base load,[4] and other systems to be described later will be better for these and competitive for intermediate loads.

GASOLINE (SPARK IGNITION, SI) ENGINE—OTTO CYCLE

The theoretical "air standard cycle" which approximates the processes of a conventional gasoline engine is the Otto cycle, shown in Fig. 13.14. Starting with the piston down (bottom dead center, BDC), the cylinder is assumed to be full of air at atmospheric pressure. The piston rises compressing the air isentropically along 1–2 to the top dead center (TDC) position. Heat is then added instantaneously, raising the temperature and pressure along 2–3. This corresponds to spark ignition in the actual engine. The hot air is then expanded isentropically along 3–4 to the BDC position. Finally,

Table 13.3. Cost of Electricity About 1980 with Simple and Recuperated Gas Turbines (1966 Basis)

	Simple	Recuperated
Capacity factor of 0.12		
Best compressor ratio	12	12
COE (mills per kWh)	58	60
Capacity factor of 0.45		
Best compressor ratio	20	8–12
COE (mills per kWh)	36.3	34.3
Capacity factor of 0.65		
Best compressor ratio	16–20	8–12
COE (mills per kWh)	32.7	30.4

Table 13.4.   Air Standard Otto Cycle. Interpolations from the Gas Tables[2,3]

Cardinal Point	P (psia)	v (ft³/lb)	T (°R)	u (Btu/lb)	v_r (relative volume)
1	14.7	13.515[a]	537	91.53	146.34
					↓ b
2				210.07	18.292
				↓ c	
3				944.47	0.2917
					↓ d
4				458.73	2.3338

$^a v_1 = RT_1/P_1 = (1545 \times 537)/(29 \times 14.7 \times 144) = 13.515$ ft³/lb.

$^b v_{r2} = v_{r1} (v_2/v_1) = 146.34/8 = 18.292$.

^{c}The heat addition of 10 Btu to $318/(1728 \times 13.515)$ or 2.01362 lb air amounts to 734.4 Btu/lb; $u_2 = u_3 + 734.4 = 210.07 + 734.4 = 944.47$ Btu/lb.

$^d v_{r4} = v_{r3} (v_4/v_3) = 0.2917 \times 8 = 2.3338$.

heat is removed at constant volume along 4–1, restoring the air to its initial condition. This corresponds to the opening of the exhaust valve in the actual engine.

The four-stroke cycle normally used in automotive engines includes also an upward exhaust stroke, which clears out the exhaust gases, and a downward intake stroke, which draws in fresh air and fuel. However, the small energy required is included in losses. These strokes are not included in the "air standard cycle."

Otto Cycle Calculations. It can be shown[2] that the thermal efficiency of the Otto air standard cycle, with the approximation of a perfect gas with constant specific heats, $C_p/C_v = k$, is given by

$$\eta = 1 - \frac{1}{r^{k-1}} \qquad (13.38)$$

which is plotted in Fig. 13.14c. For air, k is 1.4 at 100°F and 1.3 at 3000°F.

Greater theoretical accuracy is obtained by the use of the gas tables,[2,3] which use the actual specific heats. For example, determine the theoretical thermal efficiency of an Otto air standard cycle under the following conditions:

Piston displacement	318 in.³
Compression ratio	8:1
Heat added per stroke	10 Btu
Inlet air	14.7 psia, 77°F
	(1 atm, 25°C)

The necessary interpolations from the gas table are shown in Table 13.4 for the four cardinal points of the cycle. Starting with the initial temperature, $T_1 = 537$°R, the 8:1 compression ratio leads to point 2, the heat added to point 3, and the compression ratio to point 4. The simple calculations for various entries are indicated by letters. Note that the compression ratio in a gasoline or diesel engine refers to the volume ratio, whereas in the gas turbine it refers to the pressure ratio.

Once the internal energy u has been determined for each cardinal point, the efficiency and mean effective pressure can be calculated as follows:

Heat rejected 4–1 $= u_4 - u_1 = 458.73 - 91.53$
$= 367.2$ Btu/lb

Net work $=$ heat added $-$ heat rejected
$= 734.4 - 376.2 = 367.2$ Btu/lb

$$\text{Efficiency } \eta = \frac{\text{net work}}{\text{heat added}} = \frac{367.2}{734.4} = 50\%$$

As can be seen, this corresponds to a value of k between 1.3 and 1.4 in Fig. 13.14c. The mean effective pressure on the piston is

$$\text{MEP} = \frac{\text{net work}}{v_1 - v_2}$$

$$= \frac{367.2 \times 778}{13.515 \times 144 \times 7/8} = 167.8 \text{ psi}$$

This effective pressure during the power stroke

would produce the same work as the actual pressures during the complete cycle.

Because of losses in the engine, auxiliaries, and transmission, the actual efficiency varies from somewhat less than the theoretical at the best speed and power to zero at idling. When the efficiencies at different speeds are weighted for relative times spent at these speeds and corrected for the power required at various speeds for a particular vehicle, there result the miles-per-gallon ratings for city and highway driving, which have become an important criterion of overall gasoline consumption performance. These considerations are beyond the scope of this chapter, which is limited to "how the engine works," that is, to the basic conversion cycle. An overall annualized efficiency of 10% is typical of cars with SI engines, and this has been used in Table 4.11.

THE DIESEL CYCLE

The air standard diesel cycle is shown in Fig. 13.15. Air is compressed isentropically from 1 to 2, as in the Otto cycle. However, at this point heat is added at constant pressure along 2–3 during the first part of the expansion stroke. In the actual engine, constant pressure combustion is approximated by spraying the fuel in at a controlled rate, the temperature being high enough for spontaneous combustion throughout the cylinder.

At 3, the fuel is "cut off," and the remaining expansion is isentropic. The heat removal from 4 to 1 is at constant volume, as in the Otto cycle. This corresponds to the opening of the exhaust valve in the actual engine. The automotive diesel uses a four-stroke cycle, with two strokes being used to clear out the exhaust gas and take in fresh air.

As explained in Chapter 11, the high-octane properties of a SI engine fuel, to prevent spontaneous combustion and knocking, are just opposite to the high-cetane properties of a diesel fuel, which depends on spontaneous combustion for ignition and smooth operation.

It can be shown[2] that the theoretical thermal efficiency of the air standard diesel cycle is

$$\eta = 1 - \frac{1}{r^{k-1}} \times \frac{r_c^k - 1}{k(r_c - 1)} \qquad (13.39)$$

where the cut-off ratio r_c is defined as v_3/v_2 and the compression ratio r as v_1/v_2. The variation of efficiency with cut-off ratio is shown in Fig. 13.15c for

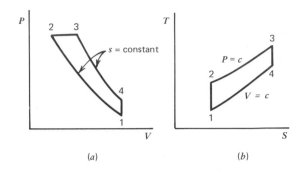

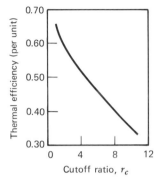

Figure 13.15. The air-standard Diesel cycle: (a) $P–V$ diagram. (b) $T–S$ diagram; (c) thermal efficiency versus cutoff ratio for a compression ratio of 15 and $k = 1.4$. From Wark, *Thermodynamics*.[2] Used with the permission of the McGraw Hill Book Co.

a compression ratio of 15 and a specific heat ratio $k = 1.4$.

Diesel Cycle Calculations. For greater accuracy as well as determination of temperatures, pressures, and work, the gas tables[2,3] can be used. Suppose, for example, it is desired to calculate the temperature and pressure at each of the cardinal points of the cycle, and also the thermal efficiency and mean effective pressure, under the following conditions:

Piston displacement	270 in.³
Compression ratio	15 : 1
Heat added during the constant pressure expansion	9 Btu
Entering air	14.4 psia, 60°F

The gas table interpolations and the connecting links from each cardinal point to the next are shown in Table 13.5. Letters indicate the simple calculations required.

Table 13.5. Air Standard Diesel Cycle. Interpolations from the Gas Table[2,3]

Cardinal Point	p (psia)	v (ft³/lb)	Gas Table T (°R)	h (Btu/lb)	u (Btu/lb)	Relative volume, v_r
1	14.4	13.36[a]	520	—	88.62[b]	158.58[b]
2	609.9[l]	0.891[c]	1468.4[e]	360.9[o]	—	10.572[d]
3	609.9[l]	2.511[h]	4139.0[g]	1130.4[f]	—	0.4101[g]
4	68.7[m]	13.36[i]	2479[k]	—	469.86[k]	2.1839[j]

[a] $v_1 = R(T_1/p_1) = (1545 \times 520)/(29 \times 14.4 \times 144) = 13.36$ ft³/lb.

[b] From the table corresponding to $T_1 = 520°R$.

[c] $v_2 = v_1/r = 13.36/15 = 0.891$ ft³/lb.

[d] $v_{r2} = v_{r1}/r = 158.58/15 = 10.572$ (isentropic).

[e] From the table corresponding to $v_{r2} = 10.572$.

[f] Heat added $= 9 \times (1728/270)$ Btu/ft³ $\times 13.360$ ft³/lb $= 769.54$ Btu/lb. Since it is added at constant pressure, $h_3 = h_2 + 769.54 = 360.9 + 769.54 = 1130.4$ Btu/lb.

[g] From the table corresponding to $h_3 = 1130.4$.

[h] $v_3 = v_2(T_3/T_2) = 0.891 \times 4139.0/1468.4 = 2.511$ ft³/lb (constant pressure).

[i] $v_4 = v_1 = 13.36$ ft³/lb.

[j] $v_{r4} = v_{r3}(v_4/v_3) = 0.4101 \times (13.36/2.511) - 2.1839$ (isentropic).

[k] From the table corresponding to $v_{r4} = 2.1839$.

[l] $p_2 = p_1(T_2/T_1)(v_1/v_2) = 14.4 \times (1468.4/520) \times (13.36/0.891) = 609.9$ psia. $p_3 = p_2$.

[m] $p_4 = p_1(T_4/T_1) = 14.4 \times (2479/520) = 68.65$ psia (constant volume).

Once the internal energy u has been determined at points 1 and 4, the work, efficiency, and MEP can be readily determined as follows:

$$\text{Heat discharged} = u_4 - u_1 = 469.86 - 88.62 = 381.24 \text{ Btu/lb}$$

$$\text{Work} = \text{heat added} - \text{heat discharged}$$

$$= 796.54 - 381.24 = 388.32 \text{ Btu/lb}$$

$$\text{Efficiency } \eta = \frac{\text{work}}{\text{heat added}} = \frac{388.32}{769.54} = 50.5\%$$

$$\text{MEP} = \frac{\text{work}}{v_1 - v_2} = \frac{388.32 \times 778}{13.36 - 0.8907}$$

$$= 168.25 \text{ psi}$$

$$\text{Cut-off ratio } r_c = \frac{v_3}{v_2} = \frac{2.5106}{0.8907} = 2.819$$

Efficiency η from Eq. 13.39 for $r = 15$, $r_c = 2.819$ is

For $k = 1.4$ 56.6%

For $k = 1.3$ 46.6%

Thus, the more accurate value, 50.5% from the gas table calculation, corresponds to a value of k between 1.3 and 1.4.

THE STIRLING AND ERICSSON CYCLES

Two additional cycles are of special interest, since they are equivalent to the Carnot cycle in theoretical efficiency. These are the Stirling cycle, devised by Robert Stirling prior to 1878, and the Ericsson cycle, stemming from the same period. The Stirling cycle has received considerable development[5] in recent years (1950s to 1970s) and may become important in hybrid electric cars and other applications. Efficiencies up to 45% have been achieved.[5] The Ericsson cycle is currently (1981) of theoretical interest only.

The $P–V$ and $T–S$ diagrams of the Stirling cycle are shown in Fig. 13.16. Heat is added at a constant high temperature T_H as the gas expands along 1–2. It is then withdrawn at constant volume along 2–3. Heat is then discharged at the constant low temperature T_L as the gas contracts along 3–4. Finally, heat is added at constant volume along 4–1 to complete the cycle.

The heat withdrawn along 2–3 is exactly equal to

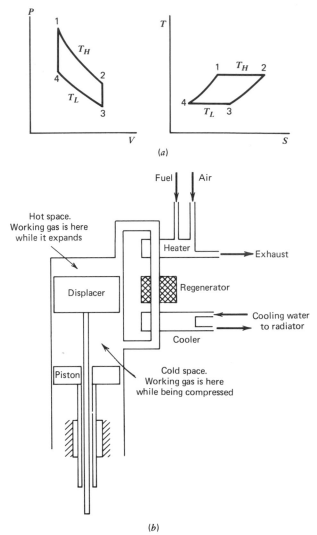

Figure 13.16. Principle of the Stirling engine—displacer type: (*a*) *P–V* and *T–S* diagrams; (*b*) schematic.

demonstrated on 10 different fuels, ranging from gasoline to salad oil, without any change of adjustment. It simply burns the fuel and discharges the products of combustion after heating the hot space. Objectionable emissions are thus relatively low. Waste heat from the cylinder is rejected through a heat exchanger. The cylinder is closed.

If the piston is thought of as fixed for the moment, the displacer can move down so that essentially all the gas above the cylinder is in the hot space, having moved up through the regenerator and heater. Or, the displacer can move up, in which case the gas has stored its heat in the regenerator, has been further cooled, and is in the cold space.

Ideally, the four processes of the cycle may be viewed as follows:

1. *Constant-temperature expansion* at the high temperature T_H. Start with the displacer down so that the gas above the piston is at T_H. It is maintained essentially at T_H as the piston moves down.

2. *Constant-volume cooling* from T_H to T_L. The displacer moves up with no change in total volume. But now, cool gas is above the piston.

3. *Constant-temperature compression* at the low temperature T_L. The piston moves up, compressing the cool gas, which is maintained close to T_L.

4. *Constant-volume heating* from T_L to T_H. The displacer moves down, forcing the gas through the regenerator and heater into the hot space, the starting point.

In practice, both the displacer and piston are connected to the same crankshaft by connecting rods and execute simple harmonic motion. The piston is phased 90° behind the displacer and so follows it as described in the foregoing steps, though not in a discontinuous manner as described. In fact, the engine is extremely smooth and quiet in operation, as there are no explosions and no valves.

Since all heat is supplied at the temperature T_H and rejected at the temperature T_L, the theoretical efficiency is $(T_H - T_L)/T_H$, as with the Carnot cycle. In practice, the efficiency is high (up to 45%) and remains relatively high at light loads, since the temperatures T_H and T_L are not changed. The control of power is by pressure of the gas in the cylinder.

The Stirling cycle can be implemented in other ways. In a four-cylinder, double-acting engine, for

that supplied along 4–1, since ΔT is the same and C_v is a function of temperature only for a perfect gas. This heat is stored in a regenerator during 2–3 and resupplied to the working fluid during 4–1. The regenerator has a temperature gradient from hot side to cold side so that all heat storage and recovery is close to reversible. Since this heat transfer is internal to the "system," the heat supplied from the heat source or rejected to the heat sink is at constant temperature, conforming to the Carnot condition.

The Philips Stirling engine,[5] shown schematically in Fig. 13.16*b*, operates on this cycle. Note that it is not an internal combustion engine. Any convenient heat source can be used to heat the "hot space" through a heat exchanger. A Philips engine has been example,[5] each cylinder contains only one piston, but both top and bottom spaces are used. The pistons are phased 90° apart. There is a movement back and forth, from the bottom of one cylinder to the top of the next, through heater, regenerator, and cooler to provide hot gas for the expansion stroke and cold gas for the compression stroke. This is the only requirement for power from the engine.

Prototype engines of the displacer type from 10 to 500 Hp per cylinder have been built, with thermal efficiencies of 30 to 45% and specific powers up to 115 Hp per liter swept cylinder volume.

Heat rejection is to water, as with the usual automobile engine. However, the amount of heat to be discharged is considerably greater since none leaves the cylinder through an exhaust.

Advantages include the ability to burn a wide variety of fuels, burning conditions that lead to low emissions, constant torque over a wide speed range, high efficiency maintained to low power, no mixing of combustion products with lubricating oil, and hence no oil consumption or contamination.

The Ericsson cycle has constant-temperature expansion and compression, similar to the Stirling cycle, but the heating and cooling are at constant pressure instead of constant volume.[5] In either of these cycles, the energy to heat the gas from T_L to T_H is the same as that given up when it cools from T_H to T_L, since both C_p and C_v are functions of T only for a perfect gas. This heat is exchanged internally by a storage mechanism in a reciprocating engine or by a heat exchanger in a steady-flow process. It is internal to the "system." Thus, all heat is supplied to the "system" at T_H and rejected at T_L, in either cycle, and the theoretical efficiency is the same as for the Carnot cycle.

THE RANKINE CYCLE

The basic cycle of the steam power plant is known as the Rankine cycle and is composed of four processes:

1. Isentropic compression of the liquid in a pump to the boiler pressure.
2. Constant-pressure heat addition which brings the liquid up to the boiling temperature, evaporates it, and optionally superheats it (in a separate section of the boiler).
3. Isentropic expansion in a turbine or engine.

4. Constant-pressure heat removal, usually in a condenser.

A practical Rankine cycle is similar, except that there is some increase in entropy in steps 1 and 3. Alternatively, in step 4 the steam may be exhausted to atmosphere, as in a steam locomotive, or into a process steam reservoir at some desired pressure.

The basic system and the T–S plot of the ideal Rankine cycle are shown in Fig. 13.17. Starting with the condensate water at temperature T_1, Fig. 13.17b, the isentropic compression to the boiler pressure is along 1–2. This is highly exaggerated for illustration. Pressurizing water to 2400 psi increases its temperature very little, 1–2, but does raise its boiling point to about 663°F as shown at 3.

The constant-pressure heat addition to the liquid raises its temperature along 2–3 until it becomes "saturated liquid" at 3, at the saturation temperature for that pressure. Further heat addition evaporates the water at constant temperature and pressure until it becomes "saturated vapor" at 4.

Actually, nearly dry steam is emitted from the boiler; and if the process is viewed as applying to 1 lb working fluid, the process 3–4 increases the amount of dry steam from 0 to 1 lb. Or, conceptually it may be viewed as transforming 1 lb water to "wet" steam and successively increasing its "quality" along 3–4 until it is "dry steam" or "saturated vapor" at 4.

If the boiler has no superheater, the saturated vapor is expanded isentropically through the turbine, becoming increasingly wet as it approaches the discharge pressure and temperature at 5. At 5, the remaining vapor is condensed to water by removal of heat at temperature T_5 and the corresponding saturation pressure P_5.

While the process 3–4 is only conceptually wet steam, the processes 4–5 and 5–1 actually do involve wet steam, the quality being defined as the proportion by weight that is vapor. In the expansion process 4–5, if the quality drops below 90% (over 10% moisture), the water droplets impinging on the vanes and blades cause excessive losses and erosion and require remedial measures, such as moisture removal, reheating, special turbine design, or some combination.

In the process 5–1, the quality drops to 0. The fluid becomes water.

Critical Point Limitation and Metallurgical Limit. It will be noted in Fig. 13.17b that at a

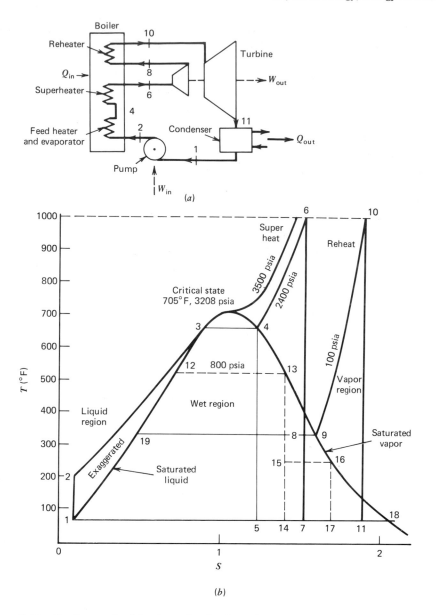

Figure 13.17. Rankine cycle: (*a*) schematic; simple cycle—evaporator direct to turbine superheat cycle, skip reheater; full cycle—as shown; (*b*) theoretical Rankine cycle with superheat and reheat, 2400 psia, 1000°F, 1000°F, 0.74 in. Hg vacuum, 0.35 psia (PWR Rankine cycle with moisture separation shown dotted).

"critical temperature" of 705°F, the "saturated liquid" and "saturated vapor" boundaries merge. Above this temperature, liquid and vapor have the same density and are indistinguishable.

At a pressure of 2400 psia, the saturation temperature is 662.3°F, which is close to critical, and the enthalpies[6] are:

Saturated liquid   718.8 Btu/lb

Enthalpy of evaporation   383.0 Btu/lb
Saturated vapor   1101.8 Btu/lb

Nearly two thirds of the heat is added to the liquid at temperatures between the condensate (about 70°F) and saturation (663°F). This is close to the upper practical limit of the basic cycle, and the critical temperature thus imposes a serious limit on the steam cycle.

The metallurgical limit is about 1000°F, although slightly higher temperatures have been used (1050°F, Eddystone Plant, Philadelphia Electric Co., 1960), and temperatures of 1200 and 1400°F have been considered.[4] While gas turbine blades and vanes operate up to 1650°F uncooled, in the steam system this top temperature applies to an extensive part of the boiler, and the materials required for such temperatures are uneconomic.

For reference, the temperature of combustion gases is of the order of several thousand degrees Fahrenheit, even with low-Btu gas.

Superheating. Some of the heat in the combustion gases above the critical temperature of water can be used to superheat the steam in a separate section of the boiler. Typically, the temperature is raised to about 1000°F, the metallurgical limit. The cycle with superheat is shown in the *T–S* plot of Fig. 13.17*b*. The steam is heated along 4–6 at constant pressure, is expanded isentropically in the turbine along 6–7, and is condensed along 7–1. In addition to increasing the average temperature at which heat is added to the system, and thereby the efficiency, much of the expansion through the turbine, 6–7, is now dry steam.

Reheating. After the steam has expanded through the high-pressure turbine, it is advantageous to reheat it to approximately the superheat temperature. This is done in a further section of the boiler in parallel with the superheater, that is, in the hottest part of the boiler. This is shown in Fig. 13.17*b*, along 8–9–10, with expansion through the low-pressure turbine along 10–11. This is virtually dry steam, and thus the moisture problem is minimized.

Supercritical Units. Somewhat higher efficiency can be obtained using a supercritical cycle, typically 3500 psia, 1000°F, with reheat to 1000°F. Feed water is heated at 3500 psia and superheated as indicated in Fig. 13.17*b*.

Condenser Temperature. The condenser temperature of 70°F, shown in Fig. 13.17*b*, would be obtained with once-through cooling. The ECAS study[4] assumed 49°F (283°K) cooling water available from rivers. With a 5°F terminal temperature difference in the condenser and a 16°F range in cooling water temperature, this would yield 70°F (49 + 5 + 16 = 70).

With wet, forced-draft cooling towers (see Chapter 15, Cooling Ponds and Towers) and the ISO ambient[4] of 51.4°F wet-bulb temperature, an approach of 22°F in the tower, a water range of 23°F, and a condenser terminal difference of 5°F would yield a condenser temperature of 101.4°F (51.4 + 22 + 23 + 5 = 101.4).

There is of course a considerable variation in these conditions at various sites, but these values are representative. The 70°F corresponds to a condenser pressure of about 0.36 psia, or 0.74 in. Hg. The 101.4°F temperature corresponds to about 0.98 psia, or 2.0 in. Hg.

Calculation of the Theoretical Rankine Cycle Efficiency

The efficiencies of the theoretical Rankine cycles illustrated in Fig. 13.17*b* can be calculated with the aid of steam tables.[6] Heat input is along 2–3–4 for the simple cycle, along 2–3–4–6 for the superheat cycle, and along 2–3–4–6 and 8–9–10 for the reheat cycle. The turbine work is along 4–5, 6–7, or 6–8 and 10–11 respectively. The pump work to compress the water, 1–2, highly exaggerated, is the same for all of these cycles.

While process 4–5 is too wet to be practical and 6–7 would represent very wet steam, in the actual process losses increase the discharge enthalpy and raise the quality somewhat. However, the calculations are made to illustrate the method and to show the successive improvements in efficiency and quality with superheat and reheat. In a practical system, with losses in each stage, optimum reheat would be at several hundred psia and expansion close to point 18.

The Carnot efficiency with source and sink temperatures of 1000 and 70°F (1460 and 530°R) is

$$\eta_c = \frac{1460 - 530}{1460} = 63.7\%$$

Theoretical Efficiency of the Simply Cycle (1–2–3–4–5–1). The specific volume of water at 70°F, 0.3632 psia is 0.01605 ft³/lb. It is practically the same when pumped to 2400 psia isentropically. Thus,

Pump work = 0.01605(2400 − 0.362)

$$\times \frac{144}{778} = 7.2 \text{ Btu/lb}$$

Heat input

$$h_2 = h_1 + \text{pump work} = 38.09 + 7.2 = 45.3 \text{ Btu/lb}$$

$$h_4 = 1101.8 \text{ Btu/lb (sat. vapor at 2400 psia)}$$

$$s_4 = 1.2441 = s_5$$

$$\text{Heat input} = h_4 - h_2 = 1101.8$$

$$- 45.3 = 1056.5 \text{ Btu/lb}$$

Turbine output To get the enthalpy at 5, note that along 1–18 the change in enthalpy and the change in entropy are proportional, since T is constant ($\Delta s = \Delta q/T$):

$$s_{18} - s_1 = 1.9896$$

(Note: h = Btu/lb, s = Btu/lb °R)

$$s_1 \qquad = 0.07463$$

$$h_1 = 38.09, h_{18} = 1092$$

$$s_5 - s_1 = 1.1695$$

$$\text{Quality}_5 = \frac{s_5 - s_1}{s_{18} - s_1} = \frac{h_5 - h_1}{h_{18} - h_1} = \frac{1.1695}{1.9896}$$

$$= 58.9\% \text{ (under 90\% in an actual turbine is serious)}$$

$$h_{18} - h_1 = 1054$$

$$h_5 - h_1 = 1054 \times 0.589 = 621$$

$$h_5 = 621 + 38 = 659$$

$$\text{Turbine output} = h_4 - h_5$$

$$= 1101.8 - 659 = 442.8 \text{ Btu/lb}$$

The theoretical efficiency of the simple cycle (1–2–3–4–5–1) is therefore

$$\eta_{\text{sim}} = \frac{\text{turbine output} - \text{pump work}}{\text{heat input}}$$

$$= \frac{442.8 - 7.2}{1056.5} = 41.2\%$$

Theoretical Efficiency of Superheat Cycle (1–2–3–4–6–7–1).

$$h_6 = 1460.6 \qquad s_6 = 1.5326 = s_7$$

$$\text{Heat input} = h_6 - h_2 = 1460.6 - 45.3 = 1415.3$$

Turbine Output

$$s_7 - s_1 = 1.5326 - 0.0746 = 1.4580$$

$$\text{Quality}_7 = \frac{s_7 - s_1}{s_{18} - s_1} = \frac{1.4580}{1.9896} = 73.5\%$$

$$(x_7 = 0.735)$$

$$h_7 - h_1 = (h_{18} - h_1)x_7 = 1054 \times 0.735 = 776$$

$$h_7 = 776 + 38 = 814$$

$$\text{Turbine output} = h_6 - h_7 = 1460.6 - 814 = 646.6$$

Thus, the theoretical efficiency of the superheat cycle is

$$\eta_{\text{sup}} = \frac{\text{turbine output} - \text{pump work}}{\text{heat input}}$$

$$= \frac{646.6 - 7.2}{1415.3} = 45.1\%$$

Theoretical Efficiency of the Reheat Cycle (1–2–3–4–6–8–9–10–11–1) (Assume that reheat is at 100 psia)

$$h_{19} = 298.61 \qquad s_{19} = 0.47439$$
$$\text{(sat. liquid at 100 psia)}$$

$$h_9 = 1187.8 \qquad s_9 = 1.6034$$
$$\text{(sat. vapor at 100 psia)}$$

$$h_{10} = 1532.1 \qquad s_{10} = s_{11} = 1.9204$$

$$s_8 = s_6 = 1.5326$$

$$s_9 - s_{19} = 1.6034 - 0.4744 = 1.1290$$

$$s_8 - s_{19} = 1.5326 - 0.4744 = 1.0582$$

$$\text{Quality}_8, x_8 = \frac{s_8 - s_{19}}{s_9 - s_{19}} = \frac{1.0582}{1.1290} = 0.937$$

$$h_9 - h_{19} = 1187.8 - 298.6 = 889.2$$

$$h_8 - h_{19} = (h_9 - h_{19})x_8 = 889 \times 0.937 = 833$$

$$h_8 = 833 + 299 = 1132$$

$$\text{Heat input} = \text{(heat input of superheat cycle)} + (h_{10} - h_8)$$

$$= 1415.3 + (1532.1 - 1132) = 1815$$

$$\text{Turbine output} = (h_6 - h_8) + (h_{10} - h_{11})$$

$$s_{11} - s_1 = 1.9204 - 0.0746 = 1.8458$$

$$x_{11} = \frac{s_{11} - s_1}{s_{18} - s_1} = \frac{1.8458}{1.9896} = 0.929$$

$$(h_{11} - h_1) = (h_{18} - h_1)x_{11} = 1054 \times 0.929 = 980$$

$$h_{11} = 980 + 38 = 1018$$

$$h_6 - h_8 = 1460.6 - 1132.0 = 328.6$$

$$h_{10} - h_{11} = 1532.1 - 1018 = \underline{514.1}$$

Turbine output = 842.7

Thus, the efficiency of the theoretical reheat cycle is

$$\eta_{Rc} = \frac{\text{turbine output} - \text{pump work}}{\text{heat input}}$$

$$= \frac{842.7 - 7.2}{1815} = 46.0\%$$

Comparison of Cycles. Note the increasing approach to the Carnot efficiency in the more complex cycles:

Cycle	Efficiency (%)	Quality of Exhaust (%)
Simple	41.1	58.9
Superheat	45.1	73.5
Reheat	46.0	92.9
Carnot	63.7	

Note also the improved quality of the exhaust steam. A quality much below 90% in an actual turbine is not practical. Note also the small amount of pump energy necessary to pump the water against the 2400 psi head. This may be compared with gas in the Brayton cycle of Fig. 13.7, where it is nearly half of the turbine output, and the compressor is larger than the turbine, Fig. 13.9.

The actual realized thermal efficiency of a modern oil-fired unit (1978) operating at 2400 psia, 1000°F, 1000°F is about 38.4%.[4] This includes all the losses in the unit and its auxiliaries as well as in the boiler and is based on the higher heating value of the fuel.

OTHER WORKING FLUIDS

In a similar manner, the theoretical cycle with any working fluid can be determined from its thermodynamic properties. These have been measured and catalogued for nearly all working substances now in practical use, particularly for single-component substances such as those shown in Fig. 13.18. The properties of some commonly used mixtures such as air[3] and combustion products[3] are also catalogued. For some, such as air, the components have like characteristics and can be treated as a

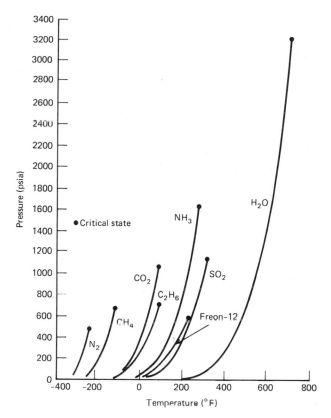

Figure 13.18. Vapor pressure curves of some common substances.

single component, as shown under Brayton Cycle earlier. The steam tables,[6] the gas tables,[3] and refrigerant tables[7] provide much of this information. Data on water (steam) is of course most complete.

Thermodynamic data on the liquid metals mercury, lithium, sodium, potassium, and cesium are also available from manufacturers. Data on mercury have been published.[8] See also the last section of the chapter.

Phases of Working Fluids. We have noted that water has both liquid and vapor phases below a critical temperature and pressure. Actually *all* single chemical compounds or elements (single-component systems) have solid, liquid, and vapor phases. They all have "critical points" or states above which liquid and vapor are indistinguishable.

They all have "triple points," or states, at a temperature and pressure at which solid, liquid, and gaseous states coexist in equilibrium. Below the triple point, the solid sublimes to a gas without going through the liquid state.

The vapor pressure curves for some common

substances are shown in Fig. 13.18. Of these, the critical pressures vary over an exceedingly wide range, from 490 psia and −250°F for nitrogen to 3205 psia and 705°F for water. The critical temperature of mercury is 2800°F. Thus, these and other working fluids provide for the operation of heat engines and refrigeration machines in the Rankine cycle (liquid and vapor phases) over a very wide range of temperatures and pressures.

In the solid, liquid, or vapor phases the thermodynamic properties, such as specific volume, enthalpy, and entropy, are functions of the two independent variables temperature and pressure. (Enthalpies of perfect gases are a function of temperature only up to 10 or 20 atm.)

At the boundary between two phases, where both phases such as water and steam can coexist, there is only one independent variable. For example, at 1 atm pressure, water boils at 212°F. Temperature and pressure are not independent.

Triple Point and Critical Point. At the triple point, there are no independent variables. Both temperature and pressure are fixed. The triple point of water is at 0.089 psia and 32.02°F (0.0061 atm, 0.01°C). This temperature is reproducible to a high degree of accuracy, of the order of one millionth of a degree Fahrenheit, and thus forms an ideal reference point for thermometry and thermodynamic properties. In the ASME Steam Tables,[6] both entropy and enthalpy are assigned the value zero at the triple point of water.

At the critical point also there are no independent variables. Both temperature and pressure are fixed, as indicated in Fig. 13.18.

Plasma—the Fourth Phase. In passing from solid to liquid to gas, the particles are less and less bound together. In the solid, the particles may be held in a crystalline lattice or in an amorphous (random) array. While held fixed in this array, each molecule is capable of moving about and having vibration energy. In the liquid phase, groups of molecules are mobile, and there is translational energy as well.

In the gas or vapor phase, the molecules completely fill the available volume. The individual molecules have acquired sufficient energy to overcome the attractive forces that held them in a well-defined space in the liquid phase. A large part of the energy is now in the translational or rotational energy of the molecules.

As the temperature of the gas is increased, the molecules tend to break apart or dissociate into atoms. This occurs at temperatures of 3000 to 5000°F, which are common in many combustion reactions. The individual atoms may then have translational energy.

A further increase in temperature and energy results in excitation of the electrons in the atoms. With sufficient energy, some of the electrons leave the atoms and the gas becomes ionized. If a significant fraction of the gas phase consists of electrons and positive ions, the gas is termed a *plasma*. Plasmas are sometimes referred to as a fourth phase of matter.

MERCURY–STEAM DUAL CYCLE

Even with superheat and reheat, it is clear from Fig. 13.17b, that the Rankine cycle, using water as the working fluid, falls far short of the theoretical Carnot efficiency between the temperatures of 70 and 1000°F. The relatively low critical temperature of water, 705°F, is partly responsible for this.

Mercury has a critical temperature of 2800°F, which would not be any restriction in working up to the metallurgical limit of 1000°F. However, it is too diffuse at low pressures to expand down to a vacuum, as with steam. The turbine would be too large and costly.

A combination of mercury–Rankin and steam–Rankine cycles, operating together, has, however, been developed and was used in a number of plants from 1922 to 1949, all under 50 MWe. As shown in Fig. 13.19, the mercury cycle operates from 1000°F down to about 450°F, where the heat is transferred to a steam cycle (used to generate steam). The steam cycle operates from this point down to a vacuum of about 0.74 in. Hg (70°F). The saturation pressure of mercury at 1000°F is 185 psia. At 450°F, the saturation pressure of mercury is 0.1 psia. The saturation pressure of steam at 430°F is 344 psia. As shown in Fig. 13.19, this makes much better use of the higher temperature heat and, in fact, results in a constant-temperature heating process at the metallurgical limit of 1000°F.

While the efficiency of these plants was about 15% higher than that of steam plants with similar top temperatures, the last of them was phased out after 1961 with the advent of larger, more efficient steam plants, with higher inlet temperatures and pressures. The value of the mercury inventory was also a factor.

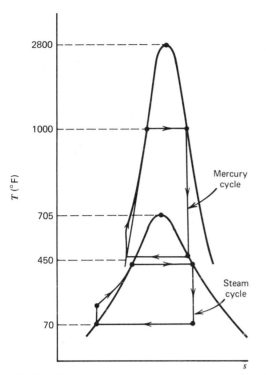

Figure 13.19. The mercury–steam binary vapor cycle. From Wark, *Thermodynamics*,[2] 1st ed., Used with the permission of the McGraw Hill Book Co.

POTASSIUM (OR CESIUM)–STEAM DUAL CYCLE[4]

Liquid-metal Rankine topping for steam plants has been studied more recently, using potassium or cesium, with considerably greater sophistication.[4] A few of the properties of these metals are given in Table 13.6, along with the properties of mercury.

The general system studied is shown in Fig. 13.20 and includes gas, liquid metal, and steam loops.

Typically, several of the liquid metal loops would be used to top one conventional steam loop, for example, eight 25 MWe metal vapor turbines with one 1000 MWe steam turbine. The size of the metal vapor turbine is limited by available technology for large disc forgings of superalloys and refractory alloys. Power output from the gas turbine generator is minor.

The potassium cycle operates from 1400°F down to 1100°F, topping a conventional steam unit at 1000°F, in contrast to the earlier mercury cycle which operated from 1000°F down to 450°F, topping a 430°F steam cycle.

The metal vapor is generated in a pressurized vessel, either a pressurized furnace burning low-Btu gas from an integral gasifier or a pressurized fluidized bed combustor, burning coal directly, with 90% of the sulfur removed by dolomite in the fluidized bed. A recirculation loop *d* is used to improve the heat transfer. Thus, the flow through the vapor generator is about $2\frac{1}{2}$ times that through the metal vapor turbine.

The pressurized combustion requires a gas turbine, shown "simple," but a recuperator and other heat recovery schemes were examined, as were feed water heaters, reheating, and steam induction, not shown. The temperatures at critical points are shown in Fig. 13.20.

Such a 1200 MWe plant, using either potassium or cesium, was estimated to have a cost of electricity of 29.5 mills/kWh for base loads, under the ECAS conditions,[4] coal at 85¢/million Btu, 18% fixed charges, and 1980 time period. However, its potential was limited in view of the simpler and more fully developed gas turbine-topped, steam-bottomed plant.

The inventories of potassium and cesium, with recirculation, were 396,000 and 1,381,000 lb, re-

Table 13.6. Properties of Potassium, Cesium, and Mercury[9]

	K	Cs	Hg
Atomic weight	39.0983	132.9054	200.59
Melting point	146.5°F,	83.1°F	−37.97°F
	63.65°C	28.40°C	−38.87°C
Boiling point	1425°F	1252°F	673.84°F
	774°C	678.4°C	356.58°C
Specific gravity (20°C)	0.862	1.873	13.546
Saturation	15.3 psia at		185 psia at
pressures (approx.)	1400°F		1000°F
	2.50 psia at		1.0 psia at
	1100°F		450°F

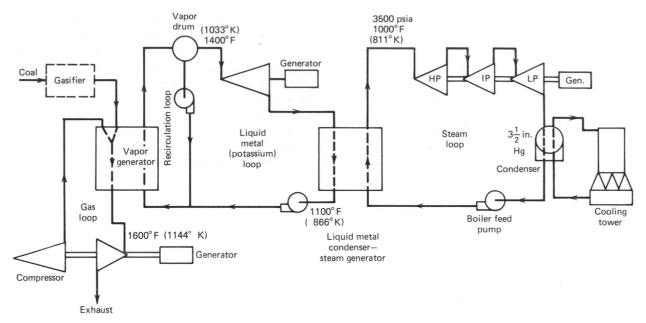

Figure 13.20.   Liquid metal-topping, steam-bottoming plant (general system used in ECAS study), 400 to 1200 MWe. (Several liquid-metal and gas loops used with one steam loop.)

spectively. At $1.68 and $18.00, respectively, these fluids cost $0.67 and $25 million, neither being a large part of the cost of a 1200 MWe plant—about $800 million.

Other dual cycles will be described later, for example, the gas turbine-topping, steam-bottoming cycle. Gas turbine inlet temperatures up to 2200°F are state of the art in 1978. Fuel costs approximately quadrupled during the 1970s, doubled in constant dollars, favoring the more efficient dual cycles, even with their greater complexity.

THE OPTIMUM USE OF HEAT: HEATERS, ECONOMIZERS, STEAM INDUCTION, MOISTURE SEPARATION

There are two major differences between a practical heat engine and the perfect Carnot engine. First of all, there is no heat source reservoir at a maximum temperature, T_1. The heat instead is generally in a combustion gas or fluid that gives up heat only as it cools. Portions of the heat are available at successively lower temperatures.

Secondly, the working substance is not compressed to the maximum temperature T_1 when heat is to be added. Instead, heat has to be added all the way from the condenser temperature up to the

maximum, in the Rankine cycle, except for a few Btu of pump work. About half of the enthalpy of superheated steam at 2400 psia, 1000°F, is added to the feed water.

Thus, while the cycle efficiency is important, the overall thermal efficiency of the process depends quite as much on matching the heating line of the working fluid with the cooling line of the heat source, as shown in Fig. 13.6. It depends on using the high-temperature heat only when needed to heat the working fluid at high temperatures and using the remaining low-temperature heat to heat the working fluid at low temperatures.

It was already demonstrated in the Brayton cycle that the best thermal efficiency was obtained not with the highest cycle efficiency at a compressor ratio of 20, but at a compressor ratio of 10, with lower cycle efficiency, in which most of the waste heat could be recuperated.

The highest temperatures that can be used economically in power machines today (1981) are still far below the stochiometric combustion temperatures of fossil fuels, and much of their low entropy and ability to do work is lost in using this heat at lower temperatures. Every study of increasing the inlet temperatures of gas turbines or Rankine cycle machines shows the prize in thermal efficiency that awaits the person who can do this economically.

Many current developments have this as a goal (1981).[4]

Thus, in the Rankine cycle there are a number of devices for optimizing the use of heat, which will be briefly described:

Economizer. After the combustion gases have given up most of their heat in evaporating water and superheating the steam, much of the remaining low-temperature heat is transferred to the feedwater in a gas-to-water heat exchanger called an "economizer."

Extraction Feedwater Heater. After the high-temperature, low-entropy working fluid has expanded through part of the turbine, producing useful work energy, some of it is extracted to heat feed water. This may be in an "open" feedwater heater in which the vapor is simply mixed with the liquid. In this case, the liquid is first pumped to the extraction pressure. Or, it may be done in a "closed" feedwater heater, a vapor-to-liquid heat exchanger.

Ideally, an infinite number of extraction points would permit each portion of vapor to produce the maximum amount of work before being used to raise the feedwater temperature a small amount. An economic analysis reveals the optimum number of extractions, heaters, and pumps and the type of heater for each location, and determines the best extraction points. Usually, this results in a small number of heaters and pumps.

Condensate Pump, Boiler Feed Pump. Both are in series. The condensate pump raises the liquid to approximately atmospheric pressure, and the boiler feed pumps, usually several in series, raise the liquid to the boiler pressure. Economizer and feedwater heaters are interspersed between pumps.

Deaerator. To avoid corrosion, it is necessary to remove air dissolved in the water. This is done in an open heater or evaporator, and while a necessary function, it does not enter into the efficiency of the cycle.

Steam Induction. In some of the combined cycles to be described, such as the gas turbine–steam turbine cycle, the gas turbine exhaust which generates the steam will be at a much lower temperature than the combustion gases in a conventional boiler. Hence, a much larger quantity of it will be required to evaporate and superheat the steam. As a result, there will be an excess of low-temperature heat. To use it effectively, further steam is generated at a low pressure, such as 30 psia, and "inducted" into the last stages of the turbine to produce additional useful work. This is called "steam induction."

Moisture Separation. Heat from the non-breeder water reactors, which currently (1981) supply most of the nuclear energy, is at a comparatively low temperature of 600°F or less, as shown in Fig. 6.3. It is too low to superheat the steam, which therefore becomes quite wet as it expands through the turbine. It is therefore necessary to use a moisture separator after the high-pressure turbine to extract most of the moisture before the steam is expanded through the low-pressure turbine. This is illustrated in principle by the dotted path 13–15–16–17 in Fig. 13.17b. Of course, with the moisture removed, less than 1 lb steam is expanded at 16–17, and this must be taken into account in the calculations.

Figure 6.8 shows the general arrangement of heaters, pumps, and moisture separator in a nuclear station. While but one boiler feed pump and two heaters are illustrated, there are usually several of each. The general arrangement is the same in a fossil fuel-fired station, with the omission of the moisture separator.

GAS TURBINE-TOPPING, STEAM RANKINE-BOTTOMING SYSTEM

As mentioned earlier, the quadrupling of fuel prices in the 1970s has made the more complex but more efficient, dual cycle plants relatively more economic. In addition the increase of gas-turbine inlet temperatures from well below 2000°F to about 2200°F (1978) have substantially increased the efficiency of the gas turbine, and have made the gas turbine–steam turbine dual cycle particularly attractive. Maximum practical steam turbine inlet temperatures are at the 1000 to 1050°F metallurgical limit.

From the ECAS[4] study of 1976, the combined gas–steam plants were found to have efficiencies several points higher than conventional steam plants. The coal-using combined cycle plant with an integrated low-Btu gasifier had an overall efficiency, from the HHV of the coal to the electric output from

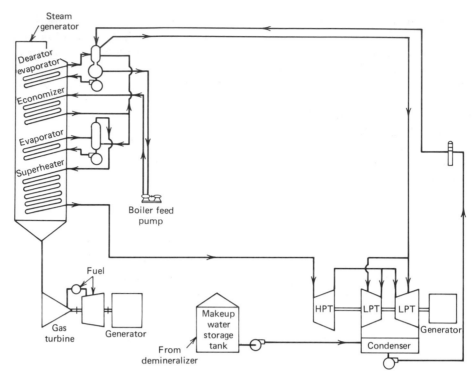

Figure 13.21. Schematic of typical combined-cycle gas-turbine, steam-turbine waste heat system. Courtesy NSF–ERDA–NASA.

the HV transformer, of 43.6% and a capitalization of $497/kW. Under the ECAS conditions,[4] of coal at 85¢/million Btu, 18% fixed charges, and 1980 time period, this yielded a cost of electricity of 24.3 mills/kWh, indicating that this plant should be considered for base load power generation (65% capacity factor).

A simplified schematic of a typical combined-cycle plant is shown in Fig. 13.21. The gas turbine is simple. No recuperator is needed since all the exhaust heat is to be used to heat steam. The boiler diagram at the left, with the highest temperature at the bottom, shows how the gas turbine exhaust is used first in the superheater, then at successively lower temperatures in the evaporator, economizer, and deaerator–evaporator, which produces steam for induction into the low-pressure turbine. More complex arrangements studied include reheaters and multiple stages of evaporator, economizer, and superheater.

Also, if a low-Btu gasification plant is used to provide fuel in place of the clean distillate from coal, the process steam required by the gasifier is taken from the steam bottoming cycle, and the compressed air for the gasifier is taken from the gas

turbine compressor. Other advantageous heat transfers are effected among the three basic components, the gas turbine, the steam turbine, and the gasifier. (See also Chapter 4, Gasification Combined-Cycle (GCC) Power Plant, in which a medium-Btu gasifier is used.)

Waste Heat Boiler Versus Supercharged Boiler. The arrangement just described is known as a "waste heat boiler," the steam cycle using the "waste heat" from the gas turbine. Earlier, with lower gas turbine inlet temperatures, the "supercharged boiler" was more economic, and many such installations have been made. In this arrangement, the hot combustion gases are first passed through the superheater and evaporator, as shown in Fig. 13.22, before being passed through the gas turbine and economizer. Instead of being burned with much excess air to limit the temperature to what the gas turbine could stand, the fuel could be burned with near stochiometric air and thus provide high temperatures for good heat transfer in the superheater and evaporator. However, the combustion must be at high pressure to supply the gas turbine, whence "supercharged."

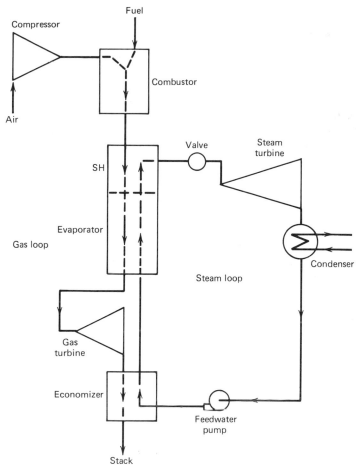

Figure 13.22. Supercharged boiler, combined cycle, gas turbine—steam turbine.

As gas turbine inlet temperatures have increased, the waste heat boiler arrangement has become more economic. In the ECAS study,[4] it is stated that "the thermodynamic transition where the more efficient system changes from supercharged to exhaust (waste heat) boiler cycle is at a gas-turbine firing temperature of about 1200°K (1700°F)." Currently (1978), state of the art is about 2200°F gas turbine inlet temperature.

However, these studies were made without provision for drastic reductions in NO_x, which may be necessary, and a reassessment may be required.

Closed-Cycle Gas Turbines. A few closed-cycle gas turbines, using air and helium as the working fluid, have been installed in Europe[4] in combined electric generation and heating plants. The ECAS study[4] included systems comprised of closed-cycle gas turbines with Rankine bottoming.

The gas turbine fluid was helium. For the Rankine cycle fluid, both steam and organic fluids were considered, the latter having considerably less exhaust volumetric flow and hence potentially smaller turbine machinery. Three organic fluids were selected for study from a list of 39 such fluids.[4] These were R-12 (dichlorodifluoromethane, the common refrigerant), methylamine, and sulfur dioxide.

These systems did not appear to be economic for coal- and oil-fired plants, due largely to the high cost of the high-temperature gas-to-gas heat exchangers to heat the helium.

MAGNETOHYDRODYNAMICS (MHD)

Nothing could be simpler in principle than MHD, and *almost* nothing more difficult to achieve practically. The idea of simply blowing a plasma through a duct and generating electricity with no moving parts is enough to excite the imagination. The calculated thermal efficiencies of well over 50% have led to truly gigantic efforts to overcome the almost insuperable materials problems that immediately present themselves. Slow but steady progress is being made, and it is the general belief (1978) that MHD will become commercially viable within the next decade. MHD is considered as a topping cycle to a conventional steam plant, supplying about half the power.

State of the Art. While an MHD duct does not require the close tolerances of rotating machinery nor have to stand the high pressures of steam and gas turbines, it does require temperatures far above those used in rotating machines. It must stand the corrosive and erosive effects of these very high-temperature combustion gases from coal, at extremely high velocities. The temperatures of over 4000°F may be compared with the 2200°F inlet temperatures of gas turbines (1978) or the 1000°F metallurgical limit of steam equipment. Thus, while runs of 1 hr continuous with direct coal firing and longer runs with cleaner fuels,[4] all in relatively small units, have been achieved, MHD is only economic in large sizes. It will require much development, particularly of materials and techniques for using them, before it reaches a commercial operating status.

While the necessary superconducting magnets can probably be built, the costs are uncertain.[4] Solid-state inverters can be assembled from multiples of quite fully developed units. The air heater is the most expensive unit and will require extensive development.

Principle of MHD

From Fig. 13.23 it can be seen that if a conductor of length l moves at a velocity v at right angles to a magnetic field of strength B, it has generated in it a voltage e proportional to the rate of cutting lines of force:

$$e = Blv \qquad (13.40)$$

where e is in volts, B is in webers/m^2 or teslas, l is in meters, and v is meters per sec. A change in flux linkages of 1 weber/sec generates 1 volt.

Assume now that the conductor shown in Fig. 13.23 is sliding on contact surfaces at top and bot-

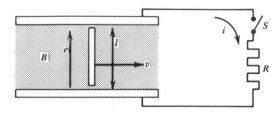

Figure 13.23. Principle of MHD.

tom. Then a dc voltage e exists between these two surfaces as long as the conductor is moving in the field.

Assume now that there are many such conductors in parallel, with some moving into the field as others move out. Then, the dc voltage is continuous as long as conductors are available.

Now assume that instead of metallic conductors there is a conducting plasma moving through the duct. The result is exactly the same, and we now have an MHD generator. The ions of the plasma are carried along with the gas with only a small slip.

Typical MHD Duct

For one of the MHD ducts considered in the ECAS[4] study, Fig. 13.24, the entering section was 1.82 m square, the entering velocity of the gas stream was 775 m/sec, and the magnetic field strength in this region was about 6 teslas. Thus, from Eq. 13.40, the no-load voltage was

$$e_0 = 6 \times 1.82 \times 775 = 8470 \text{ V}$$

The plasma has an internal resistance; and when an external load is connected, there is an internal voltage drop. This internal resistance was such that at full load the terminal voltage was about 82% of the no-load voltage:

$$e_1 = 8470 \times 0.82 = 6950 \text{ V} \qquad \text{(6930 in Fig. 13.24)}$$

Note that this duct is 22 m long and 3.81 m square at the outlet. The gas slows down as it expands along the duct, and the field is tapered downward somewhat toward the exit.

The electrodes are segmented, and, as shown, the voltage reaches a maximum at the sixth segment and then tapers down to the exit. The voltage has a longitudinal as well as a transverse component (Hall effect), the dashed lines being equipotentials connecting segments at the same potential. Thus, the

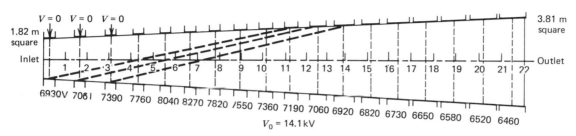

Figure 13.24. Layout of an open-cycle MHD duct. Shown is typical arrangement for connecting electrodes in series. The heavy lines are electrodes; diagonal dash lines are equipotential lines. V_0 is output voltage for two electrodes in series. Courtesy NSF–ERDA–NASA.

pair of electrodes 11 m down the duct can be connected in series with the entering pair and results in a series voltage of 14.1 kV. Similarly, the other pairs of electrodes can be connected in series and result in voltages within 5% of the 14.1 kV.

Solid-state inverter units to change the dc to ac are available for this voltage at 5 kA (1978 limit of development), and 18 of these would serve the duct shown. Each 10 kA electrode segment is divided into two 5 kA electrodes. Such inverters have remarkably low loss, 1 to 2%. The output power is thus

$$P_{out} = 14.1 \text{ kV} \times 5 \text{ kA}$$
$$\times 18 \text{ electrodes} = 1260 \text{ MWe}$$

From this must be deducted the energy required by the air compressor, about 290 MW, leaving about 970 MWe net. This unit is a topper for a 2000 MWe MHD–steam system. The values are approximate as other losses and auxiliaries must be considered as well.

Brayton Cycle

The combustion gas–MHD system operates on the Brayton cycle, the same as the open-cycle gas turbine described earlier. Air is pressurized to about 6.5 atm. Heat is added at this pressure in the air preheater and combustor. The "preheater" corresponds to the "recuperator" in the gas turbine.

Part of the work of the isentropic expansion goes into accelerating the gas from the combustor to a high velocity in a nozzle, so that it enters the duct at about 4 atm. The remaining pressure as well as that converted to velocity head are both used up in driving the "conductor" through the field and producing electrical output.

The cycle is "open" with exhaust to the atmosphere and picking up fresh air each cycle.

Seed. Even at the high temperatures in the MHD duct, 4186 to 3460°F, as shown in Fig. 13.24, the ionization is not sufficient to provide the necessary conductivity. The gas must be "seeded," for example, with potassium or cesium. Thus, "seed" is added in amount about 1% of the mass flow.

General Arrangement of MHD System

In the ECAS study,[4] three principal alternates of the open MHD cycle were considered, as well as a liquid-metal, closed-cycle system. The open-cycle alternates were (1) direct coal-burning system, (2) the use of char and product gas, and (3) a system with integral coal gasifier. Most of the principles are illustrated by the direct coal-burning system, which is shown in Fig. 13.25.

In addition to the severe materials problems common to all MHD systems, there are four interrelated problems whose solutions vary with the type of system. These are:

1. Seed and its recovery.
2. Ash removal.
3. Sulfur removal (to meet EPA requirements).
4. Nitrogen limitation (to meet EPA requirements).

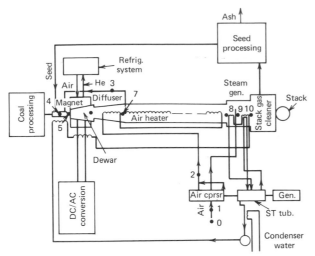

Figure 13.25. Schematic diagram and state points for open-cycle MHD, Base Case 2. Courtesy NSF–ERDA–NASA.

Location	Point No.	Pressure (psia)	Temperature (°F)	Flow (lb/sec)
Ambient	0	14.696	59.0	2768.5
Compressor inlet	1	14.40	59.0	2768.5
Compressor outlet	2	95.36	465.2	2768.5
Preheater outlet	3	92.58	2398.4	2768.5
Combustor outlet	4	88.18	4414.4	3144.3
MHD duct entrance	5	59.18	4185.8	3144.3
MHD duct exit	6	13.09	3460.4	3144.3
Diffuser exit	7	17.00	3644.0	3144.3
Preheater exit	8	16.52	2538.0	3144.3
Air quench chamber exit	9	15.58	1881.0	3435.8
Steam generator exit	10	14.696	305.0	3435.8

The particular solutions adopted for the ECAS direct coal-burning case are the following:

Referring to Fig. 13.25, powdered coal is burned with compressed preheated air, using 95% of stochiometric air to limit nitrous oxide formation at the high temperature of the flame. Some 80% of the ash is then tapped off as slag before the seed is admitted in a mixer. At the high temperature of about 4414°F, part of the ash is gaseous and cannot be tapped off as slag. About 20% of the ash passes on down the duct. However, the removal of 80% of the ash before the seed is mixed in does minimize the problem of separating the seed from the ash later.

The seed is admitted as potassium carbonate. The sulfur combines with part of the seed, converting it to a sulfate. The use of 1% mass flow of seed is adequate even for high-sulfur coal, with effluent well within the EPA requirements. For a viable system, at least 95% of this seed must be recovered and recirculated. With an integral gasifier, in which ash

and sulfur are removed in the gasifier, the use of a cesium seed, admitted as an ore, results in a viable system with the seed expendable.

The hot gas from the combustor passes in turn through the accelerating nozzle, the duct, the diffuser, the air preheater, the steam generator, and the stack gas cleaner, to the stack. Gas enters the MHD duct at 775 m/sec, 2540 ft/sec, or over 1700 mph. The corrosive and erosive effect of this 4186°F gas, containing seed and some ash, presents the principal materials problem in the construction of the duct walls.

The coils of the magnet are enclosed in a Dewar (a vacuum heat-insulated vessel) containing liquid helium, which has a boiling point within 4°C of absolute zero. At this temperature, the coils are superconducting, and the extremely high excitation to produce the 6 tesla field is possible (iron is completely saturated at 2 teslas). The refrigerating system to provide the liquid helium is shown (Fig. 13.25).

After the MHD duct, the gases are expanded in an 80% efficient diffuser to just above atmospheric pressure and enter the air preheater, a large and very expensive gas-to-gas heat exchanger operating at very high temperatures. Note: For the 2000 MWe plant, the duct component costs $652,000, the magnet $69 million, the air preheater $93 million, the seed recovery, $44.5 million, and the inverter $57.8 million, all F.O.B. factory, not installed, 1976.

Exhaust from the preheater enters the steam generator at 2538°F. When it drops to about 1880°F, the slag–seed combination starts to condense, and the gas stream is quenched with compressed air at point 9. The ash and seed then become fluffy solids that are removed by the electrostatic precipitator and sent to the seed recovery system. The sulfur is removed with the seed.

The nitrogen has been controlled by the combination of substochiometric combustion at the high temperature and final full burning with 105% stochiometric air after the air quench. The temperature is then too low for the formation of nitrous oxides and clean air passes up the stack.

Seed recovery was by hydrogen treatment of the sulfate. The turbine generator, of conventional design, drove the compressor and provided about half the output power.

Cost of Electricity. The direct coal-burning system described resulted in the lowest calculated cost of electricity of the MHD alternates consid-

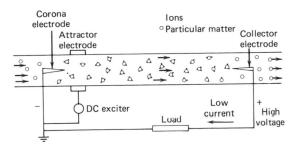

Figure 13.26. Electrogasdynamic generator.

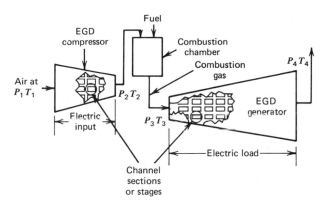

Figure 13.27. EGD generating system.

ered, 27 mills/kWh at 65% capacity factor, 18% fixed charges, coal at 85¢/million Btu, and 1980 time frame. Both magnet costs and air preheater costs were however based on very speculative assumptions of cost reductions (magnet) or successful future developments (heater).

ELECTROGASDYNAMICS (EGD)[10,11]

The basic principle of electrogasdynamic generation of electricity is shown in Fig. 13.26. Air or gas is forced under pressure through the duct from left to right. The gas stream carries an aerosol, either flyash from combustion of coal or an intentionally supplied aerosol, particles thousands of times larger than molecules. These particles are indicated by small circles. The corona electrode sprays ions, indicated by dots, on these aerosol particles, hundreds or thousands of unit charges per droplet. These droplets with their attached charges are carried along with the gas stream against the opposing field, and the charges are drawn off at the high-voltage end by the collector electrode. This is similar to the operation of a Van de Graaff generator, in which a belt carries the positive ions to the high-voltage electrode. The energy in the gas stream goes into elevating the charges to a high voltage as the gas is slowed down. Thus, the mechanical energy of the gas stream is converted to electric energy. The EGD generator is characteristically a high-voltage, low-current device (100 kW at 100 kV is typical).

The slip of the aerosol particles with respect to the air stream is very small, of the order of 2%, due to the large size of these particles relative to the gas molecules. This corresponds to a near-perfect gas turbine in which the blades slip very little with respect to the gas flowing through.

This system is obviously reversible. By reversing the polarity of the electrodes in Fig. 13.26 and pro-

viding a dc power source instead of the load, the forces on the particles toward the right will cause a compression of the gas in that direction. The duct then becomes an EGD compressor.

A complete EGD generation system is shown in Fig. 13.27. Ambient air is compressed in either an EGD or turbine-type compressor. Fuel is added and burned at constant pressure ($P_3 = P_2$). The hot gas is then expanded through the EGD generator and exhausted to the atmosphere.

This is the Brayton cycle as used in the gas turbine. As with the gas turbine, the EGD compressor and generator are each composed of many stages. This is indicated in Fig. 13.27. Each stage is suitable to the pressure and temperature range within which it operates. Channel sections in parallel are necessary to conform to the "slender duct" principle of EGD ($L \gg D$).

The ducts, being stationary, are of ceramic materials and suitable for somewhat higher inlet temperatures than gas turbines (without air cooling). However, they are exposed to the erosion effects of flyash or other aerosol and have the added requirement of maintaining their high insulating qualities under these adverse conditions.

A thermal efficiency of about 50% was calculated[12] for a pressure ratio of 20, an inlet temperature of 2300°F (1260°C), and the following assumptions:

Fuel coal 13,000 Btu/lb	HHV
Compressor efficiency	85%
Conversion efficiency	90%
Intercooling efficiency	75%
Regeneration	80%
Air temperature	85°F (29°C)
Cooling water	75°F (24°C)

"Regeneration" was called "recuperation" earlier under gas turbines. For "intercooling," see Recuperated Gas Turbines.

Advantages. The attraction of EGD generation stems from the following foreseen advantages (1965): The direct burning of pulverized coal without moving parts and with no boiler or working fluid other than the flue gases themselves, and at a higher efficiency than with the Rankine cycle, was of interest to OCR, the initial sponsors of the development. Small or zero cooling water requirements and the possible combination with the gas-cooled nuclear reactor were other significant factors. The anticipated low capital cost was yet to be proved. Current EPA standards (1981) would of course require sulfur removal and nitrous oxide limitation, as with all new plants. These were not considered in 1965.

Status of EGD. EGD is little beyond the conceptual stage (1981), and the "foreseen" advantages are still to be demonstrated. Some of the difficulties have been cited. In the meantime dual cycles have been developed raising the thermal efficiency with more conventional technology.

Principal development of this system has been by Gourdine Systems, Livingston, New Jersey, the originators, working with Foster Wheeler, and with General Public Utilities, as the projected site of a pilot installation. The project was being supported by the Office of Coal Research (OCR) now part of DOE. As of June 1967, OCR had contracted with Dynatech, Cambridge, Massachusetts, to evaluate the project.[13]

FUEL CELLS[14,15]

A fuel cell is an electric cell that converts the chemical energy of a fuel directly into electric energy in a continuous process. It differs from a battery in being a "continuous process." Both convert the energy released by chemical reactions directly into electricity. A cell that works spontaneously, a battery or fuel cell, is called a galvanic cell. One that is driven, such as an electrolysis process cell, is called an electrolytic cell.

Principle of Operation

The principle of a typical alkaline fuel cell is shown in Fig. 13.28, in which hydrogen and oxygen are supplied through two porous electrodes. The general principles illustrated by this example apply to all fuel cells.

At the cathode, or oxidizer electrode, of the alkaline fuel cell, the following reaction takes place:

$$\tfrac{1}{2} O_2 + H_2O + 2\,e^- \rightarrow 2\,OH^- \qquad (13.41)$$

A molecule of water is oxidized into two hydroxyl ions, the two needed electrons being drawn from the circuit. The hydroxyl ions migrate through the electrolyte to the other electrode.

At the anode, or fuel electrode, the reactions are

$$H_2 \rightarrow 2\,H^+ + 2\,e^- \qquad (13.42)$$

$$2\,H^+ + 2\,OH^- \rightarrow 2\,H_2O \qquad (13.43)$$

A hydrogen molecule supplies two electrons to the circuit, and two hydrogen ions are formed. These combine with the two hydroxyl ions to form two molecules of water. Thus, overall,

$$H_2(g) + \tfrac{1}{2} O_2(g) \rightarrow H_2O(l) \qquad (13.44)$$

Two electrons circulate around the circuit, and one molecule (net) of liquid water is formed. At 25°C (77°F) and 1 atm, this reaction takes place[16] with a Gibbs free energy change, ΔG, of -56.69 kcal/g mole liquid water formed, or 2.46 eV per molecule water formed. That is,

$$\frac{56.69 \text{ kcal/g mole water} \times 4186 \text{ J/kcal}}{1.6 \times 10^{-19} \text{ J/eV} \times 6.02 \times 10^{23} \text{ molecules/g-mole}}$$
$$= 2.46 \text{ eV/molecule water}$$

Since there are two electrons circulated per molecule water formed, the no-load voltage is 1.23 V. As the load current increases, the terminal voltage drops due to internal resistance. However, efficiencies as high as 90% are possible theoretically (83% for the particular cell described; see Fuel Cell Theory). The current range of cell efficiencies (1978) is 50 to 70% with 80% projected.

Thus, the entire fuel cell development is aimed at increasing the efficiency, the specific output, and the useful life and reducing the cost, all with the goal of lower cost per kWh. The rate of activity is being increased by improved catalysts and higher temperatures. Internal resistance and polarizations are being reduced. New materials and construction techniques lead to longer life. Ways are being de-

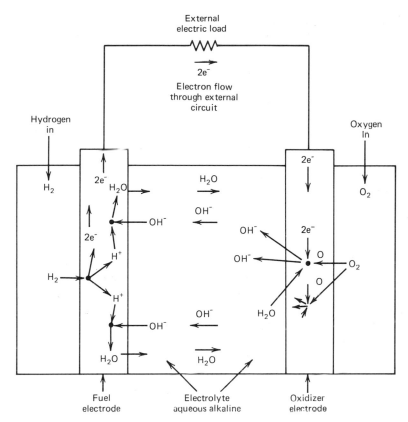

Figure 13.28. Hydrogen and oxygen each start sequences of reactions in this typical low-temperature alkaline fuel cell.

veloped to use "practical fuels"—coal, oil, gas, or methanol—with air, instead of hydrogen with oxygen.

For military and space applications, the goals are size, weight, and life; for mobile applications, size, weight, and cost; and for stationary power sources, primarily reduced overall cost of electricity.

Advantages of the Fuel Cell

The high efficiency of this direct conversion from chemical to electric energy, even in relatively small units, is the principal advantage of the fuel cell. However, its absence of moving parts, its reliability (comparable to a battery) and its negligible polluting emissions are also very important assets. Since the energy does not first have to be converted to heat, the efficiency is not limited to the Carnot efficiency. For many cells, the efficiency rises as the temperature goes down, not up, and is frequently better at light loads than at full load.

Fuel cells are also quiet and compact. To supply loads of longer duration than 50 to 100 hr, both the weight and volume are less than for batteries or engine-generator sets.

Types of Fuel Cells

In order to secure the most economical fuel cell power sources, it is desirable that the reactions proceed rapidly, with a minimum of adverse "polarizations" or charge buildups. To this end, a variety of acid and alkali electrolytes as well as "molten salts" and "solid electrolytes" have been tried. Various electrode materials have been used. A number of catalysts have been used to accelerate the activity, particularly for the lower-temperature cells. Also higher temperatures have been used which minimize or eliminate the use of catalysts.

Typical examples of systems employing these various types of cells were evaluated in the ECAS study (1976),[4] for central power applications. The fuel cell systems considered in this study were to have a minimum capacity of 25 MWe, over 100

Table 13.7. Fuel Cell Classification by Electrolyte Type

Electrolyte Type	Variations	
Aqueous acid	Phosphoric acid	(302–374°F, 423–463°K)
	Solid polymer	(167°F with air, 348°K)
		(302°F with oxygen, 423°K)
	Sulfuric acid	(140°F, 333°K)
Aqueous alkali	30 wt% Potassium hydroxide	(158°F, 343°K)
	75 wt% Potassium hydroxide	(392°F, 473°K)
	Saturated carbonate–bicarbonate	(140°F, 333°K)
Molten salt	Molten carbonate	(1202°F, 923°K)
High-temperature solid electrolyte	Stabilized zirconia	(1832°F, 1273°K)

Source: Courtesy NSF–ERDA–NASA.

times the capacity of any fuel cells yet built (1976). Thus, the evaluation was largely theoretical, based on data and experience with small units. Acid fuel cell units up to 40 kW had been tested, and 26 MWe systems were being built for a number of utilities.[4,17] However, the 4.5 MWe phosphoric acid fuel cell,[17] described later, is expected to be the first large fuel cell to be demonstrated (1981).

Fuels for Fuel Cells

Whereas pure hydrogen or hydrazine (a liquid NH_2NH_2) and oxygen can be used for small cells for space or military purposes, cells to provide central station power must be based on the conventional hydrocarbon fuels coal, oil, gas, or methanol. These must usually go through a preparation step before entering the cell, although gas can be used directly in the higher-temperature cells. Oil and gas must generally be reformed into H_2 and CO and the CO "shifted" into H_2 (see Water Gas Shift, Eq. 4.5). For alkaline cells, carbon must be practically eliminated to avoid carbonizing the electrolyte and plugging the porous electrodes with precipitated carbonates.

Coal must be gasified to CO and H_2 and the CO shifted to H_2, as with oil or gas. Thus, except in some high-temperature cells, the fuel ends up as a hydrogen-rich gas.

The use of air as the oxidant is generally most economical for large power uses, even though it entails dilution with nitrogen.

Classification of Fuel Cells

Fuel cells may be classified by electrolyte type into four categories, each of which has several varia-

tions, as shown in Table 13.7. Since no one type was unquestionably superior, the ECAS parametric studies were made of four systems. The first variation of each type in Table 13.7 was the one selected for the ECAS evaluation because of its greater promise in large power applications, such as 25 MWe.

The factors varied in the parametric studies are of prime importance in fuel cell design and application. These were:

1. Useful life and rating of the fuel cell subsystem.
2. Fuel cell power density.
3. Electrolyte thickness.
4. Fuel and oxidant types.
5. Performance degradation over the useful life.
6. Anode and cathode catalyst loadings.
7. Temperature of operation.
8. Recovery of waste heat.

Of these factors, the useful life, the power density, the fuel type, and the recovery of waste heat were found to have the principal effect, both on the efficiency and on the cost of electricity. The range of performance found is shown in Table 13.8.

After improvements deemed possible by limited developments indicated by the ECAS studies, it was projected that the cost of electricity in mills/kWh with the four systems of Table 13.8 would be: for the phosphoric acid—mid- to high-30's; for the potassium hydroxide, low 40's; for the molten carbonate, low 30's; and for the solid electrolyte, stabilized zirconia, in the high 20's (1976 basis) (see Chapter 1 for updating indices).

Table 13.8. Estimated Performance of 25 MWe Fuel Cell Systems, 1976[4]

	Capital Cost ($/kWe)	Thermal Efficiency[a] (%)	Cost of Electricity[b] (mills/kWh)
Phosphoric acid	350–450	24–29	42–50
Potassium hydroxide	450–700	26–31	46–61
Molten carbonate	480–650	32–46	38–70
Stabilized zirconia	420–950	26–53	35–61

[a]Based on the HHV of the high-Btu gas fuel used.
[b]Mills/MJ = mills/kWh × 0.2778.
Source: Courtesy NSF–ERDA–NASA.

Phosphoric Acid Cell

The phosphoric acid cell was chosen over the sulfuric acid or solid polymer electrolyte (SPE) because it "best addresses the problems of tolerance toward carbon monoxide. (It functions with CO < 0.5%), and minimization of noble metal loadings, while maintaining high levels of anodic and cathodic activity."[4] The higher operating temperature of 374°F was chosen.

The first electric utility fuel cell to be demonstrated[17] uses a phosphoric acid electrolyte and generates about $\frac{2}{3}$ of a volt dc per cell at 2500 A/m² (230 A/ft²) while operating at a cell temperature of about 190°C (375°F). This is a system supplying 4.5 MWe ac.

This system is divided into three sections—fuel processor, fuel cell, and power conditioner. The fuel processor reforms the fuel (natural gas, naphthas, or selected kerosenes) into a hydrogen rich gas, which is supplied to the fuel cells. Water and heat from the fuel cells are in turn fed to the fuel processor and utilized. The power conditioner converts the dc to ac and supplies reactive power. The fuel cells are of filter press construction resembling a storage battery. The hydrogen-rich gas is supplied to the anode, where hydrogen is dissociated into hydrogen ions, releasing electrons to the circuit. However, unlike the alkaline cell of Fig. 13.28, the hydrogen ions migrate through the electrolyte to the cathode in the acid cell. Here, they react with oxygen from the air, drawing electrons from the circuit to produce water in the form of steam. Two electrons are circulated around the circuit per molecule of water formed. While the reaction involved is that of Eq. 13.44, the free energy change is slightly less at 375°F because the water is a gas at the higher temperature. Consequently, the no-load voltage is slightly less than 1.23 V, dropping to about $\frac{2}{3}$ V at full load.

The expected heat rate of this demonstration system is 9300 Btu/kWh (36.7% thermal efficiency) at rated power and 9000 Btu/kWh (37.9% thermal efficiency) at 30% of rated power. This is the overall system heat rate from the HHV of the fuel input to the ac power at the 13.8 kV output terminals. Note that the efficiency is higher at part load.

This system is designed for a 20 yr life with scheduled overhaul and maintenance, and the cell stack assemblies are expected to last 1 yr at rated power. The load response time of 0.5 sec from 35% load to full load is of exceptional importance in providing the equivalent of "spinning reserve."

The acid cell requires a noble metal catalyst at the electrode–electrolyte interface. A loading of 1 mg Pt/cm² was used in the ECAS base case, with loadings of 0.3 and 0.1 mg/cm² considered as well. An electrolyte thickness of 0.5 mm was used with 0.25 mm as a parametric variation. The state-of-the-art performance as of 1975 is shown in Table 13.9.

Aqueous Alkali Fuel Cell—35% by Weight Potassium Hydroxide at 158°F

The 30% by weight potassium hydroxide alkaline cell was chosen (for study) over the 75% by weight potassium (Bacon) cell because of the "well-known severe corrosion problems of the latter system."[4] It was chosen over the saturated carbonate–bicarbonate cell because of low cathodic activity in the latter, due mainly to excessive concentration polarization. The significance of these terms will become clear in the ensuing discussion (see Fuel Cell Theory).

Table 13.9. State-of-the-Art Performance, Acid Fuel Cells, 1975[4]

| Oxidant | Temperature | | Power Density of Active Area | | Voltage/Cell at Full Power (V) | % of HHV |
	°K	°F	mw/cm²	w/ft²		
Air	348	167	180	167	0.66	44
Oxygen	423	302	470	437	0.75	51

Source. Courtesy NSF–ERDA–NASA.

A principal problem with alkali fuel cells is carbonation of the electrolyte by CO_2 in the air or fuel gas (when carbonaceous fuels are used). Solid carbonates also plug the porous electrodes. Long-lived air cathodes in alkaline solution operate with much better polarization characteristics than they do in acid solution. Furthermore, noble metal catalysts are not required. Silver can be used. Also nickel, at moderate cost, is stable in the cell environment, where graphite is required in acid cells. However, the efficiency advantage of the alkaline cell, resulting from lower cathodic polarization, is almost totally negated by the the efficiency penalty resulting from the need to eliminate CO_2 from the air or fuel streams. Thus overall efficiencies of 35 to 40% result for either acid or alkali systems operating on air and carbonaceous fuels.

Molten Carbonate Fuel Cells

While no large systems of this type have been built, the concept employed in the ECAS study[4] involved a filter press design with a porous nickel anode and a lithiated nickel cathode. The electrolyte was a semisolid paste of alkali aluminate powder and molten alkali carbonates (Li_2CO_3, 43.5 mole %; Na_2CO_3, 31.5 mole %; K_2CO_3, 25 mole %, mp 670°K, 747°F).

It was assumed that the high-Btu gas fuel could be reformed on the internal cell surfaces. Thus, the heat from the exothermic cell reaction would be used in part for the endothermic reforming process.

The base case operating temperature was 923°K (1202°F). At this high temperature, regeneration is necessary for economy, and both anode and cathode streams were recycled through heat exchangers to transfer the heat to incoming fuel and air streams. The remaining surplus heat would be used in heat recovery boilers for process steam needed in the area or for a bottoming steam turbine.

Solid Electrolyte Fuel Cell

The solid electrolyte operates on a semiconductor principle. The material, $(ZrO_2)_{0.85}(CaO)_{0.15}$ or $(ZrO_2)_{0.90}(Y_2O_3)_{0.10}$, (stabilized zirconia) is an impervious ceramic which has the ability to conduct a current by the passage of O^{2-} ions through the mechanism of holes in the crystal lattice. It is impervious to all other gases such as H_2, O_2, and H_2O and also to electrons. The material thus acts as an electrolyte.

Oxygen is ionized at the cathode drawing electrons from the circuit:

$$\tfrac{1}{2} O_2 + 2 e^- \rightarrow O^{2-} \qquad (13.45)$$

The oxygen ions pass through the electrolyte to the anode. For a hydrogen fuel, the hydrogen is ionized at the anode, providing electrons to the circuit. Hydrogen and oxygen ions combine to form water:

$$H_2 \rightarrow 2 H^+ + 2 e^- \qquad (13.46)$$

$$2 H^+ + O^{2-} \rightarrow H_2O \qquad (13.47)$$

The overall reaction is thus

$$H_2(g) + \tfrac{1}{2} O_2(g) \rightarrow H_2O(g) \qquad (13.48)$$

At the operating temperature of 1000°C (1832°F), the change in Gibbs free energy for this reaction is 44.789 kcal/g mole water formed, or 1.947 eV per molecule water. The open circuit voltage is thus $1.947/2 = 0.97$ V.

The Westinghouse solid electrolyte cell[18,19,20] is of bell and spigot construction. Each cell is a thin tube of ceramic with porous platinum electrodes on the inner and outer surfaces. These cells are joined in series as pipes are joined in a bell and spigot manner, with the tapered end of one section into the flared end of the next. Such series strings of cells, arranged vertically, are manifolded at top and bottom, with the fuel gas mixture passing through the inside of the tubes and the air on the outside. Fuel gas mixtures as from a coal gasifier can be handled, although the use of the water–gas shift reaction to provide a hydrogen-rich gas minimizes polarization losses at the anodes. Electrical connections are made at the upper and lower ends.

Fuel cell power systems based on stabilized-zirconia solid electrolyte offer the highest achievable efficiency.[20] Such cells, using gas from coal gasification, are under development for power generation and for high-temperature, industrial cogeneration systems.[20]

Fuel Cell Theory

It has been pointed out that, if the electrons have to pass from anode to cathode around the electric circuit to complete the chemical reaction, then in the reversible limit, of zero losses in the circuit or in the electrolyte, the Gibbs free energy change of the reaction, ΔG, at the temperature and pressure involved is all transferred to the electric circuit output. When the fuel and oxidant combine in a fuel cell or any other process, the enthalpy of combustion, ΔH, is given up. Thus, the maximum theoretical efficiency of a fuel cell is

$$\eta = \frac{\Delta G}{\Delta H} \qquad (13.49)$$

where both ΔG and ΔH are functions of temperature and pressure. Thus, for the hydrogen–oxygen cell at 25°C (77°F) and 1 atm, with the reaction of Eq. 13.44, the maximum theoretical efficiency is

$$\frac{-56.69}{-68.317} = 83\%$$

At the higher temperature of Eq. 13.48, it is

$$\frac{-44.789}{-68.317} = 66\%$$

In addition to this loss of energy which is unavailable, 17 to 34% in these cases, there are polarizations at the electrodes and resistance drops in the electrolyte and the leads that result in further losses. Thus, in Table 13.9 the cell voltage at full load

is given as a "% of HHV." This assumes that the proportion of the HHV that is converted to output is the same as the cell voltage, as a proportion of what it would be were all the HHV converted to electric output. The expression "% of HHV" is thus synonymous with "thermal efficiency" and includes both the thermodynamically irrecoverable loss and the further losses due to imperfections.

Polarization Losses. The further losses known as polarization losses are as follows:

1. Chemical activation: energy consumed in preparing the reactants and causing them to combine.
2. Concentration polarization: stemming from a decrease in the concentration of the reactants relative to products in the electrode–electrolyte film interface.
3. Resistance polarization: electrical resistance in the electrodes and ion flow resistance in the electrolyte.

Electrodes. Because of the low solubility of H_2 and O_2 in the electrolyte, the reactions take place at the electrode–electrolyte interface, requiring a large area of contact. Thus, the electrodes are porous materials which:

1. Provide a contact between the electrolyte and the gas over a wide area.
2. Catalyze the reaction.
3. Maintain the electrolyte in a very thin layer at the surface of the electrode.
4. Act as leads for current conduction.

Catalysts. The principal catalysts used are platinum, silver, nickel, cobalt, and palladium. Catalysts are being developed which are less susceptible to carbon monoxide poisoning.

Molten Carbonate Cells. At temperatures around 1202°F, these cells can use reformed hydrocarbon fuels. The carbon monoxide reaction is then as follows:

Cathode

$$CO_2 + \tfrac{1}{2} O_2 + 2\,e^- \rightarrow CO_3^{2-} \qquad (13.50)$$

Anode

$$CO + CO_3^{2-} \rightarrow 2\ CO_2 + 2\ e^- \qquad (13.51)$$

Overall

$$CO(g) + \tfrac{1}{2}\ O_2(g) \rightarrow CO_2(g) \qquad (13.52)$$

At the temperature of 923°K (1202°F), the change in Gibbs free energy for reaction 13.52 is -52.76 kcal/g mole CO_2 formed, or 2.29 eV per molecule CO_2 formed. The no-load voltage is thus $2.29/2 = 1.14$ V.

For hydrogen, the no-load voltage is 1.26 V at 77°F and 0.97 V at 1832°F. At 1202°F, it is intermediate; and with mixed CO and H_2 fuel, the effective no-load voltage would be between this value and 1.14 V, depending on the mixture.

Summary. The relation between the Gibbs free energy change for a reaction, the equilibrium constant, and the cell no-load voltage when the reaction is completed through an electrolyte may be summarized as

$$-\Delta G = RT \ln K = \frac{n\mathscr{F}\mathscr{E}}{4.1840} \qquad (13.53)$$

where, in a consistent set of units:

ΔG is the change in Gibbs free energy for the reaction in question, at the specified temperature and 1 atm, in cal/g mole (positive into the system).

R is the gas constant, 1.9872 cal/g mole °K.

T is the temperature, in °K.

K is the equilibrium constant for the reaction in question, at the specified temperature and 1 atm.

n is the number of moles of electrons carried around the circuit for the reaction as written.

$\mathscr{F}$ is the Faraday number, the charge on 1 mole of electrons, that is, on the Avogadro number of electrons, 96,500 C/g mole.

$\mathscr{E}$ is the no-load voltage, in V.

Since $n\mathscr{F}\mathscr{E}$ is in coulombs $\times$ volts = watt-seconds, or joules, it must be divided by 4.1840 J/cal to obtain calories, all per mole.

Thus, for the reaction of Eq. 13.44, from the thermochemical tables[21] $\ln K = 95.730$. Thus, as given earlier,

$$-\Delta G = \frac{1.9872 \times 298 \times 95.730}{1000} = 56.69 \text{ kcal/g mole}$$

Or for the condition of Eq. 13.52, solving Eq. 13.53 for $\mathscr{E}$:

$$\mathscr{E} = \frac{-4.184 \times \Delta G}{n\mathscr{F}}$$

$$= \frac{-4.184\ (-52,760)}{2 \times 96,500} = 1.14 \text{ V}$$

as calculated earlier.

STORAGE BATTERIES[22,23]

Strictly speaking, a battery is an assembly or "battery" of cells connected in series and/or parallel. In the simplest case, it is one cell. However, the terms battery and cell are often used interchangeably, with battery denoting a cell or battery of cells to supply electricity.

Batteries, in fact, all electrochemical cells, whether they be galvanic (spontaneous) or electrolytic (driven), such as cells for the electrolysis of water, are governed by the same basic equations. Equation 13.53 relates the Gibbs free energy change ΔG for the reaction involved to the equilibrium constant K for the reaction and to the no-load voltage $\mathscr{E}$ and the charge transferred, $n\mathscr{F}$, for the reaction.

Faraday's Law

In addition Faraday's law, which governs the amount of material converted in the process, is of great importance. It determines the size required for a given capacity. Faraday discovered that the material converted was directly proportional to the dc charge passed through the cell. The law may be stated as follows[16]:

If n is the number of electrons participating in the reaction of one molecule, or ion, then the reaction of one mole of the substance involves n faradays of electricity. In terms of coulombs, the faraday is $\mathscr{F} = $ 96,487 C/g mole.

In a fuel cell or electrolytic process, the material does not all have to be stored in the cell at any one time, and the power per unit area, or conversion rate per unit area, is a principal determinant of size. However, in a battery all of the material must be stored (except in zinc–air or iron–air cells), and this determines the size for a given capacity. The capacity is usually expressed in ampere hours at the battery voltage, under the specified conditions of rate, temperature, and so on.

Primary (Disposable) and Secondary (Storage) Batteries

Primary or disposable batteries are used as small convenience power sources for flashlights, radios, and the like, but are uneconomic for the larger power sources considered in this book. Secondary or storage batteries typically reuse the same active materials hundreds or thousands of times. They are now in use for limited-range electric cars and for driving submarines under water, and have been used in the past as emergency supplies for the downtown sections of cities, driving elevators, street cars, and large motors for a matter of hours. Some 10,000 battery-powered automobiles were operating in the United States before gasoline "took over."

Batteries are now widely used in autos, railway cars, and planes to supply lights and auxiliaries while the charging devices are at a standstill and as the emergency supply and load equalizer for dc control systems in industrial systems of all kinds. In smaller sizes, as sealed units, they now power a wide range of portable appliances, tools, and electronic gear.

It is the further development of storage batteries that is now looked to for the large electric storage jobs of the future—powering electric vehicles and providing temporary storage for electric utility load leveling and emergency. This is the area discussed primarily in the following paragraphs.

Classification of Storage Batteries

The storage batteries in common use today (1978) may be classified as follows:

1. Lead–acid, using sulfuric acid electrolyte: Pos.–PbO_2–H_2SO_4–Pb–Neg. It is by far the most common and versatile.

2. Nickel–iron (Edison cells) using potassium hydroxide electrolyte: Pos.–NiOOH–KOH–Fe–Neg. Its principal use is in cycled service in industrial trucks, railway cars, and electric automobiles.

3. Nickel–cadmium, using potassium hydroxide electrolyte: Pos.–NiCOH–KOH–CD–Neg. It is in increasing use as a premium, high-quality battery for consumer products (sealed type) and where expanded cycle life is needed in heavier duty applications (vented type).

Advanced versions of the lead–acid and nickel–iron batteries are the principal candidates for the near-term powering of electric cars (1978).

In addition, five other systems are being developed with DOE and EPRI support to provide the large, low-cost, high energy-density batteries required for electric vehicles and for electric utility load leveling in the future. These are:

4. Nickel–zinc, using potassium hydroxide electrolyte. This is a high energy-density nickel battery. Its in–out efficiency is 65 to 70%.

5. Zinc–air. Development of high-rate air electrodes for fuel cells has renewed interest in this old, low-rate system. It has relatively low in–out efficiency, 30 to 40%.

6. Iron–air. Similar to zinc–air.

7. Lithium–metal sulfide, using a molten salt electrolyte. This is a class of high-temperature batteries operating at 400 to 450°C (752 to 842°F) and having high energy and power densities. Typical examples are: Pos.–FeS or FeS_2–(LiCl–KCl eutectic,* mp 325°C, 617°F)–LiAl–Neg.,[24] Or Pos.–FeS–(LiCl–KCl eutectic)–Li_5Si–Neg.

8. Sodium–sulfur, using a solid beta-alumina ceramic electrolyte (charge carrier sodium ions). It is a high-temperature battery operating at 300°C (572°F) and having high energy and power density. Typical is: Pos.–sulfur†–(NaO 0.11, AlO_3 0.89)–sodium–Neg.[22] (see Fig. 13.29).

Several other systems are being developed privately or for other uses, or on a longer time schedule, including the following[24]:

9. Zinc–chlorine system. More like a chemical plant than a battery. Pumps and refrigera-

---

*Eutectic is the mix proportion resulting in the lowest melting point.

†A sulfur–graphite matte is used to obtain conductivity.

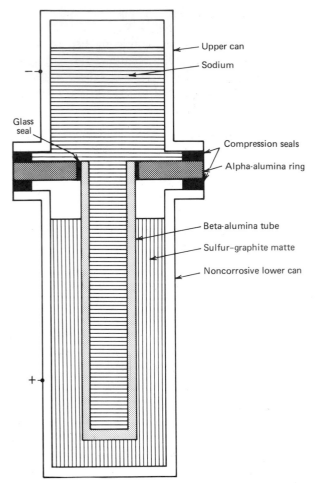

Figure 13.29. Principle of the high-temperature sodium–sulfur storage cell.

tion are used and "yellow ice" (chlorine hydrate) is stored.[22]

10. Nickel–hydrogen. A high energy-density and cycle-life system for aerospace use, as in satellites.[22]

11. Lithium–chlorine, being developed by the Electric Storage Battery Co. for fork lift trucks. The molten salt electrolyte is a KCl–LiCl eutectic at 450°C (842°F). Approximately 65 Wh/kg (30 Wh/lb).

12. Zinc–bromine.

13. Lead–manganese oxide.

14. Zinc–manganese oxide.

15. Sodium–antimony trichloride. A moderately high-temperature battery operating at 200°C (392°F).[22]

Lead–Acid Cell

The active material of the positive plates is lead peroxide, PbO_2, and that of the negative plates, sponge lead, Pb. The electrolyte is dilute sulfuric acid.

During discharge, the reactions taking place at the positive electrode are

$$PbO_2 + H_2SO_4 + 2\,e^- \rightarrow PbSO_4 + H_2O + O^{2-} \tag{13.54}$$

$$2\,H^+ + O^{2-} \rightarrow H_2O \tag{13.55}$$

Lead peroxide and sulfuric acid combine to form lead sulfate, water, and oxygen ions, drawing two electrons from the circuit. The oxygen ions combine with hydrogen ions in the electrolyte to form water.

At the negative electrode, the reaction is

$$P_b + SO_4^{2-} \rightarrow PbSO_4 + 2\,e^- \tag{13.56}$$

Lead combines with an SO_4^{2-} ion to form lead sulfate and supply two electrons to the circuit. This leaves two hydrogen ions, $2\,H^+$, to migrate. Thus overall, adding Eqs. 13.54, 13.55, and 13.56,

$$PbO_2 + Pb + 2\,H_2SO_4 \underset{\text{charge}}{\overset{\text{discharge}}{\rightleftarrows}} 2\,PbSO_4 + 2\,H_2O \tag{13.57}$$

Thus, on discharge, lead peroxide and lead on the positive and negative plates are replaced by lead sulfate, and H_2SO_4 molecules in the electrolyte are replaced by water.

Since the molecular weights of H_2O and H_2SO_4 are respectively 18 and 98, a specific gravity of 1.3 implies approximately 6.75% H_2SO_4 molecules. A specific gravity of 1.1 implies 2.25% H_2SO_4 molecules in the electrolyte. This correlation between the specific gravity and the amount of H_2SO_4 present makes it a good guide to the state of charge once the fully charged and discharged gravities are known.

Material Involved. Since two electrons traverse the circuit to produce the reaction among molecules shown in Eq. 13.57, then from Faraday's law, two faradays will produce the corresponding gram mole reaction. Two faradays are 192,974 coulombs, or 53.6 ampere hours. The weights of active material which must be converted for this output are, from Eq. 13.57:

Substance	Molecular Weight	Weight	
		g	lb
PbO_2	239.2	239.2	0.53
Pb	207.2	207.2	0.46
H_2SO_4	98	196	0.43
Total		642.4	1.42
Converted to			
$PbSo_4$	303.2	606.4	1.34
H_2O	18	36	0.08
Total		642.4	1.42

These weights apply to a single cell or 2 V battery. For a 12 V battery, the required weights will of course be six times as much. A 53.6 amp hr 12 V battery must convert $0.99 \times 6 = 6.0$ lb PbO_2 and Pb to $PbSO_4$. A 12 V car battery of about this capacity weighs about 50 lb. Thus, the structure to hold the active material, including the electrolyte, is much heavier than the material to be converted.

Construction. The plates of lead–acid batteries have in the past been of lead–antimony castings. By changing to lead–calcium castings, the gassing has been reduced sufficiently that the batteries could be "sealed." A self-sealing pressure relief vent prevents pressure buildup, and the small loss of liquid is compensated by an increased electrolyte charge at the start. This results in a "maintenance free" battery without the familiar filler caps.

Smaller lead–acid batteries, which also use the calcium–lead castings, make use of a "gelled" electrolyte and can be used in any position. These cells are suitable for 200 to 500 cycles or even 1000 partial cycles, and are used in rechargeable lanterns, appliances, tools, and electronic gear.

There are other approaches[22] to reduced gassing or disposing of the gas internally that serve the same purpose.

Nickel–Cadmium Cell

The G.E. Co. nickel–cadmium cell is made up of nickel plated steel plates on which have been sintered nickel grains, so that the surface is 80% open volume and 20% solid grains. These plates are soaked in nickel or cadmium salts and formed by an electric current into the positive and negative plates. The electrolyte is potassium hydroxide, KOH. The reactions taking place during discharge are

At the positive plates

$$2 H_2O + 2 NiOOH + 2 e^- \rightarrow 2 Ni(OH)_2 + 2 OH^-$$
$$(13.58)$$

At the Negative Plates

$$Cd + 2 OH^- - 2 e^- \rightarrow Cd(OH)_2 \quad (13.59)$$

Overall Reaction

$$Cd + 2 H_2O + 2 NiOOH \underset{\text{charge}}{\overset{\text{discharge}}{\rightleftharpoons}} 2 Ni(OH)_2$$
$$+ Cd(OH)_2 \quad (13.60)$$

The nominal voltage of these cells is 1.25 V and varies from 1.2 to 1.3 V.

While more expensive than the lead–acid battery for the same energy storage, the nickel–cadmium battery has greater cycle life, especially under deep discharge conditions, as well as long shelf life. A life of 1000 full charge–discharge cycles is typical. It therefore finds its place as a premium battery for a host of small- and medium-power applications.

These two cells, lead–acid and nickel–cadmium, as well as the fuel cells described earlier illustrate the principles of all batteries. The reaction must be completed by ions through the electrolyte and electrons around the circuit in order that the free energy change of the reaction be converted to electric energy. As with fuel cells, the limiting efficiency for a perfect battery with any given reaction is $\Delta G/\Delta H$. The real efficiency is less due to various losses. For a storage battery the final efficiency of interest is the storage efficiency, energy out/energy in. In–out efficiencies of 50 to 75% are typical (60% used in Table 4.11), 30 to 40% with an air electrode.

Batteries Under Development

What is needed in the batteries of the future? If they are to make electric vehicles competitive and be attractive for load leveling in electric utility systems,

Table 13.10.   **Storage Battery Characteristics and Interim Goals for Electric Vehicle Batteries**

| | | Energy Density | | | | |
| | | By Weight | | By Volume | | |
Type	Cost ($/kWh)	Wh/lb	Wh/kg	kWh/ft³	kWh/m³	Life Cycles[a]
1976 Batteries						
Nickel–iron	400	15	33	1.4	49.4	3000
Lead–acid						
Motive power[b]	50	10	22	2.6	91.7	1500/3000
Submarine	80	12.7	28	2.0	70.6	400
Golf cart	35	15.9	35	2.2	77.6	300
Elec. vehicle	100	15.9	35	2.8	98.8	500/800
Interim Goal						
Elec. vehicle	25–35	>31.8	>70			>1000 (3–10 yr)

Source.   Courtesy A.R. Landgrebe, K. Klunder, and N.P. Yao.[24]
[a]Range, severe to modest duty.
[b]Industrial truck and so on.

they must be improved by almost every measure: $/kWh of storage, kWh/lb, kWh/ft³, kW/lb, cycle life, years of useful service, and storage efficiency. A 4:1 improvement in performance at one fourth the price would be in the ballpark.

The battery of 1981 may be characterized as follows.[23] Automotive batteries may generally be expected to give 300 cycles or to last 2 yr. Industrial truck sizes may be expected to give 1500 to 3000 cycles in 5 to 10 yr. Standby sizes for large dc supplies may be expected to float across the dc bus for 8 to 30 yr. Generally the most costly, largest, and heaviest cells are the longest-lived.

A typical 12 V "car battery" of 1978, 7 × 10 × 8 in., or 0.32 ft³, weighs about 50 lb and costs about $50. It may be rated at 2.5 amp, 20 hr (50 amp hr) or 25 amp, 112 minutes (47 amp hr). As a reference point this is roughly 0.6 kWh of storage and results in $83/kWh, 12 Wh/lb (26 Wh/kg), 1.9 kWh/ft³ (67 kWh/m³) and 6 W/lb (13.2 W/kg) at the 112 minute rate.

Published[24] average values for 1976, based on a 6 hr rate, are given in Table 13.10. In comparison with the 1976 batteries, the tentative goals for an interim battery for electric vehicles are shown in the last line of the table. In addition, the interim battery should provide sustaining power of over 20 W/kg (9.1 W/lb) and peak power (15 sec) of over 100 W/kg (45.5 W/lb). This would correspond to currents of over 38 amp sustaining and over 190 amp peak from a 50 lb, 12 V battery.

The targeted performance of high-temperature batteries for the far term is 150 to 170 Wh/kg (68 to 77 Wh/lb), at $40/kWh, over twice the 70 Wh/kg shown as the interim goal. This explains the interest in lower electrode costs, such as for sulfur, and the higher energy densities obtainable at high temperatures and with molten salt and solid electrolytes.

As a specific example, the DOE- and EPRI-sponsored development program for the lithium–iron sulfide (Li–Al/FeS) battery has as its goal the performance shown in Table 13.11.[24]

The Future of Batteries

While a large part of golf carts, industrial trucks, and small airport vehicles are battery operated, the number of battery-powered cars in the United States is about 1800 (1976). From this beginning, it is postulated that if the battery potential, as viewed today (1980), were fully developed, the number of battery-operated cars could rise to 25,000,000, or 10 to 12% of all cars registered in the United States by the year 2000. Both full battery power and hybrid in conjunction with a small heat engine (Stirling engine) or a flywheel are being considered.

The interim battery development could make inroads on the "second car" and "third car" catego-

Table 13.11. Current (1976) Goals for LiAl–FeS Batteries

Goals	Utility Energy Storage	Electric Automobile
Cell-specific energy		
(Wh/kg)	80–150	120–160
(Wh/lb)	36–68	55–73
Cost ($/kWh)	20–30	20–40
Peak power (15 sec)		
(W/kg)		150–200
(W/lb)		68–91
Sustaining power		
(W/kg)	10–30	60–80
(W/lb)	4.5–14	27–36
Lifetime		
Cycles, equivalent deep discharge	3000	1000
Years of use	12	8–10
Annual duty (hr)	1250–2500	
Equivalent miles		100,000

Source. Courtesy R.O. Ivens, E.C. Gay, W.J. Walsh, and A.A. Skilenskas.[24]

ries, primarily for intracity use (80 mile range). There is a considerable gap between this and the intercity range (150 to 300 miles), which would be possible only with the most advanced high-temperature batteries under consideration. "The lithium–metal sulfide and sodium–sulfur systems appear to be good candidates for electric vehicles in the long run."[24] These systems are also top contenders for electric utility load-leveling service. Development of the sodium–sulfur system (see Fig. 13.29) is funded by DOE, EPRI, and NSF. A national (DOE–EPRI) battery test facility located at the Public Service Company of New Jersey, Newark, New Jersey, will be the major test site for utility batteries.

In terms of car performance, the 1976 goals for maximum daily driving range to be available at different stages were as given in Table 13.12. The attainment of these goals depends in part on attaining battery performance goals, such as the interim performance goals of Table 13.10. A long-life battery of about four times the Wh/lb and one fourth the $/kWh of present (1978) batteries would come close to meeting the requirements.

THERMOELECTRICITY

Seebeck, Peltier, and Thomson made the three discoveries of thermoelectricity, and have been honored by having the coefficients of the three phenomena named for them (or three evidences of one phenomenon).

Seebeck Effect. Seebeck (1821) observed the deflection of a compass needle near a junction of two different conductors in a closed circuit. This was later found to be due to a current flow, the emf set up in the junction being a function of temperature. The two junctions in the circuit must have been at different temperatures. The Seebeck coefficient α is, for example, volts per °C for the junction involved. Strictly, it is defined by

$$(\Delta V)_{\substack{I=0, \\ \Delta T \to 0}} = \alpha \, \Delta T \qquad (13.61)$$

It is the open circuit voltage per unit of temperature difference between the two junctions. The voltage is called the Seebeck voltage.

Table 13.12. Electric Car Daily Driving-Range Goals

Stage	Available Driving Range			Use
	Kilometers (Max.)	Miles		
		Normal	Max.	
Present 1976	80	30–40	50	Very limited
Near term 1978	80	30–40	50	Limited intracity[a]
Midterm (interim) 1982	161	60–100	100	More intracity
Far term 1986	322	120–180	200	Some intercity

[a]Electric vans and buses could use advanced lead–acid batteries in the near term, and a limited performance passenger vehicle could be introduced with a range of 40 to 60 miles for urban driving.[24]

Peltier Effect. Peltier (1834) observed that when a dc current was passed through a junction of two different conductors, heat was liberated or absorbed, depending on the direction of current flow. The Peltier coefficient π_T is, for example, watts per ampere (or volts) for the junction involved. It is a function of temperature. The heat is the "Peltier heat." Lenz (1838) demonstrated that water placed on a bismuth–antimony junction could be frozen by a current flow in one direction and melted by a current flow in the other direction.[25] These two coefficients depend only on the material of the two conductors involved, not on their size or shape.

Thomson Effect. Later, Thomson (Lord Kelvin) concluded that there must be a third phenomenon for the two previous ones to be possible thermodynamically. A current flowing in the direction of a thermal gradient, in a homogeneous conductor, must absorb or liberate heat other than the I^2R, depending on the relative directions of current flow and thermal gradient. This effect was confirmed by experiment. The Thomson coefficient τ is measured in volts per degree of temperature rise along the conductor for the particular material, and is a function of temperature.

Semiconductor Theory of Thermoelectricity. Experimentally, these three effects constitute the phenomenon of thermoelectricity, the basis of thermocouples, thermopiles, thermoelectric heating and cooling, and thermoelectric generation of electricity. The materials used should have high Seebeck coefficients, high electrical conductivity to minimize I^2R loss, and low thermal conductivity to reduce heat transfer through the devices. Semiconductors have the best combination of these properties, although the phenomenon occurs to some extent with any conductors.

The Seebeck effect arises because, if there is a temperature difference between the two ends of a metal or semiconductor, electrons at the hot end will have higher kinetic energy and speed than those at the cold end. This will lead not only to a transfer of energy, that is, to thermal conductance, but also to a diffusion of electrons toward the cold end. This will set up an emf between the two ends, which on open circuit will build up until there is no net flow along the material. The open circuit voltage so established is the Seebeck voltage[26] (actually the difference between two such voltages in dissimilar conductors).

In terms of semiconductor theory, the phenomenon is explained much more clearly. It occurs because at a semiconductor junction energy is absorbed or liberated as electrons or holes are forced to change their energy levels because of the continuity of the Fermi level (see Refs. 25 and 26 on direct energy conversion.)

Some of the materials used by Seebeck and Peltier were semiconductors, although they were not known as such in those days. Actually, materials were available then that would have produced thermoelectric generators of 3% thermal efficiency, as good as any steam engines of that day.

Thermoelectric Generators

A thermocouple is a thermoelectric generator, albeit small. It is usually used at no load so that the Seebeck or "internal voltage" of the generator can be measured to indicate temperature. A thermopile is many thermocouples in series, with alternate junctions bundled together at the same temperature. It thus multiplies the voltage by the number of pairs of junctions in series for the same temperature difference.

In a thermoelectric generator, the conductors are large and of low resistance. There may be two junctions or several, as in a thermopile, electrically in series, thermally in parallel. Junction pairs may also be thermally in series. Many different arrangements are used to adapt to different heat sources and to use the thermoelectric materials best suited to each temperature range. A simple arrangement is shown in Fig. 13.30a.

In general, a heat source is applied to the hot junctions and a heat sink to the cold junctions. Some heat flow from the hot junctions to the cold junctions is inevitable, but the configuration can be optimized to obtain the maximum efficiency, or maximum power output, as desired. Figure 13.30b shows the range of efficiencies theoretically obtainable as a function of heat source temperature, and a "figure of merit" that covers the range of available materials and other controllable variables. The heat sink is taken as 300°K (80°F).

The *n*-type alloys (75% Bi_2Te_3, 25% Bi_2Se_3) and (75% PbTe, 25% SnTe), and the *p*-type alloys (25% Bi_2Te_3, 75% Sb_2Te_3, 0.0–1.75% Se) and (AgSbTe$_2$), have high figures of merit for thermoelectric generators.[25] A generator using the first-named *n* and *p* materials and operating with a 300°K (572°F) temperature drop may have a thermal efficiency

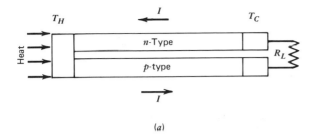

(a)

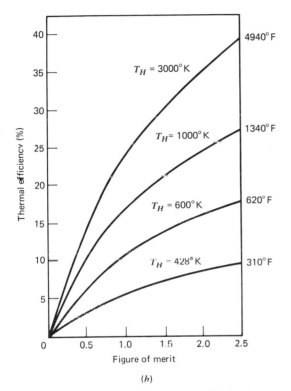

(b)

Figure 13.30. Thermoelectric generator: (a) simple generator; (b) thermal efficiency of a thermoelectric generator as a function of the source temperature and the figure of merit; the sink temperature is assumed to be 300°K (80°F). Part (b) from Stanley W. Angrist, DIRECT ENERGY CONVERSION, Second Edition. Copyright © 1971 by Allyn and Bacon, Inc., Boston. Reprinted with permission.

near 12%.[25] These materials are tellurides and selenides.

Application of Thermoelectric Generators. Thermoelectric generators have no moving parts and are ideal for small electric power sources at remote locations which must operate unattended for long periods. For example, some of the remote weather stations in the artic are supplied by thermoelectric generators that draw their heat from a supply of decaying strontium 90 (half-life 28 yr).

The generator trickle charges a sealed nickel–cadmium battery and operates unattended for 10 yr.

On a somewhat larger scale, 200 W thermoelectric power supplies a tiny irrigation pump in remote areas, with collected sun power as a source. It will supply, over a 20 ft head, the equivalent of 24 in. of rainfall per year to 4 acres of land, or 5 gallons water per day per person to 1200 people. It is assumed that there are 250 days per year with 10 hr of sunshine adequate for the 200 W output. An 8 ft parabolic mirror focuses the sun on the converter, which is 8 in. square and 2 in. deep.

The thermoelectric generator makes it possible to use any indigenous heat source to obtain small amounts of electricity almost anywhere in the world. Radioisotope power supplies with thermoelectric conversion from a fraction of a watt to several kilowatts have been used or planned in the space program.

PHOTOVOLTAIC ELECTRICITY

Early in the century, the fact that light falling on selenium or cuprous oxide produced an emf was known and used in exposure meters. About 1954, after semiconductors were better understood and the transistor had been invented (1947), researchers turned to the problem of using the photovoltaic effect as a source of power. The direct conversion of light to electricity was particularly attractive since heat energy was not an intermediary and the Carnot limitation was avoided. The unlimited supply of solar energy was also attractive.

Early efforts led to conversion efficiencies of about 6%, using p–n junctions of cadmium–sulfide and silicon. Subsequently, efficiencies near 15% were obtained using improved silicon p–n junctions. Needed energy for satellites was a strong stimulus for development as the space program got underway.

The phenomenon may be briefly described with reference to Fig. 13.31, showing the potentials in the vicinity of a p–n junction, with light photons of energy $h\nu$ impinging on the electrons of the junction region. If the photons are of sufficient energy to excite electrons from the valence band to the conduction band, as shown, the resulting free electrons and holes move under the influence of the built-in field that a p–n junction possesses. If the circuit is open, this generated current I_g builds up a charge, and the resulting field causes a recombination current I_r that

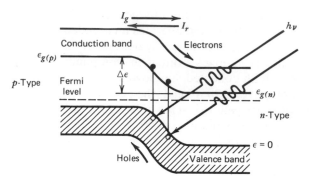

Figure 13.31. A p–n junction schematic showing the junction, the energy band scheme, and the density of donors and acceptors as a function of position in the junction. From Stanley W. Angrist etc. see above.

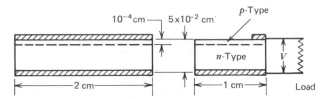

Figure 13.32. Typical configuration and dimensions of a p-on-n silicon solar cell. Characteristics: $V_{oc} \approx 0.58$ V, $I_{sc} \approx 45$ mamp, $V_m \approx 0.44$ V, $P_{out} \approx 12$ mW/cm^2, $\eta \approx 12\%$, where V_{oc} is the open circuit voltage, proportional to the band gap of the semiconductor material; I_{sc} is the short circuit current, a function of the magnitude and frequency of the incident radiation; V_m is the voltage for maximum power; P_{out} is the power output $kV_{oc}I_{sc}$; and η is the efficiency.

just balances the generated current. However, if the circuit is closed through a load, part of the generated current flows through the load. The junction thus acts as a converter of light energy directly into electrical energy.[25-27]

The electrical output is directly proportional to the number of photons having energy above the gap. The energy gap is of the order of 1 to 2 eV (1.2 eV for Si, 2.24 eV for GaP, 0.7 eV for Ge). Higher-energy photons are selectively absorbed by the atmosphere. The *average* photon energy from the sun varies from 1.48 eV outside of the atmosphere, to 1.28 eV on a clear day at sea level with the sun 60° from zenith, to 1.18 eV late in the day in bad weather. Since only photons having energy above the energy gap "count," this factor as well as the total insolation are important.

Construction and Cost. Physically, a photovoltaic cell is made with a very thin layer of one semiconductor formed on top of the other, so that the light can "get through," that is, can produce electron-hole pairs very close to the junction. Thus, a solar cell has the appearance of Fig. 13.32.

Pure single-crystal silicon is used, doped to optimum with desired impurities in the melt from which the crystal is drawn. Usually, about 1 part arsenic to 10^6 parts Si is used for n-type material. Performance with polycrystalline material is generally about one tenth of that obtained with single-crystal cells. Carefully prepared slices of this crystal form the n-type material. Boron trichloride (BCl$_3$) is diffused into the surface to a depth of 10^{-3} to 10^{-4} cm (1 to 10 μ), to form a thin p-type layer on top and the junction between. By controlled etching and

plating, connections are affixed to the p and n layers as shown.

An n-on-p type junction is made with boron in the melt and phosphorus diffused into the wafer to form the n-type material[25]; n-on-p cells are less susceptible to radiation damage.[27]

Photovoltaic cells have been relatively expensive in the past, and earth applications have usually required concentrating the solar energy by reflectors to get good utilization of the cells. For satellites, they are used directly. However, the cost has been dramatically reduced (1981) and flat panels of solar cells are now being applied directly, without concentrators, for most applications.

The solar cell is currently (1981) the most used of all direct-conversion techniques, generally under 1 kW. Applications and costs are discussed in Chapter 7.

THERMIONIC CONVERSION

Thermionic emission is the emission of electrons from a material at high temperatures. The basic principle of thermionic conversion is shown in Fig. 13.33. The cathode is heated by a heat source, emitting electrons with sufficient energy to cross the space to the anode and return through the load to the cathode. Some heat energy has thereby been converted to electric energy.

In order that electrons be boiled off from the cathode, they must be given at least the energy ϕ_c, called the surface "work function," over and above the Fermi level, the energy of the most energetic electrons within the material. Electrons leaving the

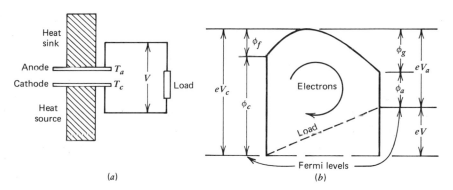

Figure 13.33. (a) Thermionic converter. (b) Potential diagram of electrons.

material form a space charge which opposes the flow of further electrons. Thus, for electrons to pass to the cathode, they must be given a further energy ϕ_f to overcome the space charge. Altogether, they must have the energy $\phi_c + \phi_f$ to rise to the potential eV_c and pass to the anode.

As they "fall" into the anode, they give up the space charge energy ϕ_g and the anode work function energy ϕ_a in arriving within the anode, which then has a Fermi level at a potential eV (a voltage V) above the cathode. This voltage V is available to force the electrons through the load impedance to the cathode, thus completing the circuit.

The heat source must supply the energy eV_c per electron at the high temperature T_c. The heat sink must be able to absorb the energy eV_a per electron and maintain the low temperature T_a. The efficiency can obviously not exceed V/V_c and will be less depending on the proportion of the total heat input that is effective in boiling off electrons of energy eV_c or higher.

An anode material of lower work function than the cathode is obviously desirable to secure high efficiency, and elimination or neutralization of space charge provides further improvement.

To minimize the space charge effect, vacuum converters must have extremely close electrode spacing (less than 10 μ). Typical emitter temperatures are 1200 to 1400°K (1700 to 2060°F). Prototype vacuum-type thermionic converters have been built[27] and tested having an output of 1 W/cm² (of emitter or collector), 5% thermal efficiency, and a life of tens of hours.

Low-pressure converters have been built in which positive ions are used to decrease the electron space charge effect. Cesium vapor is ordinarily used. This not only permits more reasonable elec-

trode spacing but also coats the anode, providing a low anode work function. Prototype low-pressure converters have yielded[27] 10 W/cm², at 10% thermal efficiency, with emitter temperatures of 2300°K (3680°F) and electrode spacings up to 1 mm (about 100 times that in the vacuum converter).

High-pressure thermionic converters (1 to 10 torr, 1 to 10 mm Hg) have yielded more than 40 W/cm², with emitter temperatures of 2200°K (3500°F) and electrode spacing of 100 μ.[27] Here again, cesium vapor is used. At these higher pressures, collisions with ions and atoms as well as space charge limit the transport of electrons. These effects are minimized by operation at high emitter temperatures, or in the "ignited" mode, to provide a copious supply of electrons and good neutralization of space charge. Because of the superior performance of the high-pressure design, this design is of greatest promise for the conversion of substantial amounts of energy.

Thermionic conversion is considered primarily for remote, inaccessible locations—outer space, deep under sea, or polar regions. Heat sources seriously considered are the sun and nuclear energy. The latter is considered for 10 kWe in outer space. However, thermionic conversion is also under consideration as a silent, light-weight replacement for the conventional engine-driven generator using liquid fuel, and has been proposed[25] as a topping unit for fossil fuel-fired central stations and nuclear plants.

REFRIGERATION, AIR CONDITIONING, AND HEAT PUMPING

Refrigeration and heat pumping differ only in purpose. In refrigeration or air conditioning, the purpose is removal of heat from a cold region to keep it cold. In heat pumping, the purpose is to get heat into a warm region to keep it warm. In either case, this is accomplished mechanically by the same two steps. First raise a fluid to a high temperature by compressing it and discharge heat to the warm region. Then cool it to a low temperature by expansion and absorb heat from the cold region to complete the cycle. In this way, heat is transferred from a cold region to a hot region at the expense of the net work of compression and expansion.

As shown earlier the maximum thermodynamically possible efficiency or "coefficient of performance" is that using the perfect reversible Carnot cycle. The corresponding heat pump and refrigeration efficiencies between two temperatures T_1 and T_2 are given by Eqs. 13.4 and 13.5, respectively.

In small systems such as domestic refrigeration and air conditioning, recovery of the expansion work is not worthwhile, and expansion by throttling is used (isenthalpic expansion). In very large systems, such as the liquefaction of natural gas, the benefits of recovering expansion work in heat engines or turbines is marginal, and both isentropic (heat engine) and isenthalpic (throttling) expansions are used in different competitive systems (see Chapter 4, Liquefied Natural Gas). At very low temperatures, throttling expansion is always used, since moving parts are not feasible.

The elevation of heat more than 100 to 140°F (55 to 78°C) in one stage is not usually practical. Thus, single-stage systems are used for domestic and commercial refrigeration, air conditioning, and heat pumping, and multistage systems are used for LNG (usually three-stage).

A two-phase liquid–vapor cycle is ordinarily used, as will be illustrated.

Working Fluids. A refrigerant must be chosen which is relatively compact over the working range to minimize the size of machinery. Fluids for the two-phase cycles must have saturation temperatures below the cold region and above the hot region at reasonable pressures. The lower pressure should be above atmospheric pressure to avoid leakage into the system. The higher pressure should not exceed 150 to 200 psia to be "reasonable" for a small compressor. The fluid should be stable, nontoxic, and nonexplosive in air. It should be of low cost and, to be "compact," should have a high enthalpy of vaporization. The thermodynamic properties should of course lead to the most economical overall solution for the particular application.

These requirements are met by a variety of working fluids suitable to different temperature ranges and applications. Freon-12 (dichlorodifluoromethane, CCl_2F_2) is a representative fluid for normal domestic and commercial applications, and the hydrocarbons of successively lower boiling points for the various stages of the cascade used in the liquefaction of natural gas. There are, however, a large number of refrigerants to choose from.[7]

In early refrigeration systems, ammonia was used in the manufacture of ice, but CO_2 was used in commercial and institutional applications because it was nontoxic. SO_2 was used in domestic refrigerators because it was easily handled in small light-weight compressors. However, the new made-to-order refrigerants, such as the several freons, have largely displaced the early working fluids.

Two Phase, Liquid-Vapor Refrigeration Cycle. Examples are given below of Freon-12 and ammonia systems. The arrangement of multistage cascades is given in Chapter 4 under Liquefied Natural Gas.

Example 13.1. Refrigeration Cycle Using Freon-12. A refrigerator using Freon-12 as the working fluid maintains a cold region temperature of 0°F. Due to thermal drops, the evaporator temperature must be at −10°F and the condenser temperature at 90°F. The liquid–vapor, two-phase refrigeration cycle shown in Fig. 13.34 is used. The refrigerant flow around the cycle is 10 lb min.

Note: In this system the refrigerant is compressed adiabatically along 1–2. It discharges heat to the warm region at constant pressure along 2–3. It is then expanded isenthalpically (throttled) along 3–4. It absorbs heat from the cold region at constant temperature and pressure along 4–1 to complete the cycle. The heat discharge along 2–3 is largely at constant temperature. The heat transfer to and from the working fluid thus *approaches* the ideal constant-temperature heat transfer of the Carnot cycle. The thermal drops, throttling expansion, and so on, are however deviations. The starting point, 1, may also be in the wet or superheat region, usually the latter.[28]

Determine the theoretical values of the coefficient of performance (C.O.P)—heat removed/work done, tons of refrigeration, and Hp input per ton of refrigeration. Determine also the coefficient of performance of a Carnot cycle operating between (a) the cold region and ambient (warm region), and (b) the

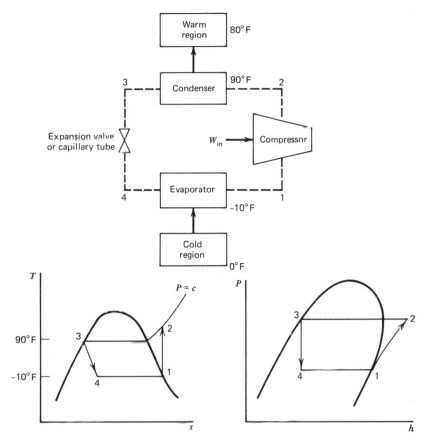

Figure 13.34. Two-phase liquid–vapor refrigeration cycle (temperatures shown for Example 13.1).

evaporator and condenser temperatures. Note: One ton of refrigeration is defined as a heat removal rate of 200 Btu/min. This heat rate would melt or freeze 1 ton of ice per day. The unit is a carryover from the days of ice refrigeration.

Solution. Referring to Fig. 13.34, the heat removed per pound of refrigerant is $h_1 - h_4$. The work done by the compressor is $h_2 - h_1$. These quantities can be obtained from tables of thermodynamic properties[2,7] for Freon-12.

At the evaporator temperature, $T_1 = -10°$ F,

$$h_1 = h_g = 76.196 \text{ Btu/lb}$$

$$s_1 = s_g = 0.16989 = s_2$$

At the condenser temperature, $T_3 = 90°F$,

$$p_3 = p_2 = 114.49 \text{ psia}$$

At the compressor outlet, 2, the vapor is super-

heated at $p_2 = 114.49$ psia, $s_2 = 0.16989$. Hence, from the superheat table,

$$T_2 = 108.6°F \quad h_2 = 89.70 \text{ Btu/lb}$$

The enthalpy at point 3, saturated liquid at 90°F, is

$$h_3 = 28.713 \text{ Btu/lb}$$

Since the throttling process is isenthalpic,

$$h_4 = 28.713 \text{ Btu/lb also.}$$

Therefore, the theoretical coefficient of performance is

$$\text{C.O.P.} = \frac{h_1 - h_4}{h_2 - h_1}$$

$$= \frac{76.196 - 28.713}{89.70 - 76.196} = 3.52$$

The tons of refrigeration are

$$\frac{10(76.196 - 28.713)}{200} = 2.37 \text{ tons}$$

The power required is

$$\frac{10 \text{ lb fluid/min} \times (89.70 - 76.19) \text{ Btu work/lb fluid}}{42.416 \text{ Btu/min per Hp}}$$
$$= 3.19 \text{ Hp}$$

Therefore, the Hp/ton is $3.19/2.37 = 1.35$.

The Carnot refrigeration efficiencies or C.O.P.s are:

Cold region to warm:

$$\frac{T_{\text{cold}}}{\Delta T} = \frac{460}{80} = 5.75$$

Evaporator to condenser:

$$\frac{T_{\text{evap}}}{\Delta T} = \frac{450}{100} = 4.50$$

Note that the C.O.P. of the cycle of Fig. 13.34 falls short of the Carnot efficiency from cold to warm regions for two reasons. First, the actual elevation of heat is 20°F higher to take care of the thermal drops; and second, the cycle used is theoretically less efficient than the Carnot cycle. The C.O.P. of the actual cycle is of course much less than the theoretical cycle. However, the theory is useful in understanding and comparing cycles.

In the actual cycle, there are losses in the compressor, motor, and auxiliaries. Also, in addition to the irreversibilities in the processes, the evaporator output is usually somewhat superheated and the condenser output subcooled, instead of points 1 and 3 being exactly saturated vapor and liquid as shown. Thus, the calculation is simply an approximation to the actual process.

Example 13.2. Refrigeration Cycle Using Ammonia. Two tons of refrigeration is needed for a refrigerated space to be kept at 30°F. Ammonia is to be used as the refrigerant. Heat is to be discharged to ambient air at a temperature of 80°F. Assume 10°F temperature drops in the evaporator and condenser heat exchangers. A liquid–vapor cycle similar to Fig. 13.34 is to be used.

What flow of refrigerant will be needed? What

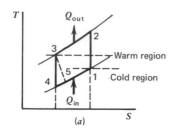

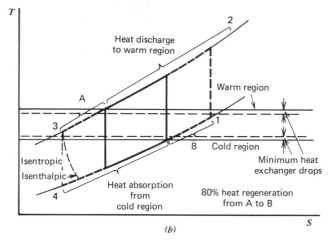

Figure 13.35. Gas refrigeration cycle using a reversed Brayton cycle.

compressor Hp? What will be the theoretical C.O.P. and the Hp/ton? What maximum temperature at the compressor output?

Solution. The evaporator and condenser saturation temperatures must be 20 and 90°F, respectively. The enthalpies at the four cardinal points, at these temperatures, can be obtained from the thermodynamic tables[2,7,29] as before. From these, the heat removed by the ammonia is 474.3 Btu/lb. The flow required for 2 tons (400 Btu/min) is thus 0.843 lb/min. The compressor work is 81.13 Btu/lb, or 68.39 Btu/min, 1.61 Hp, or 0.805 Hp/ton. The theoretical C.O.P. is thus 5.85. The compressor output temperature is 188.8°F.

Gas Refrigeration Cycle. In the gas, single-phase refrigeration cycle, the heat absorption and rejection are at varying temperature, as shown in Fig. 13.35a, since there is no phase change. The gas is compressed isentropically along 1–2. Heat is rejected to the warm region along 2–3. The gas is then expanded isentropically along 3–4, and heat is absorbed from the cold region along 4–1. This is the Brayton cycle in reverse (closed cycle).

Alternatively, the expansion may be by throttling (isenthalpic), as shown dotted along 3–5. This is less efficient but much simpler.

As shown in Fig. 13.35b, after the compressed gas has transferred the maximum heat to the warm region, it can be cooled further by the heat exchange A to B before expanding it along 3–4, and thus increase the heat absorption from the cold region. This is similar to a recuperator in reverse.

A simple open cycle is used in some aircraft. Compressed air from one of the engine compressors is cooled to near ambient by a heat exchanger. It is then expanded to cabin pressure, cooling in the process, and is admitted directly to the cabin.

Absorption-Type Refrigeration.[30] The absorption-type refrigeration system illustrated in Fig. 13.36 is activated by a heat source instead of a mechanical compressor. It is the basis of the gas refrigerator and of solar cooling systems (absorption coolers). Ammonia boiled off from a water–ammonia solution by a small gas flame or other heat source goes to a condenser at about 125 psia, where it is condensed to liquid ammonia at atmospheric temperature of about 70°F. This liquid flows through a restriction to the evaporator, the cooling coils of the refrigerator, which are maintained at 40 psia or below. At this pressure, the ammonia is evaporated at the low temperature of about 12°F, absorbing heat from the cool region. This low pressure is maintained because the ammonia vapor, as it is formed, is absorbed by water flowing down through the absorber, hence the name "absorption type."

Two unique features add to the efficiency and complete absence of moving parts. Water circulation through the absorber is maintained by a "lift tube" that acts exactly like the water lift tube in a coffee percolator. As the ammonia is boiled off by the gas flame, it carries water along with it up the lift tube to a separator. The water goes down through the absorber and returns to the generator. The ammonia vapor goes on to the condenser as described earlier.

The other unique feature is the inclusion of hydrogen in the evaporator–absorber flow circuit. Hydrogen has relatively low solubility in water compared with ammonia. Thus, the "up" fluid is light hydrogen, and the "down" fluid is the heavier ammonia–hydrogen mixture being carried to the absorber. This increases the flow rate and enhances the evaporation.

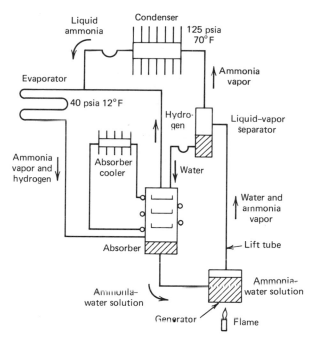

Figure 13.36. Absorption-type gas refrigerator. Pressures and temperatures shown are typical.

The absorption process is exothermic, and a cooling coil is required around the absorber, discharging heat through a cooler by convection.

Air Conditioning and Heat Pumping. By common usage, the term air conditioning has come to mean air cooling, although air conditioning units frequently do include humidity control, air cleaning and heating coils as well. More and more frequently, central units are installed to pump heat in winter and cool air in summer. We avoid confusion in this chapter by discussing only the basic principles that are used in these systems. Thus, Fig. 13.37a shows a simple air-to-air cooler such as might be used in a window air conditioner or a central unit. Two fans circulate outside air through the condenser, where it is heated, and inside air through the evaporator, where it is cooled.

By reversing the air flow ducts of a central unit, so that the room air passes the condenser, it is heated and the outside air is cooled.[28] The unit is then a heat pump, as shown in Fig. 13.37b. The heat pumped into the house in this way may be two to three times the electrical energy expended (about 1.7 times annualized, Table 4.11).

Somewhat greater efficiency is obtained by using a source of water, since it is neither as hot in sum-

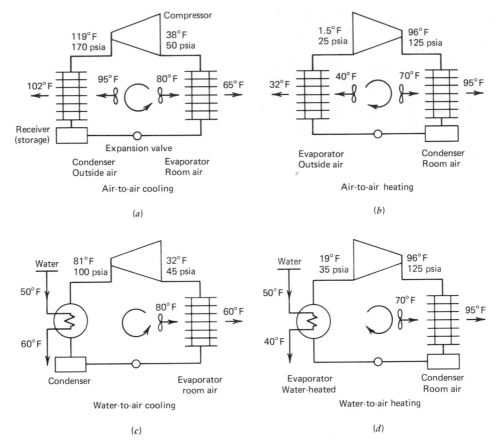

Figure 13.37. Heating and cooling with ambient air or water and liquid–vapor refrigerant. Temperatures and pressures are typical. (Refrigerant saturation temperature, Freon-12.) (*a*) Air-to-air cooling. (*b*) Air-to-air heating. (*c*) Water-to-air cooling. (*d*) Water-to-air heating.

mer nor as cold in winter as the outside air. This arrangement is shown in Figs. 13.37*c* and 13.37*d*.

The earth may also be used as a source and sink for heat. Several feet down, the earth temperature varies very little from winter to summer. Water is circulated directly to wells or to coils or tanks buried in wells or in the ground. Otherwise, the system appears the same as in Figs. 13.37*c* and 13.37*d*.

Any method of securing a colder heat sink in summer or a warmer source in winter reduces the energy required to transfer the heat. For example, the Solar Augmented Heat Pump[31] uses a relatively small solar installation (about half size) to provide a relatively warm water supply for the heat pump. Both the solar equipment and the heat pump are then more efficient, the combination being considerably more economical than a solar installation alone.

EQUILIBRIUM IN A GASIFIER OR CHEMICAL REACTOR

Throughout this book, reference is made to chemical processes taking place in various chemical reactors, such as gasifiers, shift reactors, and methanators. This section is a brief introduction to the physics of these processes. Much will have to be stated without proof, although the logic of the principles stated should be evident. The full development can be found in texts on thermodynamics.[1,2]

For a closed system in which spontaneous changes are taking place, the entropy always increases until equilibrium is reached. Then, no further change can take place. If both temperature and pressure are fixed, or known, the maximum entropy criterion for equilibrium is identical with the minimum Gibbs energy function, or Gibbs free energy, $H - TS$, which is easier to calculate.

The maximum entropy at equilibrium is a direct consequence of the Second Law, which stipulates that the entropy increases in *any* real process. In combustion reactions, it is frequently assumed, with good reason, that the reaction goes to completion or until the available oxygen is used up. By the First Law, the heat liberated is equal to the loss of chemical energy, plus the loss of sensible heat energy. However, the First Law gives no indication of whether the reaction will "go" or how far. It simply tells what will happen if it does.

The Second Law prescribes whether the process will "go" and how far. In other words, it specifies the equilibrium condition. In particular, if the temperature and pressure are fixed, it specifies the equilibrium condition at that temperature and pressure. It does not, however, tell how fast the reaction will proceed to the equilibrium condition. This is a matter of reaction kinetics, a subject that we will skip, except to say that with the reaction times allotted in typical processes described in this book, the reaction proceeds essentially to equilibrium. At least this is an excellent first approximation.

The First Law, conservation of energy, and the Second Law, entropy criterion for equilibrium, are of course perfectly general. However, to give illustrative examples we confine our attention for the moment to perfect gases.

For perfect gases, the pressure is the sum of the partial pressures of the constituent gases, and the Gibbs function is the sum of the Gibbs functions of the constituents. Thus, for a generalized chemical reaction represented by the equation

$$aA + bB \rightleftharpoons eE + fF \qquad (13.62)$$

it can be shown[2] that at equilibrium

$$\frac{(N_E)^e (N_F)^f}{(N_A)^a (N_B)^b} \left(\frac{P}{N_m} \right)^{\Delta \nu} = K_p \qquad (13.63)$$

where N_E is the number of moles of E at equilibrium; e is the stochiometric coefficient of E in the reaction equation, 13.62. (N_F and f, N_A and a, and N_B and b are defined similarly); P is the total pressure, the sum of the partial pressures in atm; $\Delta \nu$ is the difference of numerator and denominator stochiometric coefficients,

$$\Delta \nu = e + f - a - b \qquad (13.64)$$

N_m is the number of moles of gas at equilibrium, and the equilibrium constant K_p is

$$K_p = e^{-(\Delta G_T^0 / RT)} \qquad (13.65)$$

Generally, $\log_{10} K_p$ is tabulated as a function of T for various reactions (see Table 13.13). The table is based on perfect gases under standard conditions (1 atm). ΔG_T^0 is the Gibbs free energy change for the complete reaction, Eq. 13.62. Its units must be consistent with those of R and T. For example, ΔG_T^0 is in Btu/lb mole for R in Btu/lb mole °R, and T in °R.

Example 13.3. An example will make this clear. Suppose 1 mole CO and 1 mole O_2 are allowed to reach equilibrium at 2800°K (5040°R) and 1 atm pressure. What is the composition of the resulting mixture? Neglect dissociation of O_2 into O.

Solution. The basic equation involved is

$$CO(g) + \tfrac{1}{2} O_2(g) \rightleftharpoons CO_2(g) \qquad (13.66)$$

Thus, Eq. 13.63 reads

$$\frac{(N_{CO_2})^1}{(N_{CO})^1 (N_{O_2})^{1/2}} \left(\frac{P}{N_m} \right)^{-1/2} = K_p \qquad (13.67)$$

From Table 13.13 of equilibrium constants, at 2800°K, $\log K_p = 0.825$ (note that Eq. 13.66 is the reverse of that in the table, whence $+0.825$) and $K_p = 6.68$. Actually, the reaction may not go to completion, so that

$$CO + O_2 \rightarrow x\ CO + y\ O_2 + z\ CO_2 \qquad (13.68)$$

where $x = N_{CO}$, $y = N_{O_2}$, and $z = N_{CO_2}$. Substituting these in Eq. 13.67 and noting that

$$N_m + x + y + z \qquad (13.69)$$

and $P = 1$ atm, then

$$\frac{z}{xy^{1/2}} \left(\frac{1}{x + y + z} \right)^{-1/2} = 6.68 \qquad (13.70)$$

It is required to find the three unknowns x, y, and z. Mass balance equations on C and O from Eq. 13.68 give the needed additional equations:

$$1 = x + z \qquad (13.71)$$

$$3 = x + 2y + 2z \qquad (13.72)$$

Table 13.13.   **Logarithms to the Base 10 of the Equilibrium Constant $K_p{}^a$**

T (°K)	$H_2 \rightleftharpoons 2H$	$O_2 \rightleftharpoons 2O$	$N_2 \rightleftharpoons 2N$	$H_2O(g)$ $\rightleftharpoons H_2 + \frac{1}{2}O_2$	$H_2O(g)$ $\rightleftharpoons OH + \frac{1}{2}H_2$	CO_2 $\rightleftharpoons CO + \frac{1}{2}O_2$	$\frac{1}{2}O_2 + \frac{1}{2}N_2$ $\rightleftharpoons NO$	$CO_2 + H_2$ $\rightleftharpoons CO + H_2O(g)$
298	−71.228	−81.208	−159.600	−40.048	−46.054	−45.066	−15.171	−5.018
500	−40.318	−45.880	− 92.672	−22.886	−26.130	−25.025	− 8.783	−2.139
1000	−17.292	−19.614	− 43.056	−10.062	−11.280	−10.221	− 4.062	−0.159
1500	− 9.514	−10.790	− 26.434	− 5.725	− 6.284	− 5.316	− 2.487	+0.409
1800	− 6.896	− 7.836	− 20.874	− 4.270	− 4.613	− 3.693	− 1.962	+0.577
2000	− 5.582	− 6.356	− 18.092	− 3.540	− 3.776	− 2.884	− 1.699	+0.656
2200	− 4.502	− 5.142	− 15.810	− 2.942	− 3.091	− 2.226	− 1.484	+0.716
2400	− 3.600	− 4.130	− 13.908	− 2.443	− 2.520	− 1.679	− 1.305	+0.764
2500	− 3.202	− 3.684	− 13.070	− 2.224	− 2.270	− 1.440	− 1.227	+0.784
2600	− 2.836	− 3.272	− 12.298	− 2.021	− 2.038	− 1.219	− 1.154	+0.802
2800	− 2.178	− 2.536	− 10.914	− 1.658	− 1.624	− 0.825	− 1.025	+0.833
3000	− 1.606	− 1.898	− 9.716	− 1.343	− 1.265	− 0.485	− 0.913	+0.858
3200	− 1.106	− 1.340	− 8.664	− 1.067	− 0.951	− 0.189	− 0.815	+0.878
3500	− 0.462	− 0.620	− 7.312	− 0.712	− 0.547	+ 0.190	− 0.690	+0.902
4000	+ 0.400	+ 0.340	− 5.504	− 0.238	− 0.011	+ 0.692	− 0.524	+0.930
4500	+ 1.074	+ 1.086	− 4.094	+ 0.133	+ 0.408	+ 1.079	− 0.397	+0.946
5000	+ 1.612	+ 1.686	− 2.962	+ 0.430	+ 0.741	+ 1.386	− 0.296	+0.956

Source.   JANAF Interim Thermochemical Tables, Dow Chemical Company, 1965. From Wark, *Thermodynamics.*[2] Used with the permission of the McGraw-Hill Book Co.

$^a K_p = p_E^e\, p_F^f / p_A^a\, p_B^b$ for the reaction $aA + bB \rightleftharpoons eE + fF$.

Solving Eqs. 13.70 to 13.72 for x, y, and z yields 0.197, 0.598, and 0.803, respectively. Hence, the actual reaction is

$$CO + O_2 \rightarrow 0.20\ CO + 0.60\ O_2 + 0.80\ CO_2 \tag{13.73}$$

The CO_2 formed is 80% of that for complete combustion. At a lower temperature, for example, 2000°K, the equilibrium constant would be very much higher (766), and practically all the carbon would go to CO_2.

Example 13.4.   As another example, suppose that air is raised to 3000°K (5400°R) at 1 atm. How much nitrous oxide, NO, is formed? Neglect the dissociation of O_2 into O.

Solution.   The basic reaction is

$$\tfrac{1}{2}O_2 + \tfrac{1}{2}N_2 \rightleftharpoons NO \tag{13.74}$$

From Table 13.13, the equilibrium constant is 0.122. Thus, Eq. 13.63 becomes

$$\frac{(N_{NO})^1}{(N_{O_2})^{1/2}\ (N_{N_2})^{1/2}} = 0.122 \ \text{(since } \Delta\nu = 0)\quad (13.75)$$

The actual reaction is

$$0.21\ O_2 + 0.79\ N_2 \rightarrow x\ O_2 + y\ N_2 + z\ NO \tag{13.76}$$

Equation 13.75 then becomes

$$\frac{z}{x^{1/2} y^{1/2}} = 0.122 \tag{13.77}$$

The additional equations resulting from mass balances on O and N are

$$0.42 = 2x - z \tag{13.78}$$

$$1.58 = 2y - z \tag{13.79}$$

Solving for x, y, and z, Eq. 13.76 becomes

$$0.21\ O_2 + 0.79\ N_2 \rightarrow 0.187\ O_2$$
$$+\ 0.767\ N_2 + 0.046\ NO \tag{13.80}$$

At 3000°K (4940°F) at equilibrium, the air contains 4.6% NO.

These examples each involve a single basic reaction and two chemical elements. The equations are easily solved and illustrate the principle involved. In actual situations, the fuel analysis will contain a number of elements. Several basic reactions will all be possible simultaneously. But there will always be enough equations from mass balances on the elements involved to solve for all the unknowns. A computer is normally used for the solution,[41] although important cases have been solved without computer.[32] This is a very powerful tool for predetermining the stack gas analysis with different fuel compositions, fuel–air ratios, and combustion temperatures. This permits designing to meet EPA limits on SO_2 and NO_x and comparing theoretically the various systems that have been discussed in this chapter.

WORKING FLUIDS—GENERAL

A number of working fluids have been discussed in this chapter: air, steam, mercury, potassium, ammonia, Freon-12, methane, propane, sodium, and others. These are but a few of the many (over 100) working fluids that have been used from time to time. In fact, every single-component liquid boils at a saturation temperature that depends on pressure and could therefore be used in a liquid–vapor, or Rankine, cycle, either direct or reversed. All gases or vapors can undergo thermodynamic processes. All fluids, liquid or gas, transfer heat by convection or forced flow. What then are the factors that lead to the selection of a "best" working fluid for each application?

A great many factors influence the selection or motivate the creation of new fluids. These include not only the thermodynamic properties, but also the cost and availability. Fluid for a domestic refrigerator should be nontoxic, nonirritating, nonexplosive, odorless, and inexpensive. On the other hand, pure hydrogen can be safely handled in cooling large electrical machines. Hydrocarbon refrigerants are used in LNG production.

Electrical insulating properties are important in coolants for electrical machinery and in refrigerants for use in hermetically sealed units. Miscibility with oil, suitability for centrifugal compressors, density,

viscosity, surface tension, and freezing temperature are important factors in particular instances.

Historically, air and water were first used because of their ready availability. At first in windmills and water wheels, only the kinetic energy or the potential energy of elevation was exploited. Later in steam engines and then in internal combustion engines, the thermodynamic properties of the working fluids were exploited as well as the energy in fuels. For early mechanical refrigeration, commonly available substances such as ammonia, sulfur dioxide, and carbon dioxide were used.

Air and water still predominate for heat engines, although metal vapors, such as mercury, potassium, and cesium have been added in order to use the Rankine cycle more efficiently above the critical temperature of water, 705°F, up to the metallurgical limit of about 1000°F and higher. At low pressures the use of a hydrocarbon vapor, such as isobutane in a vapor turbine, greatly reduces the size and cost.

The list of refrigerants now include over 50 fluids,[7] some old and well known, some newly created by the physical chemists to meet particular requirements, thermodynamic and other. These include the halocarbons, in which the hydrogen atoms of hydrocarbons have been replaced by halogens such as chlorine and fluorine, for example, Freon-12, CCl_2F_2, from methane, CH_4.

Published Characteristics. A long list of gases and liquids have been used for heat transfer by convection or forced flow in different situations. The physical properties and thermodynamic characteristics of most of these working fluids have been published, some of the more common references being the following:

1. *The Thermodynamic Properties of Steam.*[6]
2. *Steam Tables in S.I. Units.*[33] These (1 and 2) include very complete properties of steam.
3. *Gas Tables.*[3] This includes air, oxygen, nitrogen, and products of combustion of hydrocarbon fuels with 200 or 400% of theoretical air, CO, CO_2, water vapor, and monatomic gases. These are treated as perfect gases (internal energy and enthalpy a function of temperature only), and are thus tabulated with the single independent variable, temperature. However, relative pressures and volumes are given for isentropic processes (see Brayton cycle calculations).

4. *ASHRAE Handbook and Product Directory—Fundamentals.*[7] This handbook gives properties of 50 refrigerants and full thermodynamic characteristics of 30 of them. Tabulations of saturation characteristics and charts of superheat data are given.

5. *U.S. National Bureau of Standards Circular 142.*[29] Thermodynamic properties of ammonia.

6. *U.S. National Bureau of Standards. Circular 500.*[34] This compilation gives heat of formation, ΔH_f^0; free energy, ΔF_f^0; equilibrium constant of formation, $\log K_f$; heat content function, $(H_0 - H_0^\circ)/T$; entropy, S°; heat content or enthalpy, $H_0 - H_0^\circ$; and heat capacity, C_p°, for a number of elements and compounds.

7. *JANAF Thermochemical Tables.*[21] Information similar to item 6 on a large number of elements and compounds.

8. *Handbook of Chemistry and Physics,*[9] p. E-37. Miscellaneous properties of common refrigerants.

9. *Selected Values of Physical and Thermodynamic Properties of Hydrocarbons and Related Compounds,* API Project 44.[35]

10. U.S. National Bureau of Standards. Circular 461.[36] Selected values of the properties of hydrocarbons.

11. U.S. National Bureau of Standards. Circular 564.[37] Thermal properties of gases.

12. L. A. Sheldon,[8] "Thermodynamic Properties of Mercury Vapor."

In addition, a number of texts on thermodynamics include tables adequate for working out illustrative problems. Some of these are:

1. K. Wark,[2] *Thermodynamics.* Included are tables for water, air, ammonia, Freon-12, H_2, N_2, O_2, CO, CO_2, and water vapor, as well as combustion products with 400% air, equilibrium constants for the reactions involving H, N, and O, and enthalpies of formation and combustion for a number of substances.

2. C. L. Brown,[38] *Basic Thermodynamics.* Includes water (steam), ammonia, Freon-12, CO_2, and saturated mercury.

3. V. W. Young,[39] *Basic Engineering Thermodynamics.* Includes steam, saturated mercury vapor, ammonia, CO_2, and Freon-12.

4. W. H. Severns,[40] *Air Conditioning and Refrigeration.* Includes comparative refrigerant characteristics, p. 372.

REFERENCES

1. T. A. Brzustowski, *Introduction to the Principles of Engineering . Thermodynamics*, Reading, Mass.: Addison-Wesley, 1969.

2. K. Wark, *Thermodynamics*, New York: McGraw-Hill, 1966.

3. J. H. Keenan, J. Chau, and J. Kaye, *Gas Tables*, 2nd ed., New York: Wiley, 1979.

4. *Energy Conversion Alternatives Study (ECAS)*, NASA/ERDA/NSF, Report NSF/RA Nos. as listed:, 1976. 760131: Introduction and Summary, General Assumptions; 760133: Materials Considerations; 760134: Combustors, Furnaces, and Low-Btu Gasifiers; 760135: Open Recuperated and Bottomed Gas Turbine Cycles; 760136: Combined Gas-Steam Turbine Cycles; 760137: Closed-Cycle Gas Turbine Systems; 760138: Metal-Vapor Rankine Topping—Steam Bottoming Cycles; 760139: Open-Cycle MHD; 760140: Closed-Cycle MHD; 760141: Liquid-Metal MHD Systems; 760142: Advanced Steam Systems; 760143: Fuel Cells.

5. "Stirling Engine," in *McGraw-Hill Encyclopedia of Science and Technology*, New York: McGraw-Hill, 1977.

6. J. H. Keenan and F. G. Keyes, *The Thermodynamic Properties of Steam*, New York: Wiley, 1936.

7. C. W. MacPhee, Ed. *ASHRAE Handbook and Product Directory—Fundamentals*, New York: Amer. Soc. of Heating, Refrig., and Air Cond. Engrs., Inc., 1977.

8. L. A. Sheldon, "Thermodynamic Properties of Mercury Vapor," *Transact. ASME*, Paper 49A30, 1949.

9. R. C. Weast and M. J. Astle, Eds., *Handbook of Chemistry and Physics*, 61st ed., Boca Raton, Fla.: CRC Press, 1980–81.

10. "Electrogasdynamics—A Bold New Power Source," *Product Engineering*, March 28, 1966, p. 90.

11. M. C. Gourdine and D. H. Malcolm, "Feasibility of an EGD High-Voltage Power Source," *Proc. 19th Annual Power Sources Conf.*, PSC Publ., Box 891, Red Bank, N.J., 1965.

12. E. L. Daman, *Memo—Performance EGD System*, March 2, 1966 (unpublished).

13. *Air-Water Pollution Rept. Weekly News Letter*, June 26, 1967, Silver Springs, Md.

14. N. P. Chopey, Asst. Ed., "What You Should Know About Fuel Cells," *Chem. Eng.*, May 25, 1964, p. 125.

15. "Fuel Cells," in *McGraw-Hill Encyclopedia of Science and Technology*, New York: McGraw-Hill, 1968.

16. J. Waser, *Basic Chemical Thermodynamics*, New York: W. A. Benjamin, 1966.

17. E. P. Barry, R. L. A. Fernandez, and W. A. Messner, "A

Giant Step Planned in Fuel Cell Plant Test," *Spectrum*, November 1978, p. 47.

18. J. Weissbart and R. Ruka, "A Solid Dielectric Fuel Cell," *Electrochemical Society Journal*, August 1962, p. 723.

19. D. H. Archer, J. J. Alles, W. A. English, L. Elikan, E. F. Sveerdrup, and R. L. Zahradnik, "Westinghouse Solid-Electrolyte Fuel Cell," *Advances in Chemistry Series*, No. 47, 1965, p. 322.

20. W. Feduska and A. O. Isenberg, *High-Temperature, Solid Oxide Electrolyte Fuel Cell Power Generating System*, Quarterly Report, December 1980–February 1981, C-17089-4, Springfield, Va.: National Technical Information Service, 1981, Appendix A.

21. *JANAF Thermochemical Tables*, Washington, D.C.: U.S. Bureau of Standards, 1970.

22. "Batteries," *Spectrum*, March 1976, p. 36.

23. W. W. Smith, "Storage Batteries," *McGraw Hill Encyclopedia of Science and Technology*, New York: McGraw-Hill, 1977.

24. Sessions on Secondary Batteries and Fuel Cells, *Proceedings of the 27th Power Sources Symposium*, Red Bank, N.J.: PSC Publ., Box 891, 1976.

25. S. W. Angrist, *Direct Energy Conversion*, 2nd ed., Boston: Allyn and Bacon, 1971, p. 123.

26. K. H. Spring, Ed., *Direct Generation of Electricity*, New York: Academic Press, 1965.

27. G. W. Sutton, Ed., *Direct Energy Conversion*, New York: McGraw-Hill, 1966.

28. E. R. Ambrose, *Heat Pumps and Electric Heating*, New York: Wiley, 1966, p. 18.

29. U.S. National Bureau of Standards, *Thermodynamic Properties of Ammonia*, Circ. 142, Washington, D.C.: Supt. of Documents, 1923.

30. F. W. Sears, *An Introduction to Thermodynamics*, 2nd ed., Reading, Mass.: Addison Wesley, 1952, p. 194.

31. *Solar Augmented Heat Pump System, SP-SAHP-1*, Falls Church, Va.: Westinghouse Electric Corp., 1978.

32. F. Osterle, A. Impink, M. Lipner, and A. Candris, *Pressurized Gasification Systems Coupled to Combined Steam–Gas Power Cycles for the Generation of Clean Electric Power from Penna Coal*, Harrisburg, Pa.: Penna Sci. and Engr. Fdn., 1975.

33. J. H. Keenan, F. G. Keyes, P. G. Hill, and J. G. Moore, *Steam Tables in S.I. Units*, New York: Wiley, 1978.

34. *Selected Values of Thermodynamic Properties*, U.S. National Bureau of Standards Circ. 500, Washington, D.C.: Supt. of Documents, 1952.

35. *Selected Values of Physical and Thermodynamic Properties of Hydrocarbons and Related Compounds*, API Project 44, Pittsburgh, Pa.: Carnegie Press, 1953.

36. *Selected Values of Properties of Hydrocarbons*, U.S. Bureau of Standards, Circ. 461, Washington, D.C.: Supt. of Documents, 1947.

37. *Thermal Properties of Gases*, U.S. Bureau of Standards, Circ. 564, Washington, D.C.: Supt. of Documents, 1955.

38. C. L. Brown, *Basic Thermodynamics*, New York: McGraw-Hill, 1957.

39. V. W. Young, *Basic Engineering Thermodynamics*, New York: McGraw-Hill, 1952.

40. W. H. Severns and J. R. Fellows, *Air Conditioning and Refrigeration*, New York: Wiley, 1958.

41. F. L. Stasa, *The Thermodynamic Performance of Two Combined Cycle Power Plants Integrated with Two Coal Gasification Systems*, PhD Dissertation, Carnegie Mellon Univ., Pittsburgh, Pa., 1978.

14

The Transportation and Storage of Energy

INTRODUCTION

Nature stores most of its energy in nuclear form converts it to heat energy, and transmits it by electromagnetic radiation, as from the sun to the earth. We may come to this, hopefully, in a less wasteful manner. But for now we must store it mostly in nuclear, chemical, or mechanical form and transport it in all sorts of land, water, and air vehicles, or in guides such as pipelines or electrical transmission. Most of these forms of transportation and storage are well known. We seek in this chapter to express their characteristics and relative costs in a more quantitative way.

Our principal transportation problem lies in fossil fuels. Nearly a quarter of these are converted to electricity before being used. Then we have the problem of transmitting and distributing electricity. With coal, we may have a choice of shipping fuel and generating at the load center, or generating at the source and transmitting electricity. With hydro, we have little choice. With nuclear, we may *choose* to locate plants distant from population centers and transmit the energy electrically.

Most cities have overlapping gas and electric distribution systems. The economy of this may be challenged as natural gas becomes scarce. At least half the cost of each service is in the distribution system.

Nuclear fuel represents the least transportation expense. After it is mined and milled nearby into "yellow cake" (70 to 90% uranium oxide, U_3O_8), some 200 tons of it (150 tons of U_3O_8) will run a 1000 MWe nuclear plant for a year. It is thus equivalent to about 2.5 million tons of coal. About 12,000 times as much coal is needed as yellow cake. The transportation cost for nuclear fuel in any form, from yellow cake to finished fuel rods, is thus negligible compared with all other fuels and will not be considered further. Moreover, its storage possibilities

are unbelievable. The depleted uranium stored in our enrichment plants now (1980) is equivalent to all U.S. energy needs for the next century, at the present rate of use, if used in breeder reactors.

Our principle energy storage problems are fourfold: (1) storage aboard vehicles to provide a reasonable range before refueling, (2) daily storage for load leveling on electric power systems, (3) annual storage of oil and gas to provide for seasonal peaks, and (4) strategic, or national defense storage of fuel reserves which may be held for many years.

The technologies, most of which are covered in this chapter or earlier in the book, are indicated in the following lists, with the newer technologies under study and development in 1980 indicated by asterisks. Many current technologies are also undergoing extensive *further* development, for example, storage batteries and pumped storage. This book covers mainly civilian aspects. Explosives are not covered

Energy Transportation Technologies

1. Ship.
2. Barge.
3. Railroad.
4. Pipeline.
5. Truck.
6. Electrical transmission.
7. Conveyor.
8. Air transport.
9. Space transport.*
10. Microwave.*

Energy Storage Technologies
1. Gas storage above ground.
2. Gas storage under ground.
3. Liquefied natural gas (LNG).
4. Liquefied petroleum gas (LPG).

5. Liquefied hydrogen.

6. Oil storage—crude and products.

7. Power plant storage: (a) tank farms, (b) coal piles.

8. Vehicle storage: (a) coal, (b) diesel oil, (c) gasoline, (d) jet fuel, (e) battery, (f) hydride, (g) flywheel.

9. Pumped storage hydro.

10. Storage battery.

11. Compressed air.*

12. Metal hydride.*

13. Magnetic energy—superconducting magnet.*

14. Thermal (water, rock, molten salt, steam).

15. Flywheel.*

16. Radioisotope.

17. Electric storage, capacitor.

18. Chemical storage—as used with solar.*

THE TRANSPORTATION OF ENERGY

With moving vehicles, water transportation is cheaper than land, and land is cheaper than air transportation. For general, break-bulk overseas cargo, the cost per ton-mile was about 0.5¢ by sea and 18 to 19¢ by air about 1975, a ratio of 40:1. Liquid or dry-bulk cargo was much less than 0.5¢ per ton mile by sea. Oil from the Persian Gulf to the U.S. East Coast, about 23,900 nautical miles, cost 83¢ per barrel of 298 lb in 1972, using supertankers and superports. This is 0.26 mills per sea ton mile, or 0.20 mills per land ton mile. With normal tankers over the same route, the cost was $1.30 per barrel, or 0.32 mills per land ton mile.

Inland waterways,[1] where available, are considerably cheaper than land vehicles and are comparable with pipelines in cost. While canals were largely replaced by railroads at the end of the Civil War (1865), canalized rivers and intracoastal waterways are highly economic traffic arteries today (1980), and are undergoing major improvements and extensions. The entire eastern United States is spanned by a vast waterway network (Fig. 14.1).

For 1000 miles, from New Orleans to St. Louis, the Mississippi is an "open river," with no dams or locks. Barge flotillas of 50,000 to 60,000 tons can be pushed by a single 6000 Hp "towboat." A 50,000 ton coal tow is equivalent to $8\frac{1}{3}$ 6000 ton coal trains, each with a 6000 Hp locomotive.

Even in the upper canalized sections of the rivers, 20,000 ton tows are not uncommon. An average coal tow (1980) consists of 15 1500 ton barge loads (22,500 tons) pushed by a 5600 Hp towboat.

Altogether, the average costs per land ton mile by various carriers, about 1975, were as shown in Table 14.1.

Energy Transport Pattern

Given these costs, and assuming that the cheapest available transport is always used for large bulk movements, the following energy transportation emerges:

1. There are five principal transportation modes available for bulk energy transport, namely, water, pipeline, railroad, truck, and electrical transmission.

2. There are three prime energy materials to be transported, namely, coal, oil, and gas.

3. *Overseas.* Oil, LNG, and a little coal are transported entirely by ship.

4. *Domestic Water Transportation.*[3] In addition to the inland waterways shown in Fig. 14.1, domestic water transportation includes the Domestic Ocean Trade to Alaska, Hawaii, and U.S. possessions as well as ocean-going trade along the coasts and through the Panama Canal. It also includes Great Lakes Trade. The Domestic Ocean and Great Lakes transport is similar to overseas, except with many smaller and specially designed vessels and non-self-propelled vessels. Coastal ocean-going ships are limited to about 60,000 tons by the depths of many of the harbors. Great Lakes ships are specialized to these waters and port facilities, as will be explained later.

The intracoastal waterways along the Atlantic and Gulf coasts are protected from the sea by long bars or islands which have formed along these coasts. In some sections they pass through sounds or bays, such as Chesapeake Bay.

Domestic water transportation altogether accounts for most of the ton miles of oil and coal transport between points it connects. Inland water transportation accounted for 15% of the ton miles of coal transport, and 42% of the ton miles of oil and oil products transport in 1970.

Water transport of coal usually involves an initial

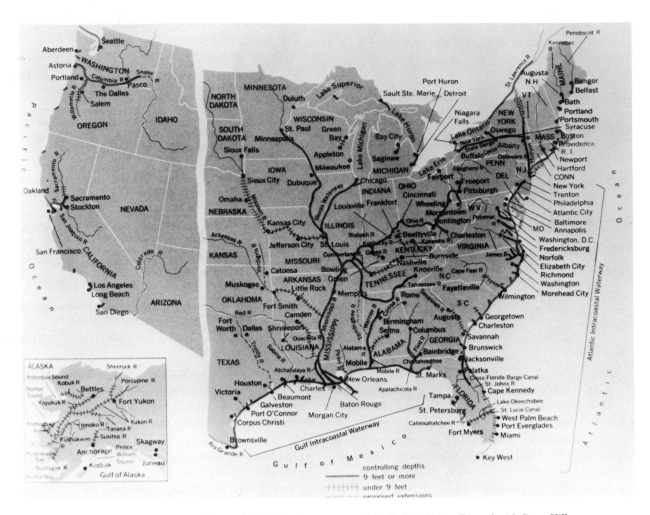

Figure 14.1. Commercially navigable inland waterways of the United States. From the *McGraw Hill Encyclopedia of Science and Technology.*[1] Copyright 1977. Used with the permission of the McGraw Hill Book Co.

truck or rail link to get it to the water. The destinations are generally along the waterways.

5. *Crude Oil and Oil-Product Pipelines.* These connect the heavily populated and industrial areas with the sources where water transport is not available. They interconnect with the water transport at many points. Most oil pipelines are interior, the coastal cities and New England being well served by water transport.

Where both pipeline and water connections are available, since costs are often comparable, many other factors enter to determine the best choice. The pipeline, being a large fixed investment, requires assured loading for many years. Ships and barges use public waterways and are much more flexible as to product and route.

Pipelines accounted for about 57% of the crude oil and oil products transported in 1970. About 75% of crude oil and 20% of products are transported by pipeline.

6. *Trucks.* These, and large off-highway vehicles, account for most of the coal gathering and short-haul delivery to shipping points and nearby plants. Trucks account for most of the delivery of oil products to retailers and to residential and commercial users.

7. *Electrical Transmission.* The maximum distance at which electrical transmission is competitive with coal by train has been increasing with the increase in the maximum transmission voltage and, recently, due to increases in rail costs. It was about 200 miles with 220 kV in 1922, 300 miles at 345 kV in

Table 14.1. Approximate Large-Bulk Transportation Costs, 1975[1]

Carrier	Cost per Ton Mile
Inland, United States	
Barge	3 mills (1.75 –7)
Pipeline	Comparable to barge
Great Lakes ship	Comparable to barge
Coastal ship	Comparable to barge
Rail	15 mills (1.5¢), unit trains 8–10 mills
Motor freight	65 mills (6.5¢)
Air freight	200 mills (20¢)
Electrical transmission at 500–765 kV was cheaper than coal by train for 500–600 miles in many places	
Overseas[a,2]	
General cargo	5 mills
Oil tanker	0.24–0.44 mills
Supertanker	0.20–0.29 mills

[a]Multiply by 1.29 for cost per long ton—nautical mile.

1952, 450 to 600 miles at 500 to 765 kV in the period 1963–1973, and is 800 to 1000 miles with 500 to 765 kV lines in 1980. Thus, much coal conversion to electricity is currently (1980) in coal fields 300 to 600 miles from the load centers, with plans for much greater distances.

The economic power blocks increase also with voltage and distance, from a few hundred MW at 220 kV, 200 miles, to 1 to 2 GW at 500 kV, 450 miles, to about 10 GW at 1100 kV, 1000 miles.

If nuclear plants must be located hundreds of miles from population centers, electrical transmission provides the most economical way of getting the energy to market. One alternative, conversion to hydrogen, transmission by pipeline, and reconversion to electricity, is several times more expensive (see Chapter 4, The "Hydrogen Economy" Concept).

8. *Coal.* Coastal shipping of coal is mainly Alabama coal via the Warrior River to Gulf Coast ports and Appalachian coal to East Coast ports. With the "clean fuel era" of the 1960s, collier traffic on the East Coast dwindled to near zero as the utilities switched to "clean" oil. This was prior to the "oil shortage era" starting in 1974. The future is still in the making (1981).

Coal accounts for about 19% of the ton miles of railroad transportation in the United States (33% of the tonnage, 16% of the revenue, 1979). However,

some railroads are primarily coal-hauling roads, and the coal traffic provides over 50% of their revenue.

9. Natural gas is transmitted almost 100% by pipeline.

The following sections provide an introduction to each of the five principal energy transportation modes over 50 miles and their important characteristics.

Transport by Ship

Nature has provided a marvelous medium for moving heavy loads slowly, namely, water. Over 5000 yr ago, the Chinese discovered that even a light breeze on sails would move a ship. Today, the term "ship" is reserved for the larger vessels that ply the oceans and the great inland lakes. The smaller craft used on rivers and inland waterways are not usually called ships. For thousands of years ships have carried most of the coastal and overseas cargo, and even today (1980) they carry 90% of all overseas cargo. Airplanes carry the other 10% and most of the passengers.

Ship Performance.[4] The empirical curve of Fig. 14.2 (full line) tells us almost everything we need to know about the power required at various

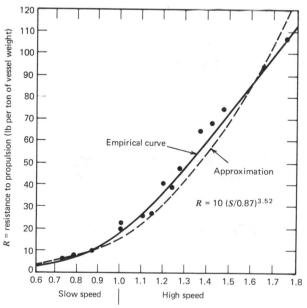

Figure 14.2. Ship performance versus specified speed; S = speed in knots divided by the square root of the ship waterline length in feet.

speeds by modern ships. It may be roughly approximated by the dashed curve:

$$R = 10(S/0.87)^{3.52} \text{ lb/ton*} \qquad (14.1)$$

For example, a cargo/passenger ship of 19,000 displacement tons (9000 deadweight tons), having a 500 ft waterline length and cruising at 20 knots ($S = 0.89$), requires 11 lb push per ton of ship weight, or 209,000 lb push. At 20 knots, the required Hp is

$$\text{Hp} = 209,000 \frac{(20 \times 6080)}{(3600 \times 550)} = 12,836 \text{ Hp}$$

This ship is powered by engines of 19,800 max. shaft Hp on a single propeller. This provides for (a) the efficiency of propulsion, normally 75 to 80%; (b) efficient operation below maximum Hp; (c) higher emergency speeds.

At a propulsion efficiency of 80%, and using the dashed approximation of Fig. 14.2, it could make

$$\left(\frac{19,800 \times 0.8}{12,836}\right)^{1/3.52} \times 20 = 21.2 \text{ knots}$$

fully loaded. With $S = 0.89$, it is relatively faster than most cargo ships ($S = 0.5$ to 1.0), but relatively slow compared with passenger ships, naval vessels, or power yachts ($S = 1.0$ to 1.8, see dots on curve).

Tanker. On the other hand, a tanker of 180,000 displacement tons (150,000 dwt), designed for operation at 16.2 knots, with a waterline length of 950 ft ($S = 0.525$) requires

$$R = 10\left(\frac{0.525}{0.87}\right)^{3.52} = 1.69 \text{ lb/ton}$$

only 15% of that required by the cargo/passenger ship. It requires a push of $1.69 \times 180,000 = 304,000$ lb. Its Hp requirement is

$$\frac{(304,000 \times 16.2 \times 6080)}{(3600 \times 550)} = 15,123 \text{ Hp}$$

This ship is equipped with 30,000 Hp engines, about the largest normally used on cargo ships, and nor-

*All tons in "Transport by Ship" are long tons unless otherwise labeled.

mally cruises at 60 to 70% power. At 80% efficiency, it could make

$$\left(\frac{30,000 \times 0.8}{15,123}\right)^{1/3.52} \times 16.2 = 18.5 \text{ knots}$$

With $S = 0.525$, it is a relatively slow-speed ship.

Typical Cargo Ships. The specifications[4] and relative speeds S for 10 typical cargo ships are given in Table 14.2. Lacking the waterline length, the bow-to-rudder dimension L_{pp} has been used as an approximation, this factor entering as a square root. All ships listed use one propeller unless otherwise noted. Figure 14.3 shows the tanker, item 9.

Nautical Terms. From long tradition, the sea has a language all its own. On land, we might "ship" 1000 tons of material 1000 miles at 20 mph. The same operation at sea would be 893 (long) tons for 868 (nautical) miles, at 17.4 knots. The "long" and "nautical" would not be mentioned. In round numbers, a knot is 1.15 mph. A nautical mile is 6080 ft. And a long ton is 2240 lb. Knots are nautical miles per hour. The principal nautical conversions are shown in Table 14.3.

When aboard ship, facing forward, "port" is on your left, "starboard" on your right. The port lights are red, the starboard lights green. The saying "Red left port" ties these terms together. At the front or "forward" part of the ship is the "bow." At the rear or "aft" is the "stern." In between is "amidships."

A ship may "list" (tip) to port or starboard due to uneven loading, or to damage. It may "trim by the bow" if the forward end is down, or "by the stern" if the aft end is down. Somewhere between level and 2 ft trim by the stern is normal in operation, any greater unbalance restricting the ship's movements in harbors near the limiting depth.

A ship's displacement is its total weight loaded in long tons. Its deadweight (dwt) is its cargo-carrying capacity in long tons. "Tonnage" is a measure of enclosed volume, less ballast tanks, in ft³/100. It is used for registry, mooring, and lockage charges, and the like. Gross tonnage is the total. Net tonnage is available for cargo; it derives from the number of tuns of wine that could be carried.

A ship's "depth" is the vertical distance from keel to main or "strength" deck. Its "draft" is the vertical distance from the waterline, loaded, to the

Table 14.2. Specifications of Typical Cargo Ships[4]

No.	Type of Ship	Length Overall (ft)	Length Between Perp's L_{pp} (ft)	Beam (ft)	Depth (ft)	Draft (ft)	Displacement (long tons)	Dead Weight (long tons)	Speed (knots)	Max. Shaft (Hp)	Normal Crew	Remarks	Relative Speed S
1	Break-bulk cargo ship	574	544.5	82	45.5	30.5	21,235	12,932	23	24,000	45		0.99
2	Combination passenger, reefer, container ship	547	508.5	79	48	29	19,799	9,234	20	19,800	121	119 passengers	0.89
3	Container ship	752	705.8	100.5	57	31.6	33,924	19,206	27	60,000	40	Two props.	1.02
4	Roll-on/roll-off ship	700	643	105	60	27	24,100	13,100	25	32,000	39		0.99
5	Barge-carrying ship	820	724	100	60	28	32,800	18,760	22.5	32,000	39		0.84
6	Ocean-going ore carrier	865	826.7	124.7	68.9	50.7	124,000	104,500	15.8	24,300	40		0.55
7	Ocean-going combination bulk carrier	815	775	104.5	61.5	41.1	78,938	63,410	15.8	18,000	41		0.57
8	Great Lakes self-unloader	730	724	75	45	27.9	38,980	29,550	13.0	7,400	32		0.48
9	Tanker	1005	951.4	155.8	78.7	52.6	179,700	151,300	16.2	30,000	29		0.53
10	LNG Ship	660	617.7	81.6	54	24.7	25,700	13,400	17	15,000	44		0.68

bottom of the keel amidships. "Freeboard" is the distance from waterline to strength deck amidships, that is, to the top of the enclosed side of the ship.

Ship Structure. Partitions across the ship are "bulkheads." The spaces between are "holds." The strength of a steel ship comes from the hull, which is covered by steel plates and framed, either by ribs or by longitudinal members. Thus, the plat-

ing of the hull, the bulkheads, and the main or strength deck, together with the framing, give the ship the necessary strength to withstand the forces imposed by waves and loading. Other in-between decks provide for storage of cargo and may add some to the strength.

The bulkheads are watertight and limit the flooding by seawater in event of damage. In cargo vessels, they are usually spaced about 50 to 60 ft apart,

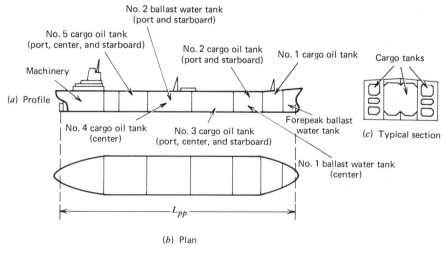

Figure 14.3. Tanker.

Table 14.3. Nautical Conversions

To Convert From	To	Multiply by
Nautical miles	Statute miles	1.1508
Knots	Miles per hour	1.1508
Long tons	Short tons	1.12
Metric tons	Short tons	1.1023
Short tons	Pounds	2000
Long tons	Pounds	2240
Metric tons	Pounds	2205

for example, four bulkheads in a 250 ft ship, eight bulkheads in a 500 ft ship.

Superstructures above the main deck are the "forecastle" at the "forward end" and the "poop" "aft." Above-deck superstructures that do not extend to the sides are "deck houses."

The "bridge" or pilot house is usually above the machinery space on ocean-going ships. Except for bulk cargo ships, the machinery and bridge are about three quarters of the way back from bow to stern. For bulk carriers, for convenience in loading and unloading the cargo, the machinery is at the stern, and the bridge at one end or the other, aft on ocean-going ships and forward on Great Lakes ships.

The Cargo. This is classified as (1) break-bulk, or general cargo, (2) unitized cargo, or (3) bulk (dry or liquid) cargo. The latter is of primary interest for the transportation of energy. However, since these ships are members of a family, a brief description of the whole family is given here.

Break-bulk or general cargo is miscellaneous goods packed in boxes, bales, crates, cases, bags, cartons, barrels, or drums, as well as lumber, motor vehicles, pipe, steel, and machinery. Break-bulk ships may also carry some refrigerated cargo in specially insulated "refrigerator holds."

Unitized cargo is handled in pallets, containers, railroad cars, trailers, barges, and lighters. Typical pallets are 24 × 32 in. to 88 × 108 in., or squares of 36 to 48 in., and are handled by fork lift trucks. "Pallett ships" carry pallets exclusively.

"Containers" are boxes of aluminum, plywood, or steel, typically 8 × 8 × 20 ft or 8 × 8 × 40 ft. "Full container ships" are specialized to this class of cargo, which makes possible fast loading and unloading, as well as reducing loss from damage or pilferage.

Roll-on, roll-off ships are designed specifically to accommodate wheeled vehicles such as railroad cars and trailers and are equipped with side or end ramps.

Barge-carrying ships are similarly equipped for loading barges or lighters, using shipboard gantry cranes.

Bulk-dry cargo includes ore, sugar, cement, and grain, as well as coal. A coal ship is a "collier," and an ore ship is an "ore carrier."

Bulk-liquid cargo, in addition to petroleum and products, includes LNG, LPG, liquid chemicals, wine, fruit juices, and molten asphalt and sulfur.

A *"reefer"* carries primarily refrigerated products. A general cargo ship may also have one or more "reefer holds."

LNG tankers, or cryogenic ships, carry liquefied natural gas, having about $\frac{1}{600}$ of the volume of natural gas. These ships carry LNG at atmospheric pressure and a temperature of about −260°F, in tanks that are insulated from all structural parts of the ship. There is also a secondary barrier that would prevent any leakage from reaching the structure to embrittle it and endanger the ship.

A typical LNG tanker (1975) has a capacity of 125,000 m³. At 0.424 sp. gr., this represents about 53,000 long tons. It would have a displacement of about 90,000 tons, considerably larger than one of the earlier LNG ships listed in Table 14.2.

The net cost to the owner is about $80,000,000. At 24,000 Btu/lb, the cargo contains 2,800,000 million Btu of energy; and at $1/million Btu, it is worth $2,800,000 delivered.

At 18% capital charge, the annual capital cost of the ship is $14.4 million. At 20 knots, and including turnarounds, this ship might make 10 trips from Algeria to the East Coast U.S.A. per year. Thus, the capital cost for the ship is $1.44 million per trip. The capital cost includes maintenance, depreciation, taxes, insurance, and return. Operating expenses for crew and fuel must be added, and about 3% boiloff (0.3% per day), so that the voyage adds about $1.6 million to the cost of the cargo. Note that over half of the delivered cost is in shipping. The costs and facilities of a complete LNG system are discussed in Chapter 4.

Ships up to 165,000 m³ were on order in 1972 for delivery about 1980, and 200,000 m³ ships were planned. The cost varies about as the 0.6 power of the capacity, that is, $106 million for a 200,000 m³ ship.

Ship Proportions. Theoretically, the size and weight of a ship could be doubled either by multiplying each linear dimension by the cube root of 2 or by holding the cross section fixed and doubling all lengths. In the first case, starting with $S = 1$ (19 lb/ton), the increase of waterline length by 1.260 would make $S = 0.891$, at the same speed, and the resistance 11 lb/ton.

If, instead, the waterline length is doubled, S becomes 0.707 at the same speed and the resistance is 4.5 lb/ton. The long, narrow ship requires less than half as much power to drive it through the water at the same speed. There is thus an economic incentive to make ships long and narrow.

However, in addition to many practical factors, such as length of locks, space required at wharves, and maneuverability, there are limits to shape determined by stability and strength. These limits have been determined by long experience, combined with tests and computations. The ship must not capsize even when damaged. It must not break in two in high waves. Passenger ships must be even safer than cargo vessels.

The baseline for safe proportions is the "length between perpendiculars," L_{pp}, shown on Fig. 14.3. It is the bow-to-rudder distance, but not less than 95% of the waterline length. The limiting proportions of ships are given in Table 14.4. Thus, a 600 ft L_{pp} cargo ship might have a breadth of $0.11 \times 600 + 15 = 81$ ft. Most of the ships in Table 14.2 have much greater breadths than given by these limits.

Ships on the Great Lakes are not subjected to as great waves as ocean-going ships and can be much more slender and hence more economical to operate, for the same size. The long ship also requires less speed reduction in high waves.

Locks, canals and harbors also introduce limitations to ship dimensions, as shown in Table 14.5.

Table 14.4. Limiting Proportions of Ships as a Percent of $L_{pp}{}^a$

Dimension	Ocean Going	Great Lakes
Breadth not less than	11% + 10–20 ft	"Somewhat less"
Depth not less than (keel to strength deck)	10–14%	5.5%
Freeboard at least		
500–700 ft dry cargo	1.6%	
Tankers	1.3%	

$^a L_{pp}$ is the distance between perpendiculars.

The three larger ships of Table 14.2 exceed the 35 ft draft limitation of the Panama and Suez canals. All exceed the 25.5 ft draft of the St. Lawrence Seaway.

A cargo ship usually has a double bottom. A bottom deck extends from one of the forward bulkheads to the stern bulkhead. The space between is used as tanks for fuel, fresh water, and seawater ballast, and also limits flooding in event of damage to the bottom of the ship.

In a tanker, the double bottom is dispensed with. However, the arrangement of tanks for cargo and ballast provides equivalent subdivision (Fig. 14.3c). While the bulkheads are more widely spaced—six in the 1005 ft tanker of Fig. 14.3—the tank walls provide the necessary subdivision and structural strength.

Propulsion. Most cargo vessels of the past have been powered by engines of 30,000 shaft Hp or less. However, there are an increasing number of fast container ships such as that listed in Table 14.2 (60,000 Hp, two propellers) and LNG ships of high capital cost, warranting increased speed.

Power plants under 300 Hp are usually diesel. Above 300 Hp, diesel and steam are used. Steam

Table 14.5. Locks, Canals, and Harbors Limitations

Waterway	Limiting			
	Length (ft)	Breadth (ft)	Draft (ft)	Clearance Under Bridges (ft)
Panama Canal	900	104	35	—
Suez Canal	—	—	35	—
St. Lawrence Seaway	730	75	25.5	120
Enlarged Soo Lock	1000	105	—	—
Many harbors	—	—	40	—

turbines are usual in U.S. ships. Some diesels as large as 25,000 Hp are used in other ships.

There are also limited applications of nuclear/ steam, gas turbine, free-piston engines, and other power plants. Sails are almost extinct for cargo vessels, with a few romantic exceptions[5] or as auxiliaries. For most ships, the increased capital cost per voyage exceeds the reduced fuel cost.

Barges and Towboats

Inland Waterways (1968). The Inland Waterways of the United States are shown in Fig. 14.1. They include the intracoastal waterways along the East Coast and the Gulf Coast and canalized portions of many of the great rivers. Some 15,000 miles of waterway are suitable for vessels of 9 ft draft, (at least 12 ft deep), and an additional 10,000 miles are suitable for vessels of less than 9 ft draft. A few stretches are deep enough for ocean-going ships—the Hudson River to Albany, the Houston Ship Canal, the Columbia River to Portland, and the Mississippi River to Baton Rouge.

Great improvements are in progress throughout the system (1980). Congress has authorized 4000 miles of extensions. The multiplicity of small locks on the upper reaches of all the great rivers is being replaced by fewer larger locks. Thus, water transportation is continually being made more economic, and its use is being expanded. Some 10% of the nation's commerce is now (1980) moved over these waterways.

Types of Vessels. Cargo vessels may be self-propelled or nonself-propelled. They may be dry cargo vessels or tankers. In the three categories of Domestic Trade, the proportions of each type of vessel in use in 1976 were as given in Table 14.6.[3]

Most cargo on inland waterways in the United States is transported by nonself-propelled vessels,

that is, by barges. A surprising amount of Domestic Oceans transport is also by large ocean-going barges, 30,000 tons and over, either towed on a hawser or with rigidly affixed propulsion unit, forming an integrated tow. These operate close to the safety of ports at all times.

Self-propelled vessels or ships have been described earlier. The next sections cover typical barges and towboats used on the Inland Waterways.

Towboats. In spite of the name, towboats are push boats. A flotilla of barges can be controlled infinitely better by a rigidly affixed towboat than by a tugboat and hawser. Towboats and tugboats are almost invariably diesel powered. Standard sizes of towboats are 1800 to 10,500 Hp, with the 5600 Hp size, shown in Fig. 14.4, being the most popular. At 8 tons/Hp, it can handle tows up to 45,000 tons. It has ample manoevering power for the smaller tows, typically 15 to 20 coal barges, 22,000 to 30,000 tons, or a double string of oil barges, 25,000 tons.

With the two propellers well under the shaped hull and rudders both fore and aft of the propellers, a high steering torque can be exerted either forward or backing.

Fuel tanks of 94,000 gal capacity, with enough fuel for 18 to 20 days of full power, are balanced by ballast water tanks of 66,000 gal capacity, so that the boat can be trimmed for maximum thrust at all times.

Barges. These are of two general types— hopper barges and tank barges. Hopper barges (Fig. 14.5) are open for coal and other materials not requiring protection. For grain and other bulk cargo requiring protection, there are lift-off covers. The barges are usually loaded through openings and unloaded by suction hoses, after lifting off the covers. General cargo barges have rolling covers that

Table 14.6.   Vessels in Domestic Trade (1976)[3]

	Self-Propelled			Nonself-Propelled			Total Trade (thousands short tons)
	Dry Cargo	Tankers	Total	Dry Cargo	Tankers	Total	
Domestic oceans	5%	68%	73%	8%	19%	27%	232,507
Great Lakes	92%	3%	95%	3%	2%	5%	140,313
Inland waterways	1%	3%	4%	54%	42%	96%	605,292
							978,212

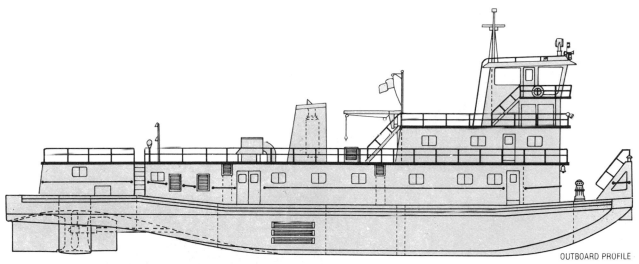

Figure 14.4. 5600 Hp towboat. Specifications: Vertical clearance above design draft (approx.)—44′6″, Height of headlog above molded baseline—12′11⅝″, Height of transom above molded baseline—12′10⅛″, Height of towing knee above design WL (approx.) —13′3″, Width of towing knee—5′11″, Outside span of towing knees—31′, Fuel oil capacity, 98% full (approx. gal.)—94,000, Potable water capacity, 100% full—14,200, Ballast water capacity, 100% full (approx. gal.)—66,000, Lube oil capacity, 98% full (approx. gal.)—1,485, Length—140′, Width—42′, Hull depth—11′, Draft—8′6″, Eye level—33′8½″. Courtesy Dravo Corp.

telescope one under the other. All hopper barges have a single hold, with no bulkheads, which makes for maximum efficiency in loading and unloading.

A typical coal unloader is shown in Fig. 14.6. The scoop bucket wheels lift the coal to a conveyor which carries it to the storage pile or tipple. The bucket wheels can be spaced horizontally and unload in several passes of the barge under the unloader.

The strength of the barge is built into its sides, ends, and bottom. The double sides are about 3 ft 3 in. apart, and the double bottoms about 1 ft 4 in. apart, as shown in Fig. 14.5b. The space between is divided into six or seven watertight compartments to limit flooding in event of damage. Both ends are heavily braced for pushing or towing. The decking at the top of the double-sided hull is the "strength deck" of the barge, and the hull depth D is measured from this deck to the bottom. Standard hull depths are 12, 13, and 14 ft (12 ft for coal barges).

Above the hull is the "coaming" E, which may be 1¼, 3, 4, or 5 ft, depending on the normal cargo. A 1½ ft coaming is usual for a coal barge. Thus a coal barge with 12′ hull depth, and 1½ ft coaming, total, 13½ ft, would have a hopper depth of 12 ft 2½ in. and a double bottom, including plating, of 1 ft 3½ in., total 13½ ft.

Standard hopper barges are 195 or 200 ft in length; 195 ft is usual, but 200 ft is the maximum for three in a 600 ft lock. The standard width for hopper

barges is 35 ft overall, or a hopper width of 28 ft 6 in. The hopper length is 175.6 ft in the 195 ft coal barge, volume 63,811 ft³ in box barge, 58,950 ft³ in "semi."

Since they operate in waters of varying depth, the draft may be 9 ft for standard inland waterways or more or less for deeper or shallower waters. A large amount of barge traffic is in lakes, bays, and sounds that are suitable for more than the standard 9 ft draft. Some 10,000 miles of waterways are substandard and suitable only for less than 9 ft draft.

Some barges are made with one end "raked," with a fixed curvature RR about 41 ft, as shown in Fig. 14.5c, and one square "box end." These are known as "rake barges" or "semis." Others are made with two box ends and are known as "box barges." In operation, they are arranged with a semi at each end and box barges in between, which provides minimum water resistance and the efficiency of a long narrow ship.

With the standard 9 ft draft, the 195 ft coal barge will carry loads as follows:

Rake barge—1598 tons   Box barge—1610 tons

(Box barge 1397 tons at 8 ft draft to 1929 tons at 10 ft 6 in. draft.) In round numbers, these are 1500 ton barges. General open hopper barges vary slightly from these figures depending on hull depth, coaming height, and type of covers.

With 8 ft draft, the box barge carries 1397 tons. It

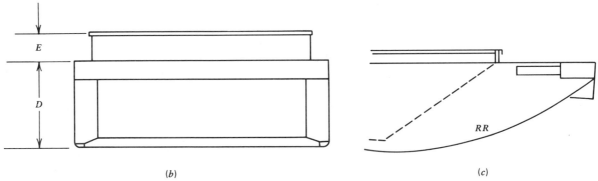

Figure 14.5. Open hopper barge. Courtesy Dravo Corp: (*a*) Open hopper rake barge or "semi"; (*b*) box end; (*c*) rake end.

displaces 1610 − 1397 = 213 tons per ft of draft, or 1917 tons total at 9 ft draft. Deducting the cargo, 1610 tons, the barge itself weighs 307 tons. In shipping language it has 1917 displacement tons and 1610 deadweight tons (dwt) when loaded to a 9 ft draft.

Tank barges are used for a variety of liquid products, including crude oil and oil products. As with open-hopper barges, there are box and rake barges fitted together to form an integrated tow. However, the individual barges are very much larger. All are 54 ft wide, just 2 ft less than the width of the smaller locks. In the larger, 110 ft wide locks, two tank barges abreast just fit with a 2 ft clearance.

The lead and trailing rake barges and the "long box barges" are all 297 ft 6 in. long, and hence just fit two in a 600 ft lock, with 5 ft spare. There is also a "short box barge" of 146 ft 6 in. length, which

together with a long rake unit and the towboat will also fit in a 600 ft lock. Normally, a five-unit integrated tank barge fleet is arranged, fore to aft—lead rake, long box, short box, trail rake, and towboat. It divides in two for locking through the 600 ft locks in the upper rivers and can be locked as a unit in the 1200 ft locks of the lower rivers.

Most 1200 ft locks are 110 ft wide and can accommodate an eight-barge tanker fleet intact, two of the four-barge units abreast, plus the towboat.

Some of the upriver locks are 110 ft wide but 600 ft long and lock the two-abreast tow in two units. Where tows are broken in length for locking, the locks are equipped to pull the forward barges out of the lock after raising or lowering.

Most tank barges are double skinned and have a uniform hull depth of 12 ft. (Some are single-skin or independent tank units.) The lead and trail rake

Figure 14.6. Coal unloader. Courtesy Dravo Corp.

units have a 2 to 3 ft shear on a 150 ft rake radius, which makes them extremely efficient in operation. Each barge is divided into three approximately equal tanks. The total tons shown in Table 14.7 correspond to the bbls at 350 lb/barrel, or at 1.0 sp. gr.

Note that at 9 ft draft the tanks could be filled to only about three quarters full, with heavy crude, of 1.0 sp. gr. Or they could be completely filled with a lighter product of 0.75 sp. gr. or less, such as gasoline.

In addition to the typical barges described, a wide variety of other sizes and shapes are used, as well as special barges for many purposes.

Barge traffic averages about 6 mph on the inland waterways, but may be much less on the upper riv-

Table 14.7.   Tank Barge Capacities

| Unit | Total Tankage | | DWT Tons 9 ft Draft | Displacement (tons, approx.) |
	Barrels	Tons		
Lead Rake	25,945	4,545	3,359	
Long Box	29,406	5,151	3,780	
Short Box	13,712	2,402	1,812	
Trail Rake	27,572	4,831	3,584	
Four-Barge	96,635	16,929	12,535	
Eight-Barge	193,270	33,858	25,070	30,080
Towboat, 5600 Hp				1,200
Eight-Barge tow				31,280

ers if the locks are congested. The typical 25,000 ton tows, with 5600 Hp towboat, would normally travel about 8 to 9 mph.

As shown in Table 14.1, the average transportation cost by barge in 1975 was about 3 mills per ton mile, compared with 15 mills by rail (8 to 10 mills by unit train), or 65 mills by truck. The cost was comparable to pipeline, and waterways provided most of the coal and oil transport between points along the water. In some large petroleum movements, the average cost was 1.75 mills per ton mile.

In 1980, the cost of a 195 ft coal barge was about $275,000; a 297½ ft tank barge, $950,000; and a 5600 Hp towboat, about $4,000,000.

Inland waterways are maintained by the Federal Government at a cost of about $525,000,000 annually.

The Barge Fleet. In 1973, there were about 4400 towboats and tugboats on the inland waterways of the United States, totalling 3,546,000 Hp, an average of 806 Hp per unit. There were 15,400 dry-cargo barges of 16,070,000 aggregate tons, or an average of 1044 tons per barge. There were 2780 tank barges of 5,120,000 aggregate tons, or an average of 1842 tons per barge.

Petroleum and products represented 38% of all barge tonnage in 1976, followed by coal, 22%. The other 40% was minerals, farm products, chemicals, and general cargo.

Transport by Rail

Railroads in the United States expanded slowly during the first 30 yr of the twentieth century, from 193,000 miles to about 260,000 miles. Since then, they have declined to about 220,000 miles in 1970, where they appear to have leveled off (1980). Railroads transported about 773 billion ton miles of freight in 1970. The average rail haul was about 283 miles.

Railroads are largely private-enterprise, taxpaying operations, which own and maintain their own right-of-ways, in contrast to trucks or waterway transports, which use public right-of-ways.

About 1976, a number of railroads in the Northeast, which were in financial difficulties, were made into a single corporation, Conrail, with government assistance. Passenger traffic in the United States is handled by Amtrack, which leases facilities and services from the various railroads.

Locomotives. These were predominantly steam until 1935 and predominantly diesel by 1965 (1935, 49,998 steam, 89 diesel; 1965, 27,888 diesel, 52 steam). Electric locomotives reached a peak of 857 in 1945 and declined to 236 by 1975. There were no further main-line railroad electrifications in the United States after 1936. Main-line diesel locomotives are normally 3000 Hp/unit, and two or more units can be operated in tandem by a single locomotive crew.

Rolling Stock. There are open-hopper cars for coal, ore, and other cargo not requiring protection from the weather, closed-hopper cars for grain, cement or other dry bulk cargo requiring protection, specialized flat cars for trailers or containers, rack cars for automobiles, closed freight cars for general cargo, tank cars for petroleum and other liquid cargo, and refrigerator cars.

Unit Trains. Paralleling the trend in water transportation, there is a trend to "unit trains," with associated highly efficient loading and unloading facilities. Unit trains, with a single cargo and a single destination, bypass the classification yards, with their associated costs and delays, and lead to the lowest-cost rail transportation. However, they can only be used for large, regularly scheduled movements, where the terminal facilities are designed to accommodate them.

Coal Hauling. About 19% of all the ton miles of railroad transport in the United States is coal haulage (33% of tonnage, 16% of revenue, 1979). The railroads account for 78% of all domestic coal transport, the balance (over 50 miles) being by waterway. Practically all transport under 50 miles is by truck.

Next to trucks, trains are the most flexible means of coal transport. Most mines are not adjacent to waterways. Slurry pipelines are fixed and economic only for very large movements. The relative costs are given in Fig. 14.13 and Table 14.12. Trains require about 1 Hp/ton, as compared with ⅛ Hp/ton by waterway, or 11 Hp/ton by truck. Trains are about four times as fuel efficient as highway vehicles. An average coal unit train uses five six-axle, 3000 Hp locomotive units to pull a 10,000 gross ton trailing train (7000 tons of coal).

Older 50 to 70 ton coal cars are being replaced by 80 to 100 ton cars, with 100 tons becoming standard

(1980). The gross weight of a loaded 100 ton car is about 140 tons.

Gas Pipelines

As an initiation to the mysteries of gas pipelines, consider a 60 mile length of 36 in. diameter pipe. Natural gas enters the pipe compressed to 1000 psia. At the receiving end, the pressure has dropped to 800 psia, where it will be recompressed to start the next leg of its journey. How much gas flows? How much energy is transferred? How much does it cost for the pipeline and the compressor? How much of the gas energy is used up in making it flow? How much energy is stored within the pipeline?

How Much Gas Flows?

The flow is a function of the difference in pressure squared at the two ends. It depends on $P_1^2 - P_2^2$. The gas is compressible. Its volume expands as the pressure drops. Hence, the mass flow depends on the drop in P^2 instead of the drop in pressure, as with water or oil. The mass flow is proportional to $(P_1^2 - P_2^2)^{0.5394}$.

The mass flow Q is measured in standard ft³/day (SCFD), expressed at a base temperature and pressure, T_0 and P_0, ordinarily 520°R (60°F) and 15.15 psia for natural gas.

Other factors that enter in are the average temperature of the flowing gas, T_f, in °R, the specific gravity of the gas, G, relative to air, the diameter and length of the pipe, and the efficiency of flow, E. Natural gas is a mixture, mostly methane of sp. gr. 0.55, but with some heavier constituents—ethane, and the like. An average sp. gr. is 0.60.

The flow efficiency varies from 0.88 to 0.94, depending on the condition of the pipe. It will be taken as 0.91.

The inside diameter of a 36 in. pipe depends on the wall thickness, which must be sufficient to support the working pressure within. Materials of different yield strength and cost are used. For a 36 in. pipe, with 1000 psi design pressure, a 0.42 in. thickness of 60,000 psi yield material would be adequate, resulting in an inside diameter d of approximately 35 in.

Compressor stations are required every 40 to 80 miles. The 60 mile spacing is near maximum economy, 80 miles for a lightly loaded pipeline and 40 miles for a heavily loaded pipeline.

Most pipeline companies use the "panhandle formula" for obtaining the flow Q in SCFD from these parameters. The panhandle formula and the parameters used in the example are

$$Q = 435.87E \left(\frac{T_0}{P_0}\right)^{1.0788} \left(\frac{P_1^2 - P_2^2}{G^{0.8359}T_f L}\right)^{0.5394} d^{2.6182} \text{ SCFD}$$

(14.2)

where:

Q = rate of flow in standard cu. ft per day (SCFD)

d = ID = 35 in.

E = pipe flow efficiency (0.88 to 0.94) = 0.91

G = sp. gr. (air = 1) = 0.6

L = length in miles = 60

P_0 = pressure base in psia = 15.15

P_1 = inlet pressure in psia = 1000

P_2 = outlet pressure in psia = 800

T_0 = temperature base in °R = 520

T_f = average flow temperature in °R = 510

$$Q = 435.87 \times 0.91 \left(\frac{520}{15.15}\right)^{1.0788}$$
$$\left(\frac{1000^2 - 800^2}{0.6^{0.8359} \times 510 \times 60}\right)^{0.5394} \times 35^{2.6182}$$

= 943 million SCFD, or 0.344×10^{12} SCF/yr

At 1000 Btu/SCF, this is 0.344×10^{15} Btu/yr, or 0.344 quad/yr. This may be compared with the U.S. use of gas, 20 quads in 1978. Some 20/0.344 = 58 such pipelines could supply the United States. One of them could supply 218/58 = 3.76 million people.

The largest LNG tankers (1980), 200,000 m³, carry $200,000 \times 3.28^3 \times 600 = 4230$ million SCF of gas. (LNG occupies about 1/600 of the volume of natural gas.) This ship would supply the 36 in. pipeline of the example for $4230/943 \cong 4\frac{1}{2}$ days.

Maximum Flow. At a sacrifice in economy, from Eq. 14.2, the flow could be increased to 1403 million SCFD, a 49% increase, by dropping P_2 to 500 psia. It could be increased to 1169 million SCFD, a 24% increase by compressor stations every 40 miles, or to 1746 million SCFD, an 85% increase, by both measures.

How Much Energy Is Transferred?

The power flow is

$$0.344 \times 10^{15} \text{ Btu/yr/8760 hr/yr/3413 Btu/kWh}$$
$$= 11.5 \text{ million kW, or } 11.5 \text{ GW}$$

EHV lines have nominal ratings (given later) of:

500 kV, 0.9 GW
765 kV, 2.3 GW

Thus, the 36 in. pipeline in this example is carrying the equivalent energy of 12.8 500 kV lines, or five 765 kV lines.

How Much Does It Cost for the Pipeline and The Compressor?

At the compressor ratio of 1.25 used in the example, the flow of 943 million SCFD would require about 10,000 Hp.[6] At a cost of $300/Hp (1970), this is $3 million per station, or $50,000 per mile.

To run it for a year requires

$$10,000 \text{ Hp} \times 8760 \text{ hr/yr} \times 0.746 \text{ kW/HP}$$
$$\times 3413 \text{ Btu/kwhr/0.75 eff} = 0.30 \times 10^{12} \text{Btu/yr},$$
$$\text{or } 0.00030 \times 10^{12} \text{ SCF/yr}$$

A fraction 0.00030/0.344 = 0.0009 of the gas is used each 60 miles, or 0.9% in 600 miles. Energy for the compressors consumes roughly 1% of the gas every 600 miles, and proportionally more if closer spacing or higher compressor ratio is used.

The pipeline itself is of the order of $4000/mile per in. OD (1970),[6] or $144,000 per mile. Altogether, pipe plus compressor station are $194,000 per mile.

Annual Cost. At 18% for maintenance, depreciation, taxes, insurance, and return, the capital cost charge is

$$\$194,000 \times 0.18 = \$35,000/\text{mile/yr}$$

The capital cost charge is normally about 80% of the annual cost, the remaining 20% being operation and fuel loss and use. Thus, the annual cost is approximately $35,000/0.8 = $44,000/mile/yr.

Unit Cost. The unit cost is thus

$$\frac{\$44,000/\text{yr/mi}}{344 \times 10^{12} \text{ Btu/yr}} = \$128/10^{12} \text{ Btu mile}$$

or 12.8¢ per million Btu for 1000 miles (1970 basis). See Chapter 1 for updating indices.

This agrees quite well with experience. In 1970, interstate pipeline companies paid an average of 18¢/MCF* for gas and sold it at an average of 27¢/MCF to electric utilities and others. The 9¢ difference would transmit gas (9/12.8) 1000 = 703 miles via 36 in. pipeline—a reasonable average distance of transport between the gas fields and the pipeline customers.

Except for escalation of all costs in later years, this gives an idea of the cost of gas transmission.

How Much Gas Is Stored in the Pipeline?

Neglecting supercompressibility, a 5 to 10% factor, the gas can be treated as a perfect gas. At 1000 psia the pipeline holds 1000/15.15 = 66 times its actual volume. A 1000 mile 36 in. pipeline has a volume of

$$1000 \times 5280 \times \pi \times \frac{(35/12)^2}{4} = 35.3 \times 10^6 \text{ ft}^3$$

At 1000 psia, it holds $66 \times 35.3 \times 10^6 = 2330 \times 10^6$ SCF. This is 2330/943 = 2.47 day supply; when "packed" to full pressure, or approximately 90% of that with an average pressure of 900 psia (2.22 days supply).

As a corollary, its average velocity is 1000/2.22 = 450 miles/day, 19 mph, or 28 ft/sec, about 10% higher at the receiving ends and 10% lower at the sending ends.

Gas by Pipeline Versus Electrical Transmission

In the large volume of a 36 in. pipeline, the cost for 1000 miles was shown to be 12.8¢/MCF or per million Btu. If the gas was burned at 40% efficiency, in a generating station, and the 400,000 Btu of electricity transmitted, using the line costs for EHV transmission, and an 18% capital charge, the cost of electrical transmission would be as follows:

Item	500 kV	765 kV
Cost per mile ($)	370,000	473,000
SIL load	900 MW	2300 MW
Cost, 400,000 Btu, 1000 miles, electrical	99.0¢†	49.5¢
Natural gas, 1,000,000 Btu, 1000 miles	12.8¢*	12.8¢

*One MCF of natural gas contains approximately one million Btu.

$$† \frac{\$370,000 \times .18}{900 \text{ MW} \times 1000 \times 8760 \times 3413} \times 400,000 \times 1000 = \$0.990$$

Table 14.8. Capacity of Gas Pipelines with 60 Mile Compressor Spacing and 1.25 Compressor Ratios

Pipe OD (in.)	Capacity (millions SCFD)
16	113
20	202
24	326
30	585
36	943
42	1412
48	2003

Even though 1978 costs are used for transmission and 1970 costs for pipelines, it is clearly more economical to transmit natural gas, as has been done in the past.

Capacity of Other Pipe Sizes

There is about 2:1 variation in capacity depending on pressure ratio and frequency of compressor stations. Corresponding to the 943×10^6 SCFD capacity of the 36 in. line, with 60 mile spacings and 1.25 compressor ratios, the capacities of other size pipes, from Eq. 14.2, are as given in Table 14.8; d is taken as $\frac{35}{36}$ OD, as for the 36 in. pipe.

General

In 1962, some 90% of U.S. natural gas reserves were in Texas, Louisiana, Oklahoma, New Mexico, and Kansas. About 40% was used in those states and 60% exported. Altogether, in the United States there were 664,000 miles of trunk gas pipelines.

Hydrogen in Pipelines

The Hydrogen Economy, discussed in Chapter 4, would involve transmitting hydrogen by pipeline. It is stated that while hydrogen has about one third the heating value of natural gas, per ft³, it is very light and can be transmitted at three times the velocity over the same pipelines (with larger compressors), at comparable overall costs per Btu mile.

Hydrogen has a heating value of 334 Btu/SCF at 60°F and 15.15 psia. Its specific gravity is $2/29 = 0.0690$, compared with air. Thus, for the 36 in. pipeline with 60 mile spacings and 1.25 compressor ratios, $Q = 2505$ million SCFD. The energy transmitted is $(334/1000)(2505/943) = 0.89$ times that with natural gas.

The horsepower corresponding to a flow of 2505 million SCFD is about 25,000 Hp, and the cost at $250/Hp for gas turbines is about $6.25 million, or $104,000 per mile. Adding this to the cost of the pipe, $144,000 per mile, the capital cost is $248,000/mile. This is $248/196 = 1.27$ times the cost of the natural gas pipeline, and other costs are in proportion. Thus, the unit cost of transmitting energy with hydrogen is $1.27/0.89 = 1.43$ times that with natural gas, or $1.43 \times \$128 = \$183/10^{12}$ Btu mile (1970 basis).

Oil Pipelines

The throughput of an oil pipeline depends on the size of pipe, the viscosity of the oil, the spacing of pump stations, and the pressures used. However, empirically, almost all crude oil and product pipelines now in use have throughputs kD^2, in barrels/day, where $k = 300$ to 600, an average value being 500, and D is the nominal pipe size, or OD, in in.[7]

For example, the throughput of a 12 in. trunkline is about $500 \times 12^2 = 72,000$ barrels/day.

The 48 in. Alaskan pipeline should deliver about $500 \times 48^2 = 1.15$ million barrels/day. Actually, starting at 500,000 barrels/day, it was expected to reach 1.2 million barrels/day by 1980, 1.5 million barrels/day later.

A 24 in. line should deliver $500 \times 24^2 = 288,000$ barrels/day. The 24 in. "Big Inch" line of World War II was rated at 300,000 barrels/day.

Thus, the approximate carrying capacities of oil or product trunklines are as given in Table 14.9, first two columns.

Velocity of the Oil. Since the throughput varies as D^2, the mean velocity is about the same in all of these lines. It is about 4 mph or 6 fps.

Energy Transmitted. At 0.85 sp. gr., oil has 5.84×10^6 Btu/barrel. Thus, the 36 in. oil pipeline will transmit about $648,000 \times 5.84 \times 10^6 = 3.78 \times 10^{12}$ Btu/day. This compares with 0.943×10^{12} Btu/day transmitted by the 36 in. gas pipeline discussed earlier. The 36 in. oil pipeline transmits roughly four times the energy of a 36 in. gas pipeline. However, since the flow in a gas pipeline varies as the 2.618 power of the diameter, Eq. 14.2, a 12 in. oil pipeline transmits about eight times the energy of a 12 in. gas pipeline.

Table 14.9. Approximate Capacities and Costs (1970) of Crude Oil Pipelines

OD (in.)	Approx. Capacity (BPD)	Approx. Hp/Sta.	f	Miles Between Stations	Capital Cost (Thousand \$/mile)			Annual Mills/Mile (Millions)	Annual Tons (Millions)	Mills per Ton Mile
					Pumps	Pipe	Total			
12	72,000	953	0.020	28.89	5.61	48	53.61	12.06	3.92	3.08
16	128,000	1,694	0.019	40.42	5.02	64	69.02	15.53	6.96	2.23
24	288,000	3,810	0.017	66.23	5.75	96	101.8	22.89	15.67	1.46
36	648,000	8,575	0.016	106.4	8.06	144	152.1	34.21	35.3	0.97
48	1,150,000	15,220	0.015	152.5	9.98	192	202.0	45.44	62.6	0.73

Cost. The cost of a pipeline is about the same for oil or gas, approximately \$4000/mile installed, per in. OD (1970).

To determine the cost and spacing of pump stations, and thus the total cost per mile of the pipeline, the following steps are necessary:

1. Assume that each pump station raises the pressure from 100 to 800 psi. From the throughput and the 700 psi pressure differential, the pump output is determined (Table 14.9). Assume 90% pump efficiency for motor Hp.*

2. Pump costs (1970) were approximately[6]:

Installed Hp	\$/Hp, Average
250–500	290
500–750	210
750–1000	170
1000–2000	120
Over 2000	100

From these data, the costs of the pump stations can be determined.

3. The spacing of the pump stations depends on the friction drop. Determine in turn[6] (a) the kinematic viscosity of the oil v, (b) Reynolds number $R = Vd/v$, (c) the friction coefficient f. For example, with 0.85 sp. gr. oil at 70°F, $v = 10^{-4}$ ft²/sec. For the 3 ft diameter pipe and 6 ft/sec flow velocity, $R = 6 \times 3/10^{-4} = 18 \times 10^4$. Steel pipe over 1 ft in diameter is essentially "smooth pipe." At $R = 18 \times 10^4$, $f = 0.016$, as shown in the table.

4. For crude oil pipelines, the pressure drop is normally determined from the Darcy formula,[6] which can be expressed as

$$P = \frac{34.87fB^2S}{d^5} \text{ psi/mile} \qquad (14.3)$$

where f was determined in paragraph 3 above, B is throughput in barrels/hr, S is sp. gr. = 0.86 for the crude oil considered, and d is the ID in in; use OD in. $\times$ 35/36, as an approximation.

5. Knowing the pressure drop per mile, the number of miles between stations to result in 700 psi drop can be determined.

6. The pump station cost can now be expressed as cost per mile and added to the pipe cost per mile to get the total capital cost. Let the annual cost be 18% for maintenance, depreciation, taxes, insurance, and return. Most of the cost of a pipeline is capital cost. Increase this by 25% to include operating cost, as with the gas pipeline earlier, to get the total annual cost per mile as given in Table 14.9.

7. From this annual cost per mile, and the throughput, determine the mills per ton-mile to compare with other energy transportation modes, or cost of a million Btu for 1000 miles, to compare with the gas pipeline. For these calculations the weight and Btu content of a barrel of 0.85 sp. gr. crude oil are 350 × 0.85 lb/barrel, and about 5.84 × 10⁶ Btu/barrel.

8. The results of these calculations for the 5 pipe sizes are shown in Table 14.9. The throughput and unit costs are based on 100% load factor, no spares, and 1970 costs, as for the gas pipeline earlier. With 50% load factor the costs per ton mile are approximately doubled. Suitable inflation factors must be applied for later years.

9. The costs per ton mile are comparable with

---

*648,000 barrels/day × 42 gal/barrel × (231/1728) ft³/gal ÷ 24 hr/day ÷ 3600 sec/hr × 700 psi × 144 in.²/ft² ÷ 550 ft lb/sec/Hp ÷ 0.9 eff = 8575 Hp.

the costs by barge (1.75 to 7 mills/ton mile), as stated earlier.

10. These costs bear no relation to the Alaska pipeline costs, for which the conditions were extraordinary. Part of it had to be supported above ground. Heaters were required in some locations. Delays and unusual problems escalated costs to several times estimates.

11. The cost of a million Btu for 1000 miles via 36 in. crude oil pipeline, at 100% load factor, is 2.48¢.* This compares with 12.8¢ for the 36 in. gas pipeline. The corresponding unit cost is $25/$10^{12}$ Btu mile, compared with $128 for natural gas and $183 for hydrogen.

12. From Table 14.9, it may be observed that pump stations are a small part of the total cost (including the 25% added for operating cost). From Eq. 14.3, the friction drop is proportional to (flow).[2] Thus, 41% increased flow could be obtained with twice the number of pump stations, at comparable unit cost—mills per ton mile. (A small decrease in f is neglected.)

Product Pipelines. These can also be estimated by the Darcy formula, as in (4). For gasoline, sp. gr. = 0.68, at 70°F, $\nu = 4.8 \times 10^{-5}$. Reynolds number is thus 1.25×10^5 and $f = 0.017$. For a 12 in. pipeline of 11.67 in. ID and 3000 barrels/hr throughput,

$$P = 34.87 \times 0.017 \times 3000^2 \times \frac{0.68}{11.67^5} =$$

$$16.76 \text{ psi/mile}$$

However, the Williams and Hazen formula[6] is widely used for product pipelines. It is

$$P = \frac{2340 B^{1.852} S}{C^{1.852} d^{4.87}} \text{ psi/1000 ft} \qquad (14.4)$$

Substituting B = 3000 barrels/hr, S = 0.68 sp. gr., C = 150 for gasoline, 130 for No. 2 Furnace Oil, 134 for kerosene, d = 11.67, yields for gasoline

$$P = 2.60 \text{ psi/1000 ft, or } 13.72 \text{ psi/mile}$$

about 18% less than the Darcy formula.

*$34,210/mi/yr ÷ 365 ÷ 648,000 barrels/day ÷ (5.84 × 10⁶Btu/barrel) × 1000 × 10⁶ = $0.0248.

Statistics. Some 75% of crude oil and 20% of products are transported domestically by pipeline. The sizes of lines vary from $3\frac{1}{2}$ in. OD for 0 to 2000 barrels/day product lines to the 48 in. Alaska pipeline with over a million barrels/day capacity.

In 1976, there were 135,000 miles of trunk oil pipelines. These transported 1609 billion barrel miles of crude oil (5824 million barrels, average 276 miles). They transported 1307 billion barrel-miles of products (3813 million barrels, average 342 miles).

Coal Slurry Pipelines

As of 1980, slurry pipelines are in considerable use for the transport of various ores and mineral products, but only one coal slurry pipeline is in use in the United States. This is the Black Mesa pipeline in Arizona, which transports about 1.4 billion ton miles of coal per year. The combined capacities of various coal slurry pipelines proposed or under construction are about 100 billion ton miles per year.

For transport as a slurry, coal is ground to 8-mesh size, about the size of granulated sugar or coarse sand. It is mixed with water to form a 45 to 55% concentration of coal. This requires about 250 gal water per ton coal, that is, 1 ton water per ton coal. Obtaining water rights for this amount of water is one of the major problems.

The slurry flows at about 5 fps, about the same as oil in a pipeline. The flow is thus turbulent and keeps the coal in suspension. Flows of 3 to 6 fps are practical, but lower velocities result in settling out of the coal. This means a practical operating range of 60 to 100% of design rating. For less than 60%, intermittent flow must be used. Pumps are located every 40 to 80 miles. The considerations are the same as for oil pipelines (Table 14.9).

Coal slurry causes some erosion of the pipes and pumps, and some extra wall thickness is necessary to provide for this loss of metal over the lifetime of the project.

While uncleaned coal can be transported, cleaned coal is preferable. With the enormous capacity of a slurry pipeline, coal must be gathered from a great area to feed it, and cleaned coal is more uniform and can be mixed. The cleaning is less costly than for other forms of transport, since the cleaned coal does not have to be dried.

At a power plant, separation and drying are required, or a loss in efficiency of about 5% accepted with undried coal. However, part of the coal handling facilities are saved with slurry delivery.

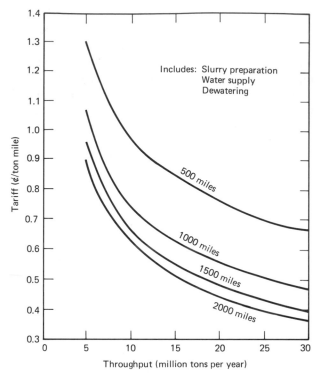

Figure 14.7. Coal slurry pipeline—transportation costs. *Source.* Reference 8.

Costs. The approximate costs of coal slurry pipelines (1975) are shown in Fig. 14.7.[8] To compete with unit train costs of 8 to 10 mills/ton mile, a slurry pipeline would need to be in the 500 mile, 10 megaton/yr range. To compete with water transport at 4 to 6 mills/ton mile, a very unlikely case, the slurry pipeline would need to be about 1000 miles and 20 megatons/yr, or equivalent. Note: Average barge cost is 3 mills/ton mile, Table 14.1; 4 to 6 mills/ton mile includes some truck or rail transport to the water.

Increases in rail tariffs since 1974 somewhat improved the economics of the slurry pipelines versus rail. However, slurry pipelines have no right of eminent domain. They cannot cross railroads. Securing the necessary right of way is a limitation in some cases.

If the coal is intended for electrical generation, the mine mouth plant and electric transmission must also be considered (see Transmission Lines and Cables). Electric transmission is competitive with coal by unit train up to 800 miles in some cases, and hence with the 500 mile slurry pipeline. However, each case must be studied individually, as relative costs and other influencing factors vary widely (see also Fig. 14.13).

Transmission Lines and Cables

Modern transmission lines serve three principal purposes:

1. They provide bulk transmission from the generators to the load centers.
2. They provide interchange flow between different parts of the interconnected system. This makes it possible to operate all generators in the most economical manner. They provide great flexibility in the location of base-load, intermediate-load, and peaking generating capacity and reliable emergency supply to all areas.
3. They supply large loads tapped off along the lines.

Thus, line voltages are selected to best serve this combination of purposes and only rarely for a specific long-distance bulk transmission project.

Cables are simply short underground transmission links in areas where overhead transmission is impractical, such as bringing large blocks of power into the center of a metropolitan area, or under water, or through a tunnel. Lower voltage distribution feeders in densely populated areas are also predominantly cables.

A power system cost is roughly 40% generation, 10% transmission, and 50% distribution. As with coal, oil, and gas, the distribution is not covered in this book.

The following paragraphs give (1) some measures of transmission needs, (2) some information about the characteristics, capabilities, and costs of transmission lines and cables, and (3) a comparison of coal by train versus equivalent energy via electrical transmission.

The Transmission Problem

As shown in Fig. 5.1, the total generating capacity in the United States in 1980 was about 600 GW. It was divided into a number of operating power pools. For example the Pennsylvania–New Jersey–Maryland (PJM) pool, described in Chapter 5, constituted about 8% of the total in 1974 and supplied about 10% of the population. An interconnected pool of 48 GW may need many transmission ties of the order of 1 to 2 GW. Large-capacity lines are also needed to tie power pools together for economy and emergency interchange.

The largest steam-electric generating unit installed in 1972 was 1.3 GW. Two units usually constitute a modern generating station, and a reliable transmission link of 2 to 3 GW capacity is needed to connect it into the system. While the rate of increase appears to have slowed (1980), economy-of-scale inevitably favors larger units as the loads grow.

A 1.8 GW generating station in the mine fields of Western Pennsylvania has supplied loads in New Jersey and Maryland over a total of 650 miles of 500 kV lines since 1967. This is a growing trend, referred to as "mine mouth generation." In 1970 over 40% of the larger coal-fired plants being constructed (18,000 MW) were "mine mouth."

In the Pacific Northwest, much of the hydro is east of the Cascade Mountains; much of the load is to the west. In 1979, transmission capacity through these mountains was about 18 GW, primarily 500 kV lines of the order of 2 GW capacity. Prime hydro energy is insufficient for future needs, and nuclear and coal-electric plants are being built. Plans for "mine mouth" generation in the coal fields farther east would more than double the transmission through the Cascades, indicating the need for 10 GW lines in the late 1980s. These examples are replicated in all parts of United States and Canada.

In 1980, there were in operation in the United States approximately 60,000 circuit miles of 230 kV transmission, 33,000 circuit miles of 345 kV, 22,000 circuit miles of 500 kV, 3600 circuit miles of 765 kV ac, and 1700 circuit miles of 800 kV ($\pm$ 400 kV) dc transmission. Short test lines at 1100 kV ac (1200 kV max.) were in use, providing design data for the next step—1100 kV ac. UHV laboratories were developing data up to 1500 kV. The historical increase in maximum transmission voltage in North America is given in Fig. 14.8.

In 1974, there were about 2700 miles of 69 to 345 kV underground cable in the United States, with plans for 525 kV pipe-type cable (6500 ft at Grand Coulee) and 800 kV cable under development.

Characteristics of Existing Lines

The principal parameters of a selection of existing (1980) EHV and HVDC lines are given in Table 14.10. The normal rating per circuit varies widely.

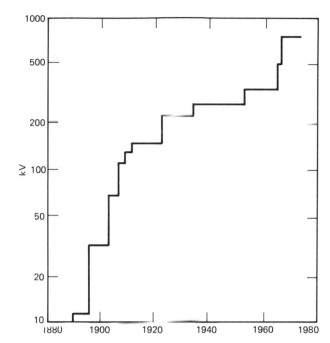

Figure 14.8. Maximum transmission voltage in North America. *Source.* Reference 9.

The upper rating is the thermal rating of the conductors. The lower rating is set by stability and other considerations as will be explained later.

Steel towers are generally used, except for single-circuit 345 kV, many of which are wood H-frame; 345 kV and 500 kV may use double-circuit towers, but above that one circuit is used per tower.

In order to keep radio and audible noise and corona loss to accepted limits, the surface voltage gradient on conductors of higher-voltage lines must be kept approximately the same as for lower voltages. This is accomplished by the use of "bundled conductors." As shown in the table, the conductor diameters are in the same range for all three EHV voltages, but the average number of conductors per bundle increases at higher voltages. This also increases the transmission capability, with the result that 1100 kV lines will have about 50 times the capability of 230 kV, rather than 23 times, as might be expected from the square of the voltage ratio.

The right of way can be compressed considerably when necessary. By going to tall towers with vertical configuration, BPA[10] places four 500 kV circuits (two high-capacity 500 kV double circuits) on a 250 ft right of way (ROW), with a total of 10 GW capacity, or one 1100 kV single circuit (planned) on a 185 ft ROW, also with 10 GW capacity.

Table 14.10.   Characteristics of Selected EHV and HVDC Transmission Lines[9]

Parameter	EHV (ac)			HVDC
	345 kV	500 kV	765 kV	800 ($\pm$ 400) kV dc
Line lengths (miles)	< 10–413	15–630 tot.	63–580	845
Years constructed	1964–74	1965–75	1971–75	1967–70
Normal rating/circuit (MVA)	352–2000	873–3500	1700–4000	1350
Circuits per structure	1 or 2	1 or 2	1	1
Single-circuit towers	Wood or steel	Steel	Steel	Steel
Double-circuit towers	Steel	Steel	—	—
Structures per mile	5 to 10	3 to 6	4	4.5
Conductor diameter (in.)	1.06–1.82	1.17–1.88	1.17–1.6	1.80
Equiv. area (MCM/cond.)	795–2493	1024–2515	954–1585	2312
Conductors per bundle	1–2	2–3	4	2
Phase separation (ft)	16–39.7	30–45	45–50	38–41
Typical right of way (single circuit) (ft)	150	200	250	

Performance of Transmission Lines

The performance of transmission lines can be calculated precisely from four physical constants: the series inductance and resistance per phase per mile, and the shunt capacitance and leakage resistance per phase per mile.[9] However, there are estimating methods of sufficient precision for this chapter which greatly simplify the calculations and improve the physical picture of the phenomenon taking place.

Because the line constants vary logarithmically with the conductor radii and spacings, the line performances indicated by these very simple approximations are within about 10% of those obtained by the most refined calculations. They are entirely adequate for an understanding of transmission line capabilities, or for estimating purposes, but certainly not for actual design.[9]

Surge Impedance Load (SIL)

If a long line is terminated in its surge impedance, conditions between the sending end and the point of termination are exactly the same as if the line continued indefinitely. There are no reflections. Thus, in the ideal case of a lossless line, the voltage is everywhere the same as at the sending end, simply delayed in time by the velocity of propagation.

If this concept is applied to a power transmission line, with the approximation that the velocity of propagation is the speed of light, some very useful relations result and are generally used for estimating the performance of lines, as mentioned earlier. These relations are as follows:

1. The surge impedance per phase is simply $\sqrt{L/C}$, where L and C are the series inductance and shunt capacitance per phase, per unit of length. It is a pure resistance. The surge impedance load is therefore of unity power factor all along the line. The SIL is about 400 MW at 345 kV, 900 MW at 500 kV, or 2300 MW at 765 kV.

2. The voltage is delayed by the speed of light. The wavelength of a 60 Hz wave is the velocity of light $\times \frac{1}{60}$, or 3105 miles. Thus, for a 100 mile line, the angle between the voltages at the two ends is 360° $\times$ (100/3105) = 11.59°. The reactive voltage drop in this 100 mile line is thus 2 sin (11.59°/2) = 0.202 per unit. This is generally taken as 20%. The reactive power losses in 100 miles of line at surge impedance loading are 20% of SIL.

3. To maintain unity power factor all along the line, the reactive power supplied by the line capacitance, per 100 miles, must also be 20% of SIL.

4. Thus, for a lossless line of any length, operated at SIL, the reactive power losses and the reactive power supplied by the line capacitance are each 20% of SIL multiplied by (line length in miles/100).

5. At 300 miles, with SIL, the line angle is about 35°. For stability reasons and other consider-

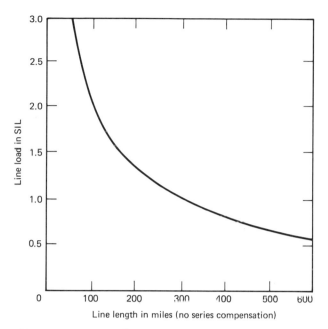

Figure 14.9. Transmission line capability. *Source.* Ref. 9.

ations, lines are not generally loaded above SIL at 300 miles, or about 0.55 SIL at 600 miles. Below 300 miles, higher loadings are used, up to approximately 3 SIL at 50 miles, or to the thermal rating of the conductors for short lines. These empirical limits are shown in Fig. 14.9.

6. If half the line reactance is canceled by series capacitors (50% compensation), these distances can be approximately doubled. Shunt reactors may also be required to absorb part of the reactive power supplied by the line capacitance.

7. Reactive losses increase as the square of the loading. At twice SIL, the reactive losses are $4 \times 20 = 80\%$ of SIL, for a 100 mile line. However, the reactive power supplied by the line capacitance remains unchanged. The difference, 60% of SIL in reactive power, must be supplied from somewhere. Similarly, at less than SIL, there is excess reactive power which must be absorbed somewhere.

8. When the line length is L miles and the load transmitted is P MW, at unity power factor near the center of the line, the two effects are superposed. The reactive power supplied is about 20% of SIL $\times$ $L/100$. The line reactive loss is about 20% of SIL $\times$ $L/100$ $\times$ (P/SIL).[2]

9. *Ferranti Effect.* At no load, the line conditions of a 100 mile line can be approximated by half of the reactive supply (10% of SIL) at each end. If the receiving end of the line is open, the 10% reactive supply flowing back through the 20% line reactance causes a voltage rise of 2% at the open end. This voltage rise increases approximately as the square of the line length, being about 8% for a 200 mile line. More precisely, it is $1/\cos [11.59°(L/100)]-1$, or 8.78% for a 200 mile line. For longer lines, this voltage rise becomes excessive, and measures must be taken to control it, such as the shunt reactors mentioned above.

The Resistive Loss. The resistive loss, which we have ignored thus far, can now be estimated. If the R/X ratio of the line is 0.4 (4/0 copper, 15 ft spacing), the loss is 40% of the reactive power, or $0.4 \times 20\% = 8\%$, for SIL and 100 miles. With $R/X = 0.2$ (1 in. diameter ACSR, 15 ft equivalent spacing), it is $0.2 \times 20\% = 4\%$ for SIL and 100 miles.

The total transmission losses on a widespread power system may be of the order of 5 to 15%. Fig. 14.10 shows how peak losses have varied on the BPA system as higher line voltages have predominated.[10]

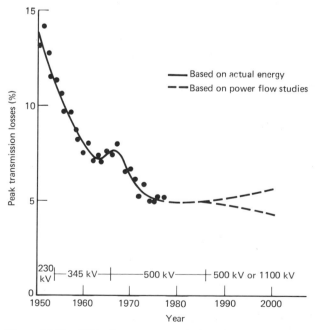

Figure 14.10. BPA System transmission peak losses. Reprinted with permission from Ref. 10. Copyright 1979 IEEE.

Estimation of Surge Impedance. From calculation of many lines, typical values of surge impedance for one, two, three, and four conductor bundle lines have been found to be approximately Z = 375, 300, 280, and 260 ohms per phase, respectively.[9] The corresponding SILs follow directly since

$$\text{SIL} = \frac{kv^2}{Z} \text{ MW} \qquad (14.5)$$

With the approximation of a lossless line and propagation velocity equal to the speed of light, it is easily shown that for a 60 Hz line

$$Z = 494.14X \qquad (14.6)$$

where $X = 2\pi fL$ is the line reactance per phase per mile, L is the inductance per phase per mile, and f = 60 Hz. Since the velocity of propagation is

$$v = \frac{1}{\sqrt{LC}} = 186{,}283 \text{ miles/sec} \qquad (14.7)$$

thus, $C = 1/(Lv^2)$, and

$$Z = \sqrt{\frac{L}{C}} = \sqrt{L^2v^2} = Lv = \left(\frac{v}{2\pi f}\right)X = 494.14X$$

as stated. With these approximations, the use of a typical surge impedance of 260 ohms for all lines of four conductors per bundle implies a typical reactance of 260/494.14 = 0.526 ohms, for all such lines. In spite of the variation in conductor size and spacing shown in Table 14.10, the logarithmic dependence of X on these variables makes this reasonable. We can also reasonably extrapolate to larger bundles.

Extrapolation to Surge Impedance of Larger Bundles. Assuming equal currents in the four conductors of a bundle, the reactance per phase with the geometry of Fig. 14.11 is

$$X = \frac{1}{4} \times 0.2794 \left(\log \frac{600}{R} + 2 \log \frac{600}{18} \right.$$
$$\left. + \log \frac{600}{25.46} \right) = 0.526 \text{ ohms} \qquad (14.8)$$

Solving, the GMR of one conductor R is 0.46 in.

The geometry of Fig. 14.11 agrees empirically with the usual estimating value of Z for a four-

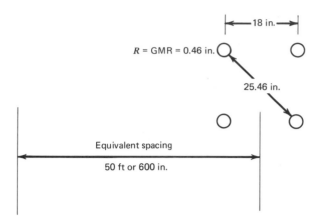

Figure 14.11. Empirical geometry corresponding to a surge impedance of 260 ohms.

conductor bundle, namely, 260 ohms. Using this geometry to extrapolate to a six-conductor bundle with 18 in. conductor spacings, and to eight-, 12-, and 16-conductor bundles of 40, 50, and 60 in. diameter, respectively, leads to the approximate values of Z and SIL for six- to 16-conductor bundles shown in Table 14.11.

These SILS can be carried for 300 miles without compensation, with amounts for other distances as shown in Fig. 14.9. With 50% compensation, the same loads can be transmitted about twice these distances.

If the voltage is firmly established by synchronous machines at each terminus, successive transmissions of the same amount are possible. Power can be shifted from system to system clear across the country, as long as the ties from system to system do not exceed the limits given.

Nominal Ratings of Transmission Lines. In summary, the most meaningful rating of a transmission line is its SIL, which it can carry for 300 miles, in round numbers

140 MW at	230 kV
400 MW at	345 kV
900 MW at	500 kV
2300 MW at	765 kV
6,000 MW at	1100 kV
10,000 MW at	1400 kV

These ratings are subject to three variations given by (1) shorter or longer lines, Fig. 14.9; (2) fewer or more conductors per bundle, Table 14.11; (3) length can be doubled by 50% compensation.

Table 14.11. Surge Impedance Loads of Typical Transmission Lines

Conductors per Bundle	Surge Impedance Z	SIL (MW)					
		HV	EHV			UHV	
		230 kV	345 kV	500 kV	765 kV	1100 kV	1400 kV
1	375	141	317	666			
2	300		397	833			
3	280			893	2090		
4	260			962	2251		
6	229				2556	5284	
8	217				2697	5576	9032
12	198					6111	9899
16	186					6505	10538

Costs of Transmission Lines

The costs of transmission lines in 1978 are shown in Fig. 14.12. The circles showing costs without right of way are given by EPRI.[11] The x values show what the cost would be with average ROW width, valued at $1500 per acre.[12] However, these land costs vary *widely*, and actual land costs must be obtained for every project.

Mine Mouth Generation Versus Coal by Railroad

At 765 kV, the highest ac transmission voltage in use in the United States in 1980, the transmission of 2.3 GW from a mine mouth plant to a load 300 miles away, at 18% capital charge, would cost

$473,000/mi × 300 mi × 0.18 = $25.54 million/yr

To generate this 2.3 GW yr of electricity at 40% efficiency would require

2.3 GW yr × 10^6 GW/KW × 8760 hr/yr × 3413 Btu/kWh/0.4 eff = $1.72 × 10^{14}$ Btu of coal

At an average heating value of 12,000 Btu/lb for Eastern high-sulfur coal, or 24 million Btu/ton, this is $1.72 × 10^{14}/24 × 10^6 = 7.16$ million tons of coal, or the transportation of 2,150 million ton miles. At a unit train rate of 1.5¢/ton mile (1980, Table 14.12) it would cost

$$2150 × 10^6 × 0.015 = \$32.25 \text{ million/yr}$$

These basic costs are in the same range, $25.54 and $32.3 million/yr. However, there are many other factors to be considered. Some of these are:

1. For reliability it might be necessary to have two lines.

2. Since transmission lines are part of a general network, all of the cost may not be allocated to this one function.[12] Also existing lines may provide part of the transmission and adequate backup so that even less than one new line is needed.

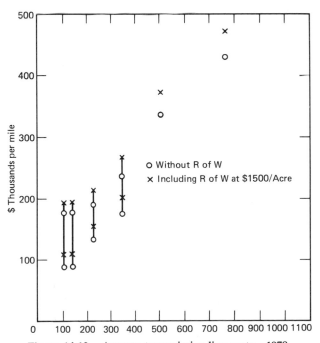

Figure 14.12. Average transmission line costs—1978.

3. The calculation assumed 100% load factor. While a new mine mouth plant is probably base load, the load factor may be under 90%. It may be as low as 50%, which would favor the unit trains.

4. The siting and water supply may be easier mine mouth than building a plant in the load area.

5. A transmission line, once built, is not subject to as great escalation as a railroad, dependent on oil fuel, labor, and replacements.

6. A coal slurry pipeline is an additional option, discussed earlier.

7. The capital charge, 18%, for maintenance, depreciation, taxes, insurance, and return may be high or low in individual cases.

The sulfur emission standards of the early 1970s caused a great shift to Western low-sulfur coal which met those standards and consequent great increase in unit trains from those areas. However, the revision of June 1979 to require at least 75% sulfur removal from *all* coal has reversed this trend. The rail revitalization act of 1976 resulted in rapid rail tariff increases. The net result of all these factors is that the maximum distance at which EHV or HVDC from a mine mouth plant is competitive with coal by unit trains increased from 500 to 600 miles in the early 1970s to the order of 800 to 1000 miles in 1980.[12]

A comparison of coal transportation alternatives is given in Fig. 14.13. Table 14.12 shows the indicated coal transport rates just after 1980. Slurry pipelines are limited to very large transport where water rights and right of way can be secured. Water carrier is accessible in only about 15% of coal transport, and trucks are limited usually to about 50 miles. Thus, unit trains and electrical transmission are the usual alternatives.

High-Voltage Transmission Cable—69 to 800 kV

About 1% of the mileage of high-voltage transmission is in underground or underwater cables. In spite of its cost, 10 to 20 times that of most overhead lines, cable is either mandatory or represents the best engineering solution in dense urban areas, underwater, or through tunnels. The narrower right-of-way requirement, 20 to 45% of that for equivalent overhead lines, and the complete ab-

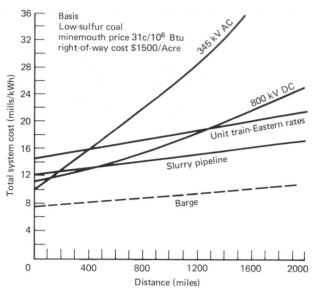

Figure 14.13. Coal transportation alternatives cost comparison. Source Ref. 13 (Barge curve added).

sence of overhead structure are the key determinants for underground cable. Its importance may be judged by the EPRI R&D budget for a typical year, 1976, $6.8 million, or 23% for overhead transmission, and $9 million, or 31% for underground transmission.

The most common system is the pipe-type cable, shown in Fig. 14.14. The three-phase conductors are individually wrapped in kraft paper, saturated with a dielectric fluid, covered with a conducting layer, and enclosed in a steel pipe. The pipe is filled with insulating oil, and pressurized to 200 to 300 psi. This high-pressure, oil-filled (HPOF) cable has been installed at voltages from 69 to 525 kV (1980).

The principal limitations are a dangerous rise in temperature with severe overloading, high dielectric losses in the paper insulation, and excessive

Table 14.12. Coal Transport Rates—Estimated Range 1980[12]

	¢/Ton Mile
Rail unit train	1.2–2.0
Slurry pipeline	0.9–1.1
Water carrier	0.6–0.8
EHV Transmission	1.2–1.5
Truck	5.0–7.0

Source. Reprinted with permission from Ref. 12, presented at the Joint ASME/IEEE/AAR Railroad Conf., April 9–10, 1980, Montreal.

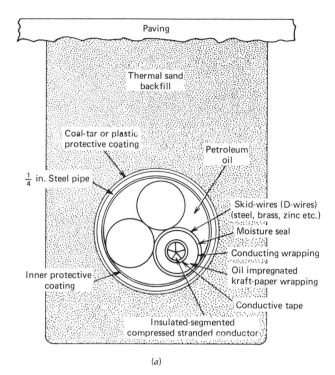

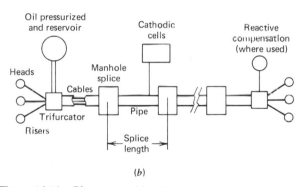

Figure 14.14. Pipe-type cable. *Source: Underground Power Transmission*, a Study prepared by Arthur D. Little, Inc., for the Edison Electric Institute, 90 Park Ave., N.Y., N.Y. October 1971 (ERC Pub. No. 1-72)

charging currents with increased voltages and distances. Cable developments currently underway are aimed at ameliorating these problems.

Forced cooling has been applied to the HPOF cable to increase its rating. The oil flows at about 6 ft/sec, as in most pipelines, and cooling stations are needed at 2 to 5 mile intervals.

The SF-6 Gas-Insulated Cable. This cable, shown in Fig. 14.15a is similar to gas-insulated bus, which has been used for some time in high-voltage stations. It eliminates the paper, with its attendant dielectric losses, high charging current, and thermal

barrier. The conductor is supported by insulators at the center of a cylindrical aluminum tube filled with SF-6 gas, at about 50 psi. The required diameter is about one eighth that which would be required with air insulation, that is, about 15 in. diameter at 345 kV. To date (1980), this has been supplied in rigid sections which have been spliced together in open trenches, but it has been very effective for short connections within stations, and its use is being extended.

Flexible gas-insulated cable is under development, with insulating discs at frequent intervals to center the conductor and with a corrugated, flexible outer pipe which can be wound on large reels. It can thus be installed in relatively long sections.

Forced cooling is also possible with SF-6-insulated cable, also known as "gas spacer" cable.

Synthetic Insulated Cables. These include synthetic tapes of lower SIC, dielectric losses, and better heat-conducting properties than kraft paper. The cable would otherwise be similar to HPOF, paper-insulated cable.

Extruded solid-dielectric insulation which would lower the fabrication cost and eliminate the oil is also under development.

The synthetics and solid dielectrics both hinge on the success of materials and fabrication developments currently underway (1980).

Cryoresistive and Superconducting Cables. These are being developed for the very high-power circuits that may soon be needed. The general concept of both systems is shown in Fig. 14.15b. Both electrical and thermal insulation are needed. The electrical insulation is shown around the individual phase conductors, all of which are at the low temperature. This system is thermally insulated from the outer pipe and ground. The coolant liquid percolates out from the small tube at the top, absorbing heat at constant temperature as it evaporates, and keeping the conductors at close to the boiling point of the fluid used.

In the cryoresistive system, the conductors operate at about 77°K, the boiling temperature of liquid nitrogen. Their resistance, at this temperature, is about 10% of normal.

In the superconducting system, the conductors operate at about 20°K, the boiling temperature of liquid hydrogen, or at about 5°K, the boiling point of liquid helium, depending on the conductor material used. In either case, the resistance is practically zero. A dc current, once set in motion in a loop, will

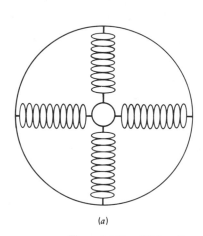

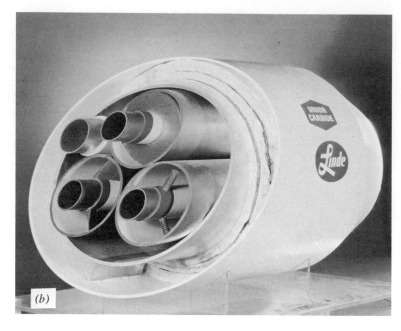

Figure 14.15.   High-voltage, high power, cable concepts. (*a*) Gas-spacer or SF-6 insulated cable. (*b*) Model of superconducting cable. (*Courtesy Linde Division, Union Carbide Corporation*)

flow undiminished almost indefinitely. An ac current incurs some small loss, since the ac field extends beyond the cryogenic envelope and causes some circulating current losses.

In either the cryoresistive or the superconducting system, the reduced losses must pay for the cryogenic system. For the cryoresistive system these costs appear to be about a standoff, although development is proceeding. However, the superconducting system appears economic for very high powers.

High-Voltage Cables in Service in the United States in 1974.   These are shown in Table 14.13. Both 800 kV ac and high-voltage dc cables were under development, although none was installed in the United States. Direct current cables at ± 100 kV

Table 14.13.   Underground and Submarine Cables in Service in the United States, 1974[14]

kV	Circuit Miles	kV	Circuit Miles
69	850	345	101
115	501	500	1[a]
138	1121	765	0
161	15	± 400 dc	0
230	96	Total	2685

[a] 1980.

have been in use between France and England under the English Channel since about 1960, with similar underwater ties elsewhere. Many dc ties are in the planning stage, and conversion of ac to dc for long high-power cable links is under consideration.

The comparative widths of right of way required by cables and overhead lines are illustrated by the two high-powered circuits, one at 345 kV and one at 500 kV, in Table 14.14.

The relative cost of overhead transmission and SF-6-insulated cable is shown in Table 14.15. The cost of SF-6 cable ranges from 2.6 to 5 times that of overhead line, considerably less than the 10 to 20 times that has been characteristic of lower-power cable circuits.

The Estimated Capabilities of Underground Transmission Systems (1971).   These are shown in Fig. 14.16. Naturally cooled pipe-type cables can be applied up to about 550 MVA at 345 kV. With forced cooling, the rating can be extended to about 1020 MVA at 345 kV, or to 1500 MVA (1981) with synthetic insulation.

Gas dielectric systems have several times the capacity of even advanced pipe-type cables, up to 2200 MVA at 500 kV, self cooled. Their capability is considered to extend to 10,000 MVA forced cooled (1981). However, these ratings are speculative for long cables since this construction has been used only for very short lengths, "with some problems."

Table 14.14.   Right-of-Way Requirements for High-Power Transmission[15]

Power Transmitted (MW)	Voltage (kV)	Overhead Line (ft)	Pipe-Type Cables No. of Cables	Pipe-Type Cables ROW Width (ft)	Gas-Insulated Transmission Line (ft)
1800	345	115	3	50	28
3000	500	150	5	70	30

Superconducting circuits are under development for capacities up to 10 GVA per circuit.

Power Density.   Superconducting dc cables result in the highest power density of any electrical transmission, 7.75 MW/cm², compared with 3.3 MW/cm² with 345 kV forced-cooled, pipe-type cable. For comparison, other attainable power densities are: 345 kV self-cooled, pipe-type cable, 1.75 MW/cm²; 500 kV gas-spacer cable, 0.36 MW/cm², self-cooled, and 1.1 MW/cm²; forced cooled, microwave wave guides, 0.38 MW/cm². Much R&D will be required before the latter can be considered.

Power Losses.   In cables up to 10 miles in length, these are relatively low, under 0.5% in 345 kV pipe-type cable or in 500 kV gas-spacer cable. However, for superconducting dc cables, they approximate $\frac{1}{100}$ of this amount, that is, 0.005% in 10 miles. This saving in loss justifies the cost of the cryogenic equipment.

Charging Current Limitation.   A typical current rating of high-voltage pipe-type cable is about 1000 amp, self-cooled. A typical capacitance is 0.5 μf/phase/mile, or a shunt reactance at 60 Hz of 5305 ohms/phase/mile. Thus, pipe-type cables of 230,

345, and 500 kV have ratings of the order of 400, 600, and 900 MVA, respectively, self-cooled. The corresponding charging currents are 25, 38, and 57 amp/mile, respectively. Thus, at lengths over 10 miles, these currents add significantly to the I^2R losses and reduce the load rating. Shunt reactor compensation can be used at the terminals, as indicated in Fig. 14.14, to absorb this reactive power; but even if balanced, half of it will flow out at each end of the cable. For this reason, the maximum length permissible without intermediate compensation is in the range of 10 to 20 miles.

ENERGY STORAGE

Energy storage may be required over time periods varying from many years to a fraction of a second. The breeder fuel now stored is enough for a century at the current U.S. rate of energy consumption. A strategic oil reserve might be held by the United States or by the Free World for many years. In Europe, a half-year supply of oil products has normally been held; in the United States, relatively little.

On a yearly cycle, large quantities of natural gas are stored each year near the load centers to provide a substantial part of the heavy winter loads. In a typical year, 1972, some 5.68 trillion SCF, or about 25% of all gas used, was stored in this way, in underground storage where feasible and in LNG

Table 14.15.   Relative Cost of Overhead Transmission and SF-6-Insulated Cable—1975[15]

Power (GVA)	Cost (Mills/kWh mile) Overhead Lines	Cost (Mills/kWh mile) SF-6 Cables
0.2	0.026–0.050	
0.5	0.013–0.028	
1.0	0.008–0.017	0.040–0.055
2.0	0.006–0.011	0.024–0.032
5.0	0.005	0.013–0.019

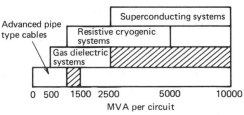

Figure 14.16.   Estimated ranges of capabilities of new underground transmission systems—1971 (crosshatched 1981).

tanks. The LNG storage has been described in Chapter 4.

The storage of liquid hydrogen is similar to LNG. Some 900,000 gal is stored in the liquid hydrogen tank at the JFK Space Center. There are proposals to store hydrogen gas underground, similar to natural gas, although this has not been demonstrated. Hydrogen is the lightest gas and most susceptible to leakage. Helium, the next lightest, is stored underground near Amarillo, Texas.

Nuclear fuel for a year is stored within a reactor.

Some hydro storage is yearly, or longer. As described in Chapter 5, the six dams in the Salt River valley can store about 2,000,000 acre ft of water. About 900,000 acre ft is released in a typical year, primarily for water supply and irrigation, but also for power. The valley could survive a severe drought lasting several years, as has occurred in the past. Several of the large up-river hydro dams have storage for over a year of full-load hydro generation, as shown in Table 5.3, although most are essentially "run-of-river," with storage for a few days of full load.

Radioisotopes may be viewed either as a form of storage or as a prime energy source. Decaying strontium 90, with a half-life of 28 yr, stores energy for 10 yr of unattended operation of remote weather stations in the artic. Radioisotopes also store energy for years of operation in space (see Chapter 13, Thermoelectricity).

Storage over the time spans of a month to a day, encompasses the important applications of energy storage for moving vehicles, and for electric-utility load leveling as well as many smaller storage needs.

For moving vehicles, energy is currently (1980) stored as fossil fuels: coal or diesel oil for ships and trains, gasoline or diesel oil for cars and trucks, and gasoline or (kerosene-like) jet fuel for planes. However, several alternatives which have "been around" for a long time are now coming into substantial use or are being studied for the future, for example, storage batteries for cars, alcohol or hydrogen (as metal hydrides) for cars, and liquified hydrogen for SSTs and very large land vehicles.

For electric utilities, storage up to a month (variable) is currently provided by the coal pile, the tank farm, or the common gas storage or as hydro pondage where available. Shorter-term storage for daily and some weekly load leveling is by hydro or pumped storage, this being an alternative to the provision of equivalent fossil-fueled peaking capacity. However, several other energy storage technologies are being actively developed or studied for this service, for example, storage batteries, compressed air, steam, magnetic (superconducting) storage, and metal hydrides.

These storage systems have been covered to a considerable extent in the foregoing chapters, hydro and pumped storage in Chapter 5, storage batteries in Chapter 13, thermal and chemical storage in Chapter 7, and LNG and LPG in Chapter 4. However, some of the newer technology applications such as metal hydrides, compressed air, or magnetic storage for load leveling are discussed here.

Natural thermal storage is all about us. However, specific devices built by man, using water, rock, steam, or phase change (molten salts), tend to be in the daily range, with many exceptions. The "ice house" kept ice from winter to summer. An oversize boiler may store energy to supply a sudden demand of 20 to 40 sec. The Stirling engine stores heat every cycle, in heat absorbing material. However, the storage units for solar heating are primarily on a daily cycle.

Large storage in the time scale of a few seconds to a few minutes can be provided by mechanical storage in flywheels or inertia of moving vehicles. A train may run "inertia grade" on a hilly terrain. Flywheels also store energy to supply repetitive loads of short duration, or to provide smoothing of fluctuating loads.

Large energy storage for time spans of microseconds to seconds is usually electric storage in capacitors. Table 14.16 shows, for the various storage technologies, either typical data or a specific case. The capacity is given in native units and in Btu. Where appropriate, the typical storage time is given. Typical unit costs are given as available. The remarks column provides data known for particular technologies.

Metal Hydride Storage

The storage of hydrogen as a solid metal hydride, in a reversible process, is being considered for automobiles, for electric utility load leveling, and for many other storage, purifying, and compressing applications. The earliest U.S. trial of this new technology was the experimental storage unit, shown in Fig. 14.17. It was built at the Brookhaven National Laboratories about 1973 and was used in a pilot load-leveling experiment by the Public Service Electric and Gas Co. of New Jersey from 1974 to 1976.

Table 14.16. Storage Characteristics (Values Tentative)

Technology	Specific	Capacity		Storage Time	Cost ($/kWh)	Remarks
		Native Units	Btu			
U-238 Stockpile	U.S.	150,000 tons	9.6×10^{18}	100 yr		
Gas underground	U.S.	5×10^{12} SCF	5×10^{15}	6 mo.		
LNG	Large tanker	200,000 m³	4.4×10^{12}			
Liquid hydrogen tank	JFK Space Center	900,000 gal	37.7×10^{9}	Mo.		11,000 MWh
One-Month 2 GW plant	Coal	521,000 tons	1.25×10^{13}	Week/mo.		
One-Month 2 GW plant	Oil	2.16 million barrels	1.25×10^{13}	Week/mo.		
Pumped storage	Luddington, Mich.	1.7×10^{7} kWh	6×10^{10}	Day/week	20–23	2000 MW 8.7 hr 60–75% eff.
Elec. util. battery	LiAl/FeS	10^{6}–10^{7} kWh	3.4×10^{9}–10^{10}	Day/week	20–30	1976 future goal
Electric car battery	Equiv. 10 gal gas	60 kWh	2×10^{5}	Day	100	1980
					20–40	Future goal
Car hydride	Equiv. 10 gal gas	1.2×10^{6} Btu	1.2×10^{6}	Day		wt = 2000 lb
Compressed air underground	Huntdorf, Germany	0.46×10^{6} kWh	1.6×10^{9}	Day		for gas turbine
Magnetic	U. Wis. Grp. proposal	10^{7} kWh	3.4×10^{10}	Day/week	50	
Thermal	Water[b]	892 gal	7.5×10^{5}	Day	5 ⎫	Home solar
	Gravel[b]	378 ft³	7.5×10^{5}	Day	5 ⎭	
Flywheel	Fusion pulse	1860 kWh	6×10^{6}	Minutes		Delivers in sec
Electric (capacitor)	0.11 farad[a] 15,000 V	3.56 kWh	1.2×10^{4}	Sec	294,000	Delivers in μsec
Hydro	Grand Coulee	1.8×10^{9} kWh	6.14×10^{12}	Mo.		
	Conowingo	6.3×10^{6} kWh	2.15×10^{10}	Days		
Radioisotope	Weather Sta. in Artic			10 yr		

[a]Princeton, plasma physics.
[b]100°F rise.

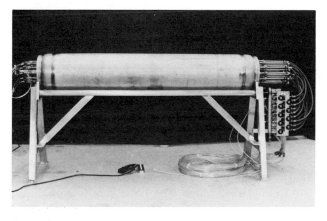

Figure 14.17. Fe–Ti metal hydride storage. Courtesy Brookhaven National Laboratory.

This experimental unit stores 12 lb hydrogen, weighs 1250 lb, and has a volume of 5.1 ft³. While great improvements are possible from this early experimental unit, it is nevertheless of interest to compare it with gasoline and storage batteries for cars (see Table 14.17).

While far from the weight efficiency of gasoline, the hydride storage compares favorably with storage batteries, and with comparable development effort may well surpass the most advanced storage batteries. It has the further advantage that it can be rapidly recharged at a hydrogen filling station.

Fifteen gal gasoline weigh 90 lb, neglecting the tank. The same energy storage as hydride, per Fig.

Table 14.17.   Storage Comparison for Cars

	Btu/lb	Engine or Motor Efficiency (%)	Output (Btu/lb)
Experimental hydride storage, Fig. 14.17	606	10	60.6
Gasoline, neglecting tank	20,400	10	2,040
Storage batteries			
1976, 16 Whr/lb	55	50	27.5
Interim goal, 32 Whr/lb	109	50	55
Advanced battery goal, 70 Whr/lb	239	50	120

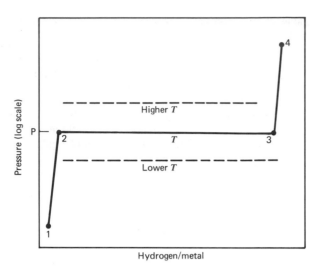

Figure 14.18.   Ideal absorption and desorption isotherm for a metal–hydrogen system.

14.17, weighs (20,400/606) 90 = 3030 lb. A 2- or 3-to-1 reduction in this weight would be needed to make it practical. This might be accomplished by the use of more weight-efficient hydrides.

Already (1978) hydrogen stored as a hydride has served as fuel for buses operated experimentally in the United States and West Germany and for an experimental automobile.[16]

Principle of Operation.   When gaseous hydrogen is brought into contact with certain metals or metal compounds, some of it combines to form a hydride, depending on the temperature and pressure. The process is reversible:

$$\text{metal} + \text{hydrogen} \rightleftharpoons \text{metal hydride}   (14.9)$$

The forward process is called absorption, and the reverse process desorption. Figure 14.18 shows the amount of hydrogen absorbed as the pressure is increased, for a fixed temperature, for an ideal metal–hydrogen system.

Starting with low hydrogen at point 1, some hydrogen is absorbed in metal solution as the pressure is raised to point 2. At this pressure, the hydriding reaction of Eq. 14.9 starts and continues at constant pressure until the metal is practically all hydrided at point 3. Further pressure increase causes a slight further addition of hydrogen in solution, to point 4.

If the pressure is now lowered, at the same constant temperature, the process is reversed and the hydrogen is given off. In practical hydrides there is a small hysteresis. The desorption path lies slightly below the absorption path.

The absorption reaction is exothermic. To keep the temperature fixed, the heat of reaction must be removed and may amount to 10 to 15% of the heat-

ing value of the hydrogen stored. Thus, the storage vessel must be built as a heat exchanger. The water tubes shown emanating from the end of the cylinder in Fig. 14.17 serve this purpose.

During desorption, the same amount of heat must be supplied. However, it is low-temperature heat, at near atmospheric temperature, and is usually available as waste heat from any process using the hydrogen. It does not mean a loss of energy in the storage cycle, as is involved in liquefying hydrogen, where about 33% of the energy is lost in the storage process.

At higher or lower temperatures, the "pressure plateaus" are correspondingly higher or lower as indicated by the dotted lines in Fig. 14.18. The hydriding reaction is very rapid at a fixed temperature and pressure. However, if the reaction heat is not carried away, the temperature rises and the reaction stops. Thus, the reaction rate is limited by the rate at which heat is carried away.

One might picture connecting the hydrogen hose at a filling station, then running cold water through the heat exchanger at a high rate to effect a rapid recharging of the hydride.

Weight and Volume Efficiencies.   As pointed out earlier, hydrogen packs three times as much energy per pound as gasoline, 63,100 versus 20,400 Btu/lb. However, on a volume basis even liquid hydrogen requires over three times as much space as gasoline for the same energy, that is, 279,000 Btu/ft³ for liquid hydrogen versus 916,000 Btu/ft³ for gasoline.

Table 14.18. Density of Hydrogen in Different Forms

Form	Weight % of H	Hydrogen Atoms per Milliliter
H_2, liquid	100	4.2×10^{22}
H_2, gas at 100 atm	100	0.5×10^{22}
MgH_2	7.6	6.7×10^{22}
VH_2	2.1	11.4×10^{22}
$FeTi\ H_{1.74} \rightarrow FeTiH_{0.14}$	1.5	5.5×10^{22}

However, the packing of hydrogen atoms in a hydride is somewhat closer than in liquid hydrogen, as shown in Table 14.18. In VH_2, the packing is nearly three times as great as in liquid hydrogen, and the energy stored per unit volume approaches that of gasoline.

Note that the volumetric packing of hydrogen in hydrides is many times that of high pressure gas. One FeTi reservoir, as shown in Fig. 14.17, containing 12 lb hydrogen would replace 12 K-size cylinders at 2000 psi, each storing 1 lb hydrogen. It would have both lower total weight (1230 lb versus 1608 lb) and lower total volume (5.1 ft³ versus 18.5 ft³). However, the hydride cylinder is much more complex, and the overall economics are yet to be determined.

The weight % of H is greater in some other hydrides than in FeTi. However, the costs are higher. For the electric utility load leveling application, weight and size are relatively unimportant compared with cost, reliability, and life.

Safety. Storage as a metal hydride is the safest form of hydrogen storage. Only small amounts are gaseous at any one time and available for catastrophic release in event of a tank rupture. The endothermic self-limiting nature of the desorbing reaction tends to limit the rate of discharge after rupture. Liquefied hydrogen, in contrast, is not considered for automobiles, although it has been considered for SSTs where adequate safety measures could be taken.

Tests have shown that hydrogen stored in hydrides is inherently less hazardous than the same amount of hydrogen stored in liquid or gaseous form. Tests have also shown that it is safer than gasoline on an equivalent energy basis.[16]

Other Hydrides. While only a few hydrides have been mentioned, there are three principal families of rechargeable hydrides, of the forms AB, AB_5, and A_2B, with several usable hydrides in each family. Altogether, these cover a very wide range of operating temperatures, pressures, and other characteristics to meet varying requirements. Examples are FeTi, $LaNi_5$, and Mg_2Ni, all developed since 1969. The technology is new and growing rapidly.

Other Uses. Because of the dependence of the pressure plateau on temperature, a hydride can be charged at low pressure at a low temperature and discharged at a higher pressure at a higher temperature. It thus acts as a compressor without moving parts, except the valves, and using low temperature waste heat for its operation.

The hydride is highly selective and will absorb only pure hydrogen from an input stream with impurities. If the remainder is exhausted, the hydride after that will deliver very pure hydrogen. It can thus act as a purifier.

While the hydride absorbs both the normal and the deuterium isotopes of hydrogen, some hydrides have different pressure plateaus for these two isotopes. This provides a basis for fractionally desorbing the deuterium at a lower pressure than the normal isotope. Deuterium, which constitutes only 0.015% of natural hydrogen, is used in heavy-water reactors and may be the fuel of future fusion reactors.

Cost. Of the various hydrides commercially available in 1978, FeTi had the lowest cost of $0.25 per gram hydrogen stored.[17] The next six materials ranged from $0.45 to $0.69 per gram hydrogen stored. FeTi is 50% hydrided at 4.2 atm at 25°C. That is, its "pressure plateau" is at 4.2 atm at 25°C. The hydride material is a large part of the total cost of a hydride storage.

One gram hydrogen contains 63,100/454 = 139 Btu, or 41 Wh (HHV). Thus, the cost of the hydride is about 0.6¢ per Wh of energy stored. It may be compared with the minimum 1976 cost of storage batteries for cars, about 3.5¢/Whr, Table 13.10. Since about five times as much hydrogen energy would need to be stored as electrical energy, and since the cost of a completed hydride storage might be twice the cost of the powder, these costs are in the same ballpark. Many other factors will determine the specific uses to which each technology is put.

Flywheels and Inertia

Flywheels and inertia are useful for storing energy for use in fractions of seconds to minutes. A moving train or an automobile may pick up speed in going down one hill and store inertial energy that helps to climb the next one. For years, freight trains have operated "inertia grade" over hilly terrain, with smaller locomotives than would be required to start the train up one of the hills.

In electric cars, an efficient flywheel storage may provide much of the extra energy needed during acceleration and thereby even the load on the battery.

Flywheels even the fluctuating power of gasoline or diesel engines, to produce smooth power drive to a vehicle or to reduce the voltage fluctuation of an engine generator.

Large flywheels have been used in the pulse supplies to particle accelerators or fusion experiments, in order to supply exceedingly high momentary loads and still draw little more than the average load from the system.

In industrial applications, large flywheels are used to even the highly fluctuating loads of large excavators or other construction equipment, so that they can be supplied by the relatively weak systems in remote locations where these operations are being carried on.

Flywheel Relations. The simplest flywheel is a solid cylinder. While the actual flywheel might be constructed somewhat differently, the solid cylinder is an adequate approximation to illustrate the energy storage capabilities of flywheels.

Let r be the radius in ft, l the axial length in ft, ω the angular velocity in radians per sec, v the peripheral velocity, $r\omega$, in ft/sec, w the specific weight of the material in lb/ft³ (about 500 for steel), g 32.2 ft/sec², W the weight in lb, I the moment of inertia in slug ft², and E the stored energy in ft lb. Then,

$$W = \pi r^2 l w \text{ lb} \tag{14.10}$$

$$I = \frac{Wr^2}{2g} = \frac{\pi r^4 l w}{2g} \text{ slug ft}^2 \tag{14.11}$$

$$E = \frac{I\omega^2}{2} = \frac{\pi r^4 l w \omega^2}{4g} \text{ ft lb} \tag{14.12}$$

For conventional, long life, waterwheel generator construction, with normal steel, the peripheral velocity, v, is limited to about 25,000 ft/min, or 417 ft/sec. Thus:

$$v = \omega r = 417 \text{ ft/sec}, r = \frac{417}{\omega}, w = 500, \text{ and}$$

$$E = \frac{\pi \, 417^4 \, l \, 500}{4 \times 32.2 \, \omega^2} = \frac{3.69 \times 10^{11} \, l}{\omega^2} \tag{14.13}$$

$$\text{and } r = \frac{417}{\omega} \tag{14.14}$$

Thus, four times as much energy can be stored per axial foot at half the speed and twice the radius.

Example 14.1.[24] The duty cycle of a typical large excavator is shown in Fig. 14.19. In a period of about 1 minute, the load varies from about 15,000 kW motoring to 15,000 kW regenerating, with an average load of 6000 kW. This total swing of 30,000 kW would result in unacceptable voltage fluctuations, as well as disastrous fluctuations of boiler load in the powerhouse as the governors continually attempted to compensate.

However, a flywheel motor generator that would store the energy in the lower shaded area and supply an equal amount in the upper shaded area would cut the power swing on the supply system to about 3500 kW, or 12% of the original swing. The energy

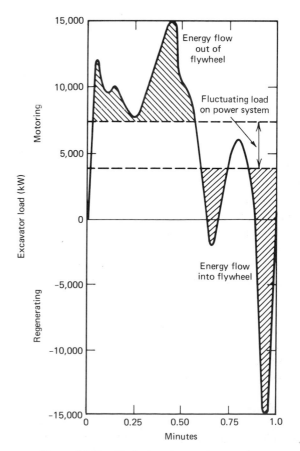

Figure 14.19. Typical excavator duty cycle.

to be stored is about 3100 kW for 30 sec, or 93,000 kW sec (25.83 kWh, 9.30 × 10⁷ J).

Solution. The flywheel to do this might be operated at ± 15% about a normal speed of 900 rpm. Its maximum speed would then be 1035 rpm, or 108.4 rad/sec, and the swing energy would be $(1.15^2 - 0.85^2)/1.15^2 = 0.454$ times the maximum energy. Thus, the maximum stored energy at 1035 rpm would need to be 93,000/0.454 = 205,000 kW sec, or 1.51 × 10⁸ ft lb, or 56,944 Whr.

From Eq. 14.13 the required length is

$$l = \frac{E\omega^2}{3.69 \times 10^{11}} = \frac{1.51 \times 10^8 \times 108.4^2}{3.69 \times 10^{11}} = 4.81 \text{ ft}$$

The radius is

$$r = \frac{417}{\omega} = \frac{417}{108.4} = 3.85 \text{ ft}$$

The weight is

$$W = \pi r^2 \, lw = \pi \, 3.85^2 \times 4.81 \times 500 = 112,000 \text{ lb},$$
$$\text{or 56 tons}$$

The specific storage at 1135 rpm is

$$\frac{56,944}{112,000} = 0.51 \text{ Wh/lb, or 4045 J/kg}$$

This calculation is highly oversimplified to illustrate the order of magnitude of the stored energy in a flywheel. An iron cylinder of 4.81 ft in length and 7.70 ft in diameter will store the desired 93,000 kW sec in speeding up from 765 to 1135 rpm, that is, in operating at 900 rpm ± 15%. It will give up this same energy in slowing down. The motor inertia has been neglected for simplicity, but of course would be considered in an actual case.

Since the motor is connected to a system of fixed frequency, its speed is regulated to the desired pattern by a variable-frequency voltage supplied to its slip rings. The cycloconverter that generates this frequency can be regulated to produce frequencies from +15% of 60 Hz to dc to −15% of 60 Hz, that is, to 9 Hz in the reverse phase rotation, causing the motor speed to vary as desired.

Example 14.2. Stored energy in flywheels was required to supply a pulse of electric power of 1860 kWh (6.7 × 10⁹ J) to a fusion experiment. This is

1860/25.8 = 72 times the storage required in Example 14.1. Obviously, a much larger diameter and slower flywheel is required. In this case, a maximum speed of 290 rpm, 30.4 rad/sec, was selected, dropping to half speed as it delivers the pulse.

Two motors, each with a 900 ton flywheel, were used, which with the motor inertia stored the requisite energy and kept the length of each flywheel within bounds. These flywheels were made of steel laminations bolted together and conservatively designed, as in a waterwheel generator application, for upward of 50 yr life. The specific storage at 290 rpm was 0.70 Wh/lb, or 5500 J/kg.

Example 14.3. At the other end of the spectrum is a very high-speed, high-energy density flywheel, designed for use on an electric automobile, to even the load on the battery. A flywheel made of Kevlar fibers, held together with an epoxy binder, was being tested (about 1977). The flywheel was designed to operate at 25,000 rpm and store 1 kWh in a 26 kg rotor (57 lb). Its specific storage was 1.4 × 10⁵ J/kg, or 17.6 Wh/lb. This is 25 times as much stored energy per unit weight as the large flywheel of Example 14.2.

Compressed-Air Storage (CAS)[18-20]

In a combustion gas turbine, about two thirds of the turbine output is used to drive the compressor. The idea of getting the compressed air from a cheaper source than the oil-fired turbine and thus tripling the size of generator that can be driven from the same turbine and the same oil is naturally very attractive. This had led to the idea of driving the compressor by a motor using low-cost off-peak power and storing the compressed air for several hours until needed during the peaks.

Underground storage of the compressed air is cheaper than above-ground storage. Thus, all such schemes are based on underground storage.

While this idea has been under consideration[21] for a number of years, the first such installation went into service in September 1978, at Huntorf, W. Germany. It was a 290 MW gas turbine plant. For 8 hr off peak some 58 MW is used to compress air which is fed into an underground cavern system. For 2 hr during the peak, 290 MW is generated, using the stored compressed air in the combustor of the gas turbine.

At Huntorf, two large caverns, solution mined in

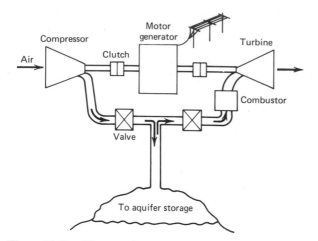

Figure 14.20. Diagram of a compressed-air storage gas turbine power station.[22] Reprinted from Jan. 1977 *Electrical World.* Copyright 1977, McGraw Hill Inc. All rights reserved.

a salt dome, are used for the storage. However, caverns mined conventionally in hard rock or the use of an aquifer as indicated in Fig. 14.20 are also being studied. While there is much experience in storing gas underground, this has been at a very slow rate compared with the daily cycling involved in this application. Studies are required to determine the consequences.

While there is no current U.S. installation of this type (1980), the three types of caverns are being studied (1979–1980) in three separate EPRI-sponsored studies.[18]

Magnetic Storage

In the superconducting region, close to absolute zero, a dc current once started in a loop circuit will flow undiminished almost indefinitely. Thus, the energy stored in its magnetic field remains until the current is intentionally diverted into an external load circuit.

Superconducting magnetic storage is under consideration[20,23] for electric utility load leveling. For this purpose, a storage of the order of 10^7 kWh (1000 MW for 10 hr) is needed. The largest pumped-storage plant, at Luddington, Michigan, stores 1.5 $\times$ 10^7 kWh. Muddy Run, on the Susquehanna, stores about 1.2 $\times$ 10^7 kWh. There have been many smaller plants in the past (Table 5.5), but most of those projected are 1000 MW or over, with pondage of 6 to 8 hr or more. That is, they are in the 10^7 kWh storage class.

A superconducting magnetic storage in this class

has been projected by a University of Wisconsin group. It would have coils 100 m in diameter and cost about $500,000,000, or $50/kWh, including the ac-to-dc conversion and the cryogenic cooling. For reference, the Luddington pumped storage cost $23/kWh, in 1973.

This scheme is in the conceptual stage (1980). Since it is only economic, even theoretically, in very large sizes, an enormous R&D commitment would be needed for the first one of economic size. The largest existing magnetic storages, incidental to other uses of the fields, are superconducting magnets for physics research—up to 280 kWh equivalent. Superconducting coils used at Los Alamos are about the size of a barrel and store 0.08 kWh.

Capacitor Storage

Capacitors are suitable for storing energy to be delivered in times of microseconds to seconds. The energy stored, in joules or watt sec, is $\frac{1}{2} CE^2$, where C is the capacitance in farads and E is the voltage in volts. If the charged capacitor is connected to a load resistance R, the initial current is E/R. This current falls off exponentially with a "time constant" $T = RC$ sec. At its initial rate of decrease, it would reach zero in T sec, as shown in Fig. 14.21. However, in each period T it decays to 36.8% of its initial value. Since the power is proportional to the square of the current, the energy is practically all delivered in the first time constant.

Small capacitors are usually measured in micro-

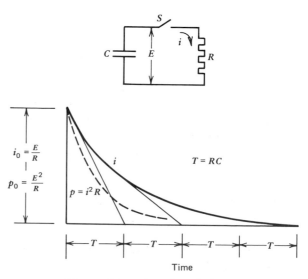

Figure 14.21. Capacitor discharge.

farads, and the time constant $T = RC$ is then in microseconds. A 100 μF capacitor connected to a 10 ohm resistor discharges with a time constant of 1000 μsec.

Capacitors are the most economic device for delivering a short pulse of energy. However, when compared with batteries or pumped storage for delivering energy over an extended period of time, they are very expensive. A 0.11$_4$ F capacitor bank at the Princeton Plasma Physics Laboratory charged to 15,000 V stores 3.564 kWh. The cost is of the order of $294,000 per kWh (8¢/J).

REFERENCES

1. B. B. Carr "Inland Waterways Transportation," in *McGraw-Hill Encyclopedia of Science and Technology,* New York: McGraw-Hill, 1977.

2. R. L. Loftness, *Energy Handbook,* New York: Van Nostrand Reinhold.

3. U.S. Dept. of Commerce, Maritime Div., *Domestic Waterborne Trade of the United States, 1972–1976,* Washington, D.C.: Supt. of Documents, 1977.

4. "Ships," in *McGraw-Hill Encyclopedia of Science and Technology,* New York: McGraw-Hill, 1977.

5. G. Ovechka, "Sail Cargo Ships to Arrive in 1980's?" *Marine Engineering,* February 1980, p. 50. "Harnessing the Wind", *National Geographic,* February 1981, p. 38.

6. T. Baumeister, Ed., *Marks Standard Handbook for Mechanical Engineers,* 7th ed., New York: McGraw-Hill, 1966.

7. *Pipeline Rules of Thumb Handbook,* Houston, Tex.: Gulf Pub. Co., 1978.

8. "Coal Slurry Pipelines," in *Transportation* Project Independence, F.E.A., Washington, D.C.: U.S. Department of Interior, 1974.

9. *Transmission Line Reference Book, 345 kV and Above,* Palo Alto, Cal.: Electric Power Research Inst., 1975.

10. R. S. Gens, E. H. Gehrig, and R. B. Eastvedt, "BPA 1100 kV Transmission System Development," *IEEE Transactions on Power Apparatus and Systems,* Nov.–Dec. 1979, p. 1916.

11. *EPRI Technical Assessment Guide,* Palo Alto, Cal.: Electric Power Research Inst., 1979.

12. B. A. Ross, "The Uncertain Future for Electric Utility Unit-Train Coal Movements," *Trans. IEEE,* 1980.

13. An *Evaluation of Regional Trends in Power Plant Siting and Energy Transport,* AA-9 (EPA-IAG-E4-0463, Proj. 1; ERDA W-31-109-Eng-38) Argonne, Ill.: Argonne National Laboratory, 1977.

14. *FPC News,* Fed. Power Comm., March 14, 1975.

15. *Guide to the Use of Gas Cable Systems,* EPRI, Rept. 78-25, Palo Alto, Cal.: Electric Power Research Inst., 1978.

16. J. J. Reilly and G. D. Sandrock, "Hydrogen Storage in Metal Hydrides," in W. H. Smith and J. G. Santangelo, Eds., *Hydrogen Production and Marketing,* ACS Symp. Ser. No. 116, Washington, D.C.: American Chemical Socy., 1980.

17. G. D. Sandrock and E. Snape, "Rechargeable Metal Hydrides: A New Concept in Hydrogen Storage, Processing, and Handling," in W. N. Smith and J. G. Santangelo, Eds., *Hydrogen Production and Marketing,* ACS Symp. Ser. No. 116, Washington, D.C.: American Chemical Socy., 1980.

18. "Eighty Atmospheres in Reserve," *EPRI J.,* April 1979, p. 14.

19. "Storage—Putting Base Load to Work on the Night Shift", *EPRI J.,* April 1980, p. 6.

20. R. E. Creagan, "Energy Storage," *McGraw-Hill Encyclopedia of Science and Technology,* New York: McGraw-Hill, 1979.

21. *Pumped Storage,* Engineering Foundation Conf., New York: ASCE, 1974.

22. W. R. Lang, "Air Stored for Peaking Power," *Electrical World,* January 1977, pp. 30–31.

23. H. A. Peterson, "Wisconsin Superconductive Energy Storage Project," Madison, Wis.: University of Wisconsin, 1974.

24. L. A. Kilgore and D. C. Washburn, Jr., "Energy Storage at Site Permits Use of Large Excavator on Small Power Systems," *Westinghouse Engineer,* November 1970, p. 162.

15

Environmental Coordination

INTRODUCTION

Early in this century, the communication companies were at the throats of the power companies for causing what they termed "telephone interference." There were angry people and expensive lawsuits.

However, by the late 1920s, reasonable engineers from the two sides had gotten together to solve the problem. They agreed to determine the "best engineering solution" in each case, whether it be in the telephone plant or in the power system, and to adopt it. Each party would bear the costs incurred in his own system, a procedure that later proved to be quite equitable and far cheaper than lawsuits. The engineering reports of the joint committee set up to solve this problem, through studies and tests over several years, are a masterpiece of engineering solution of a very complex problem.

Significantly, they changed the name of the game from "telephone interference" to "inductive coordination," which has suggested the title for this chapter.

As the world population increases, the rights of "life, liberty, and the pursuit of happiness" have to be continually redefined. One person's liberty must not deprive another of life, now or in the future. "Life" comes to mean a "healthy life." Happiness, for many, includes a world of beauty, replete with natural flora and fauna, and undisturbed natural beauties. For almost everyone, good food, clean water, pure air, good housing, and good transportation are also essential ingredients of the "good life." Most also consider the right to work and be a productive member of society as an essential ingredient.

Many of these desirable or essential things require using our natural resources. They also require energy, and require disposing of waste—human, animal, and material waste—all of which have undesirable consequences. These include pollution of the air and water, with serious health effects, despoiling the landscape, interfering with natural ecology, and exhausting our natural resources.

With this obvious conflict of goals, a compromise seems inevitable. However, given the ingenuity of man and a sufficiently strong desire, it appears that we can almost have all the desirable things—good food, water, housing, transportation, and adequate energy to run our complex society—and at the same time have air and water quality better than they have been in the recent past. In other words, the compromise is not a bad and deteriorating one but can be, if we all cooperate and will it, a fairly good and steadily improving one. The initial steps have already been taken: a beginning at understanding the problem, some public understanding and support, and some legislation and control. There is still a long way to go.

Viewed as an overall system problem, the following steps are involved:

1. Identification of all the sources of pollution of air, water, and land and all the environmentally undesirable structures, undesirable land and water uses, and undesirable activities that may have to be controlled.

2. Determine the seriousness of each one— the effect on health, property, and the ecology—establishing quantitative measures where possible. Understand the mechanisms. Some are very complex.

3. Determine the options for solution and their costs. For example, sewage *must* be disposed of. The question is simply, "How?" These costs include not only money. They may include giving up certain freedoms, accepting certain restrictions. They may mean loss of some jobs, less comfort, change of lifestyle, and accepting certain risks. The cost may include deteriorating balance of payments abroad, pressure on the dollar, and inflation,

with its impact on the poor. Large oil imports, while dragging our feet on nuclear energy, can have this effect.

4. Out of this determine a "best engineering solution," that is, a solution that results in the best environment at a cost that is judged acceptable. It is a national procedure, with the environmental aspects led by EPA but finalized into law by democratic processes. Societal motivation will also be required to stop littering. Laws are powerless against this "freedom."

This overall process has now resulted in National Air and Water Quality standards, Automobile Emission Standards (state and national), New Source Performance Standards for power plants, and Environmental Impact Statements, with their associated approval and licensing procedures, for various structures and projects. It has resulted in state and local plans for controlling existing pollution sources in all 50 states.

Air and water pollution at its source is the primary responsibility of state and local governments. The states are required to set performance standards for all existing, unmodified sources of pollution.

In general, the performance standards for new sources are based on the degree of emission limitation that can be achieved by using the best emission control system that has been adequately demonstrated, taking into account the cost of the control system. For existing sources of pollution, the limitations are also based on the cost and feasibility of "add-on" controls and the need for emission limitations. States are required to meet the National Air and Water Quality Standards in some way. The solution has to be worked out area by area.

Naturally, research, study, and action in all four phases of the problem must continue if we are to maintain and better our environment, in the face of steadily increasing world population. In this chapter, some information is given on each of the four phases of the problem, particularly as they affect energy. It is convenient to separate the considerations affecting air, water, and "the good earth," and they will be taken up in that order. Radiation will be treated separately.

AIR

The five principal air pollutants are: carbon monoxide, CO; particulates; oxides of nitrogen, NO_x; hydrocarbons, HC; and oxides of sulfur, SO_x. Table 15.1 shows the sources of these pollutants and the amounts emitted in the United States in 1970. As can be seen, the principal sources of the several pollutants are: CO—automobiles; particulates—industrial processes; SO_x—stationary fuel combustion; HC—automobiles; and NO_x—autos and stationary combustion.

As shown in Fig. 15.1, automobiles and fossil-fueled power plants together produce well over half of all the pollutants, except particulates. The principal sources of these are industrial processes such as steel, cement, plastics, and petroleum products. Formerly, bituminous coal burned in homes or plants was a large contributor. However, this source has been largely controlled or eliminated.

All of the emissions have been increasing year by year (Fig. 15.2) along with the population, except for particulates. The conversion of the railroads from coal to diesel oil, the use of precipitators in plant stacks, and the virtual elimination of coal as a domestic fuel and the resulting stagnant growth in

Table 15.1. Estimated Emissions of Air Pollutants by Weight, United States, 1970, in Millions of Tons per Year

Source	CO	Particulates	SO_x	HC	NO_x
Transportation	111.0	0.7	1.0	19.5	11.7
Fuel combustion in stationary sources	0.8	6.8	26.5	0.6	10.0
Industrial processes	11.4	13.1	6.0	5.5	0.2
Solid waste disposal	7.2	1.4	0.1	2.0	0.4
Miscellaneous	16.8	3.4	0.3	7.1	0.4
Total	147.2	25.4	33.9	34.7	22.7
Percent change 1969–1970	− 4.5	− 7.4	0	0	+ 4.5

Source. CEQ, 1972.

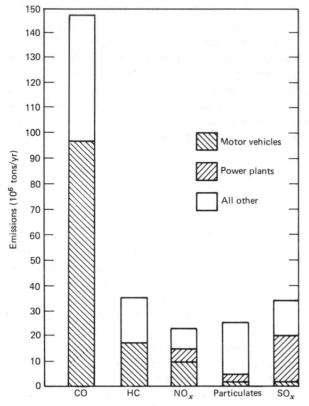

Figure 15.1. The motor vehicle and power plant contributions to total emissions in the United States, 1970 (EPA, 1973).

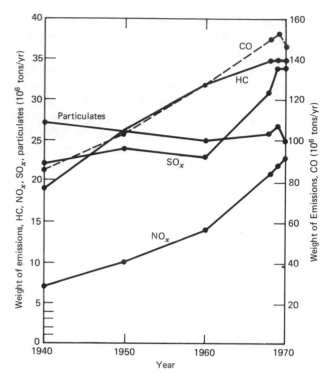

Figure 15.2. Emissions of the five major pollutants (CEQ, 1972).

coal usage (15% in 25 yr) account for the diminution in particulates, which is particularly noticeable in cities such as Pittsburgh.

Auto emission controls were made effective first in 1968 on a national basis, 1966 in California, and performance standards for new power plants somewhat later, so that their net effect is as yet unknown (1977). They should certainly produce an improvement.

Carbon Monoxide, CO

Carbon monoxide is emitted mainly by automobiles due to incomplete combustion, particularly at the low speeds of city driving. When inhaled, it reduces the capacity of the blood to carry oxygen. This interferes with body functions, accentuates pulmonary and heart ailments; and, in sufficient concentration, CO causes death.

Concentrations as low as 15 ppm for several hours cause some impairment; and by 60 ppm for several hours, patients with heart disease will be in

physiologic stress. Note: For air, the basic measurements are in parts per million (ppm), by volume for gaseous pollutants, and micrograms per cubic meter (μg/m³) for particulates.

The hemoglobin (red corpuscles) in the blood normally pick up oxygen in the lungs and carry it to the various parts of the body. With 30 ppm of CO in the atmosphere for 8 hr, some 5% of the hemoglobin is carrying CO (it becomes COHb). Considering that the oxygen concentration, 21% or 210,000 ppm, is 7000 times as great as the CO concentration, the remarkable affinity of the hemoglobin for CO is evident.

Thirty percent of the days in Chicago and 10% of the days in Los Angeles had 8-hr exposures to CO exceeding 15 ppm, according to HEW data (1962–1967). The natural background level is about 1 ppm.

The relationship between total emissions and ambient concentrations is of course very complex, involving meteorological conditions that evidently convert CO to CO_2; CO does not continually build up. However, the *actual concentrations* experienced show the need for limitation of CO emissions. As shown in Table 15.2, the National Air Quality Standard for CO has been established at 9 ppm for 8

Table 15.2. National Ambient Air Quality Standards and Recommended Federal Episode Criteria

Pollutant, Units, Averaging Time	Secondary[a]	Primary[b]	Alert[c]	Warning[c]	Emergency[c]	Significant Harm[d]
Sulfur dioxide $\mu g/m^3$ (ppm)						
1 yr	60 (0.02)	80 (0.03)				
24 hr	260 (0.10)	365 (0.14)	800 (0.3)	1,600 (0.6)	2,100 (0.8)	2,620 (1.0)
3 hr	1,300 (0.5)					
Particulate matter $\mu g/m^3$ (COH)						
1 yr	60	75				
24 hr	150	260	375 (3.0)	625 (5.0)	875 (7.0)	1,000 (8.0)
Product of sulfur dioxide and particulate matter $[\mu g/m^3]^2$ (ppm × COH)			6.5×10^4 (0.2)	2.61×10^5 (0.8)	3.93×10^5 (1.2)	4.90×10^5 (1.5)
Carbon monoxide mg/m^3 (ppm)						
8 hr	10 (9)	10 (9)	17 (15)	34 (30)	46 (40)	57.5 (50)
1 hr	40 (35)	40 (35)				144 (125)
Oxidants $\mu g/m^3$ (ppm)						
1 hr	160 (0.08)	160 (0.08)	200 (0.1)	800 (0.4)	1,200 (0.6)	1,400 (0.7)
Nitrogen dioxide $\mu g/m^3$ (ppm)						
1 yr	100 (0.05)	100 (0.05)				
24 hr			282 (0.15)	565 (0.3)	750 (0.4)	938 (0.5)
1 hr			1,130 (0.6)	2,260 (1.2)	3,000 (1.6)	3,750 (2.0)
Hydrocarbons $\mu g/m^3$ (ppm)						
3 hr (6 to 9 A.M.)	160 (0.24)	160 (0.24)				

[a] To be met within a reasonable time.
[b] To be met by 1977.
[c] The Federal Episode Criteria specify that meteorological conditions are such that pollutant concentrations can be expected to remain at these levels for 12 or more hours or increase; or, in the case of oxidants, the situation is likely to reoccur within the next 24 hr unless control actions are taken.
[d] Priority 1 regions must have a contingency plan which shall, as a minimum, provide for taking any emission control actions necessary to prevent ambient pollutant concentration at any location from reaching these levels.
[e] Corrected for methane.

hr. This will require a considerable reduction in cities such as Chicago and Los Angeles mentioned earlier.

The automobile emission standards now established (1981) are shown in Table 15.3.* Compared

*The table is much abbreviated: Contact the EPA for full conditions and exceptions.

with the CO emissions of 80 g/mile (grams per mile) prior to controls (about 1965), the emissions of CO were to be reduced to 3.4 g/mile in the 1981 models, a reduction of 23 to 1. As it takes about 10 yr to get most of the older cars off the road, it will take some time to fully assess the results with this control and some control of the industrial and miscellaneous sources.

Table 15.3. Federal New Light Duty Motor Vehicle Exhaust Emission Standards[a-c]

Model Year	Test	Standards (g/mile) HC	CO	NO$_x$
Prior to controls	7-mode	11.0	80	4.0
1970	7-mode	2.2	23	—
1971	7-mode	2.2	23	—
1972	CVS-72	3.4	39	—
1973–74	CVS-72	3.4	39	3.0
1975–76	CVS-75	1.5	15	3.1
1975 California	CVS	0.9	9	2.0
1977	CVS-75	1.5	15	2.0
1978–79	CVS-75	1.5	15	2.0
1980	CVS-75	0.41	7.0	2.0
1981	CVS-75	0.41	3.4	1.0
1985 and later	CVS-75	0.41	3.4	1.0

Source. EPA 10-8-80.

[a]CVS = Constant Volume Sampling.

[b]Particulates are also limited starting in 1982, to 6 g/mile 1985 to 2 g/mile.

[c]Evaporative hydrocarbons are also limited since 1971.

Particulates

Particulates include aerosols, fumes, dust, mist, and soot. Aerosols are particles of solid or liquid matter that can remain suspended in the air because of their small size—usually under 1 μ. Fume is liquid particles under 1 μ, formed as vapors condense or as chemical reactions take place. Dust is solid particulate matter. Mist is liquid particles up to 100 μ in diameter. Soot is finely divided carbon particles clustered together in long chains. Note: 1 μ is 1 micron, 1/1000 mm or about 1/25 mil.

In general, the eye can resolve down to 100 μ (4 mil). Particles over 10 μ settle rapidly as dust. At 10 μ, the settling velocity is 7 in./min. The large range in particle sizes leads to a problem in that the larger sizes may predominate in measurements, whereas the smaller sizes cause most of the health problems. Also, the toxicity of the particles is not accounted for in the measurements, which lump everything together. Particles smaller than about 1 μ result largely from the condensation of vaporized material after combustion. Particles larger than 10 μ result from mechanical processes such as grinding and erosion.

Particulate matter in the air correlates quite well with visibility reduction according to the relation

$$r \cong \frac{750}{c} \qquad (15.1)$$

where r is the visibility range in miles and c is the particulate concentration in μg/m^3. Thus, the National 24 hr standard (Table 15.2) of 260 μg/m^3 corresponds to roughly 3 miles visibility, and the California standard of 100 μg/m^3 to 7.5 miles visibility.

Because of the greater ease of measuring visibility, a "coefficient of haze" index, *COH,* is often used in place of the particulate concentration in μg/m^3. Unfortunately, the relationship is only approximate and varies with the character of the particulates.

In Pittsburgh, for example,[4] 1 *COH* = 150 μg/m^3. EPA/CEQ has used an average of 1 *COH* = 125 μg/m^3, as in Table 15.2 or in Ref. 4, p. 20. That is, 1 *COH* = 6 miles visibility.

Adverse health effects are noted[5] when the annual geometric mean level of particulates exceeds 80 μg/m^3. This compares with the National standard, Table 15.2, of 75 μg/m^3. Particulates act synergistically with SO$_2$ (see also discussion under Sulfur Oxides to follow).

Particulate concentrations as high as 7 *COH* (about 870 μg/m^3) along with 0.5 ppm SO$_2$ were encountered in the New York City air pollution episode of 1963, and 4500 μg/m^3, along with 1.34 ppm SO$_2$, in the worst London episode of 1952 (4000 excess deaths).

However, whenever the visibility due to London-type fog (SO$_2$ and particulates) is under 3 miles, the particulates are over 260 μg/m^3. Senior citizens in Pittsburgh well remember when this was almost a daily event during the winter, before the almost miraculous cleanup of the 1940s.

Particulates spread a layer of greasy dirt over everything, indoors and out, on the walls and furnishings of houses, on clothing, vehicles, and buildings, and require extensive cleaning. In Pittsburgh, prior to "smoke control," small boys could not climb trees without getting their clothes black.

Precipitators, both mechanical (cyclone) and electrical (Cottrell), have been in common use in connection with power plant stacks for many years. At the time given in Table 15.1 (1970), they were in use on practically all large coal-burning power plants as well as many industries. "Baghouses" are also used for the same purpose in smaller plants. Some 99% of the particulates are eliminated from the effluents by these devices. The New Source Performance Standards (NSPS) of 1979[6] limit particulate emission from new power plants to 0.03 lb/million Btu (13 ng/J). This is about 0.72 lb per ton of coal which might have 200 lb of fly ash.

The further control must be mainly in industrial processes, in restrictions in burning trash in open dumps, and in the use of wigwam burners used for lumbering wastes. This requires separate consideration of each industrial area. The states must distribute the required limitations as equitably as possible among the various polluters in order to meet the national standards as set forth in Table 15.2.

Oxides of Nitrogen, NO_x

Oxides of nitrogen are produced when the oxygen and nitrogen of air are raised to temperatures above 2000°F in the combustion processes of automobiles and stationary power plants (see Chapter 13, Equilibrium; and Chapter 4, Gasification). Some of the oxygen and nitrogen combine to form mostly NO, with a small amount of NO_2. The NO has no adverse health effects at the concentration found in the atmosphere, but it does convert to NO_2, which is corrosive, physiologically irritating, and toxic. The NO_2 also reduces the brightness and contrast of distant objects and is responsible for the reddish-brown color of photochemical smog. In fact, the creation of NO_2 in the presence of NO, light, and hydrocarbons is part of the complex reactions that produce photochemical smog.

A study in Chattanooga[2] showed that the frequency of acute bronchitis increased among school children and infants when the range of 24 hr concentrations of NO_2, measured over a 6 month period, was between 0.063 and 0.083 ppm. In 1971, the yearly average NO_2 concentrations exceeded 0.06 ppm in 54% of American cities with population between 50,000 and 500,000 and in 85% of cities with populations over 500,000, according to EPA.

As shown in Table 15.2, the national standard for NO_2 has been set at 0.05 ppm yearly average. The primary leverage on this is of course automobile and power plant emissions. By controlling the NO_x (mostly NO) emitted, the NO_2 in the atmosphere is expected to be reduced. As shown in Table 15.3, the NO_x from car emissions is being reduced from 4 g/mile before controls to 1.0 g/mile by 1981, a 4 to 1 reduction. The final reduction was in the 1981 models, and it will be some time before the results can be fully determined.

Power plants, being located outside of cities and with high stacks that afford more dilution, are less serious. Also, on the West Coast where photochemical smog is most serious, natural gas and oil are more widely used. The proportion of NO_x from power plants is far less than indicated in Table 15.1.

Automobiles produced 71% of all NO_x in California in 1972.[2] Nevertheless, stringent NO_x standards for new power plants have been established.[6] This effect will be on a longer time base. The 1979 NSPS limit NO_x emissions from new coal burning plants to 0.6 lb/million Btu (260 ng/J), oil burning plants 0.3 lb/million Btu (130 ng/J), and gas fired plants 0.2 lb/million Btu (86 ng/J).

Hydrocarbons

Hydrocarbons are compounds of hydrogen and carbon (see Chapter 11, Chemistry of Energy). Those with one to four carbon atoms are gases, the most common being methane, the principal constituent of natural gas. Natural sources, mostly biological, produce large quantities of methane and volatile terpenes and isoprenes. Nonurban air thus naturally contains 1 to 1.5 ppm methane and less than 0.1 ppm of each of a few other hydrocarbons. Some of these are exuded by conifers, and the smoky haze of the Great Smoky Mountains is from this source. Note: Terpenes, $C_{10}H_{16}$, and isoprenes, C_5H_8, both unsaturated, are from plant life.

Hydrocarbon emissions result also from the inefficient combustion of hydrocarbons and from their evaporation. Of the 35 million tons shown in Table 15.1, 16.7 million tons (48%) were emitted by motor vehicles, 5.5 million tons (15.7%) by industrial processes, and 3.1 million tons (8.9%) were from organic solvent evaporation.

Methane, CH_4, is practically inert; the next compounds ethane, C_2H_6, and ethylene, C_2H_4, are the most abundant of the hydrocarbons causing harm. It is usual to deduct the methane and use "hydrocarbons less methane" as the best index of potential harm.

While hydrocarbons in the concentrations present in the atmosphere are not known to have any adverse health effects, they are an essential ingredient of photochemical smog, which is deleterious to the health. It has been shown[2] that 0.3 ppm nonmethane hydrocarbons during the 3 hr period 6 to 9 A.M. in the Los Angeles area can be expected to result in an average 1 hr photochemical oxidant (see later paragraph) concentration of 0.1 ppm about 2 to 4 hr later, a level which has adverse health effects. Thus, federal air quality standards are designed to limit the oxidants by limiting the hydrocarbons.

In general, the unsaturated hydrocarbons such as ethylene are more active than the saturated series including ethane. Ethylene has some deleterious effects on plants at atmospheric concentrations and also causes the eye irritant formaldehyde in the photochemical reaction.

Not only are most of the hydrocarbons from cars, but they are emitted at the worst possible time, 6 to 9 A.M., during the morning rush hour, when the sunshine has all day to act on them and produce smog. Thus, the primary control is through limitation of motor vehicle hydrocarbon (HC) emissions. Compared with the 11 g/mile prior to controls, the earliest controls (California, 1966) limited HC emissions to 3.6 g/mile. The federal limitation of this effluent in the 1980 models is 0.41 g/mile, a 27 to 1 reduction from the 11 g/mile of the precontrol cars (see Table 15.3).

We have now discussed the three principle objectionable car emissions, CO, NO_x, and HC. The limitations may be summarized as follows. Note that different test methods have been used and the reductions are therefore only an order of magnitude.

Auto Emissions (g/mile)

	HC	CO_x	NO_x
Before controls (1965)	11	80	4.0
After controls (1981)	0.41	3.4	1.0
Reduction	27 to 1	23 to 1	4 to 1

In addition particulates and evaporative hydrocarbons are being controlled. Similar standards are in effect for heavy duty vehicles.

Sulfur Oxides

Sulfur, mainly from the burning of coal, has deleterious effects on health, materials, and plants. When sulfur-bearing fuels are burned, the sulfur is released, mostly as SO_2 and some SO_3. The SO_3 immediately combines with water to form sulfuric acid, H_2SO_4. Some of the SO_2 may form H_2SO_3 or other sulfates.[2]

Sulfur compounds in the atmosphere are irritating to the respiratory system and have the worst effect on individuals with chronic pulmonary or cardiac disorders, as well as the very young or old individuals. The damaging effects of sulfur on health appear to be almost directly proportional to the particulates present simultaneously. This is known as a "synergistic effect." Thus, the product of sulfur and particulate concentrations is often used as the index of health impact. The National Episode Criteria Recommendations include this product, as shown in Table 15.2. At least, the particulate level is generally stated in relating sulfur levels to health effects.

As a general rule, adverse health effects are noted when the annual mean concentration of SO_2 exceeds 0.04 ppm, or when the 24 hr mean exceeds 0.11 ppm.[2] These levels may be compared with the annual and 24 hr primary standards of 0.03 and 0.14 ppm, respectively, as shown in Table 15.2.

Along with 160 $\mu g/m^3$ particulates (5 miles visibility), the 0.04 ppm annual concentration of SO_2 causes an increase in mortality from bronchitis and lung cancer. A 24 hr SO_2 concentration of 0.25 ppm, along with 750 $\mu g/m^3$ particulates (1 mile visibility), causes an increased death rate and a sharp increase in illness rates.

From the NAS–NAE report,[7] there is good evidence that exceptional episodes of pollution (> 1000 $\mu g/m^3$ or 0.385 ppm SO_2 along with particulates) caused illness or death. There is also good evidence that sustained lower levels of pollution (> 100 $\mu g/m^3$ or 0.039 ppm SO_2 along with particulates) for a number of years affect health adversely. These two conditions are 2.7 times the primary 24 hr standard and 1.25 times the primary annual standard, respectively.

Sulfur also rapidly corrodes materials. An average annual concentration of 0.12 ppm SO_2 causes a weight loss of 16%/yr in mild steel panels. In the high-sulfur coal areas, eastern United States generally, overhead power line hardware suffers one third loss of life. Other ill effects are the loss of strength of leather, discoloration and deterioration of buildings, damage to statuary and works of art, and damage to plant life of all kinds.

Currently (1978), about 41% of U.S. electric power is derived from coal, Table 1.2, mostly high-sulfur coal. The methods of sulfur removal are discussed in Chapters 3, and 4.

Since most of the sulfur comes from fuel combustion in stationary plants, it is logical to limit these emissions to obtain the needed reduction. Coal used by electric utilities in 1975 was overwhelmingly the high-sulfur eastern coal. It averaged just under 3% sulfur and about 12,000 Btu/lb, which comes out 2.4 lb sulfur per million Btu. If all coverted to SO_2, this results in 4.8 lb SO_2 per million Btu. The New Source Performance Standards (NSPS) of June 11, 1979[6] limit the SO_2 to 1.2 lb per million Btu, with the added stipulation that at least 90% of the potential sulfur must be removed, down to 0.6 lb per million Btu. Thus, average high-sulfur coal emissions are limited to about 0.6 lb SO_2 per million Btu, an 8 to 1 average reduction from the potential of 4.8 lb per million Btu.

Even low-sulfur coal, which has an average potential of about 1.2 lb SO_2 per million Btu, is required to have 70% of its sulfur removed, a 3.3 to 1 reduction.

The NSPS covers all forms of fossil fuels, the two examples just given being the most significant in overall result. Unlike automobiles, with a 10 yr life cycle, power plants last 30 or 40 yr, so that it will be beyond the year 2000 when these controls are fully effective. In the meantime, add-on facilities are being installed as required to meet the National Ambient Air Quality Standards in individual cases.

Smog

It is necessary to differentiate between so-called "London smog," produced by high concentrations of sulfur and particulates, and photochemical smog, produced by hydrocarbons and oxides of nitrogen in the presence of sunlight.

The "London" smog, characteristic of the coal-burning regions, has been responsible for all the air pollution episodes that have thus far taken a toll of lives in a short time (few days). Nine of these have been in London, the worst, in 1952, causing 4000 "excess deaths." The SO_2 concentration reached 1.34 ppm and particulates 4500 $\mu g/m^3$ (about $\frac{1}{8}$ mile visibility).

Episodes have occurred in the United States, in Donora, Pa., 1948 (20 excess deaths), and in New York City, 1963 (200 to 400 excess deaths) and 1964 (168 excess deaths). Such "smogs" are characterized by lack of wind to blow the pollutants away and an inversion layer to trap them from vertical mixing. Donora is also in a valley.

The photochemical smog is most closely associated with Los Angeles. The phenomenon is very complex, but the inputs are NO, HC, O_2, and sunlight, and the outputs are O_3 (ozone), NO_2, PAN, and so on. The PAN (peroxoacetylnitrate, $CH_3CO_3NO_2$) and the ozone (O_3) are referred to as "photochemical oxidants." When ozone reacts with hydrocarbons, irritating substances can result, such as formaldehyde, peroxybenzoylnitrate (PB_zN), PAN, and acrolein, which are irritating to the eyes. Ozone can cause chest constriction, irritation of the mucous membrane, headache, coughing, and exhaustion. It also causes damage to organic materials such as rubber, cotton, acetate, nylon, and polyester. The oxidants have been associated with increases in asthma attacks.

Needless to say, the control of London-type smog is through reduction of its ingredients, SO_2 and particulates, as outlined earlier. The control of photochemical smog is by reduction of the principal man-made causative agents, NO_x and HC. In addition to the new source standards for cars, power plants, and industry, a number of other measures are taken, state and local, to reduce emissions.

Significant Harm

There is always a finite probability that even with normally adequate controls, very unusual meteorological and other conditions may combine to result in a dangerous amount of pollution, an "episode." In this case, EPA is charged with dealing directly with the situation and imposing any restrictions required. To this end, levels of pollutants have been established that threaten "significant harm" to health and require such emergency action. These levels are given in the final column of Table 15.2. These emergency powers have needed to be invoked only once in several years (Birmingham, 1971).

WATER

Water pollutants are divided by EPA into eight classes, only three of which are particularly energy related. The eight classes are:

1. *Oxygen-Demanding Wastes.* These are biodegradable organic compounds contained in domestic sewage and certain industrial effluents. When these are decomposed by bacteria, the dissolved oxygen in the water is used up; and, if the oxygen is insufficient, fish cannot live.

2. *Disease-Causing Agents.* These are pathogenic microorganisms which usually enter the water with human sewage. They must be controlled by sewage treatment for safe water-contact activities (swimming, water skiing) and further controlled by filtration and disinfection for drinking water.

3. *Synthetic Organic Compounds.* These include detergents, pesticides, and industrial chemicals. Many are toxic to aquatic life and harmful to humans.

4. *Plant nutrients.* These include nitrogen and phosphorous that drain from fertilized farmland, as well as discharge from sewage treatment plants. They stimulate the growth of algae and weeds. When these die and are decomposed by bacteria, the oxygen is used up as in paragraph 1. After the oxygen is used up (aerobic decay), another class of bacteria cause anaerobic decay, which leads to putrefaction and slime. While aerobic decay converts organic material to harmless CO_2 and water, anaerobic decay changes it to vile substances such as ammonia, hydrogen sulfide, and the like, which are generally foul smelling.

5. *Inorganic Chemicals.* These include acids (mine drainage) as well as heavy metals such as mercury and cadmium.

6. *Sediments.* These arise from soil erosion. They can smother bottom life and fill reservoirs and harbors.

7. *Radioactive Substances.* These can enter from mining radioactive ores, from military and civilian nuclear activities, and from medical facilities.

8. *Thermal Discharges.* The use of large quantities of water for once-through cooling in steam-electric power plants can raise the temperature of the receiving water by as much as 20°F and affects the local ecosystem. The circulated water is raised 10 to 30°F, usually about 20°F.

While not specifically included in the EPA list, trash and floating material such as foam, garbage, sewage, and oil slicks are also highly objectionable aesthetically and render the water so polluted unfit for many human activities.

Like any large industrial plant, a power plant has a variety of wastes arising from coal storage and handling, lubrication of motors, flushing of boilers or cooling towers, or disposal of ash and the particulates and sulfur removed from the stack gases. Discharges of these into public waters must be controlled as with all industries. However, electric utilities use two thirds of all coal in the United States, and the quantities dealt with are correspondingly large. (See chapters on coal and gas for methods of control.)

The three energy-related problems in the EPA list of water pollutants are thermal discharges, radioactive substances, and mine drainage. Oil spills from transportation and off-shore production of oil are also an energy problem, as are port facilities and shoreline siting.

Before going into these, there are a few characteristics of bodies of water and of the animals that live in them that need to be mentioned.

Thermal Pollution and Water Ecology

Anyone who has witnessed the prolific fish life in the warm fresh or salt waters of Florida is puzzled that warming the water could be anything but good. In fact, there have been proposals to use the water warmed by power plants for warm-water fish farms.

However, the ecosystem in any area is tuned to its climatic conditions. It is a delicate balance of plants and animal organisms that constitute the food chain (described more fully in Chapter 7). Just to review these steps, phytoplankton and attached plants grow by photosynthesis during the daylight hours. Carbon dioxide, water, and the energy from sunlight are converted into organic material. Some minerals and nitrogen are taken up. These are the producers. They are all plants. They liberate oxygen into the water in the photosynthesis process and provide food for animals. All aquatic animals are users of this oxygen, and are either users or decomposers of this food. First come the herbivores (zooplankton, snails, shrimp, and so on), that eat plants; then the carnivores, the slow and fast swimming fish that feed each on the next lower trophic level. Finally, as they die, the decomposers (bacteria) take over and return the CO_2 and minerals to the water, using oxygen in the process. Oxygen is used continuously for respiration both by the plants and the animals.

There are only two sources of oxygen: photosynthesis and surface contact with air. Both are predominantly at the surface, since the sunlight is rapidly attenuated with depth.

There is stratification in bodies of water, in lakes, or in the navigation pools of large rivers, with low summer flow. A thermocline acts similarly to the inversion layer in air to prevent mixing the upper and lower layers, except in the spring and fall turnovers. In summer, the warm water stays on top. When it becomes colder than the bottom water, there is a turnover. Thus, the oxygen in the lower water is almost always marginal and easily overloaded. Excessive algae growth from too much nutrients further restricts photosynthesis to the upper layers.

Both the biodegradable wastes and excessive growth due to nutrients have already placed an undue load on the limited oxygen supply of the lower water. When to this is added thermal pollution, two additional factors enter to worsen the situation: (1) less oxygen can be dissolved in the warmer water (about 10% less for 5°C higher temperature), and (2) the metabolic rate and use of oxygen of all organisms increase with temperature (double for a 10°C rise in temperature). Thus, waste decomposes faster in warmer water. The shortage of oxygen becomes acute.

A further factor is the thermal shock to the ecosystem when a large plant is shut down for maintenance or refueling.

Based on such considerations, EPA has recommended that the temperature of the "receiving" water not be raised above 90°F; and that the temperature rise be limited to 5°F in streams, to 3°F in lakes, and to zero in trout and salmon streams. In estuarine and marine situations, it is recommended that the temperature rise of the receiving water not exceed 4°F, September to May, or 1.5°F June to August.

Another fact to be understood about aquatic ecosystems is that the food chain includes as many as five trophic levels (as from phytoplankton to tuna) before it is eaten by man, whereas land ecosystems include only two levels before human consumption (as grass, beef, man) (see Chapter 10). Thus, the buildup of heavy metals such as mercury or of radioactive materials can be 100 or 1000 times worse in the water ecology.

When a combination of polluting factors is present, as with most waters in populous parts of the country, it is usually impossible to tell which caused the harm. Nevertheless, it is understandable that the burden of proof rests on *any new source*.

Power Plant Cooling

As a result of the foregoing considerations, the number of power plant sites where "once through" cooling is permissible have become scarce. Cooling ponds or towers are being used in many cases to dispose of waste heat to the atmosphere. While these are more costly than once-through cooling, the added cost is 1 to 2% compared with 15 times this amount to remove sulfur from stack gases.

The thermal efficiencies of modern (1981) power

Table 15.4. Relative Cooling Tower Capacities

	Nuclear Plant	Fossil Plant	Fossil with Scrubbers
Efficiency (%)	33	40	33 (lower limit)
Cooling in % of rating			
To cooling tower	185	115	153 approx.
To air or stack	15	35	47 approx.
Total	200	150	200

plants are about 40% fossil fuel, 33 to 36% fossil fuel with stack gas scrubbers, and 33% nuclear (non-breeder). For the 33% efficient plant, the heat to be disposed of is about twice the plant rating; and for the 40% efficient plant, it is 1.5 times the plant rating. However, the nuclear plant must dispose of most of its waste heat through cooling water, whereas in a fossil fuel-fired plant a part of it goes up the stack. As a consequence, the relative capacities of cooling towers are as shown in Table 15.4.

With combined-cycle plants, the heat dissipation to water will be much less. With the fast breeder, the efficiency will be about 40%, the waste heat 150% of rating, with about 140% to the cooling tower.

Cooling Ponds and Towers

In a few areas of the country, where land is available and the air is relatively dry, mostly in the southwest, shallow cooling ponds can be used to provide cooling water. These achieve their cooling by evaporation. Warm water is discharged to one end of the pond, and cool water is taken from the other end. Typically, 1 to 2 acres are required per megawatt of electric power, or 1½ to 3 square miles for a 1000 MWe plant. Evaporation can be greatly enhanced by multiple sprays over the surface of the pond, in which case it is called a spray pond.

More usual is a cooling tower having the general appearance of that shown in Fig. 15.3. There are three general types, as follows:

Wet-Cooling Tower, Natural Draft. As shown in Fig. 15.3, warm water from the condenser is sprayed over baffles. Cool air rising through the baffles cools the water both by evaporation and by direct heat transfer. The lighter warm air and vapor rise through the hyperbolically shaped tower and are discharged at the top. Many of these are highly

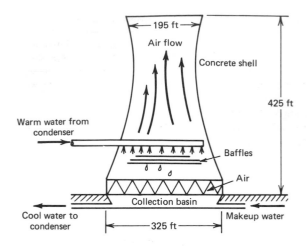

Figure 15.3. Wet, natural-draft cooling tower. Dimensions are those of the Rancho Seco tower outside of Sacramento, California. From Ref. 2 with permission of John Wiley and Sons, Publishers.

satisfactory, while in some cases the vapor plume over the tower may cause problems of local fogging and icing.

A source of makeup water, about 3%, is required to compensate for evaporation and for "blowdown." Blowdown is the continuous or periodic flushing of the cooling system to remove solids and chemicals which accumulate in the circulating cooling water. The flushing can become a water pollution problem unless special treatment is provided.

Wet-Cooling Tower, Forced Draft. Large fans are used to increase the air flow. The tower is

smaller and has lower initial cost but higher operating cost. The balance has resulted in a trend toward the natural-draft cooling tower. See Table 15.5.

Dry-Cooling Tower. Where wet-cooling towers cannot be used, for the reasons cited above or for lack of makeup water, dry towers operating on the same principle as an automobile radiator are used. Cooling is by conduction and convection between the closed water system and the air blown through. These are appreciably more expensive, as shown in Table 15.5, but are increasingly being considered (1981) where wet-cooling towers cause problems.

The Oceans

A number of the environmental concerns of the oceans are energy related. Offshore oil and gas fields introduce a number of problems. The transportation of oil and LNG, both rapidly increasing worldwide in 1981, involves oil spills and accidents. Regulation and control is made particularly difficult because of the international relations involved. Use of the precious coastline for ports, refineries, or power plants is in increased competition with aesthetic beauty, recreational activities, or downright safety.

In principle, the problem must be solved by the same four steps: (1) define the problem, (2) assess the seriousness and understand the mechanisms involved, (3) determine the available solutions or alternatives and their costs, and (4) adopt the best engineering solution at a cost that can be afforded. However, some comments are in order regarding

Table 15.5. Cooling System Costs for 40% Efficient Fossil-Fueled Steam-Electric Plant and 33% Efficient Nuclear Plant, Assuming 20°F Cooling Water Temperature Change

Cooling System	Equipment Capital Cost ($/kW)		Operation and Maintenance Costs ($/kWyr)		Total Cost ($/kWyr)		Additional[a] Cost to Consumer (%)	
	Fossil	Nuclear	Fossil	Nuclear	Fossil	Nuclear	Fossil	Nuclear
Once through	5.00	5.24	0.6	1.00	1.05	1.47	0.00	0.00
Cooling pond	6.50	7.50	0.76	1.24	1.34	1.92	0.16	0.26
Spray pond	7.60	8.10	0.90	1.50	1.58	2.23	0.30	0.43
Natural-draft wet tower	7.50	11.50	1.20	2.00	2.92	4.08	1.17	1.49
Mechanical-draft wet tower	7.20	9.40	1.88	2.66	3.23	4.20	1.25	1.56
Mechanical-draft dry tower	13.00	15.00	1.88	2.66	4.45	5.41	1.94	2.26

Source. EPA, 1969.
[a]Above cost of once-through system.

the significance of these four deceptively simple steps.

The problems now include those of international control, a problem that we have struggled with for many years in protecting whales, seals, and fishing, with only limited success.

The tragic consequences of a huge oil spill engulfing a historic beach area defies any measurement in dollars and cents. One alternative is to halt all imports of oil or gas. But the cost is too high. Here again, the cost is not measured in dollars, but in consequences to the U.S. economy and the life of the people.

On the brighter side, offshore oil and gas, huge imports of oil, and rapidly growing shipments of LNG are a relatively new phenomenon on such a large scale—almost within the last decade (See Fig. 1.1). Tremendous progress has already been made in dealing with these formidable new problems. As with our internal problems of rivers and lakes, it is conceivable that the oceans can be exploited for their transportation and resources without destroying their ecology or the vast beauty of their shorelines. With the almost limitless ingenuity of man, and with a strong national and world will to accomplish this, it should also be at an acceptable economic cost. For example, for our own coastal waters, the provisions already devised and in operation for cleaning up oil spills have not been overly expensive. Deep-water ports have been developed to avoid bringing large tankers into certain harbors, and to save coastline (see also Chapter 4, LNG).

Large refineries and other industries have already been banned from several coastal areas, and siting is an increasing problem, even though the cost of refineries has approximately doubled, due largely to environmental control requirements.

Mine Drainage

Many of the streams in the soft-coal states of eastern United States are too acid for fish to survive. They are virtually devoid of plant growth of any kind. Much of this is due to old abandoned mines that cannot be controlled by regulation. Even though the Fish Commissions have spent millions of dollars blasting shut the mouths of old mines, most of the acid pollution remains. One bad tributary may spoil a whole main stream. Currently (1978), rather expensive treatment plants are being installed on certain streams to treat *all* the water coming from the mine drainage area to make the

stream "clean" for the rest of its course. Problems have been encountered and it is too early to assess the outcome of these measures.

For presently operating mines, the effluents can be strictly controlled by regulation, as with any industrial operation. EPA interpretation of the 1972 law is:

1. Wherever possible, water that is clean enough for swimming and other recreational use and clean enough to protect fish, shellfish, and wild life—by July 1, 1983.
2. No more pollutants whatever into the nation's waters—by 1985.

A review of the list of eight pollutants given earlier shows the enormity of this task.

THE GOOD EARTH

In addition to its effect on air and water, the production of energy has some very visible effects on the face of the earth. These include strip mines and open-pit mines, deep mines and their waste disposal, overhead transmission lines, distribution circuits in the cities, power plants and substations, power dams and reservoirs, and oil fields and refineries.

Strip Mining

Strip mining has gone through three stages: (1) uncontrolled, (2) fines for failure to reclaim land, frequently less than the cost of reclamation, and (3) mandatory high-quality reclamation. In 1978, with coal at 20 to 25 $/ton, and having quadrupled in 10 yr (see Fig. 15.6) and with high-quality reclamation costing less than $1/ton, any suggestion that reclamation is expensive is utterly ridiculous.

However, coal production has been almost constant for 25 yr (15% total increase) and production almost equal to the present was at 2 to 3 $/ton in the 1930s. Large, modern earth-moving equipment had not yet been developed, and reclamation was a major cost item. No operator could afford it unless his competitors were required by law to do the same. Most of the abandoned strip mines were left in these days of no regulation or inadequate incentive.

With a 3 ft seam, 1 acre of coal contains 3 acre ft, or about 5550 tons (at 1.36 sp. gr.). An assessment of \$0.75/ton provides about \$4,200/acre for reclamation and represents the order of magnitude of cost of reclamation. Note: The U.S. General Accounting Office gives the 1977 cost per acre for reclamation as: East, \$7,877; Central, \$4,881; West, \$2,808.[13] As long as other land is available in these areas for a small fraction of this cost, there is no economic incentive to reclaim old strip-mined land. Reclamation will either have to await much higher land prices or be carried out as a public enterprise.

Nationwide, coal mining is only one of numerous operations in which the earth's surface is disrupted. The removal of stone, gravel, and limestone for building material ranks first. There are also other types of open-pit mines—copper, iron, clay, phosphate, and gold. Of the 3.2 million acres which had been disturbed by surface mining up to 1967, 1.3 million acres were related to coal production.

Overhead Transmission Lines

The transmission lines of the country are part of a vast electrical system which has been optimized to bring electric power to users at the lowest overall cost. There is no economic alternative. The transport of energy is discussed in detail in Chapter 14.

Transmission lines are objected to primarily because of their appearance on the landscape and the use of land. Rights of way are increasingly difficult to procure because of other land uses. Underground transmission of hydrogen as an alternative is fully discussed in Chapter 4, under Hydrogen Economy. In a word, it is completely uneconomic and highly wasteful of energy.

Since the power transmission ability increases about as the square of the voltage, higher voltage lines use the available rights of way more efficiently. Thus, one 765 kV line will carry about 2.5 times the power of a 500 kV line, with only slightly more right of way (see Table 14.10).

Distribution Circuits

For many years, there has been a trend to underground distribution of electricity in residential and commercial areas, both here and abroad. This is now completely feasible at a modest increase in cost. Most of the newer communities are now using this system and avoiding unsightly wires down the streets. To be effective, telephone and street-light wires must also be underground. Small substations

have also gone underground, or have been so camouflaged as to blend in with the surroundings.

Oil and gas pipelines buried in the ground and buried electrical cable are generally more acceptable than overhead lines. However, the possible environmental impact of the Alaska pipeline, mostly above ground, delayed its construction for some years.

Power Dams and Storage Reservoirs

These are frequently multipurpose, involving irrigation, public water supply, and recreation as well as power. However, in flooding a large area, one is always trading land for water, and evaluations differ. Especially difficult is a comparison of the benefits of a reservoir such as Hetch-Hetchy, in the San Francisco water supply, and the flooding of this beautiful valley just north of Yosemite.

In very large-scale projects, the ecologic effects are very difficult to predict. Construction of the high dam on the upper Nile in Egypt had completely unpredicted consequences. Lack of the annual flood eliminated the annual replenishment of nutrients in farmland all along the Nile, and virtually eliminated sardine fishing in the eastern Mediterranean. Still worse, through a complex chain of events, involving snails in the still waters of the reservoir and canals, there spread a severe disease, bilharzia, which now affects about half the population of Egypt.

Thus, while it accomplished its goal of 7 billion kWh electricity annually and 900,000 acres of land under cultivation (1400 square miles), it had completely unpredicted side effects. Environmental statements and more complete analysis prior to approval of construction are intended to avoid costly unexpected consequences. In general the human benefits from the great multipurpose dams, such as TVA and the Columbia Basin (See Chap. 5) far outweigh any disadvantages. However, good stewardship requires that we develop these great natural resources with the utmost care and with full consideration of possible consequences.

Power Plants

The power plants themselves in the past were invariably situated along rivers, lakes, or the ocean in order to take advantage of "once-through" cooling. Advent of the cooling tower provided the freedom to use other locations. For example, the Keystone

Figure 15.4. Connecticut Yankee Atomic Power Plant. Located at Haddam Neck on the east bank of the Connecticut River, about 20 miles southeast of Hartford. Pressurized-water reactor, 462,000 kW capacity, designed by Westinghouse Electric Corp. Operation of the plant, owned by Connecticut Yankee Atomic Power Company, began in 1967. Courtesy Westinghouse Electric Corp.

Plant in western Pennsylvania, when completed in 1967, was one of the largest steam plants, with 1872 MW installed capacity. It uses for makeup water a part of the flow of Plum Creek, which flows into Crooked Creek, which is one of the smaller tributaries of the Allegheny River. This water is impounded during flood periods, and the normal flow of Plum Creek is little altered. Plum Creek was, and still is, a trout-fishing stream. Thus, the prime sites along rivers and lakes need no longer be used for the comparatively unsightly power plants.

Incidentally, power plants are not necessarily unsightly, particularly nuclear or gas-fired plants that involve no coal handling or tank farms. Figure 15.4 is a view of the Connecticut Yankee Atomic Power Plant in New England.

A second reason for locating power plants along navigable rivers was of course transportation of the fuel. Many of the newer plants are being located in the coal fields, with coal transported by large trucks and by conveyors from underground mines. High-voltage lines carry the power to load areas up to 600 to 1000 miles away.

A pumped-storage hydro plant does not affect the air but has a substantial impact on the water and on the "good earth." Environmental coordination for the Luddington pumped-storage hydro plant, taken as an example in Chapter 5, involved consideration of both land and water ecology and the treatment of reservoirs and lines to lessen any adverse visual impact. Many positive features were added to make the installation an asset to the environment and to public enjoyment of the area.

RADIATION

A general discussion of radiation from nuclear plants and the hazards of plutonium fuel is given in Chapter 6. It is essentially nontechnical. The purpose of this section is to give a more quantitative picture of what is involved, so that the reader can exercise his own judgment and draw his or her own conclusions. What is the "best engineering solution," the best compromise between the need for energy that results in some radiation and the desire to avoid any serious effects on human health, now or in the future?

Radiation Exposure

The best starting point is a picture of the relative amounts of radiation from various sources (Table 15.6). The values for the year 2000 are of course estimated, but on very reasonable grounds that the reader will recognize from the figures.

The units of rems or millirems refer to radiation in the high-frequency range of X-rays and γ-rays, as shown in Fig. 15.5, especially those of 10^{-11} to 10^{-7} cm wavelength. This is the best common measure of the actual effect on man of these very penetrating radiations.

Lower-frequency ultraviolet radiations in the range of 10^{-6} to 10^{-4} cm wavelength may cause skin cancer. These are attenuated by ozone in the stratosphere. Impairment of this protective ozone layer by aerosols, with a consequent expected increase in skin cancer, has led to drastic restriction in the use of aerosols. In the following paragraphs,

Table 15.6. Whole-Body Annual Radiation Doses in the United States, 1970 and 2000 in Millirems per Person per Year

Radiation Source	1970	2000
Natural	130	130
Medical procedures	74	88
Radioactive fallout	4	5
Miscellaneous (luminous watch dials, television, air travel, etc.)	2.7	1.1
Occupational exposure	0.8	0.9
Nuclear electric power	0.05	0.5
Other AEC activities	0.015	0.012
Total[a]	211	225

Source. Klement, Miller, Minx, and Shleien, 1972.
[a]Totals may not agree due to rounding.

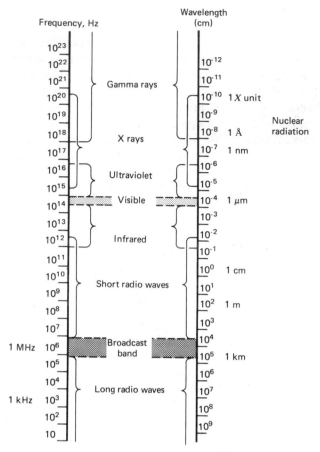

Figure 15.5. Chart of the electromagnetic spectrum.

the term radiation refers to the higher-frequency, shorter-wavelength radiation.

As shown in Table 15.6, the average natural background radiation to which the human body has been exposed since earliest man is 130 millirems (mrem) annually. This varies from an average of 100 mrem/yr in Louisiana to about 250 mrem/yr in higher-altitude Colorado.[2] This single fact enables one to make some judgment about the seriousness of 0.05 mrem/yr from nuclear power (in 1970), or 0.5 mrem/yr (in 2000). However, the other figures in the table are also revealing. Medical procedures result in an average dose of 74 mrem/yr.

A single cross-country jet flight results in 4 mrem (due to high altitude); a watch with luminous dial, 1 to 4 mrem/yr; watching TV about 1 hr/day, about 5 mrem/yr. Altogether, these miscellaneous sources cause about 2.7 mrem/yr.

Radiation does cause cancer, and there is argument whether there is any threshold below which it does not. It makes sense to minimize radiation.

But to put it in perspective, the radiation from enough nuclear plants to supply half of our electric energy in the year 2000 is about $\frac{1}{240}$ of the increased radiation the average person would get by moving to Colorado. Watching TV 1 hr a day now gives 10 times the dosage that nuclear plants will give by 2000.

Thus, compared with the total average exposure of 211 mrem/yr, the 0.05 mrem/yr contribution of the nuclear power industry is obviously insignificant. Also, as pointed out in Chapter 6, 1976 studies place the relative health effects of generating 200 billion kWh electricity ($\frac{1}{10}$ of our electricity), including all types of nuclear accidents and both industrial and civilian consequences, at:

| Coal | 600 deaths |
| Nuclear | 11 deaths |

Thus, both radiation from normal operation of nuclear plants and probable deaths from associated accidents are small compared with other well-known causes. The safeguards and controls are nevertheless being continually improved.[8]

COST OF POLLUTION AND ITS CONTROL

Again, it is convenient to divide the problem into air, water, the good earth, and radiation. Only for the first are "figures" readily available.

Air Pollution Control Costs

The EPA is required by law to submit annual estimates of the costs of carrying out the Clean Air Laws. Their estimate for the period 1971 through 1980 is given in Table 15.7.

Table 15.7. Cumulative Air Pollution Control Expenditures Required to Meet Existing Standards, 1971–1980 (in Billions of 1971 Dollars)

Sector	Capital Investment	Operating Costs	Cash Flow
Public	1.1	6.8	7.9
Private			
Automobiles	31.5	29.5	61.0
Stationary	15.6	22.0	37.6
Total	48.2	58.3	106.5

Source. CEQ, 1972.

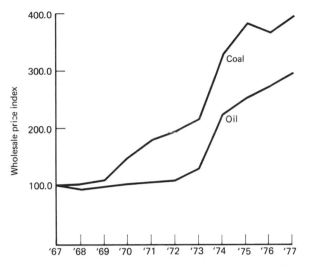

Figure 15.6. Wholesale price of coal and oil since 1967. Courtesy Harder Management Co.

Of the total costs of $106.5 billion, $61 billion, or 57%, is the increased cost and operating expense of automobiles. Private transportation, mostly automobiles, runs regularly about 13% of personal consumption expenditures, or about $126 billion in 1975. Thus, a $6.1 billion average annual cost would represent about 5% increase in personal transportation costs. Note: EPA estimated $150 to $300 for controls; manufacturers estimated $600. This might double capital investment shown in Table 15.7 and increase cost by 50%. It is still small.

The costs in stationary power plants are $37.6 billion over the 10 yr period, or about $3.8 billion/yr. One component of this, electric power plants, generate about 2000 billion kWh/yr average during this period (Fig. 5.2) at an average busbar cost of 2¢/kWh (Table 1.3), or about $40 billion worth of electricity per year at the plant bus. These plants use about two thirds of the coal (Table 3.4)

and would thus incur about two thirds of the $3.8 billion annual cost, or $2.5 billion/yr. This is a substantial increment on the plant production cost, over 6%, and its burden is concentrated on the coal-burning areas, about 40% of the total (Table 1.2). However, it is dwarfed by the *quadrupling* of the price of coal in 10 yr and nearly equal increases in all other fuels except hydro, mostly for reasons other than pollution control (see Fig. 15.6).

The public costs in Table 15.7 are primarily for administration and enforcement.

Probably for most people, cleaner air and the prospects of better health for everyone, especially the very young and old and those afflicted with cardiac and pulmonary disorders, is well worth the cost of this air pollution control. However, it is claimed that the cost of pollution on health, property, and vegetation actually exceeds the cost of eliminating it. Table 15.8 presents such an analysis, showing a cost of air pollution damage of $16.1 billion/yr to offset the $10.6 billion/yr for controls of Table 15.7.

However, these figures can be very misleading. For example, an enthusiast[9] presents them as representing a net gain of $5.5 billion/yr, for 1971 through 1980. It is now 1981. The $10.6 billion/yr has undoubtedly been spent. But where are the results? The 1981 table of damages would look surprisingly like the 1968 Table 15.8. Few of the indices have changed greatly during that time.

A far better approach is to realize that these measures are not going to be overly expensive in terms of our personal income. In 1975, $10.6 billion was 1.25% of our personal income. It will take time. It will cost something, not save money. But in the end, we shall have "saved" the air in our great American cities, an achievement that we can all be proud of. Already the disastrous rise in effluents has been halted (Fig. 15.2), except for NO_x, which was

Table 15.8. National Costs of Air Pollution Damage, by Source and Effect, 1968 (Billions of Dollars)

Effects	Stationary-Source Fuel Combustion	Transportation	Industrial Processes	Solid Waste	Miscellaneous	Total
Residential property	2.802	0.156	1.248	0.104	0.884	5.200
Materials	1.853	1.093	0.808	0.143	0.855	4.752
Health	3.281	0.197	1.458	0.119	1.005	6.060
Vegetation	0.047	0.028	0.020	0.004	0.021	0.120
Total	7.983	1.474	3.534	0.370	2.765	16.132

Source. Barrett and Waddell, 1973. See Ref. 2.

first limited nationally on new cars in 1973 and substantially in 1977 and 1981. In the meantime, *real* personal income increases 23% in 10 yr (average last 25 yr).[10] What better use for 5% of this increase?

Water Pollution

The cost of eliminating pollution from water will fall mainly on industries and public sewage systems, except for drainage from old abandoned mines, nutrients from farmland, and natural pollution such as sediments. Whatever is done will be in large units compared with the measures eventually required in over 130,000,000 motor vehicles for air pollution control. On the other hand, water polluting sources are mostly "going operations" and require retrofits, or additions (secondary sewage treatment) for control, whereas when new cars are equipped, a 10 yr delay will get most of the old ones off the road.

As a rough basis for reference, governments in the United States spent $7.4 billion on "local sanitation" in 1975, or 0.6% of personal income. Retrofits or additions to improve these sewage systems will be related to this figure. Something will have to be done quickly if the EPA interpretation of the 1972 Law is to be met: "No more pollutants whatever into the Nation's waters by 1985." The possibilities are outlined in Ref. 2 under "Sewage Treatment Fundamentals."

The use of cooling towers at all new power plants to eliminate this source of thermal pollution would increase the cost of electricity from these plants about 2% compared with once-through cooling (Table 15.6). Some 300 new nuclear plants are needed by 2000 in addition to the 63 operating in 1977.[11] These will have cooling towers wherever needed. Power plants have a lifetime of 30 to 40 yr and provide the opportunity to use cooling towers in their replacements as needed.

The Good Earth

The cost of high-quality reclamation of newly strip-mined land, at 75¢/ton, is almost lost among the other increases that have brought the cost of coal close to $25/ton (Table 1.5). Reclamation of old abandoned strip-mined land—1.3 million acres—at $4200/acre represents about $6 billion. Its reclamation is unlikely unless undertaken as a public project, possibly to provide employment and a useful result. The total acreage of disturbed land is, as mentioned, 3.2 million acres. If land is needed, the best target may not be abandoned strip mines. Nature has left much rock-strewn, unusable land which may be cheaper to reclaim. In northern Israel, where there is adequate water, much land of this type has been transformed into good farm land, using heavy machinery.

Radiation

From the beginning, the AEC has leaned over backward to ensure safety in the use of nuclear energy. Only due to such extreme measures does the nuclear industry now enjoy the safest record of all industries. Only because of this care can it be said today (1981), "No member of the American public has ever been injured, much less killed, as a result of accident or radiation from nuclear energy facilities" (Chapter 6, Ref. 2). ERDA, now DOE, and the Nuclear Regulatory Commission, are continuing these rigid safety precautions.

The shielding and protective measures are a major cost item in nuclear plant design. The delays in siting and licensing typically increase the plant cost over 20% due to interest during construction and other factors. In addition to massive shielding, each reactor has a containment building (Figs. 6.8 and 6.9). Radiation at the plant fence is limited to about 5 mrem/yr, about what one would get watching television 1 hr/day.

In addition to these measures, for the Clinch River breeder reactor, a further containment is being required by the Nuclear Regulatory Commission that would contain the postulated "worst accident." Even with these extreme burdens and increases in the price of uranium (other fuels have increased, too), nuclear energy is still the most economic for base load in all areas (1980).[12]

It would be a shame, however, to kill the goose that lays the golden eggs. It would be better to limit the delays, reduce the expense, and thus set a limit on the cost of electricity. Surely, with 72 plants operating, the factors and requirements are now known and do not require protracted research and delays for each of the 300 new plants we must build in the next 20 or 30 years.

REFERENCES

1. "Light Duty Motor Vehicle Exhaust Emission Standards," Code of Federal Register, Title 40 CFR, Part 86, 1981.

2. G. M. Masters, *Introduction to Environmental Science and Technology,* New York: Wiley, 1974.

3. "Auto Emissions Control," in *General Motors Public Information Report, 1976,* Detroit: General Motors Corp., 1977.

4. *Air Pollution Indices, U.S. and Canada,* CEQ and EPA, Washington, D.C.: Environmental Protection Agency.

5. *A Citizens Guide to Clean Air,* Washington, D.C.: The Conservation Fdn., 1972, p. 84.

6. "New Source Performance Standards," *Federal Register,* June 11, 1979, p. 33580.

7. "Health and Airborne Particles," *NAS-NAE News Report,* August 1978, p. 5.

8. "Nuclear Waste Disposal," *Wall Street Journal,* March 16, 1978, front page.

9. M. Zeldin, *The Campaign for Cleaner Air,* Public Affairs Pamphlet 494, New York: Public Affairs Pamphlets, 1973. See also "Ounce of Prevention," *Reader's Digest,* January 1981, p. 32.

10. U.S. Bureau of the Census, *Statistical Abstract of the United States,* 101st ed., Washington, D.C.: Supt. of Documents.

11. Energy Secretary Schlesinger, "Nuclear Plants" *Wall Street Journal,* June 8, 1977, front page.

12. R. A. Loth, O. D. Gildersleeve, Jr., S. A. Vejtasa, and R. W. Zeren, "Comparative Evaluation of New Electric Generating Technologies," *Proc. 7th Energy Technology Conference,* Washington, D.C., 1980.

13. *Navy Energy Fact Book,* 0584-LP-200-1420, Navy Pub. Forms Center, Philadelphia, 1979, p. 165.

Appendix

Selected Energy Relations

Physical Constants

Speed of light	c	2.997925×10^8 m/sec
Einstein relation	$E = mc^2$	8.98755×10^{16} J/kg
Electron charge	e	1.60210×10^{-19} C
Planck constant	h	6.6256×10^{-38} J/Hz
Avogadro constant	N_A	6.02252×10^{26} molecules/kg mole
Atomic mass unit	amu $= 1/N_A$	1.66043×10^{-27} kg
Faraday constant	$= \dfrac{N_A e}{1000}$	96487 C/g mole
Gravitational constant	G	6.6720×10^{-11} Nm^2/kg^2
		3.3218×10^{-11} lb ft^2/lb^2
Standard acceleration of gravity	g	980.665 cm/sec^2
		32.1740 ft/sec^2
Standard atmosphere	atm	101325 N/m^2 (Pa)
		14.6959 psi
		760 mm Hg at 0°C (torr)
		33.898 ft water at 4°C
		1.01325 bar
Molar gas constant	R	8.3143 J/g mole °K
		1.987165 cal/g mole °K
		1.987165 Btu/lb mole °R
Molar gas volume	V_m	
at 273.15°K, and 1 atm		22.4136 m^3/kg mole
		359.03 ft^3/lb mole
at 25°C, 77°F, 1 atm		391.89 ft^3/lb mole
Triple point of water		273.16°K, 6.11 millibars
Ice point		273.1500°K
Solar constant[a]		1353 W/m^2, 125.7 W/ft^2
		429.0 Btu/ft^2 hr
Mechanical equivalent of heat		3413 Btu/kWh = 777.97 ft lb/Btu[b]
		4.184 J/cal[b]

Source: Mainly JANAF Tables, Ref. 5, Chap. 11.

[a]Mean solar intensity in free space at earth distance from the sun.

[b]The thermochemical calorie is 4.814 J. However, the value 4.186 J is consistent with 3413 Btu/kWh and 777.97 ft lb/Btu used in this book:

$$\frac{(3.6 \times 10^6 \text{ J/kWh}) (1.8°F/°C)}{(3413 \text{ Btu/kWh}) (453.59237 \text{ g/lb})} = 4.185747 \text{ J/cal}$$

Energy Conversion Factors

To Convert From	To	Multiply by	Usual Approximation
Acres	Square miles	1/640	
Acres	Square ft	43,560	
Atomic mass units, amu	Millions electron volts, MeV	931.48	(931)
Angstroms, Å	Meters, m	10^{-10}	
Atmospheres, atm	Pounds per square in., psi	14.6959	(14.7)
Atmospheres, atm	Bars (10^6 dyn/cm²)	1.01325	(1.013)
Atmospheres, atm	Pascals (N/m²)	101325	
Atmospheres, atm	Torrs (mm HG)	760	
Atmospheres, atm	Inches mercury (32°F), in. Hg	29.9213	(29.9)
Barrel (petroleum), bbl	Cubic meters, m³	0.15899	
Barrel (petroleum), bbl	Gallons (U.S.), gal	42	
British thermal units, Btu	Calories, cal	251.996	(252)
British thermal units, Btu	Joules, J	1054.79	(1055)
British thermal units, Btu	Foot pounds, ft lb	777.97	(778)
British thermal units, Btu	Kilowatt hours, kWh	1/3413	
British thermal units, Btu	Ergs (dyne cm), erg	1.05479×10^{10}	
Btu/gallon (U.S.)	Megajoules per liter, MJ/l.	0.00027865	
Btu/lb	Megajoules per kilogram MJ/kg	0.0023254	
Btu/lb	Calories per gram, cal/g	1/1.8	
Btu/ft³	Megajoules/m³, MJ/m³	0.037250	
Btu/ft²	Langleys, cal/cm²,	0.27125	
Btu/hr	Watts, W	0.29300	
Celsius Degrees, °C	Fahrenheit Degrees, °F	°F = °C × 1.8 + 32	
Celsius Degrees, °C	Kelvin Degrees, °K	°K = °C + 273.15	
Calories, cal	Joules, J	4.186[a]	
Calories, cal	British thermal units, Btu	1/251.996	(1/252)
Coulombs, C	E.S. cgs units of charge	2.9979×10^9	
Dynes, dyn	Grams, g	0.00101972	
Dynes, dyn	Newtons, N	10^{-5}	
Electron charge units	E.S. cgs units of charge	4.8030×10^{-10}	
Electron volts, eV	Ergs	1.6020×10^{-12}	
Electron volts, eV	Joules, J	1.60210×10^{-19}	(1.602×10^{-19})
Electron volts, eV	Btu	1.51888×10^{-22}	
Fahrenheit Degrees, °F	Celsius Degrees, °C	°C = (°F − 32)/1.8	
Fahrenheit Degrees, °F	Rankine Degrees, °R	°R = °F + 459.67	
Ergs (dyn cm)	Joules, J	10^{-7}	
Ergs (dyn cm)	MeV	6.2418×10^5	
Feet, ft	Meters, m	0.3048 exact	
Feet, ft	Miles (statute), mi	1/5280	
Feet ft	Miles (nautical), mi	1/6076	
Feet, ft	Rods	1/16.5	
Foot pounds, ft lb	Btu	1/777.97	(1/778)
Foot pounds, ft lb	Joules (newton meters), J	1.3558	
Gallons (U.S.), gal	Barrels (petroleum), bbl	1/42	

To Convert From	To	Multiply by	Usual Approximation
Gallons (U.S.), gal	Cubic inches, in.^{3}	231	
Gallons (U.S.), gal	Liters	3.7854	(3.79)
Gauss, maxwells/cm^2, lines/cm^2	Teslas (webers/m^2), T	0.0001	
Gigawatt year, GWy	Quads	0.029898	
GWy × 85% plant factor	Megatons coal (12,707 Btu/lb)	1	
Horsepower, Hp	Foot pounds/min, ft lb/min	33,000	
Horsepower, Hp	Kilowatts, kW	0.74570	(0.746)
Horsepower, Hp	Tons of refrigeration	0.2121	
Horsepower, Hp	Btu/min	42.418	
Hectares, ha	Square meters, m^2	10,000	
Hectares, ha	Acres	2.47105	
Inches, in.	Centimeters, cm	2.54 exact	
Inches, in.	Mils	1000	
Inches mercury, in. Hg	Torrs (mm Hg), torr	25.4	
Inches mercury, in. Hg	Atmospheres, atm	0.033421	
Joules, J	Kilowatt hours, kWh	$1/(3.6 \times 10^6)$	
Kilometers, km	Miles, mi	0.62137	
Kilowatts, kW	Joules per hour, J/h	3.6×10^6	
Kelvin Degrees, °K	Rankin Degrees, °R	1.8	
Kelvin Degrees, °K	Celsius Degrees, °C	°C = °K − 273.15	
Kilograms, kg	Pounds, lb	2.2046226	
Kilograms, kg	Metric tons	0.001	
Kilograms, kg	Joules (Einstein relation), J	8.9875513×10^{16}	
Miles (statute)	Miles (nautical)	0.86898	
Miles (statute)	Feet, ft	5280	
Microns, μ	Meters, m	10^{-6}	
Newton, N	Pounds, lb	0.224809	
Pascals (N/m^2), Pa	Atmospheres, atm	1/101325	
Pascals (N/m^2), Pa	Pounds per square inch, psi	0.00014504	
Pounds, lb	Kilograms, kg	0.45359237 exact	(0.454)
Quads	Btu	10^{15}	
Quads/yr	Millions bbls oil/day	0.472	@ 5.8×10^6 Btu/bbl
Q's	Quads	1000	
Q's	Btu	10^{18}	
Rankine Degrees, °R	Kelvin Degrees, °K	1/1.8	
Rankine Degrees, °R	Fahrenheit Degrees, °F	°F = °R − 459.67	
Siemens, s	Mhos = 1/ohms	1	
Slugs	Pounds, lb	32.1740	
Tons (short)	Pounds, lb	2000	
Tons (long)	Pounds, lb	2240	
Tons (metric)	Pounds, lb	2204.6	
Tons of refrigeration	Btu/min	200	
Therms	Btu	10^5	
Webers, Wb	Maxwells (lines)	10^8	

[a]See note b under Physical Constants.

Approximate Heating Values of Fuels—1980
U.S. Average

Bituminous coal and lignite	22.14 million Btu/short ton
Petroleum	5.80 million Btu/bbl
Natural gas	1020 Btu/SCF
Gasolene	122,400 Btu/gal
Heating Oil	140,000 Btu/gal

Index